Adaptive Control Tutorial

Advances in Design and Control

SIAM's Advances in Design and Control series consists of texts and monographs dealing with all areas of design and control and their applications. Topics of interest include shape optimization, multidisciplinary design, trajectory optimization, feedback, and optimal control. The series focuses on the mathematical and computational aspects of engineering design and control that are usable in a wide variety of scientific and engineering disciplines.

Series Volumes

Shima, Tal and Rasmussen, Steven, eds., *UAV Cooperative Decision and Control: Challenges and Practical Approaches*
Speyer, Jason L. and Chung, Walter H., *Stochastic Processes, Estimation, and Control*
Krstic, Miroslav and Smyshlyaev, Andrey, *Boundary Control of PDEs: A Course on Backstepping Designs*
Ito, Kazufumi and Kunisch, Karl, *Lagrange Multiplier Approach to Variational Problems and Applications*
Xue, Dingyü, Chen, YangQuan, and Atherton, Derek P., *Linear Feedback Control: Analysis and Design with MATLAB*
Hanson, Floyd B., *Applied Stochastic Processes and Control for Jump-Diffusions: Modeling, Analysis, and Computation*
Michiels, Wim and Niculescu, Silviu-Iulian, *Stability and Stabilization of Time-Delay Systems: An Eigenvalue-Based Approach*
Ioannou, Petros and Fidan, Barış, *Adaptive Control Tutorial*
Bhaya, Amit and Kaszkurewicz, Eugenius, *Control Perspectives on Numerical Algorithms and Matrix Problems*
Robinett III, Rush D., Wilson, David G., Eisler, G. Richard, and Hurtado, John E., *Applied Dynamic Programming for Optimization of Dynamical Systems*
Huang, J., *Nonlinear Output Regulation: Theory and Applications*
Haslinger, J. and Mäkinen, R. A. E., *Introduction to Shape Optimization: Theory, Approximation, and Computation*
Antoulas, Athanasios C., *Approximation of Large-Scale Dynamical Systems*
Gunzburger, Max D., *Perspectives in Flow Control and Optimization*
Delfour, M. C. and Zolésio, J.-P., *Shapes and Geometries: Analysis, Differential Calculus, and Optimization*
Betts, John T., *Practical Methods for Optimal Control Using Nonlinear Programming*
El Ghaoui, Laurent and Niculescu, Silviu-Iulian, eds., *Advances in Linear Matrix Inequality Methods in Control*
Helton, J. William and James, Matthew R., *Extending H^∞ Control to Nonlinear Systems: Control of Nonlinear Systems to Achieve Performance Objectives*

Adaptive Control Tutorial

Petros Ioannou
University of Southern California
Los Angeles, California

Barış Fidan
National ICT Australia & Australian National University
Canberra, Australian Capital Territory, Australia

Society for Industrial and Applied Mathematics
Philadelphia

10 9 8 7 6 5 4 3 2

Library of Congress Cataloging-in-Publication Data

Ioannou, P. A. (Petros A.), 1953-
Adaptive control tutorial / Petros Ioannou, Baris Fidan.
p. cm. – (Advances in design and control)
Includes bibliographical references and index.
ISBN 978-0-898716-15-3 (pbk.)
1. Adaptive control systems. I. Fidan, Baris. II. Title.

TJ217.I628 2006
629.8'36–dc22

2006051213

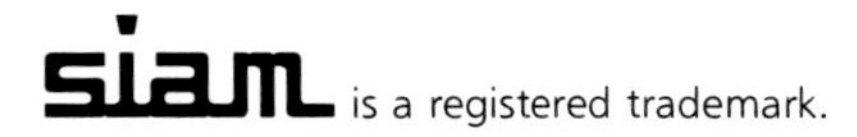

To Kira and Andreas
P.I.

To my family
B.F.

Contents

Preface

Adaptive control evolved over the years to become an accepted subject taught in many schools as an advanced control course. A good understanding of adaptive control involves a good knowledge of control design for linear time-invariant systems, basic stability theory of nonlinear systems, and some mathematical maturity. Several books and research monographs as well as numerous papers on the theory and application of adaptive control already exist. Despite the maturity of the field and the numerous publications, the field of adaptive control appears to many as a collection of unrelated techniques, intelligent tricks, and fixes, and very few researchers can really follow the long and technical stability proofs. On the implementation side, designing stable adaptive control systems and simulating them on a digital computer to demonstrate their theoretical properties could also be an adventure if one does not have a good understanding of the basic theoretical properties and limitations of adaptive control.

The purpose of this book is to present the fundamental techniques and algorithms in adaptive control in a tutorial manner making it suitable as an adaptive control textbook. The aim of the book is to serve a wider audience by addressing three classes of readers. The first class includes the readers who are simply interested to learn how to design, simulate, and implement parameter estimators and adaptive control schemes without having to fully understand the analytical and technical proofs. This class may include practicing engineers and students whose background may not be strong enough to follow proofs or who simply like to focus on the application of adaptive control. The second class involves readers who, in addition to learning how to design and implement adaptive systems, are also interested in understanding the analysis of the simple schemes and getting an idea of the steps followed in the more complex proofs. This class of readers may include the majority of students at the Masters and Ph.D. level who take adaptive control as an advanced course. The third class of readers involves the advanced students and researchers who want to study and understand the details of the long and technical proofs as training for pursuing research in adaptive control or in related topics such as nonlinear systems, etc. All of these readers may be found in the same adaptive control class consisting of students with different abilities and interests. The book is written with the objective of at least satisfying the first class of readers irrespective of how strong their theoretical background is, and at the same time serving the needs of the advanced research students on the other end without sacrificing mathematical depth and rigor. These multiple objectives and learning expectations are achieved by enriching the book with examples demonstrating the design procedures and basic analysis steps and presenting the details of the long and technical proofs in an appendix and in electronically available supplementary material. Electronically available also are additional examples and simulations using the Adaptive Control Toolbox developed by the authors, which readers can purchase separately by contacting the authors.

The material in the book is based on twenty years of experience in teaching adaptive control at the University of Southern California by the first author and on feedback from students and instructors in other universities who taught adaptive control. Our experience taught us that expecting all students to be able to understand and reproduce all technical proofs over one semester is an unrealistic goal and could often lead to confusion. The book is written in a way that allows the teacher to start from the lowest objective of teaching students how to design and simulate adaptive systems and understand their properties, and add more analysis and proofs depending on the level of the course and quality and ambition of the students involved.

The book is organized as follows. Chapter 1 presents some basic characteristics of adaptive systems and a brief history of how adaptive control evolved over the years. Chapter 2 presents the various parameterizations of plants that are suitable for parameter estimation. This is a fairly easy chapter but very important, since in subsequent chapters the expression of the unknown parameters in a parametric model suitable for estimation is the first design step in the development of parameter estimators. The design of parameter estimators or adaptive laws for continuous-time plants is presented in Chapter 3. In Chapter 4 it is shown how the continuous-time estimators developed in Chapter 3 can be discretized using Euler's method of approximating a derivative and still maintain the desired stability properties. In addition, discrete-time parameter estimators are also developed using discrete-time models of the plant. This chapter, for the first time, shows clearly the connection between continuous-time and discrete-time parameter estimators, which until now have been viewed as two completely different methodologies. The design and analysis of a class of adaptive controllers referred to as model reference adaptive control (MRAC), which has been popular in continuous-time adaptive systems, is presented in Chapter 5. Chapter 6 presents the design and analysis of adaptive pole placement (APPC) schemes, which do not require the restrictive assumptions used in the MRAC, as the control objectives are different. The discrete version of the schemes in Chapters 5 and 6 are presented in Chapter 7. In Chapters 3 to 7, the robustness of the schemes developed with respect to bounded external disturbances and dynamic uncertainties is also analyzed, and modifications are presented to enhance robustness. Finally, Chapter 8 gives a brief review, using simple examples, of the extensions of the adaptive systems developed in previous chapters to classes of nonlinear systems with unknown parameters. All mathematical preliminaries that are useful for the chapters are presented in the Appendix. The reader may go over the Appendix first before reading the chapters, as certain concepts and results presented in the Appendix are often quoted and used. Long proofs and supplementary material as well as additional examples and simulations using the Adaptive Control Toolbox are available in electronic form at http://www.siam.org/books/dc11. A solution manual is also available to instructors who can verify that they teach a class on adaptive systems using this book as textbook, by contacting the publisher or the authors.

Acknowledgments

Many friends and colleagues have helped us in the preparation of the book in many different ways. We are especially grateful to Petar Kokotovic, who introduced the first author to the field of adaptive control back in 1979 and remained a mentor and advisor for many years to follow. His continuous enthusiasm for and hard work in research has been the strongest driving force behind our research and that of our students. We thank Brian Anderson, Karl Astrom, Mike Athans, Peter Caines, Bo Egardt, Graham Goodwin, Rick Johnson, Gerhard Kreisselmeier, Yoan Landau, Lennart Ljung, David Mayne, the late R. Monopoli, Bob Narendra, and Steve Morse, whose fundamental work in adaptive control laid the foundations for our work and the work of so many Ph.D. students. We would especially like to express our deepest appreciation to Laurent Praly and former Ph.D. students in adaptive control at the University of Southern California (USC), Kostas Tsakalis, Jing Sun, Gang Tao, Aniruddha Datta, Farid Zaid, Youping Zhang, and Haojian Xu, whose work became the source of most of the material in the book. In particular, we would like to thank Jing Sun, who was the coauthor of our earlier book, *Robust Adaptive Control* (out of print), which became the main source of some of the material on adaptive control for continuous-time systems which is presented in this book in a more tutorial manner. We also thank the former USC Ph.D. students and postdocs John Reed, C. C. Chien, Houmair Raza, Alex Stotsky, Tom Xu, Elias Kosmatopoulos, Anastasios Economides, Alex Kanaris, Kun Li, Arnab Bose, Chin Liu, and Hossein Jula, whose participation in numerous discussions and research contributed to our learning. We are grateful to many colleagues for stimulating discussions at conferences, workshops, and meetings. They have helped us broaden our understanding of the field. In particular, we would like to mention Anu Annaswamy, Erwei Bai, Bob Bitmead, Marc Bodson, Karl Hedrick, David Hill, Roberto Horowitz, Ioannis Kanellakopoulos, Hassan Khalil, Bob Kosut, Miroslav Krstic, Rogelio Lozano-Leal, Iven Mareels, Rick Middleton, Romeo Ortega, Charles Rohrs, Ali Saberi, Shankar Sastry, Masayoshi Tomizuka, Pravin Varayia, and the numerous unknown reviewers of our papers whose constructive criticisms led to improvements of our results in the area of adaptive control.

We would also like to extend our thanks to our colleagues at the USC, Mike Safonov, Edmond Jonckhere, Len Silverman, Gary Rosen, and Sami Masri, and our project collaborators Anastassios Chassiakos from California State University Long Beach and Maj Mirmirani from California State University Los Angeles for many interactions and discussions. Finally, and not least, we would like to thank the Ph.D. students of the first author, Nazli Kanveci, Jianlong Zhang, Marios Lestas, Ying Huo, Matt Kuipers, Hwan Chang, and Jason Levin, for proofreading the draft chapters of the book, as well as Yun Wang, Ying Huo, Ting-Hsiang Chuang, Mike Zeitzew, and Marios Lestas for helping with the case study examples.

List of Acronyms

ACC	Adaptive cruise control
ALQC	Adaptive linear quadratic control
API	Adaptive proportional plus integral (control)
APPC	Adaptive pole placement control
ARMA	Autoregressive moving average
B–G	Bellman–Gronwall (lemma)
BIBO	Bounded-input bounded-output
B-DPM	Bilinear dynamic parametric model
B-SPM	Bilinear static parametric model
B-SSPM	Bilinear state space parametric model
CC	Cruise control
CE	Certainty equivalence (principle, approach)
CLF	Control Lyapunov function
DARMA	Deterministic autoregressive moving average
DPM	Dynamic parametric model
I/O	Input/output
LKY	Lefschetz–Kalman–Yakubovich (lemma)
LQ	Linear quadratic
LS	Least-squares
LTI	Linear time-invariant
LTV	Linear time-varying

MIMO	Multi-input multi-output
MKY	Meyer–Kalman–Yakubovich (lemma)
MRAC	Model reference adaptive control
MRC	Model reference control
PE	Persistently exciting
PI	Parameter identification/Proportional plus integral (control)
PPC	Pole placement control
PR	Positive real
RBF	Radial basis function
SISO	Single-input single-output
SPM	Static parametric model
SPR	Strictly positive real
SSPM	State space parametric model
TV	Time-varying
UCO	Uniformly completely observable
a.s.	Asymptotically stable
e.s.	Exponentially stable
m.s.s.	(In the) mean square sense
u.a.s.	Uniformly asymptotically stable
u.b.	Uniformly bounded
u.s.	Uniformly stable
u.u.b.	Uniformly ultimately bounded
w.r.t.	With respect to

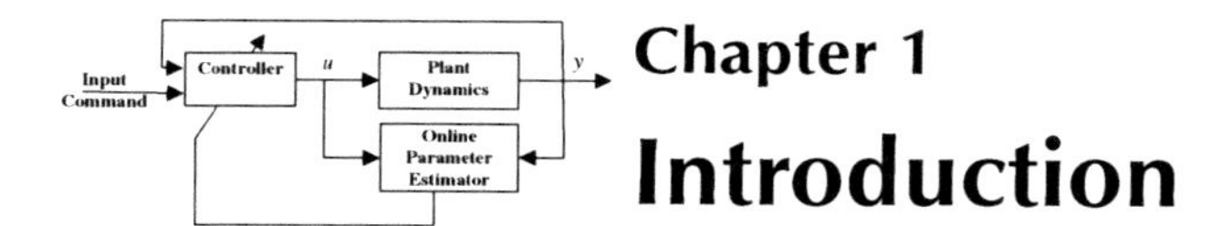

Chapter 1
Introduction

According to Webster's dictionary, to *adapt* means to "change (oneself) so that one's behavior will conform to new or changed circumstances." The words *adaptive systems* and *adaptive control* have been used as early as 1950 [1, 2]. This generic definition of adaptive systems has been used to label approaches and techniques in a variety of areas despite the fact that the problems considered and approaches followed often have very little in common. In this book, we use the following specific definition of adaptive control: *Adaptive control is the combination of a parameter estimator, which generates parameter estimates online, with a control law in order to control classes of plants whose parameters are completely unknown and/or could change with time in an unpredictable manner*. The choice of the parameter estimator, the choice of the control law, and the way they are combined leads to different classes of adaptive control schemes which are covered in this book. Adaptive control as defined above has also been referred to as *identifier-based adaptive control* in order to distinguish it from other approaches referred to as *non–identifier-based*, where similar control problems are solved without the use of an online parameter estimator.

The design of autopilots for high-performance aircraft was one of the primary motivations for active research in adaptive control in the early 1950s. Aircraft operate over a wide range of speeds and altitudes, and their dynamics are nonlinear and conceptually time-varying. For a given operating point, the complex aircraft dynamics can be approximated by a linear model. For example, for an operating point i, the longitudinal dynamics of an aircraft model may be described by a linear system of the form [3]

$$\begin{aligned} \dot{x} &= A_i x + B_i u, \quad x(t_0) = x_0, \\ y &= C_i^T x + D_i u, \end{aligned} \tag{1.1}$$

where the matrices A_i, B_i, C_i, D_i are functions of the operating point i; x is the state; u is the input; and y is the measured outputs. As the aircraft goes through different flight conditions, the operating point changes, leading to different values for A_i, B_i, C_i, D_i. Because the measured outputs carry information about the state x and parameters, one may argue that, in principle, a sophisticated feedback controller could learn the parameter changes, by processing the outputs $y(t)$, and use the appropriate adjustments to accommodate them. This argument led to a feedback control structure on which adaptive control is based. The

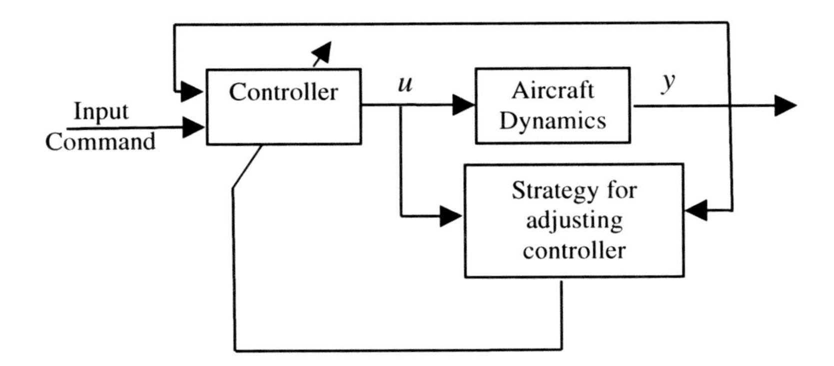

Figure 1.1. *General adaptive control structure for aircraft control.*

controller structure consists of a feedback loop and a controller with adjustable gains, as shown in Figure 1.1.

The way of adjusting the controller characteristics in response to changes in the plant and disturbance dynamics distinguishes one scheme from another.

1.1 Adaptive Control: Identifier-Based

The class of adaptive control schemes studied in this book is characterized by the combination of an online *parameter estimator*, which provides estimates of the unknown parameters at each instant of time, with a *control law* that is motivated from the known parameter case. The way the parameter estimator, also referred to as *adaptive law* in the book, is combined with the control law gives rise to two different approaches. In the first approach, referred to as *indirect adaptive control*, the plant parameters are estimated online and used to calculate the controller parameters. In other words, at each time t, the estimated plant is formed and treated as if it is the true plant in calculating the controller parameters. This approach has also been referred to as *explicit adaptive control*, because the controller design is based on an explicit plant model. In the second approach, referred to as *direct adaptive control*, the plant model is parameterized in terms of the desired controller parameters, which are then estimated directly without intermediate calculations involving plant parameter estimates. This approach has also been referred to as *implicit adaptive control* because the design is based on the estimation of an implicit plant model.

The basic structure of indirect adaptive control is shown in Figure 1.2. The plant model $G(\theta^*)$ is parameterized with respect to some unknown parameter vector θ^*. For example, for a linear time-invariant (LTI) single-input single-output (SISO) plant model, θ^* is a vector with the unknown coefficients of the numerator and denominator of the plant model transfer function. An online parameter estimator generates the estimate $\theta(t)$ of θ^* at each time t by processing the plant input u and output y. The parameter estimate $\theta(t)$ specifies an estimated plant model characterized by $G(\theta(t))$, which for control design purposes is treated as the "true" plant model and is used to calculate the controller parameter or gain vector θ_c by solving a certain algebraic equation, $\theta_c(t) = F(\theta(t))$, that relates the plant parameters with the controller parameters at each time t. The form of the control law

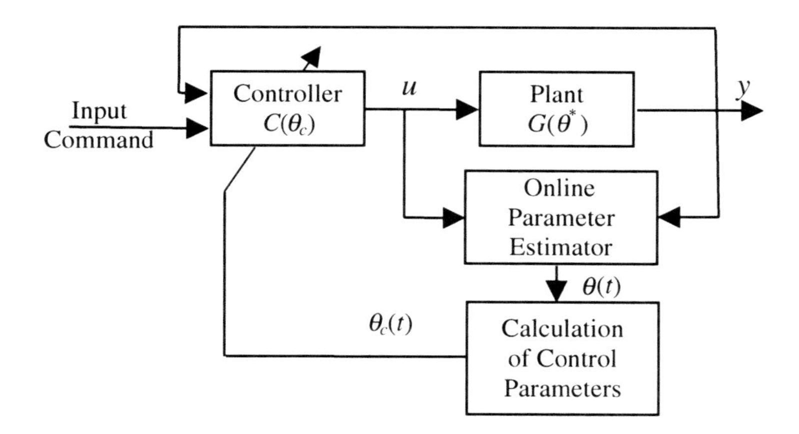

Figure 1.2. *Indirect adaptive control structure.*

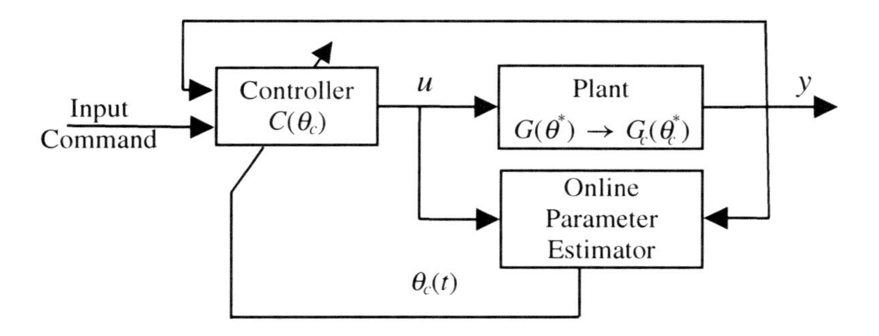

Figure 1.3. *Direct adaptive control structure.*

$C(\theta_c(t))$ and algebraic equation $\theta_c(t) = F(\theta(t))$ is chosen to be the same as that of the control law $C(\theta_c^*)$ and equation $\theta_c^* = F(\theta^*)$, which could be used to meet the performance requirements for the plant model $G(\theta^*)$ if θ^* was known. It is, therefore, clear that with this approach, $C(\theta_c(t))$ is designed at each time t to satisfy the performance requirements for the estimated plant model $G(\theta(t))$ rather than for the actual plant $G(\theta^*)$. Therefore, the main problem in indirect adaptive control is to choose the class of control laws $C(\theta_c)$ and the class of parameter estimators that generate $\theta(t)$, as well as the algebraic equation $\theta_c = F(\theta)$, so that $C(\theta_c)$ meets the performance requirements for the plant model $G(\theta^*)$ with unknown θ^*. We will study this problem in great detail in Chapters 5 and 6.

Figure 1.3 shows the structure of direct adaptive control. In this case, the plant model $G(\theta^*)$ is parameterized in terms of the unknown controller parameter vector θ_c^*, for which $C(\theta_c^*)$ meets the performance requirements, to obtain the plant model $G_c(\theta_c^*)$ with exactly the same input/output (I/O) characteristics as $G(\theta^*)$. The online parameter estimator is designed based on $G_c(\theta_c^*)$ instead of $G(\theta^*)$ to provide the direct online estimate $\theta_c(t)$ of θ_c^* at each time t,by processing the plant input u and output y. The estimate $\theta_c(t)$ is then used in the control law without intermediate calculations. The choice of the class of control laws $C(\theta_c)$ and parameter estimators that generate $\theta_c(t)$ so that the closed-loop plant meets the performance requirements is the fundamental problem in direct adaptive control. The properties of the plant model $G(\theta^*)$ are crucial in obtaining the parameterized plant model $G_c(\theta_c^*)$ that is convenient for online estimation. As a result, direct adaptive control is restricted to certain classes of plant models. In general, not every plant can be expressed in a parameterized form involving only the controller parameters, which is also a suitable form for online estimation. As we show in Chapter 5, a class of plant models that is suitable for direct adaptive control for a particular control objective consists of all SISO LTI plant models that are minimum phase; i.e., their zeros are located in $\text{Re}[s] < 0$.

In general, the ability to parameterize the plant model with respect to the desired controller parameters is what gives us the choice to use the direct adaptive control approach. Note that Figures 1.2 and 1.3 can be considered as having the exact same structure if in Figure 1.3 we add the calculation block $\theta_c(t) = F(\theta_c(t)) = \theta_c(t)$. This identical-in-structure interpretation is often used in the literature of adaptive control to argue that the separation of adaptive control into direct and indirect is artificial and is used simply for historical reasons. In general, direct adaptive control is applicable to SISO linear plants which are minimum phase, since for this class of plants the parameterization of the plant with respect to the controller parameters for some controller structures is possible. Indirect adaptive control can be applied to a wider class of plants with different controller structures, but it suffers from a problem known as the *stabilizability problem* explained as follows: As shown in Figure 1.2, the controller parameters are calculated at each time t based on the estimated plant. Such calculations are possible, provided that the estimated plant is controllable and observable or at least stabilizable and detectable. Since these properties cannot be guaranteed by the online estimator in general, the calculation of the controller parameters may not be possible at some points in time, or it may lead to unacceptable large controller gains. As we explain in Chapter 6, solutions to this stabilizability problem are possible at the expense of additional complexity. Efforts to relax the minimum-phase assumption in direct adaptive control and resolve the stabilizability problem in indirect adaptive control led to adaptive control schemes where both the controller and plant parameters are estimated online, leading to combined direct/indirect schemes that are usually more complex [4].

The principle behind the design of direct and indirect adaptive control shown in Figures 1.2 and 1.3 is conceptually simple. The form of the control law is the same as the one used in the case of known plant parameters. In the case of indirect adaptive control the unknown controller parameters are calculated at each time t using the estimated plant parameters generated by the online estimator, whereas in the direct adaptive control case the controller parameters are generated directly by the online estimator. In both cases the estimated parameters are treated as the true parameters for control design purposes. This design approach is called *certainty equivalence* (*CE*) and can be used to generate a wide class of adaptive control schemes by combining different online parameter estimators with different control laws. The idea behind the CE approach is that as the parameter estimates $\theta_c(t)$ converge to the true ones θ_c^*, the performance of the adaptive controller $C(\theta_c)$ tends to that of $C(\theta_c^*)$ used in the case of known parameters. In some approaches, the control law is modified to include nonlinear terms, and this approach deviates somewhat from the CE approach. The principal philosophy, however, that as the estimated parameters converge to the unknown constant parameters the control law converges to that used in the known parameter case, remains the same.

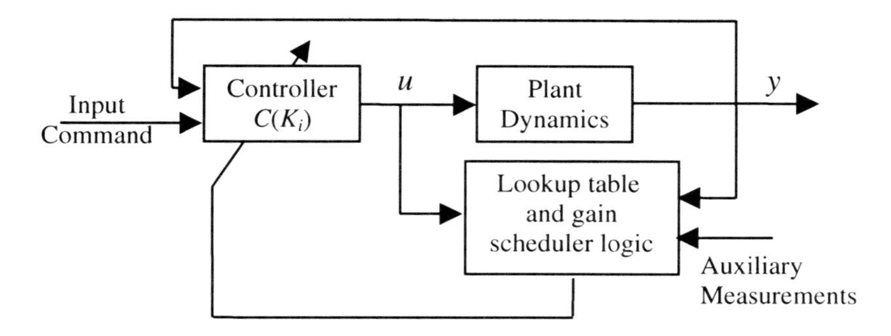

Figure 1.4. *Gain scheduling structure.*

1.2 Adaptive Control: Non–Identifier-Based

Another class of schemes that fit the generic structure given in Figure 1.1 but do not involve online parameter estimators is referred to as non–identifier-based adaptive control schemes. In this class of schemes, the online parameter estimator is replaced with search methods for finding the controller parameters in the space of possible parameters, or it involves switching between different fixed controllers, assuming that at least one is stabilizing or uses multiple fixed models for the plant covering all possible parametric uncertainties or consists of a combination of these methods. We briefly describe the main features, advantages, and limitations of these non–identifier-based adaptive control schemes in the following subsections. Since some of these approaches are relatively recent and research is still going on, we will not discuss them further in the rest of the book.

1.2.1 Gain Scheduling

Let us consider the aircraft model (1.1), where for each operating point $i = 1, 2, \dots, N$ the parameters A_i, B_i, C_i, D_i are known. For each operating point i, a feedback controller with constant gains, say K_i, can be designed to meet the performance requirements for the corresponding linear model. This leads to a controller, say $C(K_i)$, with a set of gains $K_1, K_2, \dots, K_N$ covering N operating points. Once the operating point, say i, is detected the controller gains can be changed to the appropriate value of K_i obtained from the precomputed gain set. Transitions between different operating points that lead to significant parameter changes may be handled by interpolation or by increasing the number of operating points. The two elements that are essential in implementing this approach are a lookup table to store the values of K_i and the plant measurements that correlate well with the changes in the operating points. The approach is called *gain scheduling* and is illustrated in Figure 1.4.

The gain scheduler consists of a lookup table and the appropriate logic for detecting the operating point and choosing the corresponding value of K_i from the lookup table. With this approach, plant parameter variations can be compensated by changing the controller gains as functions of the input, output, and auxiliary measurements. The advantage of gain scheduling is that the controller gains can be changed as quickly as the auxiliary measurements respond to parameter changes. Frequent and rapid changes of the controller gains,

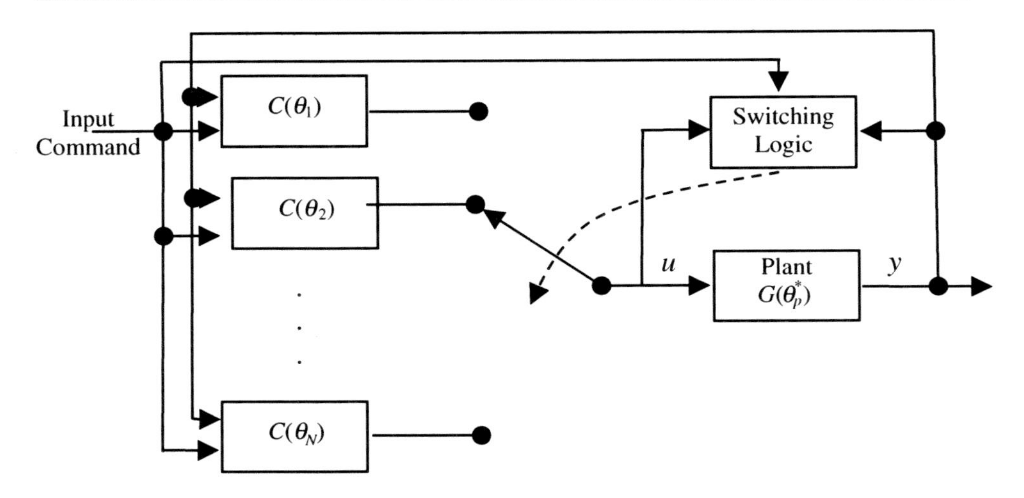

Figure 1.5. *Multiple models adaptive control with switching.*

however, may lead to instability [5]; therefore, there is a limit to how often and how fast the controller gains can be changed. One of the disadvantages of gain scheduling is that the adjustment mechanism of the controller gains is precomputed offline and, therefore, provides no feedback to compensate for incorrect schedules. A careful design of the controllers at each operating point to meet certain robustness and performance measures can accommodate some uncertainties in the values of the plant parameters A_i, B_i, C_i, D_i. Large unpredictable changes in the plant parameters, however, due to failures or other effects may lead to deterioration of performance or even to complete failure. Despite its limitations, gain scheduling is a popular method for handling parameter variations in flight control [3, 6] and other systems [7, 210, 211]. While gain scheduling falls into the generic definition of adaptive control, we do not classify it as adaptive control in this book due to the lack of online parameter estimation which could track unpredictable changes in the plant parameters.

1.2.2 Multiple Models, Search Methods, and Switching Schemes

A class of non–identifier-based adaptive control schemes emerged over the years which do not explicitly rely on online parameter estimation [8–28]. These schemes are based on search methods in the controller parameter space [8] until the stabilizing controller is found or the search method is restricted to a finite set of controllers, one of which is assumed to be stabilizing [22, 23]. In some approaches, after a satisfactory controller is found it can be tuned locally using online parameter estimation for better performance [24–26]. Since the plant parameters are unknown, the parameter space is parameterized with respect to a set of plant models which is used to design a finite set of controllers so that each plant model from the set can be stabilized by at least one controller from the controller set. A switching approach is then developed so that the stabilizing controller is selected online based on the I/O data measurements. Without going into specific details, the general structure of this multiple model adaptive control with switching, as it is often called, is shown in Figure 1.5.

In Figure 1.5, N controllers are used to control a plant whose parameters θ_p^* are unknown or could change with time. In some approaches an a priori knowledge of where the elements of θ_p^* are located, such as lower and upper bounds, is used to parameterize the plant and generate a finite set of controllers so that for each possible plant there exists at least one stabilizing controller from the set of the N controllers. This by itself could be a difficult task in some practical situations where the plant parameters are unknown or change in an unpredictable manner. Furthermore, since there is an infinite number of plants within any given bound of parametric uncertainty, finding controllers to cover all possible parametric uncertainties may also be challenging. In other approaches [22, 23], it is assumed that the set of controllers with the property that at least one of them is stabilizing is available. Once the set of controllers with the stabilizing property is available the problem of finding the stabilizing one using I/O data has to be resolved. This is achieved by the use of a switching logic that differs in detail from one approach to another. While these methods provide another set of tools for dealing with plants with unknown parameters, they cannot replace the identifier-based adaptive control schemes where no assumptions are made about the location of the plant parameters. One advantage, however, is that once the switching is over, the closed-loop system is LTI, and it is much easier to analyze its robustness and performance properties. This LTI nature of the closed-loop system, at least between switches, allows the use of the well-established and powerful robust control tools for LTI systems [29] for controller design. These approaches are still at their infancy and it is not clear how they affect performance, as switching may generate bad transients with adverse effects on performance. Switching may also increase the controller bandwidth and lead to instability in the presence of high-frequency unmodeled dynamics. Guided by data that do not carry sufficient information about the plant model, the wrong controllers could be switched on over periods of time, leading to internal excitation and bad transients before the switching process settles to the right controller. Some of these issues may also exist in classes of identifier-based adaptive control, as such phenomena are independent of the approach used.

1.3 Why Adaptive Control

The choice of adaptive control as a solution to a particular control problem involves understanding of the plant properties as well as of the performance requirements. The following simple examples illustrate situations where adaptive control is superior to linear control. Consider the scalar plant

$$\dot{x} = ax + u,$$

where u is the control input and x the scalar state of the plant. The parameter a is unknown. We want to choose the input u so that the state x is bounded and driven to zero with time. If a is a known parameter, then the linear control law

$$u = -kx, \quad k > |a|,$$

can meet the control objective. In fact if an upper bound $\bar{a} \geq |a|$ is known, the above linear control law with $k > \bar{a}$ can also meet the control objective. On the other hand, if a changes so that $a > k > 0$, then the closed-loop plant will be unstable. The conclusion is that in the absence of an upper bound for the plant parameter no linear controller could stabilize the plant and drive the state to zero. The switching schemes described in Section 1.2.2 could not solve the problem either, unless the controller set included a linear controller with $k > |a|$ or an online parameter estimator were incorporated into the switching design. As we will establish in later chapters, the adaptive control law

$$u = -kx, \quad \dot{k} = x^2,$$

guarantees that all signals are bounded and x converges to zero no matter what the value of the parameter a is. This simple example demonstrates that adaptive control is a potential approach to use in situations where linear controllers cannot handle the parametric uncertainty.

Another example where an adaptive control law may have properties superior to those of the traditional linear schemes is the following. Consider the same example as above but with an external bounded disturbance d:

$$\dot{x} = ax + u + d.$$

The disturbance is unknown but can be approximated as

$$d = \sum_{i=1}^{N} \theta_i^* \phi_i(t, x),$$

where $\phi_i(t, x)$ are known functions and θ_i^* are unknown constant parameters. In this case if we use the linear control law

$$u = -kx$$

with $k > \bar{a} \geq |a|$, we can establish that x is bounded and at steady state

$$|x| \leq \frac{d_o}{k - a},$$

where d_o is an upper bound for $|d|$. It is clear that by increasing the value of the controller gain k, we can make the steady-state value of x as small as we like. This will lead to a high gain controller, however, which is undesirable especially in the presence of high-frequency unmodeled dynamics. In principle, however, we cannot guarantee that x will be driven to zero for any finite control gain in the presence of nonzero disturbance d. The adaptive control approach is to estimate online the disturbance d and cancel its effect via feedback. The following adaptive control law can be shown to guarantee signal boundedness and convergence of the state x to zero with time:

$$u = -kx - \hat{d}, \quad \hat{d} = \sum_{i=1}^{N} \theta_i \phi_i(t, x), \quad \dot{\theta}_i = x\phi_i(t, x),$$

where $k > \bar{a} \geq |a|$, assuming of course that $\bar{a}$ is known; otherwise k has to be estimated, too. Therefore, in addition to stability, adaptive control techniques could be used to improve performance in a wide variety of situations where linear techniques would fail to meet the performance characteristics. This by no means implies that adaptive control is the most appropriate approach to use in every control problem. The purpose of this book is to teach the reader not only the advantages of adaptive control but also its limitations. Adaptive control involves learning, and learning requires data which carry sufficient information about the unknown parameters. For such information to be available in the measured data, the plant has to be excited, and this may lead to transients which, depending on the problem under consideration, may not be desirable. Furthermore, in many applications there is sufficient information about the parameters, and online learning is not required. In such cases, linear robust control techniques may be more appropriate. The adaptive control tools studied in this book complement the numerous control tools already available in the area of control systems, and it is up to the knowledge and intuition of the practicing engineer to determine which tool to use for which application. The theory, analysis, and design approaches presented in this book will help the practicing engineer to decide whether adaptive control is the approach to use for the problem under consideration.

1.4 A Brief History

Research in adaptive control has a long history of intense activities that involved debates about the precise definition of adaptive control, examples of instabilities, stability and robustness proofs, and applications. Starting in the early 1950s, the design of autopilots for high-performance aircraft motivated intense research activity in adaptive control. High-performance aircraft undergo drastic changes in their dynamics when they move from one operating point to another, which cannot be handled by constant-gain feedback control. A sophisticated controller, such as an adaptive controller, that could learn and accommodate changes in the aircraft dynamics was needed. Model reference adaptive control was suggested by Whitaker and coworkers in [30, 31] to solve the autopilot control problem. Sensitivity methods and the MIT rule were used to design the online estimators or adaptive laws of the various proposed adaptive control schemes. An adaptive pole placement scheme based on the optimal linear quadratic problem was suggested by Kalman in [32]. The work on adaptive flight control was characterized by a "lot of enthusiasm, bad hardware and nonexisting theory" [33]. The lack of stability proofs and the lack of understanding of the properties of the proposed adaptive control schemes coupled with a disaster in a flight test [34] caused the interest in adaptive control to diminish.

The 1960s became the most important period for the development of control theory and adaptive control in particular. State-space techniques and stability theory based on Lyapunov were introduced. Developments in dynamic programming [35, 36], dual control [37] and stochastic control in general, and system identification and parameter estimation [38, 39] played a crucial role in the reformulation and redesign of adaptive control. By 1966, Parks [40] and others found a way of redesigning the MIT rule-based adaptive laws used in the model reference adaptive control (MRAC) schemes of the 1950s by applying the Lyapunov design approach. Their work, even though applicable to a special class of LTI plants, set the stage for further rigorous stability proofs in adaptive control for more general classes of plant models. The advances in stability theory and the progress in control theory in the 1960s improved the understanding of adaptive control and contributed to a strong renewed interest in the field in the 1970s. On the other hand, the simultaneous development and progress in computers and electronics that made the implementation of complex controllers, such as the adaptive ones, feasible contributed to an increased interest in applications of adaptive control.

The 1970s witnessed several breakthrough results in the design of adaptive control. MRAC schemes using the Lyapunov design approach were designed and analyzed in [41–44]. The concepts of positivity and hyperstability were used in [45] to develop a wide class of MRAC schemes with well-established stability properties. At the same time parallel efforts for discrete-time plants in a deterministic and stochastic environment produced several classes of adaptive control schemes with rigorous stability proofs [44, 46]. The excitement of the 1970s and the development of a wide class of adaptive control schemes with well-established stability properties were accompanied by several successful applications [47–49]. The successes of the 1970s, however, were soon followed by controversies over the practicality of adaptive control. As early as 1979 it was pointed out by Egardt [41] that the adaptive schemes of the 1970s could easily go unstable in the presence of small disturbances. The nonrobust behavior of adaptive control became very controversial in the early 1980s when more examples of instabilities were published by Ioannou et al. [50, 51] and Rohrs et al. [52] demonstrating lack of robustness in the presence of unmodeled dynamics or bounded disturbances. Rohrs's example of instability stimulated a lot of interest, and the objective of many researchers was directed towards understanding the mechanism of instabilities and finding ways to counteract them. By the mid-1980s, several new redesigns and modifications were proposed and analyzed, leading to a body of work known as *robust adaptive control*. An adaptive controller is defined to be robust if it guarantees signal boundedness in the presence of "reasonable" classes of unmodeled dynamics and bounded disturbances as well as performance error bounds that are of the order of the modeling error. The work on robust adaptive control continued throughout the 1980s and involved the understanding of the various robustness modifications and their unification under a more general framework [41, 53–56]. In discrete time Praly [57, 58] was the first to establish global stability in the presence of unmodeled dynamics using various fixes and the use of a dynamic normalizing signal which was used in Egardt's work to deal with bounded disturbances. The use of the normalizing signal together with the switching σ-modification led to the proof of global stability in the presence of unmodeled dynamics for continuous-time plants in [59].

The solution of the robustness problem in adaptive control led to the solution of the long-standing problem of controlling a linear plant whose parameters are unknown and changing with time. By the end of the 1980s several breakthrough results were published in the area of adaptive control for linear time-varying plants [5, 60–63]. The focus of adaptive control research in the late 1980s to early 1990s was on performance properties and on extending the results of the 1980s to certain classes of nonlinear plants with unknown parameters. These efforts led to new classes of adaptive schemes, motivated from nonlinear system theory [64–69] as well as to adaptive control schemes with improved transient and steady-state performance [70–73]. New concepts such as adaptive backstepping, nonlinear damping, and tuning functions are used to address the more complex problem of dealing with parametric uncertainty in classes of nonlinear systems [66].

In the late 1980s to early 1990s, the use of neural networks as universal approximators of unknown nonlinear functions led to the use of online parameter estimators to "train" or update the weights of the neural networks. Difficulties in establishing global convergence results soon arose since in multilayer neural networks the weights appear in a nonlinear fashion, leading to "nonlinear in the parameters" parameterizations for which globally stable online parameter estimators cannot be developed. This led to the consideration of single layer neural networks where the weights can be expressed in ways convenient for estimation parameterizations. These approaches are described briefly in Chapter 8, where numerous references are also provided for further reading.

In the mid-1980s to early 1990s, several groups of researchers started looking at alternative methods of controlling plants with unknown parameters [8–29]. These methods avoid the use of online parameter estimators in general and use search methods, multiple models to characterize parametric uncertainty, switching logic to find the stabilizing controller, etc. Research in these non–identifier-based adaptive control techniques is still going on, and issues such as robustness and performance are still to be resolved.

Adaptive control has a rich literature full of different techniques for design, analysis, performance, and applications. Several survey papers [74, 75] and books and monographs [5, 39, 41, 45–47, 49, 50, 66, 76–93] have already been published. Despite the vast literature on the subject, there is still a general feeling that adaptive control is a collection of unrelated technical tools and tricks.

The purpose of this book is to present the basic design and analysis tools in a tutorial manner, making adaptive control accessible as a subject to less mathematically oriented readers while at the same time preserving much of the mathematical rigor required for stability and robustness analysis. Some of the significant contributions of the book, in addition to its relative simplicity, include the presentation of different approaches and algorithms in a unified, structured manner which helps abolish much of the mystery that existed in adaptive control. Furthermore, up to now continuous-time adaptive control approaches have been viewed as different from their discrete-time counterparts. In this book we show for the first time that the continuous-time adaptive control schemes can be converted to discrete time by using a simple approximation of the time derivative.

Chapter 2
Parametric Models

Let us consider the first-order system

$$\dot{x} = -x + ax + bu,$$

where x, u are the scalar state and input, respectively, and a, b are the unknown constants we want to identify online using the measurements of x, u. The first step in the design of online parameter identification (PI) algorithms is to lump the unknown parameters in a vector and separate them from known signals, transfer functions, and other known parameters in an equation that is convenient for parameter estimation. For the above example, one such suitable parametric representation is obtained by expressing the above system as

$$x = \frac{1}{s+1}(ax + bu) = a\frac{1}{s+1}x + b\frac{1}{s+1}u,$$

and in the compact algebraic form

$$x = \theta^{*T}\phi,$$

where

$$\theta^* = [a, b]^T, \qquad \phi = \left[\frac{1}{s+1}x, \frac{1}{s+1}u\right]^T.$$

In the general case, this class of parameterizations is of the form

$$z = \theta^{*T}\phi, \tag{2.1}$$

where $\theta^* \in \Re^n$ is the vector with all the unknown parameters and $z \in \Re$, $\phi \in \Re^n$ are signals available for measurement. We refer to (2.1) as the linear *"static" parametric model* (*SPM*). The SPM may represent a dynamic, static, linear, or nonlinear system. Any linear or nonlinear dynamics in the original system are hidden in the signals z, ϕ that usually consist of the I/O measurements of the system and their filtered values.

Another parameterization of the above scalar plant is

$$x = \frac{1}{s+1}[a, b]\begin{bmatrix} x \\ u \end{bmatrix} = \frac{1}{s+1}\theta^{*T}\phi,$$
$$\theta^* = [a, b]^T, \qquad \phi = [x, u]^T.$$

In the general case, the above parametric model is of the form

$$z = W(q)(\theta^{*T}), \tag{2.2}$$

where $z \in \Re$, $\phi \in \Re^n$ are signals available for measurement and $W(q)$ is a known stable proper transfer function, where q is either the shift operator in discrete time (i.e., $q = z$) or the differential operator ($q = s$) in continuous time. We refer to (2.2) as the linear *"dynamic" parametric model (DPM)*. The importance of the SPM and DPM as compared to other possible parameterizations is that the unknown parameter vector θ^* appears linearly. For this reason we refer to (2.1) and (2.2) as *linear in the parameters* parameterizations. As we will show later, this property is significant in designing online PI algorithms whose global convergence properties can be established analytically.

We can derive (2.1) from (2.2) if we use the fact that θ^* is a constant vector and redefine ϕ to obtain

$$z = \theta^{*T}\varphi, \qquad \varphi = W(q)\phi.$$

In a similar manner, we can filter each side of (2.1) or (2.2) using a stable proper filter and still maintain the linear in the parameters property and the form of SPM, DPM. This shows that there exist an infinite number of different parametric models in the form of SPM, DPM for the same parameter vector θ^*.

In some cases, the unknown parameters cannot be expressed in the form of the linear in the parameters models. In such cases the PI algorithms based on such models cannot be shown to converge globally. A special case of nonlinear in the parameters models for which convergence results exist is when the unknown parameters appear in the special bilinear form

$$z = \rho^*(\theta^*\phi + z_1) \tag{2.3}$$

or

$$z = W(q)\rho^*(\theta^*\phi + z_1), \tag{2.4}$$

where $z \in \Re$, $\phi \in \Re^n$, $z_1 \in \Re$ are signals available for measurement at each time t, and $\rho^* \in \Re^n$, $\theta^* \in \Re^n$ are the unknown parameters. The transfer function $W(q)$ is a known stable transfer function. We refer to (2.3) and (2.4) as the *bilinear static parametric model (B-SPM)* and *bilinear dynamic parametric model (B-DPM)*, respectively.

In some applications of parameter identification or adaptive control of plants of the form

$$\dot{x} = Ax + Bu,$$

whose state x is available for measurement, the following parametric model may be used:

$$\dot{x} = A_m x + (A - A_m)x + Bu,$$

where A_m is a stable design matrix; A, B are the unknown matrices; and x, u are signal vectors available for measurement. The model may be also expressed in the form

$$\dot{x} = A_m x + \Theta^{*T}\Phi,$$

where $\Theta^{*^T} = [A - A_m, B]$, $\Phi = [x^T, u^T]^T$. We refer to this class of parametric models as *state-space parametric models (SSPM)*. It is clear that the SSPM can be expressed in the form of the DPM and SPM. Another class of state-space models that appear in adaptive control is of the form

$$\dot{x} = A_m x + B\Theta^{*^T}\Phi,$$

where B is also unknown but is positive definite, is negative definite, or the sign of each of its elements is known. We refer to this class of parametric models as *bilinear state-space parametric models (B-SSPM)*. The B-SSPM model can be easily expressed as a set of scalar B-SPM or B-DPM.

The PI problem can now be stated as follows:

- *For the SPM and DPM*: Given the measurements $z(t)$, $\phi(t)$, generate $\theta(t)$, the estimate of the unknown vector θ^*, at each time t. The PI algorithm updates $\theta(t)$ with time so that as time evolves, $\theta(t)$ approaches or converges to θ^*. Since we are dealing with online PI, we would also expect that if θ^* changes, then the PI algorithm will react to such changes and update the estimate $\theta(t)$ to match the new value of θ^*.

- *For the B-SPM and B-DPM*: Given the measurements of z, z_1, and ϕ, generate the estimates $\theta(t)$, $\rho(t)$ of θ^*, ρ^*, respectively, at each time t the same way as in the case of SPM and DPM.

- *For the SSPM*: Given the measurements of x, u, i.e., Φ, generate the estimate Θ of θ^* (and hence the estimates $\hat{A}(t)$, $\hat{B}(t)$ of A, B, respectively) at each time t the same way as in the case of SPM and DPM.

The online PI algorithms generate estimates at each time t, by using the past and current measurements of signals. Convergence is achieved asymptotically as time evolves. For this reason they are referred to as *recursive* PI algorithms to be distinguished from the *nonrecursive* ones, in which all the measurements are collected a priori over large intervals of time and are processed offline to generate the estimates of the unknown parameters.

Generating the parametric models (2.1)–(2.4) is a significant step in the design of the appropriate PI algorithms. Below, we present several examples that demonstrate how to express the unknown parameters in the form of the parametric models presented above.

Example 2.1 Consider the mass–spring–dashpot system shown in Figure 2.1, where k is the spring constant, f is the viscous-friction or damping coefficient, M is the mass of the system, u is the forcing input, and x is the displacement of the mass M. If we assume that the spring is "linear," i.e., the force acting on the spring is proportional to the displacement, and the friction force is proportional to the velocity $\dot{x}$, using Newton's law we obtain the differential equation that describes the dynamics of the system as

$$M\ddot{x} = u - kx - f\dot{x}. \tag{2.5}$$

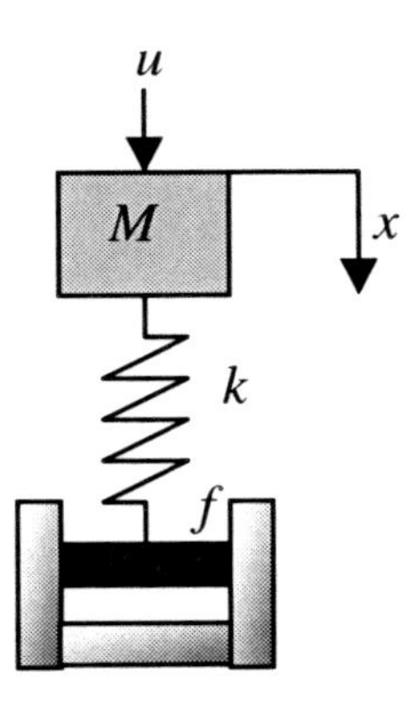

Figure 2.1. *Mass–spring–dashpot system.*

Let us assume that M, f, k are the constant unknown parameters that we want to estimate online. We can easily express (2.5) in the form of SPM by defining

$$\theta^* = [M, f, k]^T, \qquad z = u, \qquad \phi = [\ddot{x}, \dot{x}, x]^T.$$

However, in this formulation we are making the assumption that the vector $\phi = [\ddot{x}, \dot{x}, x]^T$ is available for measurement, which is true, provided that x and its first two derivatives are available for measurement. If not, the parametric model associated with $\phi = [\ddot{x}, \dot{x}, x]^T$ cannot be used for developing PI algorithms because ϕ is not available for measurement. Let us assume that only x, the displacement of the mass, is available for measurement. In this case, in order to express (2.5) in the form of the SPM, we filter both sides of (2.5) with the stable filter $\frac{1}{\Lambda(s)}$, where $\Lambda(s) = (s+\lambda)^2$ and $\lambda > 0$ is a constant design parameter we can choose arbitrarily, to obtain

$$\frac{Ms^2 + fs + k}{\Lambda(s)} x = \frac{1}{\Lambda(s)} u. \tag{2.6}$$

Using (2.6), we can express the unknown parameters in the form of (2.1) as follows:

$$z = \theta^{*T} \phi,$$

where

$$\begin{aligned} z &= \frac{1}{\Lambda(s)} u, \\ \phi &= \left[\frac{s^2}{\Lambda(s)} x, \frac{s}{\Lambda(s)} x, \frac{1}{\Lambda(s)} x \right]^T, \\ \theta^* &= [M, f, k]^T. \end{aligned}$$

In this case z, ϕ are available for measurement since they can be generated by filtering the measurements u and x, respectively. Another possible parametric model is

$$z = \theta^{*T} \phi,$$

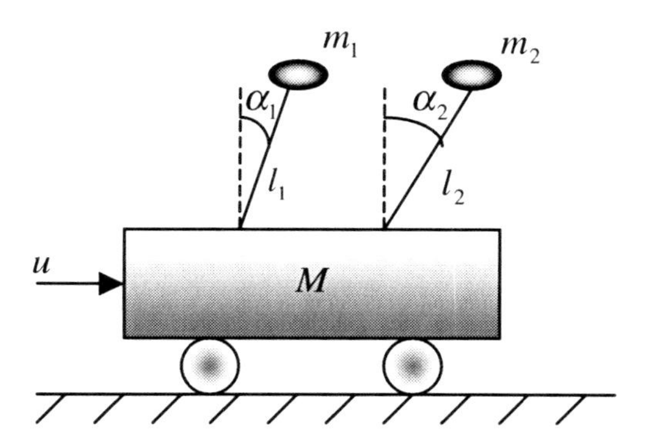

Figure 2.2. *Cart with two inverted pendulums.*

where

$$
\begin{aligned}
z &= \frac{s^2}{\Lambda(s)} x, \\
\phi &= \left[\frac{1}{\Lambda(s)} u, -\frac{s}{\Lambda(s)} x, -\frac{1}{\Lambda(s)} x \right]^T, \\
\theta^* &= \left[\frac{1}{M}, \frac{f}{M}, \frac{k}{M} \right]^T.
\end{aligned}
$$

In this model, the unknown parameters are rearranged to a different vector θ^*. ■

Example 2.2 Consider the cart with two inverted pendulums shown in Figure 2.2, where M is the mass of the cart, m_1 and m_2 are the masses of the bobs, and l_1 and l_2 are the lengths of the pendulums, respectively.

Using Newton's law and assuming small angular deviations of $|\alpha_1|$, $|\alpha_2|$, the equations of motion are given by

$$
\begin{aligned}
M\dot{v} &= -m_1 g \alpha_1 - m_2 g \alpha_2 + u, \\
m_1(\dot{v} + l_1 \ddot{\alpha}_1) &= m_1 g \alpha_1, \\
m_2(\dot{v} + l_2 \ddot{\alpha}_2) &= m_2 g \alpha_2,
\end{aligned}
$$

where v is the velocity of the cart, u is an external force, and g is the acceleration due to gravity. Letting α_1 be the output, i.e., $y = \alpha_1$, the system can be described by the differential equation

$$
y^{(4)} + a_2 y^{(2)} + a_0 y = b_2 u^{(2)} + b_0 u,
$$

where

$$
\begin{aligned}
a_2 &= -\frac{g}{M}\left(\frac{M + m_1}{l_1} + \frac{M + m_2}{l_2} \right), & a_0 &= \frac{(M + m_1 + m_2) g^2}{M l_1 l_2}, \\
b_2 &= \frac{1}{M l_1}, & b_0 &= \frac{g}{M l_1 l_2}.
\end{aligned}
$$

The above equation can be rewritten as

$$y^{(4)} = \theta^{*^T} Y_0,$$

where

$$Y_0 = [u^{(2)}, u, -y^{(2)}, -t]^T,$$
$$\theta^* = [b_2, b_0, a_2, a_0]^T.$$

In order to avoid the use of differentiators, we filter each side with fourth-order stable filter $\frac{1}{\Lambda(s)}$, e.g., $\Lambda(a) = (s+\lambda)^4$, $\lambda > 0$, to obtain the SPM model

$$z = \theta^{*T}\phi,$$

where

$$\begin{aligned} z &= \frac{s^4}{(s+\lambda)^4} y, \\ \phi &= \left[\frac{s^2}{(s+\lambda)^4}u, \frac{1}{(s+\lambda)^4}u, -\frac{s^2}{(s+\lambda)^4}y, -\frac{1}{(s+\lambda)^4}y\right]^T, \\ \theta^* &= [b_2, b_0, a_2, a_0]^T. \end{aligned}$$

If in the above model we know that a_0 is nonzero, redefining the constant parameters as $\bar{b}_2 = \frac{b_2}{a_0}$, $\bar{b}_0 = \frac{b_0}{a_0}$, $\bar{a}_1 = \frac{a_1}{a_0}$, we obtain the following B-SPM:

$$z = \rho^*(\theta^{*T}\phi + z_1),$$

where

$$\begin{aligned} z &= \frac{s^4}{(s+\lambda)^4} y, \qquad z_1 = \frac{1}{(s+\lambda)^4} y, \\ \phi &= \left[\frac{s^2}{(s+\lambda)^4}u, \frac{1}{(s+\lambda)^4}u, -\frac{s^2}{(s+\lambda)^4}y\right]^T, \\ \theta^* &= [\bar{b}_2, \bar{b}_0, \bar{a}_2]^T, \qquad \rho^* = a_0. \end{aligned}$$

∎

Example 2.3 Consider the second-order autoregressive moving average (ARMA) model

$$y(k) = -a_1 y(k-1) - a_2 y(k-2) + b_1 u(k-1) + b_2 u(k-2).$$

This model can be rewritten in the form of the SPM as

$$z(k) = \theta^{*T}\phi(k),$$

where

$$\begin{aligned} z(k) &= y(k), \\ \phi(k) &= [-y(k-1), -y(k-2), u(k-1), u(k-2)]^T, \\ \theta^* &= [a_1, a_2, b_1, b_2]^T. \end{aligned}$$

Let us assume that we know that one of the constant parameters, i.e., b_1, is nonzero. Then we can obtain a model of the system in the B-SPM form as follows:

$$z(k) = \rho^*(\theta^{*T}\phi(k) + z_1(k)),$$

where

$$\begin{aligned} z(k) &= y(k), \qquad z_1(k) = u(k-1), \\ \phi &= [-y(k-1), -y(k-2), u(k-2)]^T, \\ \theta^* &= \left[\frac{a_1}{b_1}, \frac{a_2}{b_1}, \frac{b_2}{b_1}\right]^T, \qquad \rho^* = b_1. \end{aligned}$$ ∎

Example 2.4 Consider the nonlinear system

$$\dot{x} = a_1 f_1(x) + a_2 f_2(x) + b_1 g_1(x) u + b_2 g_2(x) u,$$

where $u, x \in \Re$, f_1, g_1 are known nonlinear functions of x and a_1, b_1 are unknown constant parameters. Filtering both sides of the equation with the filter $\frac{1}{s+1}$, we can express the system in the form of the SPM

$$z = \theta^{*T}\phi,$$

where

$$\begin{aligned} z &= \frac{s}{s+1} x, \\ \phi &= \left[\frac{1}{s+1}[f_1(x)], \frac{1}{s+1}[f_2(x)], \frac{1}{s+1}[g_1(x)u], \frac{1}{s+1}[g_2(x)u]\right]^T, \\ \theta^* &= [a_1, a_, b_1, b_2]^T. \end{aligned}$$ ∎

Example 2.5 Consider the following dynamical system in the transfer function form:

$$y = \frac{s+b}{s^2 + as + c} u, \tag{2.7}$$

where the parameter b is known and a, c are the unknown parameters. The input u and output y are available for measurement. We would like to parameterize the system (2.7) in the form of the SPM. We rewrite (2.7) as

$$s^2 y + asy + cy = (s+b)u.$$

Separating the known terms from the unknown ones, we obtain

$$asy + cy = (s+b)u - s^2 y.$$

In order to avoid the use of derivatives for y and u, we filter each side with a second-order filter $\frac{1}{\Lambda(s)}$, where $\Lambda(s) = (s+\lambda)^2$, $\lambda > 0$. We then lump the unknown parameters a, c in the vector $\theta^* = [a, c]^T$ and obtain the SPM form

$$z = \theta^{*T}\phi,$$

where

$$z = \frac{s+b}{(s+\lambda)^2}u - \frac{s^2}{(s+\lambda)^2}y,$$
$$\phi = \left[\frac{s}{(s+\lambda)^2}y, \frac{1}{(s+\lambda)^2}y\right]^T.$$

Since the parameter b and the input u and output y are known, the signals z, ϕ can be generated by filtering. ∎

Example 2.6 Consider the second-order ARMA model

$$y(k+4) = -a_1 y(k+3) - a_2 y(k+2) + b_1 u(k+1) + b_2 u(k), \tag{2.8}$$

where, at instant k, $y(k)$, $u(k)$, and their past values are available for measurement. In order to express (2.8) in the form of SPM, we rewrite it as

$$y(k) = -a_1 y(k-1) - a_2 y(k-2) + b_1 u(k-3) + b_2 u(k-4)$$

by shifting each side by four in the time axis. This is equivalent to filtering each side with the fourth-order stable filter $\frac{1}{z^4}$. The shifted model where all signals are available for measurement can be rewritten in the form of the SPM as

$$z(k) = \theta^{*T}\phi(k),$$

where

$$z(k) = y(k),$$
$$\phi(k) = [-y(k-1), -y(k-2), u(k-3), u(k-4)]^T,$$
$$\theta = [a_1, a_2, b_1, b_2]^T.$$ ∎

Example 2.7 Consider the system

$$\dot{x} = -2x + \alpha_1 f(x) + \alpha_2 g(x)u,$$

where $f(x)$, $g(x)$ are known functions and α_1, α_2 are unknown constants. For identification purposes, the system may be expressed as

$$x = \frac{1}{s+2}[\alpha_1 f(x) + \alpha_2 g(x)u]$$

and put in the form of the DPM

$$z = W(s)\theta^{*T}\phi,$$

where $z = x$, $W(s) = \frac{1}{s+2}$, $\theta^* = [\alpha_1, \alpha_2]^T$, $\phi = [f(x), g(x)u]^T$.

If we want $W(s)$ to be a design transfer function with a pole, say at $\lambda > 0$, we write

$$\dot{x} = -\lambda x + (\lambda - 2)x + \alpha_1 f(x) + \alpha_2 g(x)u$$

by adding and subtracting the term λx. We then rewrite it as

$$x - \frac{\lambda - 2}{s + \lambda} x = \frac{1}{s + \lambda}[\alpha_1 f(x) + \alpha_2 g(x) u].$$

Letting $z = x - \frac{\lambda-2}{s+\lambda} x = \frac{s+2}{s+\lambda} x$, $W(s) = \frac{1}{s+\lambda}$, $\theta^* = [\alpha_1, \alpha_2]^T$, $\phi = [f(x), g(x)u]^T$, we obtain another DPM. ∎

Example 2.8 Consider the second-order plant

$$\dot{x} = Ax + Bu,$$

where $x = [x_1, x_2]^T$, $u = [u_1, u_2]^T$, and

$$A = \begin{bmatrix} a_{11} & a_{12} \\ a_{21} & a_{22} \end{bmatrix}, \qquad B = \begin{bmatrix} b_{11} & b_{12} \\ b_{21} & b_{22} \end{bmatrix}$$

are matrices with unknown elements. The SSPM is generated as

$$\dot{x} = \begin{bmatrix} -a_m & 0 \\ 0 & -a_m \end{bmatrix} x + \begin{bmatrix} a_{11} + a_m & a_{12} \\ a_{21} & a_{22} + a_m \end{bmatrix} x + \begin{bmatrix} b_{11} & b_{12} \\ b_{21} & b_{22} \end{bmatrix} u,$$

where $a_m > 0$ is a design constant. The model may be also expressed as

$$\dot{x} = \begin{bmatrix} -a_m & 0 \\ 0 & -a_m \end{bmatrix} x + \Theta^{*^T} \Phi,$$

where

$$\Theta^{*T} = \begin{bmatrix} a_{11} + a_m, a_{12}, b_{11}, b_{12} \\ a_{21}, a_{22} + a_m, b_{21}, b_{22} \end{bmatrix}, \qquad \Phi = [x_1, x_2, u_1, u_2]^T.$$ ∎

Example 2.9 (parametric model for nth-order SISO LTI system) Consider the SISO LTI system described by the I/O relation

$$y = \frac{Z(s)}{R(s)} u,$$

where

$$R(s) = s^n + a_{n-1}s^{n-1} + \cdots + a_1 s + a_0, \qquad Z(s) = b_m s^m + \cdots + b_1 s + b_0,$$

and u and y are the plant scalar input and output, respectively. We can also express the above system as an nth-order differential equation given by

$$y^{(n)} + a_{n-1} y^{(n-1)} + \cdots + a_1 \dot{y} + a_0 y = b_m u^{(m)} + \cdots + b_1 \dot{u} + b_0 u.$$

Lumping all the parameters in the vector

$$\theta^* = [b_m, \ldots, b_0, a_{n-1}, \ldots, a_0]^T,$$

we can rewrite the differential equation as

$$y^{(n)} = \theta^{*T}[u^{(m)}, \ldots, u, -y^{(n-1)}, \ldots, -y]^T.$$

Filtering both sides by $\frac{1}{\Lambda(s)}$, where $\Lambda(s) = s^n + \lambda_{n-1}s^{n-1} + \cdots + \lambda_1 s + \lambda_0$ is a monic Hurwitz polynomial, we obtain the parametric model

$$z = \theta^{*T}\phi,$$

where

$$z = \frac{1}{\Lambda(s)}y^{(n)} = \frac{s^n}{\Lambda(s)}y,$$
$$\theta^* = [b_m, \ldots, b_0, a_{n-1}, \ldots, a_0]^T \in \mathcal{R}^{n+m+1},$$
$$\phi = \left[\frac{s^m}{\Lambda(s)}u, \ldots, \frac{1}{\Lambda(s)}u, -\frac{s^{n-1}}{\Lambda(s)}y, \ldots, -\frac{1}{\Lambda(s)}y\right]^T. \quad \blacksquare$$

See the web resource [94] for examples using the Adaptive Control Toolbox.

Problems

1. Consider the third-order plant

$$y = G(s)u,$$

where

$$G(s) = \frac{b_2 s^2 + b_1 s + b_0}{s^3 + a_2 s^2 + a_1 s + a_0}.$$

   (a) Obtain parametric models for the plant in the form of SPM and DPM when $\theta^* = [b_2, b_1, b_0, a_2, a_1, a_0]^T$.

   (b) If a_0, a_1, and a_2 are known, i.e., $a_0 = 2$, $a_1 = 1$, and $a_2 = 3$, obtain a parametric model for the plant in terms of $\theta^* = [b_2, b_1, b_0]^T$.

   (c) If b_0, b_1, and b_2 are known, i.e., $b_0 = 2$, $b_1 = b_2 = 0$, obtain a parametric model in terms of $\theta^* = [a_2, a_1, a_0]^T$.

2. Consider the mass–spring–dashpot system of Figure 2.1 described by (2.5) with x, u as the only signals available for measurement. Let us assume that $M = 100$ kg and f, k are the unknown constant parameters that we want to estimate online. Develop a parametric model for estimating the unknown parameters f, k. Specify any arbitrary parameters or filters used.

3. Consider the second-order ARMA model

$$y(k) = -1.3y(k-1) - a_2 y(k-2) + b_1 u(k-1) + u(k-2),$$

where the parameters a_2, b_1 are unknown constants. Express the unknown parameters in the form of a linear parametric model. Assume that $u(k)$, $y(k)$, and their past values are available for measurement.

4. Consider the fourth-order ARMA model

$$y(k+4) = a_1 y(k+3) - a_2 y(k) + b_1 u(k) + u(k+2),$$

where a_1, a_2, b_1 are unknown constants. Express the unknown parameters in the form of a linear parametric model. Assume that only the current and the past four values of the signals u and y, i.e., $u(k), \ldots, u(k-4), y(k), \ldots, y(k-4)$, are available for measurement.

5. Consider the nonlinear system

$$\ddot{x} + 2\dot{x} + x = a_1 f_1(x) + a_2 f_2(x) + b_1 g_1(x)u + b_2 g_2(x)u,$$

where a_1, a_2, b_1, b_2 are unknown constants and x, $f_1(x)$, $f_2(x)$, $g_1(x)$, $g_2(x)$, u are available for measurement. Express the unknown parameters in the form of

   (a) the linear SPM,

   (b) the linear DPM.

6. Consider the following system described in the I/O form

$$y = K_p \frac{s+b}{s^2 + as + c} u,$$

where b, a, c, K_p are unknown constants. In addition, we know that $K_p > 0$ and only u and y are available for measurement. Express the unknown parameters in the form of the

   (a) B-SPM,

   (b) B-DPM,

   (c) linear SPM,

   (d) linear DPM.

7. Consider the nonlinear system

$$\dot{x} = f(x) + g(x)u,$$

where the state x and the input u are available for measurement and $f(x)$, $g(x)$ are smooth but unknown functions of x. In addition, it is known that $g(x) > 0 \ \forall x$. We want to estimate the unknown functions f, g online using neural network approximation techniques. It is known that there exist constant parameters W_{fi}^*, W_{gi}^*, referred to as weights, such that

$$f(x) \approx \sum_{i=1}^{m} W_{fi}^* \varphi_{fi}(x),$$

$$g(x) \approx \sum_{i=1}^{n} W_{gi}^* \varphi_{gi}(x),$$

where $\varphi_{fi}(\cdot)$, $\varphi_{gi}(\cdot)$ are some basis functions that are known and n, m are known integers representing the number of nodes of the neural network. Obtain a parameterization of the system in the form of SPM that can be used to identify the weights W_{fi}^*, W_{gi}^* online.

8. Consider the mass–spring–dashpot system of Figure 2.1 described by (2.5) and the SPM with $\theta^* = [1/M, f/M, k/M]^T$ presented in Example 2.1.

   (a) Generate the signals z, ϕ of the parametric model using the Adaptive Control Toolbox for $M = 100$ kg, $f = 0.15$ kg/sec, $k = 7$ kg/sec^2, $u(t) = 1 + \cos(\frac{\pi}{3}t)$, and $0 \leq t \leq 25$ sec.

   (b) The SPM in (a) is based on the assumption that M, f, k are unknown. Assume that M is known. Use the Adaptive Control Toolbox to generate the signals of the reduced SPM for the same values of M, f, k, $u(t) = 1 + \cos(\frac{\pi}{3}t)$, and $0 \leq t \leq 25$ sec.

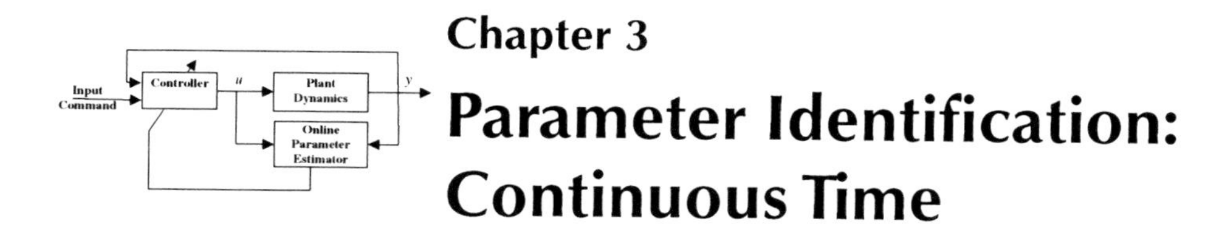

Chapter 3
Parameter Identification: Continuous Time

3.1 Introduction

The purpose of this chapter is to present the design, analysis, and simulation of a wide class of algorithms that can be used for online parameter identification of continuous-time plants. The online identification procedure involves the following three steps.

Step 1. Lump the unknown parameters in a vector θ^* and express them in the form of the parametric model SPM, DPM, B-SPM, or B-DPM.

Step 2. Use the estimate θ of θ^* to set up the *estimation model* that has the same form as the parametric model. The difference between the outputs of the estimation and parametric models, referred to as the *estimation error*, reflects the distance of the estimated parameters $\theta(t)$ from the unknown parameters θ^* weighted by some signal vector. The estimation error is used to drive the adaptive law that generates $\theta(t)$ online. The adaptive law is a differential equation of the form

$$\dot{\theta} = H(t)\varepsilon,$$

where ε is the estimation error that reflects the difference between $\theta(t)$ and θ^* and $H(t)$ is a time-varying gain vector that depends on measured signals. A wide class of adaptive laws with different $H(t)$ and ε may be developed using optimization techniques and Lyapunov-type stability arguments.

Step 3. Establish conditions that guarantee that $\theta(t)$ converges to θ^* with time. This step involves the design of the plant input so that the signal vector $\phi(t)$ in the parametric model is *persistently exciting* (a notion to be defined later on), i.e., it has certain properties that guarantee that the measured signals that drive the adaptive law carry sufficient information about the unknown parameters. For example, for $\phi(t) = 0$, we have $z = \theta^{*T}\phi = 0$, and the measured signals ϕ, z carry no information about θ^*. Similar arguments could be made for ϕ that is orthogonal to θ^* leading to $z = 0$ even though $\theta \neq \theta^*$, etc.

We demonstrate the three design steps using the following example of a scalar plant.

3.2 Example: One-Parameter Case

Consider the first-order plant model

$$y = \frac{a}{s+2}u, \tag{3.1}$$

where a is the only unknown parameter and y and u are the measured output and input of the system, respectively.

Step 1: Parametric Model We write (3.1) as

$$y = a\frac{1}{s+2}u = au_f, \tag{3.2}$$

where $u_f = \frac{1}{s+2}u$. Since u is available for measurement, u_f is also available for measurement. Therefore, (3.2) is in the form of the SPM

$$z = \theta^* \phi, \tag{3.3}$$

where $\theta^* = a$ and $z = y$, $\phi = u_f$ are available for measurement.

Step 2: Parameter Identification Algorithm This step involves the development of an estimation model and an estimation error used to drive the adaptive law that generates the parameter estimates.

Estimation Model and Estimation Error The estimation model has the same form as the SPM with the exception that the unknown parameter θ^* is replaced with its estimate at time t, denoted by $\theta(t)$, i.e.,

$$\hat{z} = \theta(t)\phi, \tag{3.4}$$

where $\hat{z}$ is the estimate of z based on the parameter estimate $\theta(t)$ at time t. It is obvious that the difference between z and $\hat{z}$ is due to the difference between $\theta(t)$ and θ^*. As $\theta(t)$ approaches θ^* with time we would expect that $\hat{z}$ would approach z at the same time. (Note that the reverse is not true, i.e., $\hat{z}(t) = z(t)$ does not imply that $\theta(t) = \theta^*$; see Problem 1.) Since θ^* is unknown, the difference $\tilde{\theta} = \theta(t) - \theta^*$ is not available for measurement. Therefore, the only signal that we can generate, using available measurements, that reflects the difference between $\theta(t)$ and θ^* is the error signal

$$\varepsilon = \frac{z - \hat{z}}{m_s^2}, \tag{3.5}$$

which we refer to as the *estimation error*. $m_s^2 \geq 1$ is a normalization signal[1] designed to guarantee that $\frac{\phi}{m_s}$ is bounded. This property of m_s is used to establish the boundedness of the estimated parameters even when ϕ is not guaranteed to be bounded. A straightforward choice for m_s in this example is $m_s^2 = 1 + \alpha\phi^2$, $\alpha > 0$. If ϕ is bounded, we can take $\alpha = 0$,

[1] Note that any $m_s^2 \geq$ nonzero constant is adequate. The use of a lower bound 1 is without loss of generality.

i.e., $m_s^2 = 1$. Using (3.4) in (3.5), we can express the estimation error as a function of the parameter error $\tilde{\theta} = \theta(t) - \theta^*$, i.e.,

$$\varepsilon = -\frac{\tilde{\theta}\phi}{m_s^2}. \tag{3.6}$$

Equation (3.6) shows the relationship between the estimation error ε and the parameter error $\tilde{\theta}$. It should be noted that ε cannot be generated using (3.6) because the parameter error $\tilde{\theta}$ is not available for measurement. Consequently, (3.6) can be used only for analysis.

Adaptive Law A wide class of adaptive laws or parameter estimators for generating $\theta(t)$, the estimate of θ^*, can be developed using (3.4)–(3.6). The simplest one is obtained by using the SPM (3.3) and the fact that ϕ is scalar to write

$$\theta(t) = \frac{z(t)}{\phi(t)}, \tag{3.7}$$

provided $\phi(t) \neq 0$. In practice, however, the effect of noise on the measurements of $\phi(t)$, especially when $\phi(t)$ is close to zero, may lead to erroneous parameter estimates. Another approach is to update $\theta(t)$ in a direction that minimizes a certain cost of the estimation error ε. With this approach, $\theta(t)$ is adjusted in a direction that makes $|\varepsilon|$ smaller and smaller until a minimum is reached at which $|\varepsilon| = 0$ and updating is terminated. As an example, consider the cost criterion

$$J(\theta) = \frac{\varepsilon^2 m_s^2}{2} = \frac{(z - \theta\phi)^2}{2m_s^2}, \tag{3.8}$$

which we minimize with respect to θ using the gradient method to obtain

$$\dot{\theta} = -\gamma \nabla J(\theta), \tag{3.9}$$

where $\gamma > 0$ is a scaling constant or step size which we refer to as the *adaptive gain* and where $\nabla J(\theta)$ is the gradient of J with respect to θ. In this scalar case,

$$\nabla J(\theta) = \frac{dJ}{d\theta} = -\frac{(z - \theta\phi)}{m_s^2}\phi = -\varepsilon\phi,$$

which leads to the adaptive law

$$\dot{\theta} = \gamma\varepsilon\phi, \quad \theta(0) = \theta_0. \tag{3.10}$$

Step 3: Stability and Parameter Convergence The adaptive law should guarantee that the parameter estimate $\theta(t)$ and the speed of adaptation $\dot{\theta}$ are bounded and that the estimation error ε gets smaller and smaller with time. These conditions still do not imply that $\theta(t)$ will get closer and closer to θ^* with time unless some conditions are imposed on the vector $\phi(t)$, referred to as the *regressor vector*.

Let us start by using (3.6) and the fact that $\dot{\tilde{\theta}} = \dot{\theta} - \dot{\theta}^* = \dot{\theta}$ (due to θ^* being constant) to express (3.10) as

$$\dot{\tilde{\theta}} = -\gamma \frac{\phi^2}{m_s^2} \tilde{\theta}, \quad \tilde{\theta}(0) = \tilde{\theta}_0. \tag{3.11}$$

This is a scalar linear time-varying differential equation whose solution is

$$\tilde{\theta}(t) = e^{-\gamma \int_0^t \frac{\phi^2(\tau)}{m_s^2(\tau)} d\tau} \tilde{\theta}_0, \tag{3.12}$$

which implies that for

$$\int_0^t \frac{\phi^2(\tau)}{m_s^2(\tau)} d\tau \geq \alpha_0 t \tag{3.13}$$

and some $\alpha_0 > 0$, $\tilde{\theta}(t)$ converges to zero exponentially fast, which in turn implies that $\theta(t) \rightarrow \theta^*$ exponentially fast. It follows from (3.12) that $\theta(t)$ is always bounded for any $\phi(t)$ and from (3.11) that $\dot{\theta}(t) = \dot{\tilde{\theta}}(t)$ is bounded due to $\frac{\phi(t)}{m_s(t)}$ being bounded.

Another way to analyze (3.11) is to use a Lyapunov-like approach as follows: We consider the function

$$V = \frac{\tilde{\theta}^2}{2\gamma}.$$

Then

$$\dot{V} = \frac{\tilde{\theta}}{\gamma} \frac{d\tilde{\theta}}{dt} = -\frac{\phi^2}{m_s^2} \tilde{\theta}^2 \leq 0$$

or, using (3.6),

$$\dot{V} = -\frac{\phi^2}{m_s^2} \tilde{\theta}^2 = -\varepsilon^2 m_s^2 \leq 0. \tag{3.14}$$

We should note that $\dot{V} = -\varepsilon^2 m_s^2 \leq 0$ implies that $\dot{V}$ is a negative semidefinite function in the space of $\tilde{\theta}$. $\dot{V}$ in this case is not negative definite in the space of $\tilde{\theta}$ because it can be equal to zero when $\tilde{\theta}$ is not zero. Consequently, if we apply the stability results of the Appendix, we can conclude that the equilibrium $\tilde{\theta}_e = 0$ of (3.11) is uniformly stable (u.s.) and that the solution of (3.11) is uniformly bounded (u.b.). These results are not as useful, as our objective is asymptotic stability, which implies that the parameter error converges to zero. We can use the properties of V, $\dot{V}$, however, to obtain additional properties for the solution of (3.11) as follows.

Since $V > 0$ and $\dot{V} \leq 0$, it follows that (see the Appendix) V is bounded, which implies that $\tilde{\theta}$ is bounded and V converges to a constant, i.e., $\lim_{t\rightarrow\infty} V(t) = V_\infty$. Let us now integrate both sides of (3.14). We have

$$\int_0^t \dot{V}(\tau) d\tau = -\int_0^t \varepsilon^2(\tau) m_s^2(\tau) d\tau$$

or

$$V(t) - V(0) = -\int_0^t \varepsilon^2(\tau) m_s^2(\tau) d\tau. \tag{3.15}$$

Since $V(t)$ converges to the limit V_∞ as $t \to \infty$, it follows from (3.15) that

$$\int_0^\infty \varepsilon^2(\tau) m_s^2(\tau) d\tau = V(0) - V_\infty < \infty,$$

i.e., εm_s is square integrable or $\varepsilon m_s \in \mathcal{L}_2$. Since $m_s^2 \geq 1$, we have $\varepsilon^2 \leq \varepsilon^2 m_s^2$, which implies $\varepsilon \in \mathcal{L}_2$. From (3.6) we conclude that $\frac{\tilde{\theta}\phi}{m_s} \in \mathcal{L}_2$ due to $\varepsilon m_s \in \mathcal{L}_2$. Using (3.10), we write

$$\dot{\theta} = \gamma \varepsilon m_s \frac{\phi}{m_s}.$$

Since $\frac{\phi}{m_s}$ is bounded and $\varepsilon m_s \in \mathcal{L}_2 \cap \mathcal{L}_\infty$, it follows (see Problem 2) that $\dot{\theta} \in \mathcal{L}_2 \cap \mathcal{L}_\infty$. In summary, we have established that the adaptive law (3.10) guarantees that (i) $\theta \in \mathcal{L}_\infty$ and (ii) $\varepsilon, \varepsilon m_s, \dot{\theta} \in \mathcal{L}_2 \cap \mathcal{L}_\infty$ independent of the boundedness of ϕ. The $\mathcal{L}_2$ property of ε, εm_s, and $\dot{\theta}$ indicates that the estimation error and the speed of adaptation $\dot{\theta}$ are bounded in the $\mathcal{L}_2$ sense, which in turn implies that their average value tends to zero with time.

It is desirable to also establish that ε, εm_s, and $\dot{\theta}$ go to zero as $t \to \infty$, as such a property will indicate the end of adaptation and the completion of learning. Such a property can be easily established when the input u is bounded (see Problem 3).

The above properties still do not guarantee that $\theta(t) \to \theta^*$ as $t \to \infty$. In order to establish that $\theta(t) \to \theta^*$ as $t \to \infty$ exponentially fast, we need to restrict $\frac{\phi}{m_s}$ to be persistently exciting (PE), i.e., to satisfy

$$\frac{1}{T} \int_t^{t+T} \frac{\phi^2(\tau)}{m_s^2} d\tau \geq \alpha_0 > 0 \tag{3.16}$$

$\forall t \geq 0$ and some constants $T, \alpha_0 > 0$. The PE property of $\frac{\phi}{m_s}$ is guaranteed by choosing the input u appropriately. Appropriate choices of u for this particular example include (i) $u = c > 0$, (ii) $u = \sin \omega t$ for any $\omega \neq 0$ and any bounded input u that is not vanishing with time. The condition (3.16) is necessary and sufficient for exponential convergence of $\theta(t) \to \theta^*$ (see Problem 4).

The PI algorithm for estimating the constant a in the plant (3.1) can now be summarized as

$$\begin{aligned}
\dot{\theta} &= \gamma \varepsilon \phi, & \theta(0) &= \theta_0, & \\
\varepsilon &= \frac{(z - \hat{z})}{m_s^2}, & \hat{z} &= \theta \phi, & \\
z &= y, & \phi &= \frac{1}{s+2} u, & m_s^2 = 1 + \phi^2,
\end{aligned}$$

where $\theta(t)$ is the estimate of the constant a in (3.1).

The above analysis for the scalar example carries over to the vector case without any significant modifications, as demonstrated in the next section. One important difference, however, is that in the case of a single parameter, convergence of the Lyapunov-like function V to a constant implies that the estimated parameter converges to a constant. Such a result cannot be established in the case of more than one parameter for the gradient algorithm.

3.3 Example: Two Parameters

Consider the plant model

$$y = \frac{b}{s+a}u, \tag{3.17}$$

where a, b are unknown constants. Let us assume that $y, \dot{y}, u$ are available for measurement. We would like to generate online estimates for the parameters a, b.

Step 1: Parametric Model Since y, $\dot{y}$ are available for measurement, we can express (3.17) in the SPM form

$$z = \theta^{*T}\phi,$$

where $z = \dot{y}$, $\theta^* = [b, a]^T$, $\phi = [u, -y]^T$, and z, ϕ are available for measurement.

Step 2: Parameter Identification Algorithm

Estimation Model

$$\hat{z} = \theta^T\phi,$$

where $\theta(t)$ is the estimate of θ^* at time t.

Estimation Error

$$\varepsilon = \frac{z-\hat{z}}{m_s^2} = \frac{z-\theta^T\phi}{m_s^2}, \tag{3.18}$$

where m_s is the normalizing signal such that $\frac{\phi}{m_s} \in \mathcal{L}_\infty$. A straightforward choice for m_s is $m_s^2 = 1 + \alpha\phi^T\phi$ for any $\alpha > 0$.

Adaptive Law We use the gradient method to minimize the cost,

$$J(\theta) = \frac{\varepsilon^2 m_s^2}{2} = \frac{(z-\theta^T\phi)^2}{2m_s^2} = \frac{(z-\theta_1\phi_1-\theta_2\phi_2)^2}{2m_s^2},$$

where $\phi_1 = u$, $\phi_2 = -y$, and set

$$\dot{\theta} = -\Gamma\nabla J,$$

where

$$\nabla J = \left[\frac{\partial J}{\partial\theta_1}, \frac{\partial J}{\partial\theta_2}\right]^T,$$

$\Gamma = \Gamma^T > 0$ is the adaptive gain, and θ_1, θ_2 are the elements of $\theta = [\theta_1, \theta_2]^T$. Since

$$\frac{\partial J}{\partial\theta_1} = -\frac{(z-\theta^T\phi)}{m_s^2}\phi_1 = -\varepsilon\phi_1, \qquad \frac{\partial J}{\partial\theta_2} = -\frac{(z-\theta^T\phi)}{m_s^2}\phi_2 = -\varepsilon\phi_2,$$

we have

$$\dot{\theta} = \Gamma\varepsilon\phi, \qquad \theta(0) = \theta_0, \tag{3.19}$$

which is the adaptive law for updating $\theta(t)$ starting from some initial condition $\theta(0) = \theta_0$.

Step 3: Stability and Parameter Convergence As in the previous example, the equation for the parameter error $\tilde{\theta} = \theta - \theta^*$ is obtained from (3.18), (3.19) by noting that

$$\varepsilon = \frac{z - \theta^T\phi}{m_s^2} = \frac{\theta^{*T}\phi - \theta^T\phi}{m_s^2} = -\frac{\tilde{\theta}^T\phi}{m_s^2} = -\frac{\phi^T\tilde{\theta}}{m_s^2} \tag{3.20}$$

and $\dot{\tilde{\theta}} = \dot{\theta}$, i.e.,

$$\dot{\tilde{\theta}} = \Gamma\phi\varepsilon = -\Gamma\frac{\phi\phi^T}{m_s^2}\tilde{\theta}. \tag{3.21}$$

It is clear from (3.21) that the stability of the equilibrium $\tilde{\theta}_e = 0$ will very much depend on the properties of the time-varying matrix $-\frac{\Gamma\phi\phi^T}{m_s^2}$, which in turn depends on the properties of ϕ. For simplicity let us assume that the plant is stable, i.e., $a > 0$. If we choose $m_s^2 = 1$, $\Gamma = \gamma I$ for some $\gamma > 0$ and a constant input $u = c_0 > 0$, then at steady state $y = c_1 \triangleq \frac{c_0 b}{a} \neq 0$ and $\phi = [c_0, -c_1]^T$, giving

$$-\frac{\Gamma\phi\phi^T}{m_s^2} = -\gamma\begin{bmatrix} c_0^2 & -c_0c_1 \\ -c_0c_1 & c_1^2 \end{bmatrix} \triangleq A,$$

i.e.,

$$\dot{\tilde{\theta}} = A\tilde{\theta},$$

where A is a constant matrix with eigenvalues 0, $-\gamma(c_0^2 + c_1^2)$, which implies that the equilibrium $\theta_e = 0$ is only marginally stable; i.e., $\tilde{\theta}$ is bounded but does not necessarily converge to 0 as $t \to \infty$. The question that arises in this case is what properties of ϕ guarantee that the equilibrium $\tilde{\theta}_e = 0$ is exponentially stable. Given that

$$\phi = H(s)u,$$

where for this example $H(s) = [1, -\frac{b}{s+a}]^T$, the next question that comes up is how to choose u to guarantee that ϕ has the appropriate properties that imply exponential stability for the equilibrium $\tilde{\theta}_e = 0$ of (3.21). Exponential stability for the equilibrium point $\tilde{\theta}_e = 0$ of (3.21) in turn implies that $\theta(t)$ converges to θ^* exponentially fast. As demonstrated above for the two-parameter example, a constant input $u = c_0 > 0$ does not guarantee exponential stability. We answer the above questions in the following section.

3.4 Persistence of Excitation and Sufficiently Rich Inputs

We start with the following definition.

Definition 3.4.1. *The vector $\phi \in \mathcal{R}^n$ is PE with level α_0 if it satisfies*

$$\int_t^{t+T_0} \phi(\tau)\phi^T(\tau)d\tau \geq \alpha_0 T_0 I \tag{3.22}$$

for some $\alpha_0 > 0$, $T_0 > 0$ and $\forall t \geq 0$.

Since $\phi\phi^T$ is always positive semidefinite, the PE condition requires that its integral over any interval of time of length T_0 is a positive definite matrix.

Definition 3.4.2. *The signal $u \in \mathcal{R}$ is called sufficiently rich of order n if it contains at least $\frac{n}{2}$ distinct nonzero frequencies.*

For example, $u = \sum_{i=1}^{10} \sin \omega_i t$, where $\omega_i \neq \omega_j$ for $i \neq j$ is sufficiently rich of order 20. A more general definition of sufficiently rich signals and associated properties may be found in [95].

Let us consider the signal vector $\phi \in \mathcal{R}^n$ generated as

$$\phi = H(s)u, \tag{3.23}$$

where $u \in \mathcal{R}$ and $H(s)$ is a vector whose elements are transfer functions that are strictly proper with stable poles.

Theorem 3.4.3. *Consider (3.23) and assume that the complex vectors $H(j\omega_1), \dots, H(j\omega_n)$ are linearly independent on the complex space $\mathcal{C}^n$ $\forall \omega_1, \omega_2, \dots, \omega_n \in \mathcal{R}$, where $\omega_i \neq \omega_j$ for $i \neq j$. Then ϕ is PE if and only if u is sufficiently rich of order n.*

Proof. The proof of Theorem 3.4.3 can be found in [56, 87]. $\square$

We demonstrate the use of Theorem 3.4.3 for the example in section 3.3, where

$$\phi = H(s)u$$

and

$$H(s) = \begin{bmatrix} 1 \\ -\frac{b}{s+a} \end{bmatrix}.$$

In this case $n = 2$ and

$$H(j\omega_1) = \begin{bmatrix} 1 \\ -\frac{b}{j\omega_1+a} \end{bmatrix}, \qquad H(j\omega_2) = \begin{bmatrix} 1 \\ -\frac{b}{j\omega_2+a} \end{bmatrix}.$$

We can show that the matrix $[H(j\omega_1), H(j\omega_2)]$ is nonsingular, which implies that $H(j\omega_1), H(j\omega_2)$ are linearly independent for any ω_1, ω_2 different than zero and $\omega_1 \neq \omega_2$.

Let us choose

$$u = \sin \omega_0 t$$

for some $\omega_0 \neq 0$ which is sufficiently rich of order 2. According to Theorem 3.4.3, this input should guarantee that ϕ is PE for the example in section 3.3. Ignoring the transient terms that converge to zero exponentially fast, we can show that at steady state

$$\phi = \begin{bmatrix} \sin \omega_0 t \\ c_0 \sin(\omega_0 t + \varphi_0) \end{bmatrix},$$

where

$$c_0 = \frac{|b|}{\sqrt{\omega_0^2 + a^2}}, \qquad \varphi_0 = \arg\left(\frac{-b}{j\omega_0 + a}\right).$$

Now

$$\phi\phi^T = \begin{bmatrix} \sin^2 \omega_0 t & c_0 \sin \omega_0 t \sin(\omega_0 t + \varphi_0) \\ c_0 \sin \omega_0 t \sin(\omega_0 t + \varphi_0) & c_0^2 \sin^2(\omega_0 t + \varphi) \end{bmatrix}$$

and

$$\int_t^{t+T_0} \phi(\tau)\phi^T(\tau)d\tau = \begin{bmatrix} a_{11} & a_{12} \\ a_{12} & a_{22} \end{bmatrix},$$

where

$$\begin{aligned} a_{11} &= \frac{T_0}{2} - \frac{\sin 2\omega_0(t+T_0) - \sin 2\omega_0 t}{4\omega_0}, \\ a_{12} &= c_0 \frac{T_0}{2} \cos\varphi_0 + c_0 \frac{\sin\varphi_0}{4\omega_0}(\cos 2\omega_0 t - \cos 2\omega_0(t+T_0)), \\ a_{22} &= c_0^2 \frac{T_0}{2} - c_0^2 \frac{\sin 2(\omega_0(t+T_0)+\varphi_0) - \sin 2(\omega_0 t + \varphi_0)}{4\omega_0}. \end{aligned}$$

Choosing $T_0 = \frac{\pi}{\omega_0}$ it follows that

$$a_{11} = \frac{T_0}{2}, \qquad a_{12} = \frac{T_0 c_0}{2} \cos\varphi_0, \qquad a_{22} = \frac{T_0}{2} c_0^2$$

and

$$\int_t^{t+T_0} \phi(\tau)\phi^T(\tau)d\tau = \frac{T_0}{2} \begin{bmatrix} 1 & c_0\cos\varphi_0 \\ c_0 \cos\varphi_0 & c_0^2 \end{bmatrix},$$

which is a positive definite matrix. We can verify that for $\alpha_0 = \frac{1}{2}\frac{(1-\cos^2\varphi_0)c_0^2}{1+c_0^2} > 0$,

$$\int_t^{t+T_0} \phi(\tau)\phi^T(\tau)d\tau \geq T_0\alpha_0 I,$$

which implies that ϕ is PE.

Let us consider the plant model

$$y = \frac{b(s^2+4)}{(s+5)^3}u,$$

where b is the only unknown parameter. A suitable parametric model for estimating b is

$$z = \theta^*\phi,$$

where

$$z = y, \qquad \theta^* = b, \qquad \phi = \frac{s^2+4}{(s+5)^3}u.$$

In this case $\phi \in \mathcal{R}$ and $H(s) = \frac{s^2+4}{(s+5)^3}$; i.e., $n = 1$ in Theorem 3.4.3. Let us use Theorem 3.4.3 to choose a sufficiently rich signal u that guarantees ϕ to be PE. In this case, according to the linear independence condition of Theorem 3.4.3 for the case of $n = 1$, we should have

$$|H(j\omega_0)| = \frac{4-\omega_0^2}{(25+\omega_0^2)^{3/2}} \neq 0$$

for any $\omega_0 \neq 0$. This condition is clearly violated for $\omega_0 = 2$, and therefore a sufficiently rich input of order 1 may not guarantee ϕ to be PE. Indeed, the input $u = \sin 2t$ leads to $y = 0$, $\phi = 0$ at steady state, which imply that the output y and regressor ϕ carry no information about the unknown parameter b. For this example $u = \sin \omega_0 t$ will guarantee ϕ to be PE, provided $\omega_0 \neq 2$. Also, $u =$ constant $\neq 0$ and $u = \sum_{i=1}^{m} \sin \omega_i t$, $m \geq 2$, will guarantee that ϕ is PE. In general, for each two unknown parameters we need at least a single nonzero frequency to guarantee PE, provided of course that $H(s)$ does not lose its linear independence as demonstrated by the above example.

The two-parameter case example presented in section 3.3 leads to the differential equation (3.21), which has exactly the same form as in the case of an arbitrary number of parameters. In the following section, we consider the case where θ^*, ϕ are of arbitrary dimension and analyze the convergence properties of equations of the form (3.21).

3.5 Example: Vector Case

Consider the SISO system described by the I/O relation

$$y = G(s)u, \qquad G(s) = \frac{Z(s)}{R(s)} = k_p \frac{\bar{Z}(s)}{R(s)}, \tag{3.24}$$

where u and y are the plant scalar input and output, respectively,

$$R(s) = s^n + a_{n-1}s^{n-1} + \cdots + a_1 s + a_0,$$
$$Z(s) = b_m s^m + \cdots + b_1 s + b_0,$$

and $k_p = b_m$ is the high-frequency gain.[2]

We can also express (3.24) as an nth-order differential equation given by

$$y^{(n)} + a_{n-1}y^{(n-1)} + \cdots + a_1\dot{y} + a_0 y = b_m u^{(m)} + \cdots + b_1\dot{u} + b_0 u. \tag{3.25}$$

Parametric Model Lumping all the parameters in (3.25) in the vector

$$\theta^* = [b_m, \ldots, b_0, a_{n-1}, \ldots, a_0]^T,$$

we can rewrite (3.25) as

$$y^{(n)} = \theta^{*T}[u^{(m)}, \ldots, u, -y^{(n-1)}, \ldots, -y]^T. \tag{3.26}$$

Filtering each side of (3.26) with $\frac{1}{\Lambda(s)}$, where $\Lambda(s) = s^n + \lambda_{n-1}s^{n-1} + \cdots + \lambda_1 s + \lambda_0$ is a monic Hurwitz polynomial, we obtain the parametric model

$$z = \theta^{*T}\phi, \tag{3.27}$$

[2]At high frequencies or large s, the plant behaves as $\frac{k_p}{s^{n-m}}$; therefore, k_p is termed high-frequency gain.

where

$$z = \frac{1}{\Lambda(s)} y^{(n)} = \frac{s^n}{\Lambda(s)} y,$$
$$\theta^* = [b_m, \ldots, b_0, a_{n-1}, \ldots, a_0]^T \in \mathcal{R}^{n+m+1},$$
$$\phi = \left[\frac{s^m}{\Lambda(s)} u, \ldots, \frac{1}{\Lambda(s)} u, -\frac{s^{n-1}}{\Lambda(s)} y, \ldots, -\frac{1}{\Lambda(s)} y \right]^T.$$

If $\bar{Z}(s)$ is Hurwitz, a bilinear model can be obtained as follows: Consider the polynomials $P(s) = p_{n-1}s^{n-1} + \cdots + p_1 s + p_0$, $Q(s) = s^{n-1} + q_{n-2}s^{n-2} + \cdots + q_1 s + q_0$ which satisfy the Diophantine equation (see the Appendix)

$$k_p \bar{Z}(s) P(s) + R(s) Q(s) = \bar{Z}(s) \Lambda_0(s),$$

where $\Lambda_0(s)$ is a monic Hurwitz polynomial of order $2n - m - 1$. If each term in the above equation operates on the signal y, we obtain

$$k_p \bar{Z}(s) P(s) y + Q(s) R(s) y = \bar{Z}(s) \Lambda_0(s) y.$$

Substituting for $R(s)y = k_p \bar{Z}(s) u$, we obtain

$$k_p \bar{Z}(s) P(s) y + k_p \bar{Z}(s) Q(s) u = \bar{Z}(s) \Lambda_0(s) y.$$

Filtering each side with $\frac{1}{\Lambda_0(s)\bar{Z}(s)}$, we obtain

$$y = k_p \left[\frac{P(s)}{\Lambda_0(s)} y + \frac{Q(s)}{\Lambda_0(s)} u \right].$$

Letting

$$z = y,$$
$$\rho^* = k_p,$$
$$\theta^* = [q_{n-2}, \ldots, q_0, p_{n-1}, \ldots, p_0]^T,$$
$$\phi = \left[\frac{s^{n-2}}{\Lambda_0(s)} u, \ldots, \frac{1}{\Lambda_0(s)} u, \frac{s^{n-1}}{\Lambda_0(s)} y, \ldots, \frac{1}{\Lambda_0(s)} y \right]^T,$$
$$z_0 = \frac{s^{n-1}}{\Lambda_0(s)} u,$$

we obtain the B-SPM

$$z = \rho^* (\theta^{*T} \phi + z_0). \tag{3.28}$$

We should note that in this case θ^* contains not the coefficients of the plant transfer function but the coefficients of the polynomials $P(s)$, $Q(s)$. In certain adaptive control systems such as MRAC, the coefficients of $P(s)$, $Q(s)$ are the controller parameters, and parameterizations such as (3.28) allow the direct estimation of the controller parameters by processing the plant I/O measurements.

If some of the coefficients of the plant transfer function are known, then the dimension of the vector θ^* in the SPM (3.27) can be reduced. For example, if a_1, a_0, b_m are known, then (3.27) can be rewritten as

$$z = \theta^{*T}\phi, \tag{3.29}$$

where

$$z = \frac{s^n + a_1 s + a_0}{\Lambda(s)} y - \frac{b_m s^m}{\Lambda(s)} u,$$
$$\theta^* = [b_{m-1}, \ldots, b_0, a_{n-1}, \ldots, a_2]^T,$$
$$\phi = \left[\frac{s^{m-1}}{\Lambda(s)} u, \ldots, \frac{1}{\Lambda(s)} u, -\frac{s^{n-1}}{\Lambda(s)} y, \ldots, -\frac{s^2}{\Lambda(s)} y\right]^T.$$

Adaptive Law Let us consider the SPM (3.27). The objective is to process the signals $z(t)$ and $\phi(t)$ in order to generate an estimate $\theta(t)$ for θ^* at each time t. This estimate may be generated as

$$\dot{\theta}(t) = H(t)\varepsilon(t),$$

where $H(t)$ is some gain vector that depends on $\phi(t)$ and $\varepsilon(t)$ is the estimation error signal that represents a measure of how far $\theta(t)$ is from θ^*. Different choices for $H(t)$ and $\varepsilon(t)$ lead to a wide class of adaptive laws with, sometimes, different convergence properties, as demonstrated in the following sections.

3.6 Gradient Algorithms Based on the Linear Model

The gradient algorithm is developed by using the gradient method to minimize some appropriate functional $J(\theta)$. Different choices for $J(\theta)$ lead to different algorithms. As in the scalar case, we start by defining the estimation model and estimation error for the SPM (3.27).

The estimate $\hat{z}$ of z is generated by the estimation model

$$\hat{z} = \theta^T\phi, \tag{3.30}$$

where $\theta(t)$ is the estimate of θ^* at time t. The estimation error is constructed as

$$\varepsilon = \frac{z - \hat{z}}{m_s^2} = \frac{z - \theta^T\phi}{m_s^2}, \tag{3.31}$$

where $m_s^2 \geq 1$ is the normalizing signal designed to bound ϕ from above. The normalizing signal often has the form $m_s^2 = 1 + n_s^2$, where $n_s \geq 0$ is referred to as the *static normalizing signal* designed to guarantee that $\frac{\phi}{m_s}$ is bounded from above. Some straightforward choices for n_s include

$$n_s^2 = \alpha\phi^T\phi, \quad \alpha > 0,$$

or

$$n_s^2 = \phi^T P\phi, \quad P = P^T > 0,$$

where α is a scalar and P is a matrix selected by the designer.

The estimation error (3.31) and the estimation model (3.30) are common to several algorithms that are generated in the following sections.

3.6.1 Gradient Algorithm with Instantaneous Cost Function

The cost function $J(\theta)$ is chosen as

$$J(\theta) = \frac{\varepsilon^2 m_s^2}{2} = \frac{(z - \theta^T \phi)^2}{2m_s^2}, \tag{3.32}$$

where m_s is the normalizing signal given by (3.31). At each time t, $J(\theta)$ is a convex function of θ and therefore has a global minimum. The gradient algorithm takes the form

$$\dot{\theta} = -\Gamma \nabla J, \tag{3.33}$$

where $\Gamma = \Gamma^T > 0$ is a design matrix referred to as the *adaptive gain*. Since $\nabla J = -\frac{(z-\theta^T\phi)\phi}{m_s^2} = -\varepsilon\phi$, we have

$$\dot{\theta} = \Gamma \varepsilon \phi. \tag{3.34}$$

The adaptive law (3.34) together with the estimation model (3.30), the estimation error (3.31), and filtered signals z, ϕ defined in (3.29) constitute the gradient parameter identification algorithm based on the instantaneous cost function whose stability properties are given by the following theorem.

Theorem 3.6.1. *The gradient algorithm* (3.34) *guarantees the following:*

(i) $\varepsilon, \varepsilon m_s, \dot{\theta} \in \mathcal{L}_2 \cap \mathcal{L}_\infty$ *and* $\theta \in \mathcal{L}_\infty$.

(ii) *If* $\frac{\phi}{m_s}$ *is PE, i.e.,* $\int_t^{t+T_0} \frac{\phi\phi^T}{m_s^2} d\tau > \alpha_0 T_0 I$ $\forall t \geq 0$ *and for some* $T_0, \alpha_0 > 0$, *then* $\theta(t) \to \theta^*$ *exponentially fast. In addition,*

$$(\theta(t) - \theta^*)^T \Gamma^{-1} (\theta(t) - \theta^*) \leq (1 - \gamma_1)^n (\theta(0) - \theta^*)^T \Gamma^{-1} (\theta(0) - \theta^*),$$

where $0 \leq t \leq nT_0$, $n = 0, 1, 2, \ldots,$ *and*

$$\gamma_1 = \frac{2\alpha_0 T_0 \lambda_{\min}(\Gamma)}{2 + \beta^4 \lambda_{\max}^2(\Gamma) T_0^2}, \quad \beta = \sup_t \left| \frac{\phi}{m_s} \right|.$$

(iii) *If the plant model* (3.24) *has stable poles and no zero-pole cancellations and the input* u *is sufficiently rich of order* $n + m + 1$, *i.e., it consists of at least* $\frac{n+m+1}{2}$ *distinct frequencies, then* ϕ, $\frac{\phi}{m_s}$ *are PE. Furthermore,* $|\theta(t) - \theta^*|$, ε, εm_s, $\dot{\theta}$ *converge to zero exponentially fast.*

Proof. (i) Since θ^* is constant, $\dot{\tilde{\theta}} = \dot{\theta}$ and from (3.34) we have

$$\dot{\tilde{\theta}} = \Gamma \varepsilon \phi = -\Gamma \frac{\phi \phi^T}{m_s^2} \tilde{\theta}. \tag{3.35}$$

We choose the Lyapunov-like function

$$V(\tilde{\theta}) = \frac{\tilde{\theta}^T \Gamma^{-1} \tilde{\theta}}{2}.$$

Then along the solution of (3.35), we have

$$\dot{V} = \tilde{\theta}^T \phi \varepsilon = -\varepsilon^2 m_s^2 \le 0, \tag{3.36}$$

where the second equality is obtained by substituting $\tilde{\theta}^T \phi = -\varepsilon m_s^2$ from (3.31). Since $V > 0$ and $\dot{V} \le 0$, it follows that $V(t)$ has a limit, i.e.,

$$\lim_{t \to \infty} V(\tilde{\theta}(t)) = V_\infty < \infty,$$

and $V, \tilde{\theta} \in \mathcal{L}_\infty$, which, together with (3.31), imply that $\varepsilon, \varepsilon m_s \in \mathcal{L}_\infty$. In addition, it follows from (3.36) that

$$\int_0^\infty \varepsilon^2 m_s^2 d\tau \le V(\tilde{\theta}(0)) - V_\infty,$$

from which we establish that $\varepsilon m_s \in \mathcal{L}_2$ and hence $\varepsilon \in \mathcal{L}_2$ (due to $m_s^2 = 1 + n_s^2$). Now from (3.35) we have

$$|\dot{\tilde{\theta}}| = |\dot{\theta}| \le \|\Gamma\| |\varepsilon m_s| \frac{|\phi|}{m_s},$$

which together with $\frac{|\phi|}{m_s} \in \mathcal{L}_\infty$ and $|\varepsilon m_s| \in \mathcal{L}_2$ imply that $\dot{\theta} \in \mathcal{L}_2 \cap \mathcal{L}_\infty$, and the proof of (i) is complete.

The proof of parts (ii) and (iii) is longer and is presented in the web resource [94]. □

Comment 3.6.2 The rate of convergence of θ to θ^* can be improved if we choose the design parameters so that $1 - \gamma_1$ is as small as possible or, alternatively, $\gamma_1 \in (0, 1)$ is as close to 1 as possible. Examining the expression for γ_1, it is clear that the constants β, α_0, T_0 depend on each other and on ϕ. The only free design parameter is the adaptive gain matrix Γ. If we choose $\Gamma = \lambda I$, then the value of λ that maximizes γ_1 is

$$\lambda^* = \left(\frac{2}{2 - \beta^4 T_0^2} \right)^{1/2},$$

provided $\beta^4 T_0^2 < 2$. For $\lambda > \lambda^*$ or $\lambda < \lambda^*$, the expression for γ_1 suggests that the rate of convergence is slower. This dependence of the rate of convergence on the value of the adaptive gain Γ is often observed in simulations; i.e., very small or very large values of Γ lead to slower convergence rates. In general the convergence rate depends on the signal input and filters used in addition to Γ in a way that is not understood quantitatively.

Comment 3.6.3 Properties (i) and (ii) of Theorem 3.6.1 are independent of the boundedness of the regressor ϕ. Additional properties may be obtained if we make further assumptions about ϕ. For example, if $\phi, \dot{\phi} \in \mathcal{L}_\infty$, then we can show that $\varepsilon, \varepsilon m_s, \dot{\theta} \to 0$ as $t \to \infty$ (see Problem 6).

Comment 3.6.4 In the proof of Theorem 3.6.1(i)–(ii), we established that $\lim_{t\to\infty} V(t) = V_\infty$, where V_∞ is a constant. This implies that

$$\lim_{t\to\infty} V(\tilde{\theta}(t)) = \lim_{t\to\infty} \frac{\tilde{\theta}^T(t)\Gamma^{-1}\tilde{\theta}(t)}{2} = V_\infty.$$

We cannot conclude, however, that $\tilde{\theta} = \theta - \theta^*$ converges to a constant vector. For example, take $\Gamma = I$, $\tilde{\theta}(t) = [\sin t, \cos t]^T$. Then

$$\frac{\tilde{\theta}^T\tilde{\theta}}{2} = \frac{\sin^2 t + \cos^2 t}{2} = \frac{1}{2},$$

and $\tilde{\theta}(t) = [\sin t, \cos t]^T$ does not have a limit.

Example 3.6.5 Consider the nonlinear system

$$\dot{x} = af(x) + bg(x)u,$$

where a, b are unknown constants, $f(x)$, $g(x)$ are known continuous functions of x, and x, u are available for measurement. We want to estimate a, b online. We first obtain a parametric model in the form of an SPM by filtering each side with $\frac{1}{s+\lambda}$ for some $\lambda > 0$, i.e.,

$$\frac{s}{s+\lambda}x = a\frac{1}{s+\lambda}f(x) + b\frac{1}{s+\lambda}g(x)u.$$

Then, for

$$z = \frac{s}{s+\lambda}x, \qquad \theta^* = [a, b]^T, \qquad \phi = \frac{1}{s+\lambda}[f(x), g(x)u]^T,$$

we have

$$z = \theta^{*T}\phi.$$

The gradient algorithm (3.34),

$$\dot{\theta} = \Gamma\varepsilon\phi,$$
$$\varepsilon = \frac{z - \theta^T\phi}{m_s^2}, \quad m_s^2 = 1 + n_s^2, \quad n_s^2 = \phi^T\phi,$$

can be used to generate $\theta(t) = [\hat{a}(t), \hat{b}(t)]^T$ online, where $\hat{a}(t)$, $\hat{b}(t)$ are the online estimates of a, b, respectively. While this adaptive law guarantees properties (i) and (ii) of Theorem 3.6.1 independent of ϕ, parameter convergence of $\theta(t)$ to θ^* requires $\frac{\phi}{m_s}$ to be PE. The important question that arises in this example is how to choose the plant input u so that $\frac{\phi}{m_s}$ is PE. Since the plant is nonlinear, the choice of u that makes $\frac{\phi}{m_s}$ PE depends on the form of the nonlinear functions $f(x)$, $g(x)$ and it is not easy, if possible at all, to establish conditions similar to those in the LTI case for a general class of nonlinearities. ∎

Example 3.6.6 Consider the dynamics of a hard-disk drive servo system [96] given by

$$y = \frac{k_p}{s^2}(u + d),$$

where y is the position error of the head relative to the center of the track, k_p is a known constant, and

$$d = A_1 \sin(\omega_1 t + \varphi_1) + A_2 \sin(\omega_2 t + \varphi_2)$$

is a disturbance that is due to higher-order harmonics that arise during rotation of the disk drive. In this case, ω_1, ω_2 are the known harmonics that have a dominant effect and A_i, φ_i, $i = 1, 2$, are the unknown amplitudes and phases. We want to estimate d in an effort to nullify its effect using the control input u.

Using $\sin(a + b) = \sin a \cos b + \cos a \sin b$, we can express d as

$$d = \theta_1^* \sin \omega_1 t + \theta_2^* \cos \omega_1 t + \theta_3^* \sin \omega_2 t + \theta_4^* \cos \omega_2 t,$$

where

$$\begin{aligned} \theta_1^* &= A_1 \cos \varphi_1, & \theta_2^* &= A_1 \sin \varphi_1, \\ \theta_3^* &= A_2 \cos \varphi_2, & \theta_4^* &= A_2 \sin \varphi_2 \end{aligned} \tag{3.37}$$

are the unknown parameters. We first obtain a parametric model for

$$\theta^* = [\theta_1^*, \theta_2^*, \theta_3^*, \theta_4^*]^T.$$

We have

$$s^2 y = k_p u + k_p \theta^{*T} \psi,$$

where

$$\psi(t) = [\sin \omega_1 t, \cos \omega_1 t, \sin \omega_2 t, \cos \omega_2 t]^T.$$

Filtering each side with $\frac{1}{\Lambda(s)}$, where $\Lambda(s) = (s + \lambda_1)(s + \lambda_2)$ and $\lambda_1, \lambda_2 > 0$ are design constants, we obtain the SPM

$$z = \theta^{*T} \phi,$$

where

$$\begin{aligned} z &= \frac{s^2}{\Lambda(s)} y - k_p \frac{1}{\Lambda(s)} u, \\ \phi &= k_p \frac{1}{\Lambda(s)} \psi(t) = k_p \frac{1}{\Lambda(s)} [\sin \omega_1 t, \cos \omega_1 t, \sin \omega_2 t, \cos \omega_2 t]^T. \end{aligned}$$

Therefore, the adaptive law

$$\begin{aligned} \dot{\theta} &= \Gamma \varepsilon \phi, \\ \varepsilon &= \frac{z - \theta^T \phi}{m_s^2}, \quad m_s^2 = 1 + n_s^2, \quad n_s^2 = \alpha \phi^T \phi, \end{aligned}$$

where $\Gamma = \Gamma^T > 0$ is a 4×4 constant matrix, may be used to generate $\theta(t)$, the online estimate of θ^*. In this case, $\phi \in \mathcal{L}_\infty$ and therefore we can take $\alpha = 0$, i.e., $m_s^2 = 1$. For

$\omega_1 \neq \omega_2$, we can establish that ϕ is PE and therefore $\theta(t) \to \theta^*$ exponentially fast. The online estimate of the amplitude and phase can be computed using (3.37) as follows:

$$\tan \hat{\varphi}_1(t) = \frac{\theta_2(t)}{\theta_1(t)}, \qquad \tan \hat{\varphi}_2(t) = \frac{\theta_4(t)}{\theta_3(t)},$$
$$\hat{A}_1(t) = \sqrt{\theta_1^2(t) + \theta_2^2(t)}, \qquad \hat{A}_2(t) = \sqrt{\theta_3^2(t) + \theta_4^2(t)},$$

provided of course that $\theta_1(t) \neq 0, \theta_3(t) \neq 0$. The estimated disturbance

$$\hat{d}(t) = \hat{A}_1(t)\sin(\omega_1 t + \hat{\varphi}_1(t)) + \hat{A}_2(t)\sin(\omega_2 t + \hat{\varphi}_2(t))$$

can then be generated and used by the controller to cancel the effect of the actual disturbance d. ∎

3.6.2 Gradient Algorithm with Integral Cost Function

The cost function $J(\theta)$ is chosen as

$$J(\theta) = \frac{1}{2}\int_0^t e^{-\beta(t-\tau)}\varepsilon^2(t,\tau)m_s^2(\tau)d\tau,$$

where $\beta > 0$ is a design constant acting as a forgetting factor and

$$\varepsilon(t,\tau) = \frac{z(\tau) - \theta^T(t)\phi(\tau)}{m_s^2(\tau)}, \qquad \varepsilon(t,t) = \varepsilon, \qquad \tau \leq t,$$

is the estimation error that depends on the estimate of θ at time t and on the values of the signals at $\tau \leq t$. The cost penalizes all past errors between $z(\tau)$ and $\hat{z}(\tau) = \theta^T(t)\phi(\tau)$, $\tau \leq t$, obtained by using the current estimate of θ at time t with past measurements of $z(\tau)$ and $\phi(\tau)$. The forgetting factor $e^{-\beta(t-\tau)}$ is used to put more weight on recent data by discounting the earlier ones. It is clear that $J(\theta)$ is a convex function of θ at each time t and therefore has a global minimum. Since $\theta(t)$ does not depend on τ, the gradient of J with respect to θ is easy to calculate despite the presence of the integral. Applying the gradient method, we have

$$\dot{\theta} = -\Gamma\nabla J,$$

where

$$\nabla J = -\int_0^t e^{-\beta(t-\tau)}\frac{z(\tau) - \theta^T(t)\phi(\tau)}{m_s^2(\tau)}\phi(\tau)d\tau.$$

This can be implemented as (see Problem 7)

$$\dot{\theta} = -\Gamma(R(t)\theta + Q(t)), \qquad \theta(0) = \theta_0,$$
$$\dot{R} = -\beta R + \frac{\phi\phi^T}{m_s^2}, \qquad R(0) = 0,$$
$$\dot{Q} = -\beta Q - \frac{z\phi}{m_s^2}, \qquad Q(0) = 0,$$

where $R \in \mathcal{R}^{n\times n}$, $Q \in \mathcal{R}^{n\times 1}$; $\Gamma = \Gamma^T > 0$ is the adaptive gain; n is the dimension of the vector θ^*; and m_s is the normalizing signal defined in (3.31).

Theorem 3.6.7. *The gradient algorithm with integral cost function guarantees that*

(i) $\varepsilon, \varepsilon m_s, \dot{\theta} \in \mathcal{L}_2 \cap \mathcal{L}_\infty$ *and* $\theta \in \mathcal{L}_\infty$.

(ii) $\lim_{t\to\infty} |\dot{\theta}(t)| = 0$.

(iii) *If* $\frac{\phi}{m_s}$ *is PE, then* $\theta(t) \to \theta^*$ *exponentially fast. Furthermore, for* $\Gamma = \gamma I$, *the rate of convergence increases with* γ.

(iv) *If* u *is sufficiently rich of order* $n + m + 1$, *i.e., it consists of at least* $\frac{n+m+1}{2}$ *distinct frequencies, and the plant is stable and has no zero-pole cancellations, then* ϕ, $\frac{\phi}{m_s}$ *are PE and* $\theta(t) \to \theta^*$ *exponentially fast.*

Proof. The proof is presented in the web resource [94]. □

Theorem 3.6.7 indicates that the rate of parameter convergence increases with increasing adaptive gain. Simulations demonstrate that the gradient algorithm based on the integral cost gives better convergence properties than the gradient algorithm based on the instantaneous cost. The gradient algorithm based on the integral cost has similarities with the least-squares (LS) algorithms to be developed in the next section.

3.7 Least-Squares Algorithms

The LS method dates back to the eighteenth century, when Gauss used it to determine the orbits of planets. The basic idea behind LS is fitting a mathematical model to a sequence of observed data by minimizing the sum of the squares of the difference between the observed and computed data. In doing so, any noise or inaccuracies in the observed data are expected to have less effect on the accuracy of the mathematical model.

The LS method has been widely used in parameter estimation both in recursive and nonrecursive forms mainly for discrete-time systems [46, 47, 77, 97, 98]. The method is simple to apply and analyze in the case where the unknown parameters appear in a linear form, such as in the linear SPM

$$z = \theta^{*T}\phi. \tag{3.38}$$

We illustrate the use and properties of LS by considering the simple scalar example

$$z = \theta^*\phi + d_n,$$

where $z, \theta^*, \phi \in \mathcal{R}$, $\phi \in \mathcal{L}_\infty$, and d_n is a noise disturbance whose average value goes to zero as $t \to \infty$, i.e.,

$$\lim_{t\to\infty} \frac{1}{t}\int_0^t d_n(\tau)d\tau = 0.$$

In practice, d_n may be due to sensor noise or external sources, etc. We examine the following estimation problem: Given the measurements of $z(\tau)$, $\phi(\tau)$ for $0 \leq \tau < t$, find a "good" estimate $\theta(t)$ of θ^* at time t. One possible solution is to calculate $\theta(t)$ as

$$\theta(t) = \frac{z(\tau)}{\phi(\tau)} = \theta^* + \frac{d_n(\tau)}{\phi(\tau)} \tag{3.39}$$

by using the measurements of $z(\tau)$, $\phi(\tau)$ at some $\tau < t$ for which $\phi(\tau) \neq 0$. Because of the noise disturbance, however, such an estimate may be far off from θ^*. For example, at the particular time τ at which we measured z and ϕ, the effect of $d_n(\tau)$ may be significant, leading to an erroneous estimate for $\theta(t)$ generated by (3.39).

A more intelligent approach is to choose the estimate $\theta(t)$ at time t to be the one that minimizes the square of all the errors that result from the mismatch of $z(\tau) - \theta(t)\phi(\tau)$ for $0 \leq \tau \leq t$. Hence the estimation problem above becomes the following LS problem: Minimize the cost

$$J(\theta) = \frac{1}{2}\int_0^t |z(\tau) - \theta(t)\phi(\tau)|^2 d\tau \tag{3.40}$$

w.r.t. $\theta(t)$ at any given time t. The cost $J(\theta)$ penalizes all the past errors from $\tau = 0$ to t that are due to $\theta(t) \neq \theta^*$. Since $J(\theta)$ is a convex function over $\mathcal{R}$ at each time t, its minimum satisfies

$$\nabla J(\theta) = -\int_0^t z(\tau)\phi(\tau)d\tau + \theta(t)\int_0^t \phi^2(\tau)d\tau = 0,$$

which gives the LS estimate

$$\theta(t) = \left(\int_0^t \phi^2(\tau)d\tau\right)^{-1}\int_0^t z(\tau)\phi(\tau)d\tau,$$

provided of course that the inverse exists. The LS method considers all past data in an effort to provide a good estimate for θ^* in the presence of noise d_n. For example, when $\phi(t) = 1$, $\forall t \geq 0$, we have

$$\lim_{t\to\infty}\theta(t) = \lim_{t\to\infty}\frac{1}{t}\int_0^t z(\tau)d\tau = \theta^* + \lim_{t\to\infty}\frac{1}{t}\int_0^t d_n(\tau)d\tau = \theta^*;$$

i.e., $\theta(t)$ converges to the exact parameter value despite the presence of the noise disturbance d_n.

Let us now extend this problem to the linear model (3.38). As in section 3.6, the estimate $\hat{z}$ of z and the normalized estimation are generated as

$$\hat{z} = \theta^T\phi, \qquad e = \frac{z - \hat{z}}{m_s^2} = \frac{z - \theta^T\phi}{m_s^2},$$

where $\theta(t)$ is the estimate of θ^* at time t, and $m_s^2 = 1 + n_s^2$ is designed to guarantee $\frac{\phi}{m_s} \in \mathcal{L}_\infty$. Below we present different versions of the LS algorithm, which correspond to different choices of the LS cost $J(\theta)$.

3.7.1 Recursive LS Algorithm with Forgetting Factor

Consider the function

$$J(\theta) = \frac{1}{2}\int_0^t e^{-\beta(t-\tau)}\frac{[z(\tau) - \theta^T(t)\phi(\tau)]^2}{m_s^2(\tau)}d\tau + \frac{1}{2}e^{-\beta t}(\theta - \theta_0)^T Q_0(\theta - \theta_0), \quad (3.41)$$

where $Q_0 = Q_0^T > 0$, $\beta \geq 0$ are design constants and $\theta_0 = \theta(0)$ is the initial parameter estimate. This cost function is a generalization of (3.40) to include possible discounting of past data and a penalty on the initial error between the estimate θ_0 and θ^*. Since $\frac{z}{m_s}$, $\frac{\phi}{m_s} \in \mathcal{L}_\infty$, $J(\theta)$ is a convex function of θ over $\mathcal{R}^n$ at each time t. Hence, any local minimum is also global and satisfies

$$\nabla J(\theta(t)) = 0 \quad \forall t \geq 0.$$

The LS algorithm for generating $\theta(t)$, the estimate of θ^*, in (3.38) is therefore obtained by solving

$$\nabla J(\theta) = e^{-\beta t}Q_0(\theta(t) - \theta_0) - \int_0^t e^{-\beta(t-\tau)}\frac{z(\tau) - \theta^T(t)\phi(\tau)}{m_s^2(\tau)}\phi(\tau)d\tau = 0 \quad (3.42)$$

for $\theta(t)$, which yields the *nonrecursive LS algorithm*

$$\theta(t) = P(t)\left[e^{-\beta t}Q_0\theta_0 + \int_0^t e^{-\beta(t-\tau)}\frac{z(\tau)\phi(\tau)}{m_s^2(\tau)}d\tau\right], \quad (3.43)$$

where

$$P(t) = \left[e^{-\beta t}Q_0 + \int_0^t e^{-\beta(t-\tau)}\frac{\phi(\tau)\phi^T(\tau)}{m_s^2(\tau)}d\tau\right]^{-1} \quad (3.44)$$

is the so-called *covariance matrix*. Because $Q_0 = Q_0^T > 0$ and $\phi\phi^T$ is positive semidefinite, $P(t)$ exists at each time t. Using the identity

$$\frac{d}{dt}PP^{-1} = \dot{P}P^{-1} + P\frac{d}{dt}P^{-1} = 0$$

and $\varepsilon m_s^2 = z - \theta^T\phi$, and differentiating $\theta(t)$ w.r.t. t, we obtain the *recursive LS algorithm with forgetting factor*

$$\begin{aligned} \dot{\theta} &= P\varepsilon\phi, & \theta(0) &= \theta_0, \\ \dot{P} &= \beta P - P\frac{\phi\phi^T}{m_s^2}P, & P(0) &= P_0 = Q_0^{-1}. \end{aligned} \quad (3.45)$$

The stability properties of (3.45) depend on the value of the *forgetting factor* β, as discussed in the following sections. If $\beta = 0$, the algorithm becomes the *pure LS* algorithm discussed and analyzed in section 3.7.2. When $\beta > 0$, stability cannot be established unless $\frac{\phi}{m_s}$ is PE. In this case (3.45) is modified, leading to a different algorithm discussed and analyzed in section 3.7.3.

The following theorem establishes the stability and convergence of θ to θ^* of the algorithm (3.45) in the case where $\frac{\phi}{m_s}$ is PE.

Theorem 3.7.1. *If $\frac{\phi}{m_s}$ is PE, then the recursive LS algorithm with forgetting factor* (3.45) *guarantees that $P, P^{-1} \in \mathcal{L}_\infty$ and that $\theta(t) \to \theta^*$ as $t \to \infty$. The convergence of $\theta(t) \to \theta^*$ is exponential when $\beta > 0$.*

Proof. The proof is given in [56] and in the web resource [94]. □

Since the adaptive law (3.45) could be used in adaptive control where the PE property of $\frac{\phi}{m_s}$ cannot be guaranteed, it is of interest to examine the properties of (3.45) in the absence of PE. In this case, (3.45) is modified in order to avoid certain undesirable phenomena, as discussed in the following sections.

3.7.2 Pure LS Algorithm

When $\beta = 0$ in (3.41), the algorithm (3.45) reduces to

$$\begin{aligned} \dot{\theta} &= P\varepsilon\phi, & \theta(0) &= \theta_0, \\ \dot{P} &= -P\frac{\phi\phi^T}{m_s^2}P, & P(0) &= P_0, \end{aligned} \tag{3.46}$$

which is referred to as the *pure LS algorithm.*

Theorem 3.7.2. *The pure LS algorithm* (3.46) *guarantees that*

(i) *$\varepsilon, \varepsilon m_s, \dot{\theta} \in \mathcal{L}_2 \cap \mathcal{L}_\infty$ and $\theta, P \in \mathcal{L}_\infty$.*

(ii) *$\lim_{t\to\infty} \theta(t) = \bar{\theta}$, where $\bar{\theta}$ is a constant vector.*

(iii) *If $\frac{\phi}{m_s}$ is PE, then $\theta(t) \to \theta^*$ as $t \to \infty$.*

(iv) *If* (3.38) *is the SPM for the plant* (3.24) *with stable poles and no zero-pole cancellations, and u is sufficiently rich of order $n + m + 1$, i.e., consists of at least $\frac{n+m+1}{2}$ distinct frequencies, then $\phi, \frac{\phi}{m_s}$ are PE and therefore $\theta(t) \to \theta^*$ as $t \to \infty$.*

Proof. From (3.46) we have that $\dot{P} \le 0$, i.e., $P(t) \le P_0$. Because $P(t)$ is nonincreasing and bounded from below (i.e., $P(t) = P^T(t) \ge 0 \; \forall t \ge 0$) it has a limit, i.e.,

$$\lim_{t\to\infty} P(t) = \bar{P},$$

where $\bar{P} = \bar{P}^T \ge 0$ is a constant matrix. Let us now consider the identity

$$\frac{d}{dt}(P^{-1}\tilde{\theta}) = -P^{-1}\dot{P}P^{-1}\tilde{\theta} + P^{-1}\dot{\tilde{\theta}} = \frac{\phi\phi^T\tilde{\theta}}{m_s^2} + \varepsilon\phi = 0,$$

where the last two equalities are obtained using $\dot{\theta} = \dot{\tilde{\theta}}$, $\frac{d}{dt}P^{-1} = -P^{-1}\dot{P}P^{-1}$, and $\varepsilon = -\frac{\tilde{\theta}^T\phi}{m_s^2} = -\frac{\phi^T\tilde{\theta}}{m_s^2}$. Hence, $P^{-1}(t)\tilde{\theta}(t) = P_0^{-1}\tilde{\theta}(0)$ and therefore $\tilde{\theta}(t) = P(t)P_0^{-1}\tilde{\theta}(0)$ and $\lim_{t\to\infty}\tilde{\theta}(t) = \bar{P}P_0^{-1}\tilde{\theta}(0)$, which implies that $\lim_{t\to\infty}\theta(t) = \theta^* + \bar{P}P_0^{-1}\tilde{\theta}(0) \triangleq \bar{\theta}$.

Because $P(t) \leq P_0$ and $\tilde{\theta}(t) = P(t)P_0^{-1}\tilde{\theta}(0)$ we have $\theta, \tilde{\theta} \in \mathcal{L}_\infty$, which, together with $\frac{\phi}{m_s} \in \mathcal{L}_\infty$, implies that $\varepsilon m_s = -\frac{\tilde{\theta}^T\phi}{m_s}$ and $\varepsilon, \varepsilon m_s \in \mathcal{L}_\infty$. Let us now consider the function

$$V(\tilde{\theta}, t) = \frac{\tilde{\theta}^T P^{-1}(t)\tilde{\theta}}{2}.$$

The time derivative $\dot{V}$ of V along the solution of (3.46) is given by

$$\dot{V} = \varepsilon\tilde{\theta}^T\phi + \frac{\tilde{\theta}^T\phi\phi^T\tilde{\theta}}{2m_s^2} = -\varepsilon^2 m_s^2 + \frac{\varepsilon^2 m_s^2}{2} = -\frac{\varepsilon^2 m_s^2}{2} \leq 0,$$

which implies that $V \in \mathcal{L}_\infty$, $\varepsilon m_s \in \mathcal{L}_2$, and therefore $\varepsilon \in \mathcal{L}_2$. From (3.46) we have

$$|\dot{\theta}| \leq \|P\|\frac{|\phi|}{m_s}|\varepsilon m_s|.$$

Since P, $\frac{\phi}{m_s}$, $\varepsilon m_s \in \mathcal{L}_\infty$, and $\varepsilon m_s \in \mathcal{L}_2$, we have $\dot{\theta} \in \mathcal{L}_\infty \cap \mathcal{L}_2$, which completes the proof for (i) and (ii). The proofs of (iii) and (iv) are included in the proofs of Theorems 3.7.1 and 3.6.1, respectively. $\square$

The pure LS algorithm guarantees that the parameters converge to some constant $\bar{\theta}$ without having to put any restriction on the regressor ϕ. If $\frac{\phi}{m_s}$, however, is PE, then $\bar{\theta} = \theta^*$. Convergence of the estimated parameters to constant values is a unique property of the pure LS algorithm. One of the drawbacks of the pure LS algorithm is that the covariance matrix P may become arbitrarily small and slow down adaptation in some directions. This is due to the fact that

$$\frac{d(P^{-1})}{dt} = \frac{\phi\phi^T}{m_s^2} \geq 0,$$

which implies that P^{-1} may grow without bound, which in turn implies that P may reduce towards zero. This is the so-called *covariance wind-up* problem. Another drawback of the pure LS algorithm is that parameter convergence cannot be guaranteed to be exponential.

Example 3.7.3 In order to get some understanding of the properties of the pure LS algorithm, let us consider the scalar SPM

$$z = \theta^*\phi,$$

where $z, \theta^*, \phi \in \mathcal{R}$. Let us assume that $\phi \in \mathcal{L}_\infty$. Then the pure LS algorithm is given by

$$\begin{aligned} \dot{\theta} &= p\varepsilon\phi, & \theta(0) &= \theta_0, \\ \dot{p} &= -p^2\phi^2, & p(0) &= p_0 > 0, \\ \varepsilon &= z - \theta\phi = -\tilde{\theta}\phi. \end{aligned}$$

Let us also take $\phi = 1$, which is PE, for this example. Then we can show by solving the differential equation via integration that

$$p(t) = \frac{p_0}{1 + p_0 t}$$

and $\tilde{\theta}(t) = \frac{\tilde{\theta}(0)}{1+p_0 t}$, i.e.,

$$\theta(t) = \theta^* + \frac{\theta(0) - \theta^*}{1 + p_0 t}.$$

It is clear that as $t \to \infty$, $p(t) \to 0$, leading to the so-called covariance wind-up problem. Since $\phi = 1$ is PE, however, $\theta(t) \to \theta^*$ as $t \to \infty$ with a rate of $\frac{1}{t}$ (not exponential) as predicted by Theorem 3.7.2. Even though $\theta(t) \to \theta^*$, the covariance wind-up problem may still pose a problem in the case where θ^* changes to some other value after some time. If at that instance $p(t) \cong 0$, leading to $\dot{\theta} \cong 0$, no adaptation will take place and $\theta(t)$ may not reach the new θ^*.

For the same example, consider $\phi(t) = \frac{1}{1+t}$, which is not PE since

$$\int_t^{t+T} \phi^2(\tau)d\tau = \int_t^{t+T} \frac{1}{(1+\tau)^2}d\tau = \frac{1}{1+t} - \frac{1}{1+t+T}$$

goes to zero as $t \to \infty$, i.e., it has zero level of excitation. In this case, we can show that

$$p(t) = \frac{p_0(1+t)}{1+(1+p_0)t},$$
$$\theta(t) = \theta^* + (\theta(0) - \theta^*)\frac{1+t}{1+(1+p_0)t}$$

by solving the differential equations above. It is clear that $p(t) \to \frac{p_0}{1+p_0}$ and $\theta(t) \to \frac{p_0\theta^* + \theta(0)}{1+p_0}$ as $t \to \infty$; i.e., $\theta(t)$ converges to a constant but not to θ^* due to lack of PE. In this case $p(t)$ converges to a constant and no covariance wind-up problem arises. ■

3.7.3 Modified LS Algorithms

One way to avoid the covariance wind-up problem is to modify the pure LS algorithm using a covariance resetting modification to obtain

$$\begin{aligned}
\dot{\theta} &= P\varepsilon\phi, & \theta(0) &= \theta_0, \\
\dot{P} &= -P\frac{\phi\phi^T}{m_s^2}P, & P(t_r^+) &= P_0 = \rho_0 I, \\
m_s^2 &= 1 + n_s^2, & n_s^2 &= \alpha\phi^T\phi, \quad \alpha > 0,
\end{aligned} \tag{3.47}$$

where t_r^+ is the time at which $\lambda_{\min}(P(t)) \leq \rho_1$ and $\rho_0 > \rho_1 > 0$ are some design scalars. Due to covariance resetting, $P(t) \geq \rho_1 I$ $\forall t \geq 0$. Therefore, P is guaranteed to be positive definite for all $t \geq 0$. In fact, the *pure LS algorithm with covariance resetting* can be viewed as a gradient algorithm with time-varying adaptive gain P, and its properties are very similar to those of a gradient algorithm analyzed in the previous section. They are summarized by Theorem 3.7.4 in this section.

When $\beta > 0$, the covariance wind-up problem, i.e., $P(t)$ becoming arbitrarily small, does not exist. In this case, $P(t)$ may grow without bound. In order to avoid this phenomenon, the following *modified LS algorithm with forgetting factor* is used:

$$\dot{\theta} = P\varepsilon\phi,$$
$$\dot{P} = \begin{cases} \beta P - \frac{P\phi\phi^T P}{m_s^2} & \text{if } \|P(t)\| \le R_0, \\ 0 & \text{otherwise,} \end{cases} \tag{3.48}$$

where $P(0) = P_0 = P_0^T > 0$, $\|P_0\| \le R_0$, R_0 is a constant that serves as an upper bound for $\|P\|$, and $m_s^2 = 1 + n_s^2$ is the normalizing signal which satisfies $\frac{\phi}{m_s} \in \mathcal{L}_\infty$.

The following theorem summarizes the stability properties of the two modified LS algorithms.

Theorem 3.7.4. *The pure LS algorithm with covariance resetting* (3.47) *and the modified LS algorithm with forgetting factor* (3.48) *guarantee that*

(i) $\varepsilon, \varepsilon m_s, \dot{\theta} \in \mathcal{L}_2 \cap \mathcal{L}_\infty$ *and* $\theta \in \mathcal{L}_\infty$.

(ii) *If* $\frac{\phi}{m_s}$ *is PE, then* $\theta(t) \to \theta^*$ *as* $t \to \infty$ *exponentially fast.*

(iii) *If* (3.38) *is the SPM for the plant* (3.24) *with stable poles and no zero-pole cancellations, and* u *is sufficiently rich of order* $n+m+1$, *then* ϕ, $\frac{\phi}{m_s}$ *are PE, which guarantees that* $\theta(t) \to \theta^*$ *as* $t \to \infty$ *exponentially fast.*

Proof. The proof is presented in the web resource [94]. □

3.8 Parameter Identification Based on DPM

Let us consider the DPM

$$z = W(s)[\theta^{*T}\psi].$$

This model may be obtained from (3.27) by filtering each side with $W(s)$ and redefining the signals z, ϕ. Since θ^* is a constant vector, the DPM may be written as

$$z = W(s)L(s)[\theta^{*T}\phi], \tag{3.49}$$

where $\phi = L^{-1}(s)\psi$, $L(s)$ is chosen so that $L^{-1}(s)$ is a proper stable transfer function, and $W(s)L(s)$ is a proper strictly positive real (SPR) transfer function.

$$\hat{z} = W(s)L(s)[\theta^T\phi].$$

We form the normalized estimation error

$$\varepsilon = z - \hat{z} - W(s)L(s)[\varepsilon n_s^2], \tag{3.50}$$

where the static normalizing signal n_s is designed so that $\frac{\phi}{m_s} \in \mathcal{L}_\infty$ for $m_s^2 = 1 + n_s^2$. If $W(s)L(s) = 1$, then (3.50) has the same expression as in the case of the gradient algorithm. Substituting for z in (3.50), we express ε in terms of the parameter error $\tilde{\theta} = \theta - \theta^*$:

$$\varepsilon = W(s)L(s)[-\tilde{\theta}^T\phi - \varepsilon n_s^2]. \tag{3.51}$$

For simplicity, let us assume that $W(s)L(s)$ is strictly proper and rewrite (3.51) in the minimum state-space representation form

$$\begin{aligned} \dot{e} &= A_c e + b_c(-\tilde{\theta}^T \phi - \varepsilon n_s^2), \\ \varepsilon &= c_c^T e, \end{aligned} \tag{3.52}$$

where $W(s)L(s) = c_c^T(sI - A_c)^{-1}b_c$. Since $W(s)L(s)$ is SPR, it follows that (see the Appendix) there exist matrices $P_c = P_c^T > 0$, $L_c = L_c^T > 0$, a vector q, and a scalar $\nu > 0$ such that

$$\begin{aligned} P_c A_c + A_c^T P_c &= -qq^T - \nu L_c, \\ P_c b_c &= c_c. \end{aligned} \tag{3.53}$$

The adaptive law for θ is generated using the Lyapunov-like function

$$V = \frac{e^T P_c e}{2} + \frac{\tilde{\theta}^T \Gamma^{-1} \tilde{\theta}}{2},$$

where $\Gamma = \Gamma^T > 0$. The time derivative $\dot{V}$ of V along the solution of (3.52) is given by

$$\dot{V} = -\frac{1}{2} e^T qq^T e - \frac{\nu}{2} e^T L_c e + e^T P_c b_c(-\tilde{\theta}^T \phi - \varepsilon n_s^2) + \tilde{\theta}^T \Gamma^{-1} \dot{\tilde{\theta}}.$$

Since $e^T P_c b_c = e^T c_c = \varepsilon$, it follows that by choosing $\dot{\tilde{\theta}} = \dot{\theta}$ as

$$\dot{\theta} = \Gamma \varepsilon \phi, \tag{3.54}$$

we get

$$\dot{V} = -\frac{1}{2} e^T qq^T e - \frac{\nu}{2} e^T L_c e - \varepsilon^2 n_s^2 \leq 0.$$

As before, from the properties of V, $\dot{V}$ we conclude that $e, \varepsilon, \theta \in \mathcal{L}_\infty$ and $e, \varepsilon, \varepsilon n_s \in \mathcal{L}_2$. These properties in turn imply that $\dot{\theta} \in \mathcal{L}_2$. Note that without the use of the second equation in (3.53), we are not able to choose $\dot{\tilde{\theta}} = \dot{\theta}$ using signals available for measurement to make $\dot{V} \leq 0$. This is because the state e in (3.52) cannot be generated since it depends on the unknown input $\tilde{\theta}^T \phi$. Equation (3.52) is used only for analysis.

The stability properties of the adaptive law (3.54) are summarized by the following theorem.

Theorem 3.8.1. *The adaptive law* (3.54) *guarantees that*

(i) $\varepsilon, \theta \in \mathcal{L}_\infty$ *and* $\varepsilon, \varepsilon n_s, \dot{\theta} \in \mathcal{L}_2$.

(ii) *If* $n_s, \phi, \dot{\phi} \in \mathcal{L}_\infty$ *and* ϕ *is PE, then* $\theta(t) \to \theta^*$ *exponentially fast.*

Proof. The proof for (i) is given above. The proof of (ii) is a long one and is given in [56] as well as in the web resource [94]. □

The adaptive law (3.54) is referred to as the adaptive law based on the SPR-Lyapunov synthesis approach.

Comment 3.8.2 The adaptive law (3.54) has the same form as the gradient algorithm even though it is developed using a Lyapunov approach and the SPR property. In fact, for $W(s)L(s) = 1$, (3.54) is identical to the gradient algorithm.

3.9 Parameter Identification Based on B-SPM

Consider the bilinear SPM described by (3.28), i.e.,

$$z = \rho^*(\theta^{*T}\phi + z_0), \tag{3.55}$$

where z, z_0 are known scalar signals at each time t and ρ^*, θ^* are the scalar and vector unknown parameters, respectively. The estimation error is generated as

$$\hat{z} = \rho(\theta^T\phi + z_0),$$
$$\varepsilon = \frac{z - \hat{z}}{m_s^2},$$

where $\rho(t)$, $\theta(t)$ are the estimates of ρ^*, θ^*, respectively, at time t and where m_s is designed to bound ϕ, z_0 from above. An example of m_s with this property is $m_s^2 = 1 + \phi^T\phi + z_0^2$.

Let us consider the cost

$$J(\rho, \theta) = \frac{\varepsilon^2 m_s^2}{2} = \frac{(z - \rho^*\theta^T\phi - \rho\xi + \rho^*\xi - \rho^* z_0)^2}{2m_s^2},$$

where

$$\xi = \theta^T\phi + z_0$$

is available for measurement. Applying the gradient method we obtain

$$\dot{\theta} = -\Gamma_1 \nabla J_\theta = \Gamma_1 \varepsilon \rho^* \phi,$$
$$\dot{\rho} = -\gamma \nabla J_\rho = \gamma \varepsilon \xi,$$

where $\Gamma_1 = \Gamma_1^T > 0$, $\gamma > 0$ are the adaptive gains. Since ρ^* is unknown, the adaptive law for θ cannot be implemented. We bypass this problem by employing the equality

$$\Gamma_1 \rho^* = \Gamma_1 |\rho^*| \operatorname{sgn}(\rho^*) = \Gamma \operatorname{sgn}(\rho^*),$$

where $\Gamma = \Gamma_1|\rho^*|$. Since Γ_1 is arbitrary any $\Gamma = \Gamma^T > 0$ can be selected without having to know $|\rho^*|$. Therefore, the adaptive laws for θ, ρ, may be written as

$$\begin{aligned} \dot{\theta} &= \Gamma \varepsilon \phi \operatorname{sgn}(\rho^*), \\ \dot{\rho} &= \gamma \varepsilon \xi, \\ \varepsilon &= \frac{z - \rho\xi}{m_s^2}, \quad \xi = \theta^T\phi + z_0. \end{aligned} \tag{3.56}$$

Theorem 3.9.1. *The adaptive law* (3.56) *guarantees that*

(i) $\varepsilon, \varepsilon m_s, \dot{\theta}, \dot{\rho} \in \mathcal{L}_2 \cap \mathcal{L}_\infty$ *and* $\theta, \rho \in \mathcal{L}_\infty$.

(ii) *If* $\frac{\xi}{m_s} \in \mathcal{L}_2$, *then* $\rho(t) \to \bar{\rho}$ *as* $t \to \infty$, *where* $\bar{\rho}$ *is a constant.*

(iii) *If* $\frac{\xi}{m_s} \in \mathcal{L}_2$ *and* $\frac{\phi}{m_s}$ *is PE, then* $\theta(t)$ *converges to* θ^* *as* $t \to \infty$.

(iv) *If the plant* (3.24) *has stable poles with no zero-pole cancellations and* u *is sufficiently rich of order* $n + m + 1$, *then* ϕ, $\frac{\phi}{m_s}$ *are PE and* $\theta(t)$ *converges to* θ^* *as* $t \to \infty$.

Proof. Consider the Lyapunov-like function

$$V = \frac{\tilde{\theta}^T \Gamma^{-1} \tilde{\theta}}{2} |\rho^*| + \frac{\tilde{\rho}^2}{2\gamma}.$$

Then

$$\dot{V} = \tilde{\theta}^T \phi \varepsilon |\rho^*| \operatorname{sgn}(\rho^*) + \tilde{\rho} \varepsilon \xi.$$

Using $|\rho^*| \operatorname{sgn}(\rho^*) = \rho^*$ and the expression

$$\begin{aligned}
\varepsilon m_s^2 &= \rho^* \theta^{*T} \phi + \rho^* z_0 - \rho \theta^T \phi - \rho z_0 \\
&= -\tilde{\rho} z_0 + \rho^* \theta^{*T} \phi - \rho \theta^T \phi + \rho^* \theta^T \phi - \rho^* \theta^T \phi \\
&= -\tilde{\rho}(z_0 + \theta^T \phi) - \rho^* \tilde{\theta}^T \phi = -\tilde{\rho} \xi - \rho^* \tilde{\theta}^T \phi,
\end{aligned}$$

we have

$$\dot{V} = \varepsilon(\rho^* \tilde{\theta}^T \phi + \tilde{\rho} \xi) = -\varepsilon^2 m_s^2 \leq 0,$$

which implies that $V \in \mathcal{L}_\infty$ and therefore $\rho, \theta \in \mathcal{L}_\infty$. Using similar analysis as in the case of the gradient algorithms for the SPM, we can establish (i) from the properties of V, $\dot{V}$ and the form of the adaptive laws.

(ii) We have

$$\begin{aligned}
\rho(t) - \rho(0) = \int_0^t \dot{\rho} d\tau \leq \int_0^t |\dot{\rho}| d\tau &\leq \gamma \int_0^t |\varepsilon m_s| \frac{|\xi|}{m_s} d\tau \\
&\leq \gamma \left(\int_0^t \varepsilon^2 m_s^2 d\tau \right)^{1/2} \left(\int_0^t \frac{|\xi|^2}{m_s^2} d\tau \right)^{1/2} < \infty,
\end{aligned}$$

where the last inequality is obtained using the Schwarz inequality (see (A.14)). Since εm_s, $\frac{\xi}{m_s} \in \mathcal{L}_2$, the limit as $t \to \infty$ exists, which implies that $\dot{\rho} \in \mathcal{L}_1$ and $\lim_{t \to \infty} \rho(t) = \bar{\rho}$ for some constant $\bar{\rho}$. The proof of (iii) is long and is presented in the web resource [94]. The proof of (iv) is included in the proof of Theorem 3.6.1. □

The assumption that the sign of ρ^* is known can be relaxed, leading to an adaptive law for θ, ρ with additional nonlinear terms. The reader is referred to [56, 99–104] for further reading on adaptive laws with unknown high-frequency gain.

3.10 Parameter Projection

In many practical problems, we may have some a priori knowledge of where θ^* is located in $\mathcal{R}^n$. This knowledge usually comes in terms of upper and/or lower bounds for the elements of θ^* or in terms of location in a convex subset of $\mathcal{R}^n$. If such a priori information is available, we want to constrain the online estimation to be within the set where the unknown parameters are located. For this purpose we modify the gradient algorithms based on the unconstrained minimization of certain costs using the gradient projection method presented in section A.10.3 as follows.

The gradient algorithm with projection is computed by applying the gradient method to the following minimization problem with constraints:

$$\begin{aligned} &\text{minimize } J(\theta) \\ &\text{subject to } \theta \in S, \end{aligned}$$

where S is a convex subset of $\mathcal{R}^n$ with smooth boundary almost everywhere. Assume that S is given by

$$S = \{\theta \in \mathcal{R}^n | g(\theta) \leq 0\},$$

where $g : \mathcal{R}^n \to \mathcal{R}$ is a smooth function.

The adaptive laws based on the gradient method can be modified to guarantee that $\theta \in S$ by solving the constrained optimization problem given above to obtain

$$\dot{\theta} = \Pr(-\Gamma\nabla J) = \begin{cases} -\Gamma\nabla J & \text{if } \theta \in S^0 \\ & \text{or } \theta \in \delta(S) \text{ and } -(\Gamma\nabla J)^T\nabla g \leq 0, \\ -\Gamma\nabla J + \Gamma\frac{\nabla g\nabla g^T}{\nabla g^T\Gamma\nabla g}\Gamma\nabla J & \text{otherwise,} \end{cases} \tag{3.57}$$

where $\delta(S) = \{\theta \in \mathcal{R}^n | g(\theta) = 0\}$ and $S^0 = \{\theta \in \mathcal{R}^n | g(\theta) < 0\}$ denote the boundary and the interior, respectively, of S and $\Pr(\cdot)$ is the projection operator as shown in section A.10.3.

The *gradient algorithm based on the instantaneous cost function with projection* follows from (3.57) by substituting for $\nabla J = -\varepsilon\phi$ to obtain

$$\dot{\theta} = \Pr(\Gamma\varepsilon\phi) = \begin{cases} \Gamma\varepsilon\phi & \text{if } \theta \in S^0 \\ & \text{or } \theta \in \delta(S) \text{ and } (\Gamma\varepsilon\phi)^T\nabla g \leq 0, \\ \Gamma\varepsilon\phi - \Gamma\frac{\nabla g\nabla g^T}{\nabla g^T\Gamma\nabla g}\Gamma\varepsilon\phi & \text{otherwise,} \end{cases} \tag{3.58}$$

where $\theta(0) \in S$.

The *pure LS algorithm with projection* becomes

$$\dot{\theta} = \Pr(P\varepsilon\phi) = \begin{cases} P\varepsilon\phi & \text{if } \theta \in S^0 \\ & \text{or if } \theta \in \delta(S) \text{ and } (P\varepsilon\phi)^T\nabla g \leq 0, \\ P\varepsilon\phi - P\frac{\nabla g\nabla g^T}{\nabla g^T P\nabla g}P\varepsilon\phi & \text{otherwise,} \end{cases} \tag{3.59}$$

where $\theta(0) \in S$,

$$\dot{P} = \begin{cases} \beta P - P\frac{\phi\phi^T}{m_s^2}P & \text{if } \theta \in S^0 \\ & \text{or if } \theta \in \delta(S) \text{ and } (P\varepsilon\phi)^T\nabla g \leq 0, \\ 0 & \text{otherwise,} \end{cases}$$

and $P(0) = P_0 = P_0^T > 0$.

Theorem 3.10.1. *The gradient adaptive laws of section* 3.6 *and the LS adaptive laws of section* 3.7 *with the projection modifications given by* (3.57) *and* (3.59), *respectively, retain all the properties that are established in the absence of projection and in addition guarantee that* $\theta(t) \in S\ \forall t \geq 0$, *provided* $\theta(0) \in S$ *and* $\theta^* \in S$.

Proof. The adaptive laws (3.57) and (3.59) both guarantee that whenever $\theta \in \delta(S)$, the direction of $\dot{\theta}$ is either towards S^0 or along the tangent plane of $\delta(S)$ at θ. This property together with $\theta(0) \in S$ guarantees that $\theta(t) \in S\ \forall t \geq 0$.

The gradient adaptive law (3.57) can be expressed as

$$\dot{\theta} = -\Gamma\nabla J + (1 - \mathrm{sgn}(|g(\theta)|))\max(0, \mathrm{sgn}(-(\Gamma\nabla J)^T\nabla g))\Gamma\frac{\nabla g \nabla g^T}{\nabla g^T \Gamma \nabla g}\Gamma\nabla J, \quad (3.60)$$

where $\mathrm{sgn}\,|x| = 0$ when $x = 0$. Hence, for the Lyapunov-like function V used in section 3.6, i.e., $V = \frac{\tilde{\theta}^T\Gamma^{-1}\tilde{\theta}}{2}$, we have

$$\dot{V} = -\tilde{\theta}^T\nabla J + V_p,$$

where

$$V_p = (1 - \mathrm{sgn}(|g(\theta)|))\max(0, \mathrm{sgn}(-(\Gamma\nabla J)^T\nabla g))\tilde{\theta}^T\frac{\nabla g \nabla g^T}{\nabla g^T \Gamma \nabla g}\Gamma\nabla J.$$

The term V_p is nonzero only when $\theta \in \delta(S)$, i.e., $g(\theta) = 0$ and $-(\Gamma\nabla J)^T\nabla g > 0$. In this case

$$\begin{aligned} V_p &= \tilde{\theta}^T\frac{\nabla g \nabla g^T}{\nabla g^T \Gamma \nabla g}\Gamma\nabla J = \frac{1}{\nabla g^T \Gamma \nabla g}(\tilde{\theta}^T\nabla g)(\nabla g^T\Gamma\nabla J) \\ &= \frac{1}{\nabla g^T \Gamma \nabla g}(\tilde{\theta}^T\nabla g)((\Gamma\nabla J)^T\nabla g). \end{aligned}$$

Since $\tilde{\theta}^T\nabla g = (\theta - \theta^*)^T\nabla g \geq 0$ for $\theta^* \in S$ and $\theta \in \delta(S)$ due to the convexity of S, $(\Gamma\nabla J)^T\nabla g < 0$ implies that $V_p \leq 0$. Therefore, the projection due to $V_p \leq 0$ can only make $\dot{V}$ more negative. Hence

$$\dot{V} \leq -\tilde{\theta}^T\nabla J, \quad (3.61)$$

which is the same expression as in the case without projection except for the inequality. Hence the results in section 3.6 based on the properties of V and $\dot{V}$ are valid for the adaptive law (3.57) as well. Moreover, since $\Gamma\frac{\nabla g \nabla g^T}{\nabla g^T\Gamma\nabla g} \in \mathcal{L}_\infty$, from (3.60) we have

$$|\dot{\theta}|^2 \leq c|\Gamma\nabla J|^2$$

for some constant $c > 0$, which can be used together with (3.61) to show that $\dot{\theta} \in \mathcal{L}_2$. The same arguments apply to the case of the LS algorithms. $\square$

Example 3.10.2 Let us consider the plant model

$$y = \frac{b}{s+a}u,$$

where a, b are unknown constants that satisfy some known bounds, e.g., $b \geq 1$ and $20 \geq a \geq -2$. For simplicity, let us assume that y, $\dot{y}$, u are available for measurement so that the SPM is of the form

$$z = \theta^{*T}\phi,$$

where $z = \dot{y}$, $\theta^* = [b, a]^T$, $\phi = [u, -y]^T$. In the unconstrained case the gradient adaptive law is given as

$$\dot{\theta} = \Gamma\varepsilon\phi, \quad \varepsilon = \frac{z - \theta^T\phi}{m_s^2},$$

where $m_s^2 = 1 + \phi^T\phi$; $\theta = [\hat{b}, \hat{a}]^T$; $\hat{b}$, $\hat{a}$ are the estimates of b, a, respectively. Since we know that $b \geq 1$ and $20 \geq a \geq -2$, we can constrain the estimates $\hat{b}$, $\hat{a}$ to be within the known bounds by using projection. Defining the sets for projection as

$$\begin{aligned} S_b &= \{\hat{b} \in \mathcal{R} | 1 - \hat{b} \leq 0\}, \\ S_a^l &= \{\hat{a} \in \mathcal{R} | -2 - \hat{a} \leq 0\}, \\ S_a^u &= \{\hat{a} \in \mathcal{R} | \hat{a} - 20 \leq 0\} \end{aligned}$$

and applying the projection algorithm (3.57) for each set, we obtain the adaptive laws

$$\dot{\hat{b}} = \begin{cases} \gamma_1\varepsilon u & \text{if } \hat{b} > 1 \text{ or } (\hat{b} = 1 \text{ and } \varepsilon u \geq 0), \\ 0 & \text{if } \hat{b} = 1 \text{ and } \varepsilon u < 0, \end{cases}$$

with $\hat{b}(0) \geq 1$, and

$$\dot{\hat{a}} = \begin{cases} -\gamma_2\varepsilon y & \text{if } 20 > \hat{a} > -2 \text{ or } (\hat{a} = -2 \text{ and } \varepsilon y \leq 0) \\ & \text{or } (\hat{a} = 20 \text{ and } \varepsilon y \geq 0), \\ 0 & \text{if } (\hat{a} = -2 \text{ and } \varepsilon y > 0) \text{ or } (\hat{a} = 20 \text{ and } \varepsilon y < 0), \end{cases}$$

with $\hat{a}(0)$ satisfying $20 \geq \hat{a}(0) \geq -2$. ∎

Example 3.10.3 Let us consider the gradient adaptive law

$$\dot{\theta} = \Gamma\varepsilon\phi$$

with the a priori knowledge that $|\theta^*| \leq M_0$ for some known bound $M_0 > 0$. In most applications, we may have such a priori information. We define

$$S = \left\{\theta \in \mathcal{R}^n | g(\theta) = \frac{\theta^T\theta}{2} - \frac{M_0^2}{2} \leq 0\right\}$$

and use (3.57) together with $\nabla g = \theta$ to obtain the adaptive law with projection

$$\dot{\theta} = \begin{cases} \Gamma\varepsilon\phi & \text{if } |\theta| < M_0 \text{ or } (|\theta| = M_0 \text{ and } \phi^T\Gamma\theta\varepsilon \leq 0), \\ \Gamma\varepsilon\phi - \Gamma\frac{\theta\theta^T}{\theta^T\Gamma\theta}\Gamma\varepsilon\phi & \text{if } |\theta| = M_0 \text{ and } \phi^T\Gamma\theta\varepsilon > 0 \end{cases}$$

with $|\theta(0)| \leq M_0$. ∎

3.11 Robust Parameter Identification

In the previous sections we designed and analyzed a wide class of PI algorithms based on the parametric models

$$z = \theta^{*T}\phi \quad \text{or} \quad W(s)\theta^{*T}\phi.$$

These parametric models are developed using a plant model that is assumed to be free of disturbances, noise, unmodeled dynamics, time delays, and other frequently encountered uncertainties. In the presence of plant uncertainties we are no longer able to express the unknown parameter vector θ^* in the form of the SPM or DPM where all signals are measured and θ^* is the only unknown term. In this case, the SPM or DPM takes the form

$$z = \theta^{*T}\phi + \eta \quad \text{or} \quad z = W(s)\theta^{*T}\phi + \eta, \tag{3.62}$$

where η is an unknown function that represents the modeling error terms. The following examples are used to show how (3.62) arises for different plant uncertainties.

Example 3.11.1 Consider the scalar constant gain system

$$y = \theta^* u + d, \tag{3.63}$$

where θ^* is the unknown scalar and d is a bounded external disturbance due to measurement noise and/or input disturbance. Equation (3.60) is already in the form of the SPM with modeling error given by (3.62). ■

Example 3.11.2 Consider a system with a small input delay τ given by

$$y = \frac{b}{s+a} e^{-\tau s} u, \tag{3.64}$$

where a, b are the unknown parameters to be estimated and $u \in \mathcal{L}_\infty$. Since τ is small, the plant may be modeled as

$$y = \frac{b}{s+a} u \tag{3.65}$$

by assuming $\tau = 0$. Since a parameter estimator for a, b developed based on (3.65) has to be applied to the actual plant (3.64), it is of interest to see how $\tau \neq 0$ affects the parametric model for a, b. We express the plant as

$$y = \frac{b}{s+a} u + \frac{b}{s+a}(e^{-\tau s} - 1)u,$$

which we can rewrite in the form of the parametric model (3.62) as

$$z = \theta^{*T}\phi + \eta,$$

where

$$z = \frac{s}{s+\lambda} y, \qquad \theta^* = [b, a]^T, \qquad \phi = \begin{bmatrix} \frac{1}{s+\lambda} u \\ -\frac{1}{s+\lambda} y \end{bmatrix}, \qquad \eta = \frac{b}{s+\lambda}(e^{-\tau s} - 1)u,$$

and $\lambda > 0$. It is clear that $\tau = 0$ implies $\eta = 0$ and small τ implies small η. ■

Example 3.11.3 Let us consider the plant

$$y = \theta^*(1 + \mu\Delta_m(s))u, \tag{3.66}$$

where μ is a small constant and $\Delta_m(s)$ is a proper transfer function with poles in the open left half s-plane. Since μ is small and $\Delta_m(s)$ is proper with stable poles, the term $\mu\Delta_m(s)$ can be treated as the modeling error term which can be approximated with zero. We can express (3.66) in the form of (3.62) as

$$y = \theta^* u + \eta,$$

where

$$\eta = \mu\theta^*\Delta_m(s)u$$

is the modeling error term. ∎

For LTI plants, the parametric model with modeling errors is usually of the form

$$\begin{aligned} z &= \theta^{*T}u + \eta, \\ \eta &= \Delta_1(s)u + \Delta_2(s)y + d, \end{aligned} \tag{3.67}$$

where $\Delta_1(s)$, $\Delta_2(s)$ are proper transfer functions with stable poles and d is a bounded disturbance. The principal question that arises is how the stability properties of the adaptive laws that are developed for parametric models with no modeling errors are affected when applied to the actual parametric models with uncertainties. The following example demonstrates that the adaptive laws of the previous sections that are developed using parametric models that are free of modeling errors cannot guarantee the same properties in the presence of modeling errors. Furthermore, it often takes only a small disturbance to drive the estimated parameters unbounded.

3.11.1 Instability Example

Consider the scalar constant gain system

$$y = \theta^* u + d,$$

where d is a bounded unknown disturbance and $u \in \mathcal{L}_\infty$. The adaptive law for estimating θ^* derived for $d = 0$ is given by

$$\dot{\theta} = \gamma\varepsilon u, \quad \varepsilon = y - \theta u, \tag{3.68}$$

where $\gamma > 0$ and the normalizing signal is taken to be 1. If $d = 0$ and $u, \dot{u} \in \mathcal{L}_\infty$, then we can establish that (i) $\dot{\theta}, \theta, \varepsilon \in \mathcal{L}_\infty$, (ii) $\varepsilon(t) \to 0$ as $t \to \infty$ by analyzing the parameter error equation

$$\dot{\tilde{\theta}} = -\gamma u^2\tilde{\theta},$$

which is a linear time-varying differential equation. When $d \neq 0$, we have

$$\dot{\tilde{\theta}} = -\gamma u^2\tilde{\theta} + \gamma d u. \tag{3.69}$$

In this case we cannot guarantee that the parameter estimate $\theta(t)$ is bounded for any bounded input u and disturbance d. In fact, for $\theta^* = 2$, $\gamma = 1$,

$$u = (1+t)^{-1/2},$$
$$d(t) = (1+t)^{-1/4}\left(\frac{5}{4} - 2(1+t)^{-1/4}\right) \to 0 \quad \text{as } t \to \infty,$$

we have

$$y(t) = \frac{5}{4}(1+t)^{-1/4} \to 0 \quad \text{as } t \to \infty,$$
$$\varepsilon(t) = \frac{1}{4}(1+t)^{-1/4} \to 0 \quad \text{as } t \to \infty,$$
$$\theta(t) = (1+t)^{-1/4} \to \infty \quad \text{as } t \to \infty;$$

i.e., the estimated parameter drifts to infinity with time even though the disturbance $d(t)$ disappears with time. This instability phenomenon is known as *parameter drift*. It is mainly due to the pure integral action of the adaptive law, which, in addition to integrating the "good" signals, integrates the disturbance term as well, leading to the parameter drift phenomenon.

Another interpretation of the above instability is that, for $u = (1+t)^{-1/2}$, the homogeneous part of (3.69), i.e., $\dot{\tilde{\theta}} = -\gamma u^2 \tilde{\theta}$, is only uniformly stable, which is not sufficient to guarantee that the bounded input γdu will produce a bounded state $\tilde{\theta}$. If u is persistently exciting, i.e., $\int_t^{t+T_0} u^2(\tau)d\tau \geq \alpha_0 T_0$ for some $\alpha_0, T_0 > 0$ and $\forall t \geq 0$, then the homogeneous part of (3.69) is e.s. and the bounded input γdu produces a bounded state $\tilde{\theta}$ (show it!). Similar instability examples in the absence of PE may be found in [50–52, 105].

If the objective is parameter convergence, then parameter drift can be prevented by making sure the regressor vector is PE with a level of excitation higher than the level of the modeling error. In this case the plant input in addition to being sufficiently rich is also required to guarantee a level of excitation for the regressor that is higher than the level of the modeling error. This class of inputs is referred to as *dominantly rich* and is discussed in the following section.

3.11.2 Dominantly Rich Excitation

Let us revisit the example in section 3.11.1 and analyze (3.69), i.e.,

$$\dot{\tilde{\theta}} = -\gamma u^2 \tilde{\theta} + \gamma du, \tag{3.70}$$

when u is PE with level $\alpha_0 > 0$. The PE property of u implies that the homogeneous part of (3.70) is e.s., which in turn implies that

$$|\tilde{\theta}(t)| \leq e^{-\gamma\alpha_1 t}|\tilde{\theta}(0)| + \frac{1}{\alpha_1}(1 - e^{-\gamma\alpha_1 t}) \sup_{\tau \leq t} |u(\tau)d(\tau)|$$

for some $\alpha_1 > 0$ which depends on α_0. Therefore, we have

$$\lim_{t\to\infty} \sup_{\tau \geq t} |\tilde{\theta}(\tau)| \leq \frac{1}{\alpha_1} \lim_{t\to\infty} \sup_{\tau \geq t} |u(\tau)d(\tau)| = \frac{1}{\alpha_1} \sup_{\tau} |u(\tau)d(\tau)|. \tag{3.71}$$

The bound (3.71) indicates that the PI error at steady state is of the order of the disturbance; i.e., as $d \to 0$ the parameter error also reduces to zero. For this simple example, it is clear that if we choose $u = u_0$, where u_0 is a constant different from zero, then $\alpha_1 = \alpha_0 = u_0^2$; therefore, the bound for $|\tilde{\theta}|$ is $\sup_t \frac{|d(t)|}{u_0}$. Thus the larger the value of u_0 is, the smaller the parameter error. Large u_0 relative to $|d|$ implies large signal-to-noise ratio and therefore better accuracy of identification.

Example 3.11.4 (unmodeled dynamics) Let us consider the plant

$$y = \theta^*(1 + \Delta_m(s))u,$$

where $\Delta_m(s)$ is a proper transfer function with stable poles. If $\Delta_m(s)$ is much smaller than 1 for small s or all s, then the system can be approximated as

$$y = \theta^* u$$

and $\Delta_m(s)$ can be treated as an unmodeled perturbation. The adaptive law (3.68) that is designed for $\Delta_m(s) = 0$ is used to identify θ^* in the presence of $\Delta_m(s)$. The parameter error equation in this case is given by

$$\dot{\tilde{\theta}} = -\gamma u^2 \tilde{\theta} + \gamma u \eta, \quad \eta = \theta^* \Delta_m(s) u. \tag{3.72}$$

Since u is bounded and $\Delta_m(s)$ is stable, it follows that $\eta \in \mathcal{L}_\infty$ and therefore the effect of $\Delta_m(s)$ is to introduce the bounded disturbance term η in the adaptive law. Hence, if u is PE with level $\alpha_0 > 0$, we have, as in the previous example, that

$$\lim_{t\to\infty} \sup_{\tau \geq t} |\tilde{\theta}(\tau)| \leq \frac{1}{\alpha_1} \sup_t |u(t)\eta(t)|$$

for some $\alpha_1 > 0$. The question that comes up is how to choose u so that the above bound for $|\tilde{\theta}|$ is as small as possible. The answer to this question is not as straightforward as in the example of section 3.11.1 because η is also a function of u. The bound for $|\tilde{\theta}|$ depends on the choice of u and the properties of $\Delta_m(s)$. For example, for constant $u = u_0 \neq 0$, we have $\alpha_0 = \alpha_1 = u_0^2$ and $\eta = \theta^* \Delta_m(s) u_0$, i.e., $\lim_{t\to\infty} |\eta(t)| = |\theta^*||\Delta_m(0)||u_0|$, and therefore

$$\lim_{t\to\infty} \sup_{\tau \geq t} |\tilde{\theta}(\tau)| \leq |\Delta_m(0)||\theta^*|.$$

If the plant is modeled properly, $\Delta_m(s)$ represents a perturbation that is small in the low-frequency range, which is usually the range of interest. Therefore, for $u = u_0$, we should have $|\Delta_m(0)|$ small if not zero leading to the above bound, which is independent of u_0. Another choice of a PE input is $u = \cos \omega_0 t$ for some $\omega_0 \neq 0$. For this choice of u, since

$$e^{-\gamma \int_0^t \cos^2 \omega_0 \tau d\tau} = e^{-\frac{\gamma}{2}\left(t + \frac{\sin 2\omega_0 t}{2\omega_0}\right)} = e^{-\frac{\gamma}{4}\left(t + \frac{\sin 2\omega_0 t}{\omega_0}\right)} e^{-\frac{\gamma}{4}t} \leq e^{-\frac{\gamma}{4}t}$$

(where we used the inequality $t + \frac{\sin 2\omega_0 t}{\omega_0} \geq 0 \ \forall t \geq 0$) and $\sup_t |\eta(t)| \leq |\Delta_m(j\omega_0)||\theta^*|$, we have

$$\lim_{t\to\infty} \sup_{r \geq t} |\tilde{\theta}(r)| \leq c|\Delta_m(j\omega_0)|,$$

where $c = \frac{4|\theta^*|}{\gamma}$. This bound indicates that for small parameter error, ω_0 should be chosen so that $|\Delta_m(j\omega_0)|$ is as small as possible. If $\Delta_m(s)$ is due to high-frequency unmodeled dynamics, then $|\Delta_m(j\omega_0)|$ is small, provided ω_0 is a low frequency. As an example, consider

$$\Delta_m(s) = \frac{\mu s}{1 + \mu s},$$

where $\mu > 0$ is a small constant. It is clear that for low frequencies $|\Delta_m(j\omega)| = O(\mu)$[3] and $|\Delta_m(j\omega)| \to 1$ as $\omega \to \infty$. Since

$$|\Delta_m(j\omega_0)| = \frac{|\mu\omega_0|}{\sqrt{1 + \mu^2\omega_0^2}},$$

it follows that for $\omega_0 = \frac{1}{\mu}$ we have $|\Delta_m(j\omega_0)| = \frac{1}{\sqrt{2}}$ and for $\omega_0 \gg \frac{1}{\mu}$, $\Delta_m(j\omega_0) \approx 1$, whereas for $\omega_0 < \frac{1}{\mu}$ we have $|\Delta_m(j\omega_0)| = O(\mu)$. Therefore, for more accurate PI, the input signal should be chosen to be PE, but the PE property should be achieved with frequencies that do not excite the unmodeled dynamics. For the above example of $\Delta_m(s)$, $u = u_0$ does not excite $\Delta_m(s)$ at all, i.e., $\Delta_m(0) = 0$, whereas for $u = \sin \omega_0 t$ with $\omega_0 \ll \frac{1}{\mu}$, the excitation of $\Delta_m(s)$ is small leading to an $O(\mu)$ steady-state error for $|\tilde{\theta}|$. ∎

The above example demonstrates the well-known fact in control systems that the excitation of the plant should be restricted to be within the range of frequencies where the plant model is a good approximation of the actual plant. We explain this statement further using the plant

$$y = G_0(s)u + \Delta_a(s)u + d, \tag{3.73}$$

where $G_0(s)$, $\Delta_a(s)$ are proper and stable, $\Delta_a(s)$ is an additive perturbation of the modeled part $G_0(s)$, and d is a bounded disturbance. We would like to identify the coefficients of $G_0(s)$ by exciting the plant with the input u and processing the I/O data.

Because $\Delta_a(s)u$ is treated as a disturbance, the input u should be chosen so that at each frequency ω_i contained in u, we have $|G_0(j\omega_i)| \gg |\Delta_a(j\omega_i)|$. Furthermore, u should be rich enough to excite the modeled part of the plant that corresponds to $G_0(s)$ so that y contains sufficient information about the coefficients of $G_0(s)$. For such a choice of u to be possible, the spectrums of $G_0(s)$ and $\Delta_a(s)u$ should be separated, or $|G_0(j\omega)| \gg |\Delta_a(j\omega)|$ at all frequencies. If $G_0(s)$ is chosen properly, then $|\Delta_a(j\omega)|$ should be small relative to $|G_0(j\omega)|$ in the frequency range of interest. Since we are usually interested in the system response at low frequencies, we would assume that $|G_0(j\omega)| \gg |\Delta_a(j\omega)|$ in the low-frequency range for our analysis. But at high frequencies, we may have $|G_0(j\omega)|$ of the same order as or smaller than $|\Delta_a(j\omega)|$. The input signal u should therefore be designed to be sufficiently rich for the modeled part of the plant, but its richness should be achieved in the low-frequency range for which $|G_0(j\omega)| \gg |\Delta_a(j\omega)|$. An input signal with these two properties is called *dominantly rich* [106] because it excites the modeled or dominant part of the plant much more than the unmodeled one.

The separation of spectrums between the dominant and unmodeled part of the plant can be seen more clearly if we rewrite (3.73) as

$$y = G_0(s)u + \Delta_a(\mu s)u + d, \tag{3.74}$$

[3] A function $f(x)$ is of $O(\mu)$ $\forall x \in \Omega$ if there exists a constant $c \geq 0$ such that $\|f(x)\| \leq c|\mu|$ $\forall x \in \Omega$.

where $\Delta_a(\mu s)$ is due to high-frequency dynamics and $\mu > 0$ is a small parameter referred to as the singular perturbation parameter. $\Delta_a(\mu s)$ has the property that $\Delta_a(0) = 0$ and $|\Delta_a(\mu j\omega)| \leq O(\mu)$ for $\omega \ll \frac{1}{\mu}$, but $\Delta_a(\mu s)$ could satisfy $|\Delta_a(\mu j\omega)| = O(1)$ for $\omega \gg \frac{1}{\mu}$. Therefore, for μ small, the contribution of Δ_a in y is small, provided that the plant is not excited with frequencies of the order of $\frac{1}{\mu}$ or higher. Let us assume that we want to identify the coefficients of $G_0(s) = \frac{Z(s)}{R(s)}$ in (3.74) which consist of m zeros and n stable poles, a total of $n + m + 1$ parameters, by treating $\Delta_a(\mu s)u$ as a modeling error. We can express (3.74) in the form

$$z = \theta^{*T}\phi + \eta,$$

where $\eta = \frac{\Delta_a(\mu s)R(s)}{\Lambda(s)}u + d$, $\frac{1}{\Lambda(s)}$ is a filter with stable poles, $\theta^* \in \mathcal{R}^{n+m+1}$ contains the coefficients of the numerator and denominator of $G_0(s)$, and

$$\phi = \frac{1}{\Lambda(s)}[s^m u, s^{m-1}u, \dots, u, -s^{n-1}y, -s^{n-2}y, \dots, -y]^T.$$

If ϕ is PE, we can establish that the adaptive laws of the previous sections guarantee convergence of the parameter error to a set whose size depends on the bound for the modeling error signal η, i.e.,

$$|\tilde{\theta}(t)| \leq \alpha_1 e^{-\alpha t}(|\tilde{\theta}(0)| + \eta_0) + \alpha_2\eta_0,$$

where $\eta_0 = \sup_t |\eta(t)|$ and $\alpha, \alpha_1, \alpha_2 > 0$ are some constants that depend on the level of excitation of ϕ. The problem is how to choose u so that ϕ is PE despite the presence of $\eta \neq 0$. It should be noted that it is possible, for an input that is sufficiently rich, for the modeled part of the plant not to guarantee the PE property of ϕ in the presence of modeling errors because the modeling error could cancel or highly corrupt terms that are responsible for the PE property ϕ. We can see this by examining the relationship between u and ϕ given by

$$\phi = H_0(s)u + H_1(\mu s, s)u + B_1 d, \tag{3.75}$$

where

$$H_0(s) = \frac{1}{\Lambda(s)}[s^m, s^{m-1}, \dots, s, 1, -s^{n-1}G_0(s), -s^{n-2}G_0(s), \dots, -sG_0(s), -G_0(s)]^T,$$

$$H_1(\mu s, s) = \frac{1}{\Lambda(s)}[0, \dots, 0, -s^{n-1}\Delta_a(\mu s), -s^{n-2}\Delta_a(\mu s), \dots, -\Delta_a(\mu s)]^T,$$

$$B_1 = \frac{1}{\Lambda(s)}[0, \dots, 0, -s^{n-1}, -s^{n-2}, \dots, -s, -1]^T.$$

It is clear that the term $H_1(\mu s, s)u + B_1 d$ acts as a modeling error term and could destroy the PE property of ϕ, even for inputs u that are sufficiently rich of order $n+m+1$. So the problem is to define the class of sufficiently rich inputs of order $n+m+1$ that would guarantee ϕ to be PE despite the presence of the modeling error term $H_1(\mu s, s)u + B_1 d$ in (3.75).

Definition 3.11.5. *A sufficiently rich input u of order $n + m + 1$ for the dominant part of the plant* (3.74) *is called* dominantly rich *of order $n + m + 1$ if it achieves its richness with frequencies ω_i, $i = 1, 2, \dots, N$, where $N \geq \frac{n+m+1}{2}$, $|\omega_i| < O(\frac{1}{\mu})$, $|\omega_i - \omega_j| > O(\mu)$, $i \neq j$, and $|u| > O(\mu) + O(d)$.*

Lemma 3.11.6. *Let $H_0(s)$, $H_1(\mu s, s)$ satisfy the following assumptions:*

(a) *The vectors $H_0(j\omega_1), H_0(j\omega_2), \ldots, H_0(j\omega_{\bar{n}})$ are linearly independent on $\mathcal{C}^{\bar{n}}$ for all possible $\omega_1, \omega_2, \ldots, \omega_{\bar{n}} \in R$, where $\bar{n} \triangleq n + m + 1$ and $\omega_i \neq \omega_k$ for $i \neq k$.*

(b) *For any set $\{\omega_1, \omega_2, \ldots, \omega_{\bar{n}}\}$ satisfying $|\omega_i - \omega_k| > O(\mu)$ for $i \neq k$ and $|\omega_i| < O(\frac{1}{\mu})$, we have $|\det(\bar{H})| > O(\mu)$, where $\bar{H} \triangleq [H_0(j\omega_1), H_0(j\omega_2), \ldots, H_0(j\omega_{\bar{n}})]$.*

(c) *$|H_1(j\mu\omega, j\omega)| \leq c$ for some constant c independent of μ and $\forall \omega \in \mathcal{R}$.*

Then there exists a $\mu^* > 0$ such that for $\mu \in [0, \mu^*)$, ϕ is PE of order $n + m + 1$ with level of excitation $\alpha_1 > O(\mu)$, provided that the input signal u is dominantly rich of order $n + m + 1$ for the plant (3.74).

Proof. Since $|u| > O(d)$, we can ignore the contribution of d in (3.75) and write

$$\phi = \phi_0 + \phi_1,$$

where

$$\phi_0 = H_0(s)u, \qquad \phi_1 = H_1(\mu s, s)u.$$

Since $H_0(s)$ does not depend on μ, ϕ_0 is the same signal vector as in the ideal case. The sufficient richness of order $\bar{n}$ of u together with the assumed properties (a)–(b) of $H_0(s)$ imply that ϕ_0 is PE with level $\alpha_0 > 0$ and α_0 is independent of μ, i.e.,

$$\frac{1}{T}\int_t^{t+T} \phi_0(\tau)\phi_0^T(\tau)d\tau \geq \alpha_0 I \tag{3.76}$$

$\forall t \geq 0$ and for some $T > 0$. On the other hand, since $H_1(\mu s, s)$ is stable and $|H_1(j\mu\omega, j\omega)| \leq c \; \forall \omega \in \mathcal{R}$, we have $\phi_1 \in \mathcal{L}_\infty$,

$$\frac{1}{T}\int_t^{t+T} \phi_1(\tau)\phi_1^T(\tau)d\tau \leq \beta I \tag{3.77}$$

for some constant β which is independent of μ. Because of (3.76) and (3.77), we have

$$\begin{aligned}
\frac{1}{T}\int_t^{t+T} \phi(\tau)\phi^T(\tau)d\tau &= \frac{1}{T}\int_t^{t+T} (\phi_0(\tau) + \mu\phi_1(\tau))(\phi_0^T(\tau) + \mu\phi_1^T(\tau))d\tau \\
&\geq \frac{1}{T}\left(\int_t^{t+T} \frac{\phi_0(\tau)\phi_0^T(\tau)}{2}d\tau - \mu^2\int_t^{t+T} \phi_1(\tau)\phi_1^T(\tau)d\tau\right) \\
&\geq \frac{\alpha_0}{2}I - \mu^2\beta I = \frac{\alpha_0}{4}I + \frac{\alpha_0}{4}I - \mu^2\beta I,
\end{aligned}$$

where the first inequality is obtained by using $(x + y)(x + y)^T \geq \frac{xx^T}{2} - yy^T$. In other words, ϕ has a level of PE $\alpha_1 = \frac{\alpha_0}{4}$ for $\mu \in [0, \mu^*)$, where $\mu^* \triangleq \sqrt{\frac{\alpha_0}{4\beta}}$. □

For further reading on dominant richness and robust parameter estimation, see [56].

Example 3.11.7 Consider the plant

$$y = \frac{b}{s+a}\left(1 + \frac{2\mu(s-1)}{(\mu s+1)^2}\right)u,$$

where a, b are the unknown parameters and $\mu = 0.001$. The plant may be modeled as

$$y = \frac{b}{s+a}u$$

by approximating $\mu = 0.001 \cong 0$. The input $u = \sin\omega_0 t$ with $1 \ll \omega_0 \ll 1000$ would be a dominantly rich input of order 2. Frequencies such as $\omega_0 = 0.006$ rad/sec or $\omega_0 = 900$ rad/sec would imply that u is not dominantly rich even though u is sufficiently rich of order 2. ∎

3.12 Robust Adaptive Laws

If the objective in online PI is convergence of the estimated parameters (close) to their true values, then dominantly rich excitation of appropriate order will meet the objective, and no instabilities will be present. In many applications, such as in adaptive control, the plant input is the result of feedback and cannot be designed to be dominantly or even sufficiently rich. In such situations, the objective is to drive the plant output to zero or force it to follow a desired trajectory rather than convergence of the online parameter estimates to their true values. It is therefore of interest to guarantee stability and robustness for the online parameter estimators even in the absence of persistence of excitation. This can be achieved by modifying the adaptive laws of the previous sections to guarantee stability and robustness in the presence of modeling errors independent of the properties of the regressor vector ϕ.

Let us consider the general plant

$$y = G_0(s)u + \Delta_u(s)u + \Delta_y(s)y + d, \tag{3.78}$$

where $G_0(s)$ is the dominant part, $\Delta_u(s)$, $\Delta_y(s)$ are strictly proper with stable poles and small relative to $G_0(s)$, and d is a bounded disturbance. We are interested in designing adaptive laws for estimating the coefficients of

$$G_0(s) = \frac{Z(s)}{R(s)} = \frac{b_m s^m + \cdots + b_1 s + b_0}{s^n + a_{n-1}s^{n-1} + \cdots + a_1 s + a_0}.$$

The SPM for (3.78) is

$$z = \theta^{*T}\phi + \eta, \tag{3.79}$$

where

$$\theta^* = [b_m, \ldots, b_0, a_{n-1}, \ldots, a_0]^T \in \mathcal{R}^{n+m+1},$$

$$z = \frac{s^n}{\Lambda(s)}y,$$

$$\phi = \left[\frac{s^m}{\Lambda(s)}u, \dots, \frac{1}{\Lambda(s)}u, -\frac{s^{n-1}}{\Lambda(s)}y, \dots, -\frac{1}{\Lambda(s)}y \right]^T,$$
$$\eta = \Delta_u(s)\frac{R(s)}{\Lambda(s)}u + \Delta_y(s)\frac{R(s)}{\Lambda(s)}y + \frac{R(s)}{\Lambda(s)}d.$$

The modeling error term η is driven by both u and y and cannot be assumed to be bounded unless u and y are bounded.

The parametric model can be transformed into one where the modeling error term is bounded by using normalization, a term already used in the previous sections. Let us assume that we can find a signal $m_s > 0$ with the property $\frac{\phi}{m_s}, \frac{\eta}{m_s} \in \mathcal{L}_\infty$. Then we can rewrite (3.79) as

$$\bar{z} = \theta^{*T}\bar{\phi} + \bar{\eta}, \tag{3.80}$$

where $\bar{z} = \frac{z}{m_s}$, $\bar{\phi} = \frac{\phi}{m_s}$, $\bar{\eta} = \frac{\eta}{m_s}$. The normalized parametric model has all the measured signals bounded and can be used to develop online adaptive laws for estimating θ^*. The bounded term $\bar{\eta}$ can still drive the adaptive laws unstable, as we showed using an example in section 3.11.1. Therefore, for robustness, we need to use the following modifications:

- Design the normalizing signal m_s to bound the modeling error in addition to bounding the regressor vector ϕ.
- Modify the "pure" integral action of the adaptive laws to prevent parameter drift.

In the following sections, we develop a class of normalizing signals and modified adaptive laws that guarantee robustness in the presence of modeling errors.

3.12.1 Dynamic Normalization

Let us consider the SPM with modeling error

$$z = \theta^{*T}\phi + \eta,$$

where

$$\eta = \Delta_1(s)u + \Delta_2(s)y.$$

$\Delta_1(s)$, $\Delta_2(s)$ are strictly proper transfer functions with stable poles. Our objective is to design a signal m_s so that $\frac{\eta}{m_s} \in \mathcal{L}_\infty$. We assume that $\Delta_1(s)$, $\Delta_2(s)$ are analytic in $\Re[s] \geq -\frac{\delta_0}{2}$ for some known $\delta_0 > 0$. Apart from this assumption, we require no knowledge of the parameters and/or dimension of $\Delta_1(s)$, $\Delta_2(s)$. It is implicitly assumed, however, that they are small relative to the modeled part of the plant in the frequency range of interest; otherwise, they would not be treated as modeling errors.

Using the properties of the $\mathcal{L}_{2\delta}$ norm (see Lemma A.5.9), we can write

$$|\eta(t)| \leq \|\Delta_1(s)\|_{2\delta}\|u_t\|_{2\delta} + \|\Delta_2(s)\|_{2\delta}\|y_t\|_{2\delta}$$

for any $\delta \in [0, \delta_0]$, where the above norms are defined as

$$\|x_t\|_{2\delta} \triangleq \left(\int_0^t e^{-\delta(t-\tau)}x^T(\tau)x(\tau)d\tau \right)^{1/2},$$

$$\|H(s)\|_{2\delta} \triangleq \frac{1}{\sqrt{2\pi}} \left(\int_{-\infty}^{\infty} \left| H\left(j\omega - \frac{\delta}{2}\right) \right|^2 d\omega \right)^{1/2}.$$

If we define

$$n_d = \|u_t\|_{2\delta_0}^2 + \|y_t\|_{2\delta_0}^2,$$

which can be generated by the differential equation

$$\dot{n}_d = -\delta_0 n_d + u^2 + y^2, \quad n_d(0) = 0,$$

then it follows that

$$m_s^2 = 1 + n_d$$

bounds $|\eta(t)|$ from above and

$$\frac{|\eta(t)|}{m_s} \leq \|\Delta_1(s)\|_{2\delta_0} + \|\Delta_2(s)\|_{2\delta_0}.$$

The normalizing signal m_s used in the adaptive laws for parametric models free of modeling errors is required to bound the regressor ϕ from above. In the presence of modeling errors, m_s should be chosen to bound both ϕ and the modeling error η for improved robustness properties. In this case the normalizing signal m_s has the form

$$m_s^2 = 1 + n_s^2 + n_d,$$

where n_s^2 is the static part and n_d the dynamic one. Examples of static and dynamic normalizing signals are $n_s^2 = \phi^T \phi$ or $\phi^T P \phi$, where $P = P^T > 0$,

$$\dot{n}_d = -\delta_0 n_d + \delta_1 (u^2 + y^2), \quad n_d(0) = 0, \tag{3.81}$$

or

$$n_d = n_1^2, \quad \dot{n}_1 = -\delta_0 n_1 + \delta_1 (|u| + |y|), \quad n_1(0) \geq 0, \tag{3.82}$$

or

$$n_d = n_\infty^2, \quad n_\infty = \delta_1 \max\left(\sup_{\tau \leq t} |u(\tau)|, \sup_{\tau \leq t} |y(\tau)| \right). \tag{3.83}$$

Any one of choices (3.81)–(3.83) can be shown to guarantee that $\frac{\eta}{m_s} \in \mathcal{L}_\infty$. Since $\phi = H(s)[\begin{smallmatrix} u \\ y \end{smallmatrix}]$, the dynamic normalizing signal can be chosen to bound ϕ from above, provided that $H(s)$ is analytic in $\Re[s] \geq -\frac{\delta_0}{2}$, in which case $m_s^2 = 1 + n_d$ bounds both ϕ and η from above.

3.12.2 Robust Adaptive Laws: σ-Modification

A class of robust modifications involves the use of a small feedback around the "pure" integrator in the adaptive law, leading to the adaptive law structure

$$\dot{\theta} = \Gamma\varepsilon\phi - \sigma_\ell \Gamma\theta, \tag{3.84}$$

where $\sigma_\ell \geq 0$ is a small design parameter and $\Gamma = \Gamma^T > 0$ is the adaptive gain, which in the case of LS is equal to the covariance matrix P. The above modification is referred to as the *σ-modification* [50, 51, 59] or as *leakage*.

Different choices of σ_ℓ lead to different robust adaptive laws with different properties, described in the following sections. We demonstrate the properties of these modifications when applied to the SPM with modeling error,

$$z = \theta^{*T}\phi + \eta, \tag{3.85}$$

where η is the modeling error term that is bounded from above by the normalizing signal. Let us assume that η is of the form

$$\eta = \Delta_1(s)u + \Delta_2(s)y,$$

where $\Delta_1(s)$, $\Delta_2(s)$ are strictly proper transfer functions analytic in $\Re[s] \geq -\frac{\delta_0}{2}$ for some known $\delta_0 > 0$. The normalizing signal can be chosen as

$$\begin{aligned} m_s^2 &= 1 + \alpha_0\phi^T\phi + \alpha_1 n_d, \\ \dot{n}_d &= -\delta_0 n_d + \delta_1(u^2 + y^2), \quad n_d(0) = 0, \end{aligned} \tag{3.86}$$

for some design constants $\delta_1, \alpha_0, \alpha_1 > 0$, and shown to guarantee that $\phi/m_s, \eta/m_s \in \mathcal{L}_\infty$.

Fixed σ-Modification

In this case

$$\sigma_\ell(t) = \sigma > 0 \quad \forall t \geq 0, \tag{3.87}$$

where σ is a small positive design constant. The gradient adaptive law for estimating θ^* in (3.85) takes the form

$$\begin{aligned} \dot{\theta} &= \Gamma\varepsilon\phi - \sigma\Gamma\theta, \\ \varepsilon &= \frac{z - \theta^T\phi}{m_s^2}, \end{aligned} \tag{3.88}$$

where m_s is given by (3.86). If some a priori estimate θ_0 of θ^* is available, then the term $\sigma\Gamma\theta$ may be replaced with $\sigma\Gamma(\theta - \theta_0)$ so that the leakage term becomes larger for larger deviations of θ from θ_0 rather than from zero.

Theorem 3.12.1. *The adaptive law* (3.88) *guarantees the following properties:*

(i) $\varepsilon, \varepsilon m_s, \theta, \dot{\theta} \in \mathcal{L}_\infty$.

(ii) $\varepsilon, \varepsilon m_s, \dot{\theta} \in \mathcal{S}(\sigma + \frac{\eta^2}{m_s^2})$.[4]

(iii) *If* $\frac{\phi}{m_s}$ *is PE with level* $\alpha_0 > 0$, *then* $\theta(t)$ *converges exponentially to the residual set* $D_\sigma = \{\tilde{\theta} \mid |\tilde{\theta}| \leq c(\sigma + \bar{\eta})\}$, *where* $\bar{\eta} = \sup_t \frac{|\eta(t)|}{m_s(t)}$ *and* $c \geq 0$ *is some constant.*

Proof. (i) From (3.88), we obtain

$$\dot{\tilde{\theta}} = \Gamma\varepsilon\phi - \sigma\Gamma\theta, \quad \varepsilon m_s^2 = -\tilde{\theta}^T\phi + \eta.$$

Consider the Lyapunov-like function

$$V(\tilde{\theta}) = \frac{\tilde{\theta}^T\Gamma^{-1}\tilde{\theta}}{2}.$$

We have

$$\dot{V} = \tilde{\theta}^T\phi\varepsilon - \sigma\tilde{\theta}^T\theta$$

or

$$\dot{V} = -\varepsilon^2 m_s^2 + \varepsilon\eta - \sigma\tilde{\theta}^T(\tilde{\theta} + \theta^*) \leq -\varepsilon^2 m_s^2 + |\varepsilon m_s| \left|\frac{\eta}{m_s}\right| - \sigma\tilde{\theta}^T\tilde{\theta} + \sigma|\tilde{\theta}||\theta^*|.$$

Using $-a^2 + ab = -\frac{a^2}{2} - \frac{1}{2}(a-b)^2 + \frac{b^2}{2} \leq -\frac{a^2}{2} + \frac{b^2}{2}$ for any a, b, we write

$$\dot{V} \leq -\frac{\varepsilon^2 m_s^2}{2} - \frac{\sigma}{2}|\tilde{\theta}|^2 + \frac{1}{2}\frac{|\eta|^2}{m_s^2} + \frac{\sigma}{2}|\theta^*|^2.$$

Since $\frac{\eta}{m_s} \in \mathcal{L}_\infty$, the positive terms in the expression for $\dot{V}$ are bounded from above by a constant. Now

$$V(\tilde{\theta}) \leq |\tilde{\theta}|^2 \frac{\lambda_{\max}(\Gamma^{-1})}{2}$$

and therefore

$$-\frac{\sigma}{2}|\tilde{\theta}|^2 \leq -\frac{\sigma}{\lambda_{\max}(\Gamma^{-1})}V(\tilde{\theta}),$$

which implies

$$\dot{V} \leq -\frac{\varepsilon^2 m_s^2}{2} - \frac{\sigma}{\lambda_{\max}(\Gamma^{-1})}V(\tilde{\theta}) + c_0\left(\sigma + \frac{\eta^2}{m_s^2}\right),$$

where $c_0 = \max\{\frac{|\theta^*|^2}{2}, \frac{1}{2}\}$. Hence for

$$V(\tilde{\theta}) \geq V_0 \triangleq c_0(\sigma + \bar{\eta}^2)\frac{\lambda_{\max}(\Gamma^{-1})}{\sigma},$$

where $\bar{\eta} = \sup_t \frac{|\eta(t)|}{m_s(t)}$, we have $\dot{V} \leq -\frac{\varepsilon^2 m_s^2}{2} \leq 0$, which implies that V is bounded and therefore $\tilde{\theta} \in \mathcal{L}_\infty$. From $\varepsilon m_s = -\frac{\tilde{\theta}^T\phi}{m_s} + \frac{\eta}{m_s}$, $\tilde{\theta} \in \mathcal{L}_\infty$, and the fact that m_s bounds

[4] As defined in the Appendix, $\mathcal{S}(\mu) = \{x : [0, \infty) \to \mathcal{R}^n \mid \int_t^{t+T} x^T(\tau)x(\tau)d\tau \leq c_0\mu T + c_1 \ \forall t, T \geq 0\}$.

ϕ, η from above, we have $\varepsilon m_s, \varepsilon \in \mathcal{L}_\infty$. Furthermore, using $\dot{\tilde{\theta}} = \Gamma\varepsilon m_s \frac{\phi}{m_s} - \sigma\Gamma\theta$ and $\theta, \varepsilon m_s, \frac{\phi}{m_s} \in \mathcal{L}_\infty$, we obtain $\dot{\tilde{\theta}} = \dot{\theta} \in \mathcal{L}_\infty$.

(ii) The inequality

$$\dot{V} \leq -\frac{\varepsilon^2 m_s^2}{2} - \frac{\sigma}{\lambda_{\max}(\Gamma^{-1})} V(\tilde{\theta}) + c_0\left(\sigma + \frac{\eta^2}{m_s^2}\right)$$

obtained above implies

$$\dot{V} \leq -\frac{\varepsilon^2 m_s^2}{2} + c_0\left(\sigma + \frac{\eta^2}{m_s^2}\right).$$

Integrating both sides, we obtain

$$\frac{1}{2}\int_0^t \varepsilon^2 m_s^2 d\tau \leq V(0) - V(t) + c_0 \int_0^t \left(\sigma + \frac{\eta^2}{m_s^2}\right) d\tau,$$

which implies that $\varepsilon m_s \in \mathcal{S}(\sigma + \frac{\eta^2}{m_s^2})$. This together with $\theta, \frac{\phi}{m_s} \in \mathcal{L}_\infty$ implies that $\varepsilon, \dot{\theta} \in \mathcal{S}(\sigma + \frac{\eta^2}{m_s^2})$.

(iii) Using $\varepsilon m_s = -\frac{\tilde{\theta}^T\phi}{m_s} + \frac{\eta}{m_s} = -\frac{\phi^T\tilde{\theta}}{m_s} + \frac{\eta}{m_s}$, we have

$$\dot{\tilde{\theta}} = -\Gamma\frac{\phi}{m_s}\frac{\phi^T}{m_s}\tilde{\theta} - \sigma\Gamma\theta + \frac{\eta}{m_s}.$$

Since $\frac{\phi}{m_s}$ is PE, it follows that the homogenous part of the above equation is e.s., which implies that

$$|\tilde{\theta}(t)| \leq \alpha_1 e^{-\alpha_0 t}|\tilde{\theta}(t)| + \alpha_1 \int_0^t e^{-\alpha_0(t-\tau)}\left(\sigma\|\Gamma\||\theta(\tau)| + \frac{|\eta|}{m_s}\right) d\tau$$

for some $\alpha_1, \alpha_0 > 0$. Using $\frac{\eta}{m_s}, \theta \in \mathcal{L}_\infty$, we can write

$$|\tilde{\theta}(t)| \leq \alpha_1 e^{-\alpha_0 t}|\tilde{\theta}(0)| + c(\sigma + \bar{\eta})(1 - e^{-\alpha_0 t}),$$

where $\bar{\eta} = \sup_t \frac{|\eta(t)|}{m_s(t)}$, $c = \max(\frac{\alpha_1}{\alpha_0}\|\Gamma\| \sup_t |\theta(t)|, \frac{\alpha_1}{\alpha_0})$, which concludes that $\tilde{\theta}(t)$ converges to D_σ exponentially fast. □

One can observe that if the modeling error is removed, i.e., $\eta = 0$, then the fixed σ-modification will not guarantee the ideal properties of the adaptive law since it introduces a disturbance of the order of the design constant σ. This is one of the main drawbacks of the fixed σ-modification that is removed in the next section. One of the advantages of the fixed σ-modification is that no assumption about bounds or location of the unknown θ^* is made.

Switching σ-Modification

The drawback of the fixed σ-modification is eliminated by using a switching σ_ℓ term which activates the small feedback term around the integrator when the magnitude of the parameter vector exceeds a certain value M_0. The assumptions we make in this case are that $|\theta^*| \le M_0$ and M_0 is known. Since M_0 is arbitrary it can be chosen to be high enough in order to guarantee $|\theta^*| \le M_0$ in the case where limited or no information is available about the location of θ^*. The switching σ-modification is given by

$$\sigma_\ell(t) = \sigma_s = \begin{cases} 0 & \text{if } |\theta| \le M_0, \\ \left(\frac{|\theta|}{M_0} - 1\right)^{q_0} \sigma_0 & \text{if } M_0 < |\theta| \le 2M_0, \\ \sigma_0 & \text{if } |\theta| > 2M_0, \end{cases} \tag{3.89}$$

where $q_0 \ge 1$ is any finite integer and M_0, σ_0 are design constants satisfying $M_0 > |\theta^*|$ and $\sigma_0 > 0$. The switching from 0 to σ_0 is continuous in order to guarantee the existence and uniqueness of solutions of the differential equation. If an a priori estimate θ_0 of θ^* is available, it can be incorporated in (3.89) to replace $|\theta|$ with $|\theta - \theta_0|$. In this case, M_0 is an upper bound for $|\theta^* - \theta_0| \le M_0$.

The gradient algorithm with the switching σ-modification given by (3.89) is described as

$$\begin{aligned} \dot{\theta} &= \Gamma\varepsilon\phi - \sigma_s\Gamma\theta, \\ \varepsilon &= \frac{z - \theta^T\phi}{m_s^2}, \end{aligned} \tag{3.90}$$

where m_s is given by (3.86).

Theorem 3.12.2. *The gradient algorithm* (3.90) *with the switching σ-modification guarantees the following:*

(i) $\varepsilon, \varepsilon m_s, \theta, \dot{\theta} \in \mathcal{L}_\infty$.

(ii) $\varepsilon, \varepsilon m_s, \dot{\theta} \in \mathcal{S}(\frac{\eta^2}{m_s^2})$.

(iii) *In the absence of modeling error, i.e., for $\eta = 0$, the properties of the adaptive law* (3.90) *are the same as those of the respective unmodified adaptive laws (i.e., with* $\sigma_\ell = 0$).

(iv) *If $\frac{\phi}{m_s}$ is PE with level $\alpha_0 > 0$, then*

(a) *$\tilde{\theta}$ converges exponentially fast to the residual set*

$$D_s = \{\tilde{\theta} \mid |\tilde{\theta}| \le c(\sigma_0 + \bar{\eta})\},$$

where $c \ge 0$ is some constant;

(b) *there exists a constant $\bar{\eta}^* > 0$ such that for $\bar{\eta} < \bar{\eta}^*$, $\tilde{\theta}$ converges exponentially to the residual set*

$$\bar{D}_s = \{\tilde{\theta} \mid |\tilde{\theta}| \le c\bar{\eta}\}.$$

Proof. We choose the same Lyapunov-like function as in the case of no modeling error, i.e.,

$$V = \frac{\tilde{\theta}^T \Gamma^{-1} \tilde{\theta}}{2}.$$

Then

$$\dot{V} = \tilde{\theta}^T \phi \varepsilon - \sigma_s \tilde{\theta}^T \theta.$$

Substituting $\varepsilon m_s^2 = -\tilde{\theta}^T \phi + \eta$, we obtain

$$\dot{V} = -\varepsilon^2 m_s^2 + \varepsilon \eta - \sigma_s \tilde{\theta}^T \theta.$$

Note that $-\varepsilon^2 m_s^2 + \varepsilon \eta \leq -\frac{\varepsilon^2 m_s^2}{2} + \frac{\eta^2}{2m_s^2}$ and

$$\sigma_s \tilde{\theta}^T \theta = \sigma_s (|\theta|^2 - \theta^{*T} \theta) \geq \sigma_s |\theta| (|\theta| - M_0 + M_0 - |\theta^*|).$$

Since $\sigma_s (|\theta| - M_0) \geq 0$ and $M_0 > |\theta^*|$, it follows that

$$\sigma_s \tilde{\theta}^T \theta \geq \sigma_s |\theta| (|\theta| - M_0) + \sigma_s |\theta| (M_0 - |\theta^*|) \geq \sigma_s |\theta| (M_0 - |\theta^*|) \geq 0,$$

i.e.,

$$\sigma_s |\theta| \leq \sigma_s \frac{\tilde{\theta}^T \theta}{M_0 - |\theta^*|}. \tag{3.91}$$

Therefore, the inequality for $\dot{V}$ can be written as

$$\dot{V} \leq -\frac{\varepsilon^2 m_s^2}{2} - \sigma_s \tilde{\theta}^T \theta + \frac{\eta^2}{2m_s^2}. \tag{3.92}$$

Since for $|\theta| = |\tilde{\theta} + \theta^*| > 2M_0$ the term $-\sigma_s \tilde{\theta}^T \theta = -\sigma_0 \tilde{\theta}^T \theta \leq -\frac{\sigma_0}{2} |\tilde{\theta}|^2 + \frac{\sigma_0}{2} |\theta^*|^2$ behaves as the equivalent fixed σ term, we can follow the same procedure as in the proof of Theorem 3.12.1 to show the existence of a constant $V_0 > 0$ for which $\dot{V} \leq 0$ whenever $V \geq V_0$ and conclude that $V, \varepsilon, \theta, \dot{\theta} \in \mathcal{L}_\infty$, completing the proof of (i).

Integrating both sides of (3.92) from t_0 to t, we obtain that $\varepsilon, \varepsilon m_s, \sqrt{\sigma_s \tilde{\theta}^T \theta} \in \mathcal{S}(\frac{\eta^2}{m_s^2})$. From (3.91), it follows that

$$\sigma_s^2 |\theta|^2 \leq c_2 \sigma_s \tilde{\theta}^T \theta$$

for some constant $c_2 > 0$ that depends on the bound for $\sigma_0 |\theta|$, and therefore $|\dot{\theta}|^2 \leq c(|\varepsilon m_s|^2 + \sigma_s \tilde{\theta}^T \theta)$ for some $c \geq 0$. Since $\varepsilon m_s, \sqrt{\sigma_s \tilde{\theta}^T \theta} \in \mathcal{S}(\frac{\eta^2}{m_s^2})$, it follows that $\dot{\theta} \in \mathcal{S}(\frac{\eta^2}{m_s^2})$, which completes the proof of (ii).

The proof of part (iii) follows from (3.92) by setting $\eta = 0$, using $-\sigma_s \tilde{\theta}^T \theta \leq 0$, and repeating the above calculations for $\eta = 0$.

The proof of (iv)(a) is almost identical to that of Theorem 3.12.1(iii) and is omitted.

To prove (iv)(b), we follow the same arguments used in the proof of Theorem 3.12.1(iii) to obtain the inequality

$$\begin{aligned} |\tilde{\theta}| &\leq \beta_0 e^{-\beta_2 t} + \beta_1' \int_0^t e^{-\beta_2 (t-\tau)} (|\eta| + \sigma_s |\theta|) d\tau \\ &\leq \beta_0 e^{-\beta_2 t} + \frac{\beta_1'}{\beta_2} \bar{\eta} + \int_0^t e^{-\beta_2 (t-\tau)} \sigma_s |\theta| d\tau \end{aligned} \tag{3.93}$$

for some positive constants $\beta_0, \beta_1', \beta_2$. From (3.91), we have

$$\sigma_s|\theta| \le \frac{1}{M_0 - |\theta^*|}\sigma_s\tilde{\theta}^T\theta \le \frac{1}{M_0 - |\theta^*|}\sigma_s|\theta||\tilde{\theta}|. \tag{3.94}$$

Therefore, using (3.94) in (3.93), we have

$$|\tilde{\theta}| \le \beta_0 e^{-\beta_2 t} + \frac{\beta_1'}{\beta_2}\bar{\eta} + \beta_1'' \int_0^t e^{-\beta_2(t-\tau)}\sigma_s|\theta||\tilde{\theta}|d\tau, \tag{3.95}$$

where $\beta_1'' = \frac{\beta_1'}{M_0 - |\theta^*|}$. Applying the Bellman–Gronwall (B–G) Lemma 3 (Lemma A.6.3) to (3.95), it follows that

$$|\tilde{\theta}| \le \left(\beta_0 + \frac{\beta_1'}{\beta_2}\bar{\eta}\right) e^{-\beta_2(t-t_0)} e^{\beta_1'' \int_{t_0}^t \sigma_s|\theta|ds} + \beta_1'\bar{\eta} \int_{t_0}^t e^{-\beta_2(t-\tau)} e^{\beta'' \int_{t_0}^t \sigma_s|\theta|ds} d\tau. \tag{3.96}$$

Note from (3.91)–(3.92) that $\sqrt{\sigma_s|\theta|} \in \mathcal{S}(\frac{\eta^2}{m_s^2})$, i.e.,

$$\int_{t_0}^t \sigma_s|\theta|ds \le c_1\bar{\eta}^2(t - t_0) + c_0$$

$\forall t \ge t_0 \ge 0$ and for some constants c_0, c_1. Therefore,

$$|\tilde{\theta}| \le \bar{\beta}_1 e^{-\bar{\alpha}(t-t_0)} + \bar{\beta}_2\bar{\eta} \int_{t_0}^t e^{-\bar{\alpha}(t-\tau)}d\tau, \tag{3.97}$$

where $\bar{\alpha} = \beta_2 - \beta_2''c_1\bar{\eta}^2$ and $\bar{\beta}_1, \bar{\beta}_2 \ge 0$ are some constants that depend on c_0 and on the constants in (3.96). Hence, for any $\bar{\eta} \in [0, \bar{\eta}^*)$, where $\bar{\eta}^* = \sqrt{\frac{\beta_2}{\beta_1''c_1}}$, we have $\bar{\alpha} > 0$, and (3.97) implies that

$$|\tilde{\theta}| \le \frac{\bar{\beta}_2}{\bar{\alpha}}\bar{\eta} + ce^{-\bar{\alpha}(t-t_0)}$$

for some constant c and $\forall t \ge t_0 \ge 0$, which completes the proof of (iv). □

Another class of σ-modification involves leakage that depends on the estimation error ε, i.e.,

$$\sigma_s = |\varepsilon m_s|\nu_0,$$

where $\nu_0 > 0$ is a design constant. This modification is referred to as the *ε-modification* and has properties similar to those of the fixed σ-modification in the sense that it cannot guarantee the ideal properties of the adaptive law in the absence of modeling errors. For more details on the ε-modification, see [56, 86, 107].

The use of σ-modification is to prevent parameter drift by forcing $\theta(t)$ to remain bounded. It is often referred to as *soft projection* because it forces $\theta(t)$ not to deviate far from the bound $M_0 \ge |\theta|$. That is, in the case of soft projection $\theta(t)$ may exceed the M_0 bound, but it will still be bounded.

In the following section, we use projection to prevent parameter drift by forcing the estimated parameters to be within a specified bounded set.

3.12.3 Parameter Projection

The two crucial techniques that we used in sections 3.12.1 and 3.12.2 to develop robust adaptive laws are the dynamic normalization m_s and leakage. The normalization guarantees that the normalized modeling error term $\frac{\eta}{m_s}$ is bounded and therefore acts as a bounded input disturbance in the adaptive law. Since a bounded disturbance may cause parameter drift, the leakage modification is used to guarantee bounded parameter estimates. Another effective way to guarantee bounded parameter estimates is to use projection to constrain the parameter estimates to lie inside a bounded convex set in the parameter space that contains the unknown θ^*. Adaptive laws with projection have already been introduced and analyzed in section 3.10. By requiring the parameter set to be bounded, projection can be used to guarantee that the estimated parameters are bounded by forcing them to lie within the bounded set. In this section, we illustrate the use of projection for a gradient algorithm that is used to estimate θ^* in the parametric model

$$z = \theta^{*T}\phi + \eta.$$

In order to avoid parameter drift, we constrain θ to lie inside a bounded convex set that contains θ^*. As an example, consider the set

$$\wp = \{\theta \mid g(\theta) = \theta^T\theta - M_0^2 \leq 0\},$$

where M_0 is chosen so that $M_0 \geq |\theta^*|$. Following the results of section 3.10, we obtain

$$\dot{\theta} = \begin{cases} \Gamma\varepsilon\phi & \text{if } \theta^T\theta < M_0^2 \\ & \text{or if } \theta^T\theta = M_0^2 \text{ and } (\Gamma\varepsilon\phi)^T\theta \leq 0, \\ \left(I - \frac{\Gamma\theta\theta^T}{\theta^T\Gamma\theta}\right)\Gamma\varepsilon\phi & \text{otherwise,} \end{cases} \tag{3.98}$$

where $\theta(0)$ is chosen so that $\theta^T(0)\theta(0) \leq M_0^2$ and $\varepsilon = \frac{z-\theta^T\phi}{m_s^2}$, $\Gamma = \Gamma^T > 0$.

The stability properties of (3.98) for estimating θ^* in (3.85) in the presence of the modeling error term η are given by the following theorem.

Theorem 3.12.3. *The gradient algorithm with projection described by* (3.98) *and designed for the parametric model* $z = \theta^{*T}\phi + \eta$ *guarantees the following:*

(i) $\varepsilon, \varepsilon m_s, \theta, \dot{\theta} \in \mathcal{L}_\infty$.

(ii) $\varepsilon, \varepsilon m_s, \dot{\theta} \in \mathcal{S}(\frac{\eta^2}{m_s^2})$.

(iii) *If* $\eta = 0$, *then* $\varepsilon, \varepsilon m_s, \dot{\theta} \in \mathcal{L}_2$.

(iv) *If* $\frac{\phi}{m_s}$ *is PE with level* $\alpha_0 > 0$, *then*

(a) $\tilde{\theta}$ *converges exponentially to the residual set*

$$D_p = \{\tilde{\theta} \mid |\tilde{\theta}| \leq c(f_0 + \bar{\eta})\},$$

where $\bar{\eta} = \sup_t \frac{|\eta|}{m}$, $c \geq 0$ *is a constant, and* $f_0 \geq 0$ *is a design constant;*

(b) *there exists a constant* $\bar{\eta}^* > 0$ *such that if* $\bar{\eta} < \bar{\eta}^*$, *then* $\tilde{\theta}$ *converges exponentially fast to the residual set*

$$D_p = \{\tilde{\theta} \mid |\tilde{\theta}| \leq c\bar{\eta}\}$$

for some constant $c \geq 0$.

Proof. As established in section 3.10, projection guarantees that $|\theta(t)| \leq M_0 \ \forall t \geq 0$, provided $|\theta(0)| \leq M_0$. Let us choose the Lyapunov-like function

$$V = \frac{\tilde{\theta}^T \Gamma^{-1} \tilde{\theta}}{2}.$$

Along the trajectory of (3.98), we have

$$\dot{V} = \begin{cases} -\varepsilon^2 m_s^2 + \varepsilon\eta & \text{if } \theta^T\theta < M_0^2 \\ & \text{or if } \theta^T\theta = M_0^2 \text{ and } (\Gamma\varepsilon\phi)^T\theta \leq 0, \\ -\varepsilon^2 m_s^2 + \varepsilon\eta - \frac{\tilde{\theta}^T\theta}{\theta^T\Gamma\theta}\theta^T\Gamma\varepsilon\phi & \text{if } \theta^T\theta = M_0^2 \text{ and } (\Gamma\varepsilon\phi)^T\theta > 0. \end{cases}$$

For $\theta^T\theta = M_0^2$ and $(\Gamma\varepsilon\phi)^T\theta = \theta^T\Gamma\varepsilon\phi > 0$, we have $\operatorname{sgn}\{\frac{\tilde{\theta}^T\theta}{\theta^T\Gamma\theta}\theta^T\Gamma\varepsilon\phi\} = \operatorname{sgn}\{\tilde{\theta}^T\theta\}$. For $\theta^T\theta = M_0^2$, we have

$$\tilde{\theta}^T\theta = \theta^T\theta - \theta^{*T}\theta \geq M_0^2 - |\theta^*||\theta| = M_0(M_0 - |\theta^*|) \geq 0,$$

where the last inequality is obtained using the assumption that $M_0 \geq |\theta^*|$. Therefore, it follows that $\frac{\tilde{\theta}^T\theta\theta^T\Gamma\varepsilon\phi}{\theta^T\Gamma\theta} \geq 0$ when $\theta^T\theta = M_0^2$ and $(\Gamma\varepsilon\phi)^T\theta = \theta^T\Gamma\varepsilon\phi > 0$. Hence the term due to projection can only make $\dot{V}$ more negative, and therefore

$$\dot{V} = -\varepsilon^2 m_s^2 + \varepsilon\eta \leq -\frac{\varepsilon^2 m_s^2}{2} + \frac{\eta^2}{2m_s^2}.$$

Since V is bounded due to $\theta \in \mathcal{L}_\infty$, which is guaranteed by the projection, it follows that $\varepsilon m_s \in \mathcal{S}(\frac{\eta^2}{m_s^2})$, which implies that $\varepsilon \in \mathcal{S}(\frac{\eta^2}{m_s^2})$. From $\tilde{\theta} \in \mathcal{L}_\infty$ and $\frac{\phi}{m_s}, \frac{\eta}{m_s} \in \mathcal{L}_\infty$, we have $\varepsilon, \varepsilon m_s \in \mathcal{L}_\infty$. Now, for $\theta^T\theta = M_0^2$, we have $\frac{\|\Gamma\theta\theta^T\|}{\theta^T\Gamma\theta} \leq c$ for some constant $c \geq 0$, which implies that

$$|\dot{\theta}| \leq c|\varepsilon\phi| \leq c|\varepsilon m_s|.$$

Hence $\dot{\theta} \in \mathcal{S}(\frac{\eta^2}{m_s^2})$, and the proof of (i) and (ii) is complete. The proof of (iii) follows by setting $\eta = 0$ and has already been established in section 3.10.

The proof for parameter error convergence is completed as follows: Define the function

$$f \triangleq \begin{cases} \frac{\theta^T\Gamma\varepsilon\phi}{\theta^T\Gamma\theta} & \text{if } \theta^T\theta = M_0^2 \text{ and } (\Gamma\varepsilon\phi)^T\theta > 0, \\ 0 & \text{otherwise}. \end{cases} \tag{3.99}$$

It is clear from the analysis above that $f(t) \geq 0 \ \forall t \geq 0$. Then (3.98) may be written as

$$\dot{\theta} = \Gamma\varepsilon\phi - \Gamma f\theta. \tag{3.100}$$

We can establish that $f\tilde{\theta}^T\theta$ has properties very similar to $\sigma_s\tilde{\theta}^T\theta$, i.e.,

$$f|\tilde{\theta}||\theta| \geq f\tilde{\theta}^T\theta \geq f|\theta|(M_0 - |\theta^*|), \quad f \geq 0,$$

and $|f(t)| \leq f_0\ \forall t \geq 0$ for some constant $f_0 \geq 0$. Therefore, the proof of (iv)(a)–(b) can be completed following exactly the same procedure as in the proof of Theorem 3.12.2. □

The gradient algorithm with projection has properties identical to those of the switching σ-modification, as both modifications aim at keeping $|\theta| \leq M_0$. In the case of the switching σ-modification, $|\theta|$ may exceed M_0 but remain bounded, whereas in the case of projection $|\theta| \leq M_0\ \forall t \geq 0$, provided $|\theta(0)| \leq M_0$.

3.12.4 Dead Zone

Let us consider the estimation error

$$\varepsilon = \frac{z - \theta^T\phi}{m_s^2} = \frac{-\tilde{\theta}^T\phi + \eta}{m_s^2} \tag{3.101}$$

for the parametric model

$$z = \theta^{*T}\phi + \eta.$$

The signal ε is used to "drive" the adaptive law in the case of the gradient and LS algorithms. It is a measure of the parameter error $\tilde{\theta}$, which is present in the signal $\tilde{\theta}^T\phi$, and of the modeling error η. When $\eta = 0$ and $\tilde{\theta} = 0$, we have $\varepsilon = 0$ and no adaptation takes place. Since $\frac{\eta}{m_s}, \frac{\phi}{m_s} \in \mathcal{L}_\infty$, large εm_s implies that $\frac{\tilde{\theta}^T\phi}{m_s}$ is large, which in turn implies that $\tilde{\theta}$ is large. In this case, the effect of the modeling error η is small, and the parameter estimates driven by ε move in a direction which reduces $\tilde{\theta}$. When εm_s is small, however, the effect of η on εm_s may be more dominant than that of the signal $\tilde{\theta}^T\phi$, and the parameter estimates may be driven in a direction dictated mostly by η. The principal idea behind the dead zone is to monitor the size of the estimation error and adapt only when the estimation error is large relative to the modeling error η, as shown below.

We consider the gradient algorithm for the linear parametric model (3.85). We consider the same cost function as in the ideal case, i.e.,

$$J(\theta, t) = \frac{\varepsilon^2 m_s^2}{2} = \frac{(z - \theta^T\phi)^2}{2m_s^2},$$

and write

$$\dot{\theta} = \begin{cases} -\Gamma\nabla J(\theta) & \text{if } |\varepsilon m_s| > g_0 > \frac{|\eta|}{m_s}, \\ 0 & \text{otherwise,} \end{cases} \tag{3.102}$$

where g_0 is a known upper bound of the normalized modeling error $\frac{|\eta|}{m_s}$. In other words, we move in the direction of the steepest descent only when the estimation error is large relative to the modeling error, i.e., when $|\varepsilon m_s| > g_0$. In view of (3.102) we have

$$\dot{\theta} = \Gamma\phi(\varepsilon + g), \quad g = \begin{cases} 0 & \text{if } |\varepsilon m_s| > g_0, \\ -\varepsilon & \text{if } |\varepsilon m_s| \leq g_0. \end{cases}$$

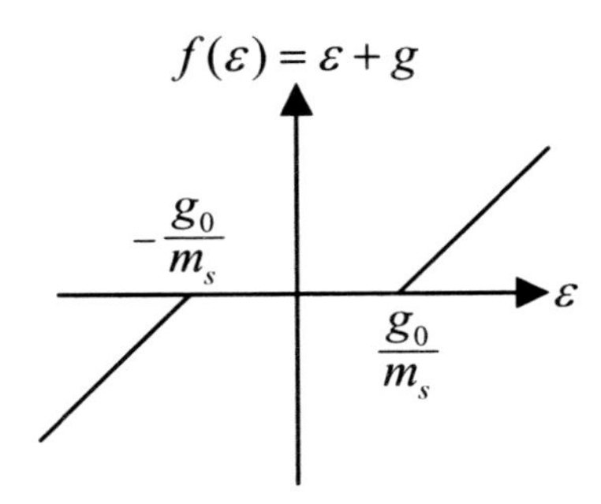

Figure 3.1. *Normalized dead zone function.*

To avoid any implementation problems which may arise due to the discontinuity in (3.102), the dead zone function is made continuous as shown in Figure 3.1; i.e.,

$$\dot{\theta} = \Gamma\phi(\varepsilon + g), \quad g = \begin{cases} \frac{g_0}{m_s} & \text{if } \varepsilon m_s < -g_0, \\ -\frac{g_0}{m_s} & \text{if } \varepsilon m_s > g_0, \\ -\varepsilon & \text{if } -g_0 \le \varepsilon m_s \le g_0. \end{cases} \tag{3.103}$$

Since the size of the dead zone depends on m_s, this dead zone function is often referred to as the *variable* or *relative dead zone*.

Theorem 3.12.4. *The adaptive law* (3.103) *guarantees the following properties:*

(i) $\varepsilon, \varepsilon m_s, \theta, \dot{\theta} \in \mathcal{L}_\infty$.

(ii) $\varepsilon, \varepsilon m_s, \dot{\theta} \in \mathcal{S}(g_0 + \frac{\eta^2}{m_s^2})$.

(iii) $\dot{\theta} \in \mathcal{L}_1 \cap \mathcal{L}_2$.

(iv) $\lim_{t\to\infty} \theta(t) = \bar{\theta}$, *where* $\bar{\theta}$ *is a constant vector.*

(v) *If* $\frac{\phi}{m_s}$ *is PE with level* $\alpha_0 > 0$, *then* $\tilde{\theta}(t)$ *converges to the residual set*

$$D_d = \{\tilde{\theta} \in \mathcal{R}^n \mid |\tilde{\theta}| \le c(g_0 + \bar{\eta})\},$$

where $\bar{\eta} = \sup_t \frac{|\eta(t)|}{m_s(t)}$ *and* $c \ge 0$ *is a constant.*

Proof. The proof can be found in [56] and in the web resource [94]. □

The dead zone modification has the following properties:

- It guarantees that the estimated parameters always converge to a constant. This is important in adaptive control employing an adaptive law with dead zone because at steady state the gains of the adaptive controller are almost constant, leading to an LTI closed-loop system when the plant is LTI.

- As in the case of the fixed σ-modification, the ideal properties of the adaptive law are destroyed in an effort to achieve robustness. That is, if the modeling error term becomes zero, the ideal properties of the adaptive law cannot be recovered unless the dead zone modification is also removed.

The robust modifications that include leakage, projection, and dead zone are analyzed for the case of the gradient algorithm for the SPM with modeling error. The same modifications can be used in the case of LS and DPM, B-SPM, and B-DPM with modeling errors. For further details on these modifications the reader is referred to [56]. One important property of these modifications is that $\varepsilon, \varepsilon m_s, \dot{\theta} \in \mathcal{S}(\lambda_0 + \frac{\eta^2}{m_s^2})$ for some constant $\lambda_0 \geq 0$. This means that the estimation error and speed of adaptation are guaranteed to be of the order of the modeling error and design parameter in the mean square sense. In other words intervals of time could exist at steady state where the estimation error and speed of adaptation could assume values higher than the modeling error. This phenomenon is known as "bursting," and it may take considerable simulation time to appear [105]. The dead zone exhibits no bursting as the parameters converge to constant values. Bursting can be prevented in all the other modifications if they are combined with a small-size dead zone at steady state. The various combinations of modifications as well as the development of new ones and their evaluation using analysis and simulations are a good exercise for the reader.

3.13 State-Space Identifiers

Let us consider the state-space plant model

$$\dot{x} = A_p x + B_p u, \tag{3.104}$$

where $x \in \mathcal{R}^n$ is the state, $u \in \mathcal{R}^m$ is the input vector, and $A_p \in \mathcal{R}^{n\times n}$, $B_p \in \mathcal{R}^{n\times m}$ are unknown constant matrices. We assume that x, u are available for measurement. One way to estimate the elements of A_p, B_p online is to express (3.104) as a set of n scalar differential equations and then generate n parametric models with scalar outputs and apply the parameter estimation techniques covered in the previous sections. Another more compact way of estimating A_p, B_p is to develop online estimators based on an SSPM model for (3.104) as follows.

We express (3.104) in the form of the SSPM:

$$\dot{x} = A_m x + (A_p - A_m)x + B_p u,$$

where A_m is an arbitrary stable matrix. The estimation model is then formed as

$$\dot{\hat{x}} = A_m \hat{x} + (\hat{A}_p - A_m)x + \hat{B}_p u = A_m(\hat{x} - x) + \hat{A}_p x + \hat{B}_p u,$$

where $\hat{A}_p(t)$, $\hat{B}_p(t)$ are the estimates of A_p, B_p at time t, respectively. The above estimation model has been referred to as the *series-parallel model* in the literature [56]. The estimation error vector is defined as

$$\varepsilon = x - \hat{x} - (sI - A_m)^{-1}\varepsilon n_s^2$$

or

$$\varepsilon = x - \hat{x} - w,$$

$$\dot{w} = A_m w + \varepsilon n_s^2, \quad w(0) = 0,$$

where n_s^2 is the static normalizing signal designed to guarantee $\frac{x}{\sqrt{1+n_s^2}}, \frac{u}{\sqrt{1+n_s^2}} \in \mathcal{L}_\infty$. A straightforward choice for n_s is $n_s^2 = x^T x + u^T u$. It is clear that if the plant model (3.104) is stable and the input u is bounded, then n_s can be taken to be equal to zero.

It follows that the estimation error satisfies

$$\dot{\varepsilon} = A_m \varepsilon - \tilde{A}_p x - \tilde{B}_p u - \varepsilon n_s^2,$$

where $\tilde{A}_p \triangleq \hat{A}_p - A_p$, $\tilde{B}_p \triangleq \hat{B}_p - B_p$ are the parameter errors.

The adaptive law for generating $\hat{A}_p$, $\hat{B}_p$ is developed by considering the Lyapunov function

$$V(\varepsilon, \tilde{A}_p, \tilde{B}_p) = \varepsilon^T P \varepsilon + \operatorname{tr}\left(\frac{\tilde{A}_p^T P \tilde{A}_p}{\gamma_1}\right) + \operatorname{tr}\left(\frac{\tilde{B}_p^T P \tilde{B}_p}{\gamma_2}\right), \tag{3.105}$$

where $\operatorname{tr}(A)$ denotes the trace of matrix A; $\gamma_1, \gamma_2 > 0$ are constant scalars; and $P = P^T > 0$ is chosen as the solution of the Lyapunov equation

$$P A_m + A_m P^T = -Q \tag{3.106}$$

for some $Q = Q^T > 0$, whose solution is guaranteed by the stability of A_m (see the Appendix). The time derivative $\dot{V}$ is given by

$$\dot{V} = \dot{\varepsilon}^T P \varepsilon + \varepsilon^T P \dot{\varepsilon} + \operatorname{tr}\left(\frac{\dot{\tilde{A}}_p^T P \tilde{A}_p}{\gamma_1} + \frac{\tilde{A}_p^T P \dot{\tilde{A}}_p}{\gamma_1}\right) + \operatorname{tr}\left(\frac{\dot{\tilde{B}}_p^T P \tilde{B}_p}{\gamma_2} + \frac{\tilde{B}_p^T P \dot{\tilde{B}}_p}{\gamma_2}\right).$$

Substituting for $\dot{\varepsilon}$, using (3.106), and employing the equalities $\operatorname{tr}(A + B) = \operatorname{tr}(A) + \operatorname{tr}(B)$ and $\operatorname{tr}(A^T) = \operatorname{tr}(A)$ for square matrices A, B of the same dimension, we obtain

$$\dot{V} = -\varepsilon^T Q \varepsilon - 2\varepsilon^T P \tilde{A}_p x - 2\varepsilon^T P \tilde{B}_p u + 2\operatorname{tr}\left(\frac{\tilde{A}_p^T P \dot{\tilde{A}}_p}{\gamma_1} + \frac{\tilde{B}_p^T P \dot{\tilde{B}}_p}{\gamma_2}\right) - 2\varepsilon^T P \varepsilon n_s^2. \tag{3.107}$$

Using the equality $v^T y = \operatorname{tr}(v y^T)$ for vectors v, y of the same dimension, we rewrite (3.107) as

$$\dot{V} = -\varepsilon^T Q \varepsilon + 2\operatorname{tr}\left(\frac{\tilde{A}_p^T P \dot{\tilde{A}}_p}{\gamma_1} - \tilde{A}_p^T P \varepsilon x^T + \frac{\tilde{B}_p^T P \dot{\tilde{B}}_p}{\gamma_2} - \tilde{B}_p^T P \varepsilon u^T\right) - 2\varepsilon^T P \varepsilon n_s^2. \tag{3.108}$$

The obvious choice for $\dot{\tilde{A}}_p = \dot{\hat{A}}_p$, $\dot{\tilde{B}}_p = \dot{\hat{B}}_p$ to make $\dot{V}$ negative is

$$\dot{\hat{A}}_p = \gamma_1 \varepsilon x^T, \qquad \dot{\hat{B}}_p = \gamma_2 \varepsilon u^T, \tag{3.109}$$

which gives us

$$\dot{V} = -\varepsilon^T Q \varepsilon - 2\varepsilon^T P \varepsilon n_s^2 \le 0.$$

This implies that V, $\hat{A}_p$, $\hat{B}_p$, ε are bounded. We can also write

$$\dot{V} \le -|\varepsilon|^2 \lambda_{\min}(Q) - 2|\varepsilon n_s|^2 \lambda_{\min}(P) \le 0$$

and use similar arguments as in the previous sections to establish that $\varepsilon, \varepsilon n_s \in \mathcal{L}_2$. From (3.109) we have that

$$\|\dot{\hat{A}}_p\| \le \gamma_1 |\varepsilon m_s| \frac{|x^T|}{m_s}, \qquad \|\dot{\hat{B}}_p\| \le \gamma_2 |\varepsilon m_s| \frac{|u^T|}{m_s},$$

where $m_s^2 = 1 + n_s^2$. Since $\frac{|x|}{m_s}, \frac{|u|}{m_s} \in \mathcal{L}_\infty$ and $|\varepsilon m_s| \in \mathcal{L}_2$, we can also conclude that $\|\dot{\hat{A}}_p\|, \|\dot{\hat{B}}_p\| \in \mathcal{L}_2$.

We have, therefore, established that independent of the stability of the plant and boundedness of the input u, the adaptive law (3.109) guarantees that

- $\|\hat{A}_p\|, \|\hat{B}_p\|, \varepsilon \in \mathcal{L}_\infty$,
- $\|\dot{\hat{A}}_p\|, \|\dot{\hat{B}}_p\|, \varepsilon, \varepsilon n_s \in \mathcal{L}_2$.

These properties are important for adaptive control where the adaptive law is used as part of the controller and no a priori assumptions are made about the stability of the plant and boundedness of the input. If the objective, however, is parameter estimation, then we have to assume that the plant is stable and the input u is designed to be bounded and sufficiently rich for the plant model (3.104). In this case, we can take $n_s^2 = 0$.

Theorem 3.13.1. *Consider the plant model* (3.104) *and assume that* (A_p, B_p) *is controllable and* A_p *is a stable matrix. If each element* u_i, $i = 1, 2, \dots, m$, *of vector* u *is bounded, sufficiently rich of order* $n+1$, *and uncorrelated, i.e., each* u_i *contains different frequencies, then* $\hat{A}_p(t)$, $\hat{B}_p(t)$ *generated by* (3.109) (*where* n_s *can be taken to be zero*) *converge to* A_p, B_p, *respectively, exponentially fast.*

The proof of Theorem 3.13.1 is long, and the reader is referred to [56] for the details.

Example 3.13.2 Consider the second-order plant

$$\dot{x} = Ax + Bu,$$

where $x = [x_1, x_2]^T$, $u = [u_1, u_2]^T$ is a bounded input vector, the matrices A, B are unknown, and A is a stable matrix. The estimation model is generated as

$$\dot{\hat{x}} = \begin{bmatrix} -a_m & 0 \\ 0 & -a_m \end{bmatrix} (\hat{x} - x) + \begin{bmatrix} \hat{a}_{11} & \hat{a}_{12} \\ \hat{a}_{21} & \hat{a}_{22} \end{bmatrix} x + \begin{bmatrix} \hat{b}_{11} & \hat{b}_{12} \\ \hat{b}_{21} & \hat{b}_{22} \end{bmatrix} u,$$

where $\hat{x} = [\hat{x}_1, \hat{x}_2]^T$ and $a_m > 0$. The estimation error is given by

$$\varepsilon = x - \hat{x},$$

where $n_s = 0$ due to the stability of A and boundedness of u. The adaptive law (3.109) can be written as

$$\dot{\hat{a}}_{ij} = \gamma_1 \varepsilon_i x_j, \qquad \dot{\hat{b}}_{ij} = \gamma_2 \varepsilon_i u_j$$

for $i = 1, 2$, $j = 1, 2$, and adaptive gains γ_1, γ_2. An example of a sufficiently rich input for this plant is

$$u_1 = c_1 \sin 2.5t + c_2 \sin 4.6t,$$
$$u_2 = c_3 \sin 7.2t + c_4 \sin 11.7t$$

for some nonzero constants c_i, $i = 1, 2, 3, 4$. ∎

The class of plants described by (3.104) can be expanded to include more realistic plants with modeling errors. The adaptive laws in this case can be made robust by using exactly the same techniques as in the case of SISO plants described in previous sections, and this is left as an exercise for the reader.

3.14 Adaptive Observers

Consider the LTI SISO plant

$$\begin{aligned} \dot{x} &= Ax + Bu, \quad x(0) = x_0, \\ y &= C^T x, \end{aligned} \tag{3.110}$$

where $x \in \mathcal{R}^n$. We assume that u is a piecewise continuous bounded function of time and that A is a stable matrix. In addition, we assume that the plant is completely controllable and completely observable. The problem is to construct a scheme that estimates both the plant parameters, i.e., A, B, C, as well as the state vector x using only I/O measurements. We refer to such a scheme as the *adaptive observer*.

A good starting point for designing an adaptive observer is the Luenberger observer used in the case where A, B, C are known. The Luenberger observer is of the form

$$\begin{aligned} \dot{\hat{x}} &= A\hat{x} + Bu + K(y - \hat{y}), \quad \hat{x}(0) = \hat{x}_0, \\ \hat{y} &= C^T \hat{x}, \end{aligned} \tag{3.111}$$

where K is chosen so that $A - KC^T$ is a stable matrix, and guarantees that $\hat{x} \to x$ exponentially fast for any initial condition x_0 and any input u. For $A - KC^T$ to be stable, the existence of K is guaranteed by the observability of (A, C).

A straightforward procedure for choosing the structure of the adaptive observer is to use the same equation as the Luenberger observer (3.111), but replace the unknown parameters A, B, C with their estimates $\hat{A}$, $\hat{B}$, $\hat{C}$, respectively, generated by some adaptive law. The problem we face with this procedure is the inability to estimate uniquely the n^2+2n parameters of A, B, C from the I/O data. The best we can do in this case is to estimate the parameters of the plant transfer function and use them to calculate $\hat{A}$, $\hat{B}$, $\hat{C}$. These calculations, however, are not always possible because the mapping of the $2n$ estimated parameters of the transfer function to the n^2+2n parameters of $\hat{A}$, $\hat{B}$, $\hat{C}$ is not unique unless (A, B, C) satisfies certain structural constraints. One such constraint is that (A, B, C) is in the observer form, i.e., the plant is represented as

$$\dot{x}_\alpha = \left[-a_p \;\middle|\; \begin{matrix} I_{n-1} \\ \hline 0 \end{matrix} \right] x_\alpha + b_p u, \qquad (3.112)$$
$$y = [1, 0, \ldots, 0] x_\alpha,$$

where $a_p = [a_{n-1}, a_{n-2}, \ldots, a_0]^T$ and $b_p = [b_{n-1}, b_{n-2}, \ldots, b_0]^T$ are vectors of dimension n, and $I_{n-1} \in \mathcal{R}^{(n-1)\times(n-1)}$ is the identity matrix. The elements of a_p and b_p are the coefficients of the denominator and numerator, respectively, of the transfer function

$$\frac{y(s)}{u(s)} = \frac{b_{n-1}s^{n-1} + b_{n-2}s^{n-2} + \cdots + b_0 s}{s^n + a_{n-1}s^{n-1} + a_{n-2}s^{n-2} + \cdots + a_0 s} \qquad (3.113)$$

and can be estimated online from I/O data using the techniques presented in the previous sections.

Since both (3.110) and (3.112) represent the same plant, we can focus on the plant representation (3.112) and estimate x_α instead of x. The disadvantage is that in a practical situation x may represent some physical variables of interest, whereas x_α may be an artificial state vector.

The adaptive observer for estimating the state x_α of (3.112) is motivated from the Luenberger observer structure (3.111) and is given by

$$\dot{\hat{x}} = \hat{A}(t)\hat{x} + \hat{b}_p(t)u + K(t)(y - \hat{y}), \quad \hat{x}(0) = \hat{x}_0, \qquad (3.114)$$
$$\hat{y} = [1, 0, \ldots, 0]\hat{x},$$

where $\hat{x}$ is the estimate of x_α,

$$\hat{A}(t) = \left[-\hat{a}_p(t) \;\middle|\; \begin{matrix} I_{n-1} \\ \hline 0 \end{matrix} \right], \qquad K(t) = a^* - \hat{a}_p(t),$$

$\hat{a}_p(t)$ and $\hat{b}_p(t)$ are the estimates of the vectors a_p and b_p, respectively, at time t, and $a^* \in \mathcal{R}^n$ is chosen so that

$$A^* = \left[-a^* \;\middle|\; \begin{matrix} I_{n-1} \\ \hline 0 \end{matrix} \right] \qquad (3.115)$$

is a stable matrix that contains the eigenvalues of the observer.

A wide class of adaptive laws may be used to generate $\hat{a}_p(t)$ and $\hat{b}_p(t)$ online. As in Chapter 2, the parametric model

$$z = \theta^{*T}\phi \qquad (3.116)$$

may be developed using (3.113), where

$$\theta^* = [b_p^T, a_p^T]^T$$

and z, ϕ are available signals, and used to design a wide class of adaptive laws to generate $\theta(t) = [\hat{b}_p^T(t), a_p^T(t)]^T$, the estimate of θ^*. As an example, consider the gradient algorithm

$$\dot{\theta} = \Gamma \varepsilon \phi, \quad \varepsilon = \frac{z - \theta^T \phi}{m_s^2}, \tag{3.117}$$

where $m_s^2 = 1 + \alpha \phi^T \phi$ and $\alpha \geq 0$.

Theorem 3.14.1. *The adaptive observer described by* (3.114)–(3.117) *guarantees the following properties:*

(i) *All signals are bounded.*

(ii) *If u is sufficiently rich of order $2n$, then the state observation error $|\hat{x} - x_\alpha|$ and the parameter error $|\theta - \theta^*|$ converge to zero exponentially fast.*

Proof. (i) Since A is stable and u is bounded, we have $x_\alpha, y, \phi \in \mathcal{L}_\infty$ and hence $m_s^2 = 1 + \alpha \phi^T \phi \in \mathcal{L}_\infty$. The adaptive law (3.117) guarantees that $\theta \in \mathcal{L}_\infty$ and $\varepsilon, \varepsilon m_s, \dot{\theta} \in \mathcal{L}_2 \cap \mathcal{L}_\infty$. The observer equation may be written as

$$\dot{\hat{x}} = A^* \hat{x} + \hat{b}_p(t) u + (\hat{A}(t) - A^*) x_\alpha.$$

Since A^* is a stable matrix and $\hat{b}_p$, $\hat{A}$, u, x_α are bounded, it follows that $\hat{x} \in \mathcal{L}_\infty$, which in turn implies that all signals are bounded.

(ii) The state observation error $\tilde{x} = \hat{x} - x_\alpha$ satisfies

$$\dot{\tilde{x}} = A^* \tilde{x} + \tilde{b}_p u - \tilde{a}_p y, \tag{3.118}$$

where $\tilde{b}_p = \hat{b}_p - b_p$, $\tilde{a}_p = \hat{a}_p - a_p$ are the parameter errors. Since for u sufficiently rich we have that $\theta(t) \to \theta^*$ as $t \to \infty$ exponentially fast, it follows that $\tilde{b}_p, \tilde{a}_p \to 0$ exponentially fast. Since $u, y \in \mathcal{L}_\infty$, the error equation consists of a homogenous part that is exponentially stable and an input that is decaying to zero. This implies that $\tilde{x} = \hat{x} - x_\alpha \to 0$ as $t \to \infty$ exponentially fast. $\square$

For further reading on adaptive observers, the reader is referred to [56, 86, 108–110].

3.15 Case Study: Users in a Single Bottleneck Link Computer Network

The congestion control problem in computer networks has been identified as a feedback control problem. The network users adjust their sending data rates, in response to congestion signals they receive from the network, in an effort to avoid congestion and converge to a stable equilibrium that satisfies certain requirements: high network utilization, small queue sizes, small delays, fairness among users, etc. Many of the proposed congestion control schemes require that at each link the number of flows, N say, utilizing the link is known. Since the number of users varies with time, N is an unknown time-varying parameter,

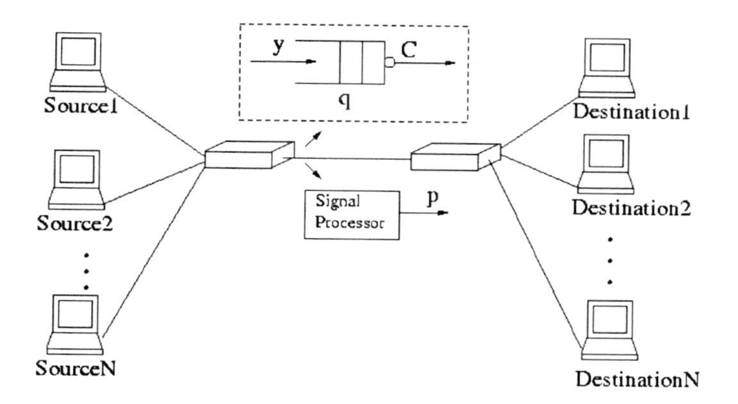

Figure 3.2. *Network topology.*

which needs to be estimated online. Estimation algorithms, which have been proposed in the literature, are based on pointwise time division, which is known to lack robustness and may lead to erroneous estimates. In this study, we consider a simple estimation algorithm, which is based on online parameter identification.

We consider the single bottleneck link network shown in Figure 3.2. It consists of N users which share a common bottleneck link through high bandwidth access links. At the bottleneck link we assume that there exists a buffer, which accommodates the incoming packets. The rate of data entering the buffer is denoted by y, the queue size is denoted by q, and the output capacity is denoted by C. At the bottleneck link, we implement a signal processor, which calculates the desired sending rate p. This information is communicated to the network users, which set their sending rate equal to p. The desired sending rate p is updated according to the following control law:

$$\dot{p} = \begin{cases} \frac{1}{\hat{N}}[k_i(C-y) - k_q q] & \text{if } 1 < p < C, \\ \frac{1}{\hat{N}}[k_i(C-y) - k_q q] & \text{if } p = 1,\ \frac{1}{\hat{N}}[k_i(C-y) - k_q q] > 0, \\ \frac{1}{\hat{N}}[k_i(C-y) - k_q q] & \text{if } p = C,\ \frac{1}{\hat{N}}[k_i(C-y) - k_q q] < 0, \\ 0 & \text{otherwise}, \end{cases} \tag{3.119}$$

where $\hat{N}$ is an estimate of N which is calculated online and k_i, k_q are design parameters. Since N is changing with time, its estimate $\hat{N}$ has to be updated accordingly. In this study we use online parameter estimation to generate $\hat{N}$. Since the sending rate of all users is equal to p, it follows that

$$y = Np. \tag{3.120}$$

Since y and p are measured at the bottleneck link, (3.120) is in the form of an SPM with N as the unknown parameter. We also know that N cannot be less than 1. Using the

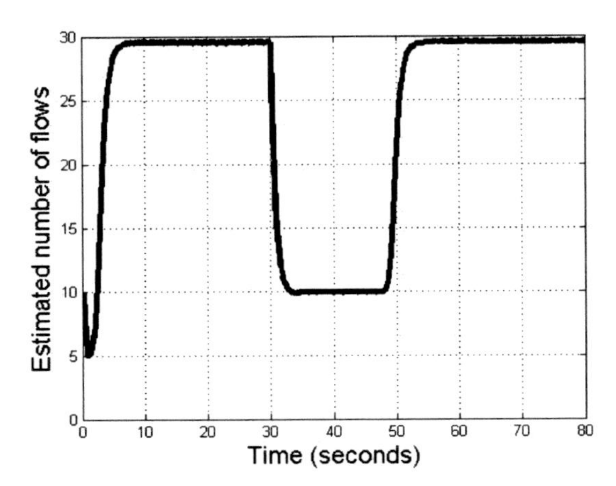

Figure 3.3. *Time response of the estimate of the number of flows.*

results of the chapter, we propose the following online parameter estimator:

$$\dot{\hat{N}} = \begin{cases} \gamma \varepsilon p & \text{if } \hat{N} > 1 \text{ or } \hat{N} = 1 \text{ and } \varepsilon p \geq 0, \\ 0 & \text{otherwise,} \end{cases}$$
$$\varepsilon = \frac{y - \hat{N} p}{1 + p^2}, \tag{3.121}$$

where $\hat{N}(0) \geq 1$. We demonstrate the effectiveness of the proposed algorithm using simulations, which we conduct on the packet-level simulator ns-2. We consider the network topology of Figure 3.2 in our simulations. The bandwidth of the bottleneck link is set to 155 Mb/s, and the propagation delay of each link is set to 20 ms. The design parameters are chosen as follows: $\gamma = 0.1$, $k_i = 0.16$, $k_q = 0.32$. Initially 30 users utilize the network. The estimator starting with an initial estimate of 10 converges to 30. After $t = 30$ seconds 20 of these users stop sending data, while an additional 20 users enter the network at $t = 45$ seconds. The output of the estimator at the bottleneck link is shown in Figure 3.3. We observe that the estimator accurately tracks the number of flows utilizing the network. In addition we observe good transient behavior as the responses are characterized by fast convergence and no overshoots. The estimator results are obtained in the presence of noise and delays which were not included in the simple model (3.120). Since the number of parameters to be estimated is 1, the PE property is satisfied for $p \neq 0$, which is always the case in this example.

See the web resource [94] for examples using the Adaptive Control Toolbox.

Problems

1. Consider the SPM

$$z = \theta^{*T} \phi$$

and the estimation model

$$\hat{z} = \theta^T(t)\phi.$$

Find values for $\theta(t)$, $\phi(t)$ such that $z = \hat{z}$ but $\theta(t) \neq \theta^*$.

2. Consider the adaptive law

$$\dot{\theta} = \gamma\varepsilon\phi,$$

where $\theta, \phi \in \mathcal{R}^n$, $\frac{\phi}{m_s} \in \mathcal{L}_\infty$, $\varepsilon m_s \in \mathcal{L}_2 \cap \mathcal{L}_\infty$, and $m_s \geq 1$. Show that $\dot{\theta} \in \mathcal{L}_2 \cap \mathcal{L}_\infty$.

3. Show that if $u \in \mathcal{L}_\infty$ in (3.1), then the adaptive law (3.10) with $m_s^2 = 1+\alpha\phi^2$, $\alpha \geq 0$, guarantees that $\varepsilon, \varepsilon m_s, \dot{\theta} \in \mathcal{L}_2 \cap \mathcal{L}_\infty$ and that $\varepsilon(t), \varepsilon(t)m_s(t), \dot{\theta}(t) \to 0$ as $t \to \infty$.

4. (a) Show that (3.16) is a necessary and sufficient condition for $\theta(t)$ in the adaptive law (3.10) to converge to θ^* exponentially fast.

   (b) Establish which of the following choices for u guarantee that ϕ in (3.10) is PE:

   (i) $u = c_0 \neq 0$, c_0 is a constant.

   (ii) $u = \sin t$.

   (iii) $u = \sin t + \cos 2t$.

   (iv) $u = \frac{1}{1+t}$.

   (v) $u = e^{-t}$.

   (vi) $u = \frac{1}{(1+t)^{1/2}}$.

   (c) In (b), is there a choice for u that guarantees that $\theta(t)$ converges to θ^* but does not guarantee that ϕ is PE?

5. Use the plant model (3.24) to develop the bilinear parametric model (3.28). Show all the steps.

6. In Theorem 3.6.1, assume that $\phi, \dot{\phi} \in \mathcal{L}_\infty$. Show that the adaptive law (3.34) with $m_s^2 = 1 + n_s^2$, $n_s^2 = \alpha\phi^T\phi$, and $\alpha \geq 0$ guarantees that $\varepsilon(t), \varepsilon(t)m_s(t), \dot{\theta}(t) \to 0$ as $t \to \infty$.

7. Consider the SPM $z = \theta^{*T}\phi$ and the cost function

$$J(\theta) = \frac{1}{2}\int_0^t e^{-\beta(t-\tau)}\frac{(z(\tau) - \theta^T(t)\phi(\tau))^2}{m_s^2(\tau)}d\tau,$$

where $m_s^2 = 1 + \phi^T\phi$ and $\theta(t)$ is the estimate of θ^* at time t.

   (a) Show that the minimization of $J(\theta)$ w.r.t. θ using the gradient method leads to the adaptive law

$$\dot{\theta}(t) = \Gamma\int_0^t e^{-\beta(t-\tau)}\frac{z(\tau) - \theta^T(t)\phi(\tau)}{m_s^2(\tau)}\phi(\tau)d\tau, \quad \theta(0) = \theta_0.$$

   (b) Show that the adaptive law in part (a) can be implemented as

$$\dot{\theta}(t) = -\Gamma(R(t)\theta(t) + Q(t)), \quad \theta(0) = \theta_0,$$

$$\dot{R}(t) = -\beta R(t) + \frac{\phi(t)\phi^T(t)}{m_s^2(t)}, \quad R(0) = 0,$$

$$\dot{Q}(t) = -\beta Q(t) - \frac{z(t)\phi(t)}{m_s^2(t)}, \qquad Q(0) = 0,$$

which is referred to as the integral adaptive law.

8. Consider the second-order stable system

$$\dot{x} = \begin{bmatrix} a_{11} & a_{12} \\ a_{21} & 0 \end{bmatrix} x + \begin{bmatrix} b_1 \\ b_2 \end{bmatrix} u,$$

where x, u are available for measurement, $u \in \mathcal{L}_\infty$, and a_{11}, a_{12}, a_{21}, b_1, b_2 are unknown parameters. Design an online estimator to estimate the unknown parameters. Simulate your scheme using $a_{11} = -0.25$, $a_{12} = 3$, $a_{21} = -5$, $b_1 = 1$, $b_2 = 2.2$, and $u = 10 \sin 2t$. Repeat the simulation when $u = 10 \sin 2t + 7 \cos 3.6t$. Comment on your results.

9. Consider the nonlinear system

$$\dot{x} = a_1 f_1(x) + a_2 f_2(x) + b_1 g_1(x) u + b_2 g_2(x) u,$$

where $u, x \in \mathcal{R}$; f_i, g_i are known nonlinear functions of x; and a_i, b_i are unknown constant parameters and $i = 1, 2$. The system is such that $u \in \mathcal{L}_\infty$ implies $x \in \mathcal{L}_\infty$. Assuming that x, u can be measured at each time t, design an estimation scheme for estimating the unknown parameters online.

10. Design and analyze an online estimation scheme for estimating θ^* in (3.49) when $L(s)$ is chosen so that $W(s)L(s)$ is biproper and SPR.

11. Design an online estimation scheme to estimate the coefficients of the numerator polynomial

$$Z(s) = b_{n-1}s^{n-1} + b_{n-2}s^{n-2} + \cdots + b_1 s + b_0$$

of the plant

$$y = \frac{Z(s)}{R(s)} u$$

when the coefficients of $R(s) = s^n + a_{n-1}s^{n-1} + \cdots + a_1 s + a_0$ are known. Repeat the same problem when $Z(s)$ is known and $R(s)$ is unknown.

12. Consider the mass–spring–damper system shown in Figure 3.4, where β is the damping coefficient, k is the spring constant, u is the external force, and $y(t)$ is the displacement of the mass m resulting from the force u.

    (a) Verify that the equations of motion that describe the dynamic behavior of the system under small displacements are

    $$m\ddot{y} + \beta\dot{y} + ky = u.$$

    (b) Design a gradient algorithm to estimate the constants m, β, k when y, u can be measured at each time t.

    (c) Repeat (b) for an LS algorithm.

    (d) Simulate your algorithms in (b) and (c) assuming $m = 20$ kg, $\beta = 0.1$ kg/sec, $k = 5$ kg/sec^2, and inputs u of your choice.

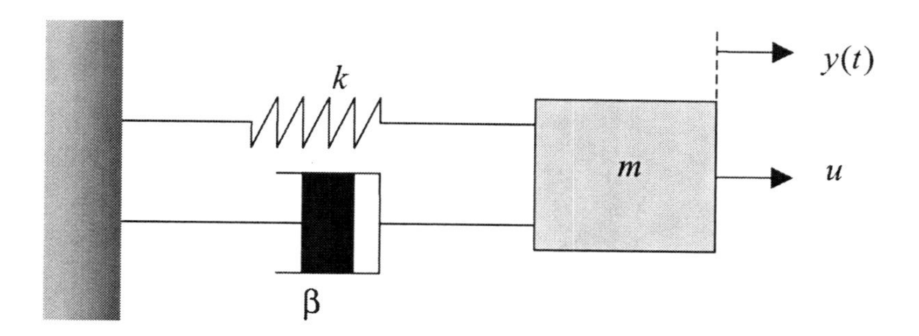

Figure 3.4. *The mass–spring–damper system for Problem* 12.

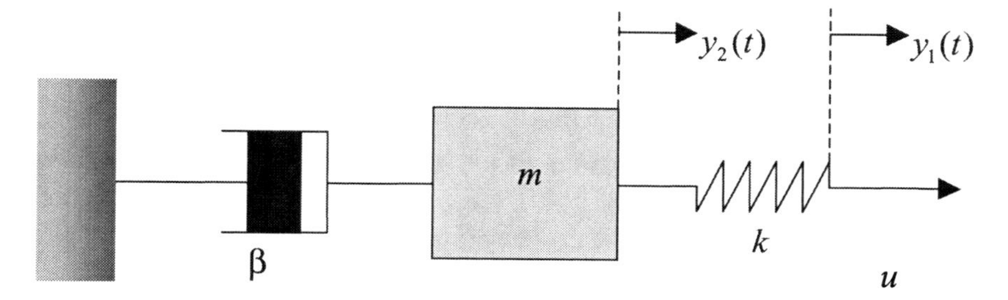

Figure 3.5. *The mass–spring–damper system for Problem* 13.

(e) Repeat (d) when $m = 20$ kg for $0 \leq t \leq 20$ sec and $m = 20(2 - e^{-0.01(t-20)})$ kg for $t \geq 20$ sec.

13. Consider the mass–spring–damper system shown in Figure 3.5.

(a) Verify that the equations of motion are given by

$$k(y_1 - y_2) = u,$$
$$k(y_1 - y_2) = m\ddot{y}_2 + \beta\dot{y}_2.$$

(b) If y_1, y_2, u can be measured at each time t, design an online parameter estimator to estimate the constants k, m, and β.

(c) We have the a priori knowledge that $0 \leq \beta \leq 1$, $k \geq 0.1$, and $m \geq 10$. Modify your estimator in (b) to take advantage of this a priori knowledge.

(d) Simulate your algorithm in (b) and (c) when $\beta = 0.2$ kg/sec, $m = 15$ kg, $k = 2$ kg/sec^2, and $u = 5 \sin 2t + 10.5$ kg $\cdot$ m/sec^2.

14. Consider the block diagram of a steer-by-wire system of an automobile shown in Figure 3.6, where r is the steering command in degrees, θ_p is the pinion angle in degrees, and $\dot{\theta}$ is the yaw rate in degree/sec.
The transfer functions $G_0(s)$, $G_1(s)$ are of the form

$$G_0(s) = \frac{k_0\omega_0^2}{s^2 + 2\xi_0\omega_0 s + \omega_0^2(1 - k_0)},$$

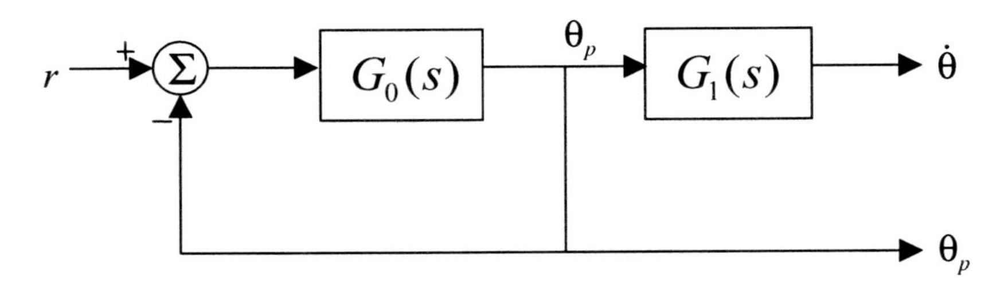

Figure 3.6. *Block diagram of a steer-by-wire system for Problem* 14.

$$G_1(s) = \frac{k_1\omega_1^2}{s^2 + 2\xi_1\omega_1 s + \omega_1^2},$$

where k_0, ω_0, ξ_0, k_1, ω_1, ξ_1 are functions of the speed of the vehicle. Assuming that $r, \theta_p, \dot{\theta}$ can be measured at each time t, do the following:

(a) Design an online parameter estimator to estimate k_i, ω_i, ξ_i, $i = 0, 1$, using the measurements of θ_p, $\dot{\theta}$, r.

(b) Consider the values of the parameters shown in the following table at different speeds:

Speed V	k_0	ω_0	ξ_0	k_1	ω_1	ξ_1
30 mph	0.81	19.75	0.31	0.064	14.0	0.365
60 mph	0.77	19.0	0.27	0.09	13.5	0.505

Assume that between speeds the parameters vary linearly. Use these values to simulate and test your algorithm in (a) when

(i) $r = 10\sin 0.2t + 8$ degrees and $V = 20$ mph.

(ii) $r = 5$ degrees and the vehicle speeds up from $V = 30$ mph to $V = 60$ mph in 40 seconds with constant acceleration and remains at 60 mph for 10 seconds.

15. Consider the equation of the motion of the mass–spring–damper system given in Problem 12, i.e.,

$$m\ddot{y} + \beta\dot{y} + ky = u.$$

This system may be written in the form

$$y = \rho^*(u - m\ddot{y} - \beta\dot{y}),$$

where $\rho^* = \frac{1}{k}$ appears in a bilinear form with the other unknown parameters m, β. Use the adaptive law based on the bilinear parametric model to estimate ρ^*, m, β when u, y are the only signals available for measurement. Since $k > 0$, the sign of ρ^* may be assumed to be known. Simulate your adaptive law using the numerical values given in (d) and (e) of Problem 12.

16. The effect of initial conditions on the SPM can be modeled as

$$z = \theta^{*T}\phi + \eta_0,$$

$$\dot{\omega}_0 = \Lambda \omega_0,$$
$$\eta_0 = C^T \omega_0,$$

where Λ is a transfer matrix with all poles in $\Re[s] < 0$ and $\omega_0 \in \mathcal{R}^n$, $\eta_0 \in \mathcal{R}$. Show that the properties of an adaptive law (gradient, LS, etc.) with $\eta_0 = 0$ are the same as those for the SPM with $\eta_0 \neq 0$.

17. Consider the system

$$y = e^{-\tau s} \frac{b}{(s-a)(\mu s + 1)} u,$$

where $0 < \tau \ll 1$, $0 < \mu \ll 1$, and a, b, τ, μ are unknown constants. We want to estimate a, b online.

   (a) Obtain a parametric model that can be used to design an adaptive law to estimate a, b.

   (b) Design a robust adaptive law of your choice to estimate a, b online.

   (c) Simulate your scheme for $a = -5, b = 100, \tau = 0.01, \mu = 0.001$, and different choices of the input signal u. Comment on your results.

18. The dynamics of a hard-disk drive servo system are given by

$$y = \frac{k_p}{s^2} \sum_{i=1}^{6} \frac{b_{1i} s + b_{0i}}{s^2 + 2\zeta_i \omega_i s + \omega_i^2} u,$$

where ω_i, $i = 1, \dots, 6$, are the resonant frequencies which are large, i.e., $\omega_1 = 11.2\pi \times 10^3$ rad/sec, $\omega_2 = 15.5\pi \times 10^3$ rad/sec, $\omega_3 = 16.6\pi \times 10^3$ rad/sec, $\omega_4 = 18\pi \times 10^3$ rad/sec, $\omega_5 = 20\pi \times 10^3$ rad/sec, $\omega_6 = 23.8\pi \times 10^3$ rad/sec. The unknown constants b_{1i} are of the order of 10^4, b_{0i} are of the order of 10^8, and k_p is of the order of 10^7. The damping coefficients ζ_i are of the order of 10^{-2}.

   (a) Derive a low-order model for the servo system. (*Hint*: Take $\frac{\alpha}{\omega_i} \cong 0$ and hence $\frac{\alpha^2}{\omega_i^2} \cong 0$ for α of the order of less than 10^3.)

   (b) Assume that the full-order system parameters are given as follows:

$$\begin{aligned}
&b_{16} = -5.2 \times 10^4, && b_{01} = 1.2 \times 10^9, && b_{02} = 5.4 \times 10^8,\\
&b_{03} = -7.7 \times 10^8, && b_{04} = -1.6 \times 10^8, && b_{05} = -1.9 \times 10^8,\\
&b_{06} = 1.2 \times 10^9, && k_p = 3.4 \times 10^7, && \zeta_1 = 2.6 \times 10^{-2},\\
&\zeta_2 = 4.4 \times 10^{-3}, && \zeta_3 = 1.2 \times 10^{-2}, && \zeta_4 = 2.4 \times 10^{-3},\\
&\zeta_5 = 6.8 \times 10^{-3}, && \zeta_6 = 1.5 \times 10^{-2}.
\end{aligned}$$

   Obtain a Bode plot for the full-order and reduced-order models.

   (c) Use the reduced-order model in (a) to obtain a parametric model for the unknown parameters. Design a robust adaptive law to estimate the unknown parameters online.

19. Consider the time-varying plant

$$\dot{x} = -a(t)x + b(t)u,$$

where $a(t)$, $b(t)$ are slowly varying unknown parameters; i.e., $|\dot{a}|$, $|\dot{b}|$ are very small.

 (a) Obtain a parametric model for estimating a, b.

 (b) Design and analyze a robust adaptive law that generates the estimates $\hat{a}(t)$, $\hat{b}(t)$ of $a(t)$, $b(t)$, respectively.

 (c) Simulate your scheme for a plant with $a(t) = 5 + \sin \mu t$, $b(t) = 8 + \cos 2\mu t$ for $\mu = 0, 0.01, 0.1, 1, 5$. Comment on your results.

20. Consider the parameter error differential equation (3.69), i.e.,

$$\dot{\tilde{\theta}} = -\gamma u^2 \tilde{\theta} + \gamma d u.$$

Show that if the equilibrium $\tilde{\theta}_e = 0$ of the homogeneous equation

$$\dot{\tilde{\theta}} = -\gamma u^2 \tilde{\theta}$$

is exponentially stable, then the bounded input $\gamma d u$ will lead to a bounded solution $\tilde{\theta}(t)$. Obtain an upper bound for $|\tilde{\theta}(t)|$ as a function of the upper bound of the disturbance term $\gamma d u$.

21. Consider the system

$$y = \theta^* u + \eta,$$
$$\eta = \Delta(s) u,$$

where y, u are available for measurement, θ^* is the unknown constant to be estimated, and η is a modeling error signal with $\Delta(s)$ being proper and analytic in $\Re[s] \geq -0.5$. The input u is piecewise continuous.

 (a) Design an adaptive law with a switching σ-modification to estimate θ^*.

 (b) Repeat (a) using projection.

 (c) Simulate the adaptive laws in (a), (b) using the following values:

$$\theta^* = 5 + \sin 0.1t,$$
$$\Delta(s) = 10\mu \frac{\mu s - 1}{(\mu s + 1)^2}$$

 for $\mu = 0, 0.1, 0.01$ and $u =$ constant, $u = \sin \omega_0 t$, where $\omega_0 = 1, 10, 100$. Comment on your results.

22. The linearized dynamics of a throttle angle θ to vehicle speed V subsystem are given by the third-order system

$$V = \frac{b p_1 p_2}{(s + a)(s + p_1)(s + p_2)} \theta + d,$$

where $p_1, p_2 > 20$, $1 \geq a > 0$, and d is a load disturbance.

(a) Obtain a parametric model for the parameters of the dominant part of the system.

(b) Design a robust adaptive law for estimating these parameters online.

(c) Simulate your estimation scheme when $a = 0.1, b = 1, p_1 = 50, p_2 = 100$, and $d = 0.02 \sin 5t$, for different constant and time-varying throttle angle settings θ of your choice.

23. Consider the parametric model

$$z = \theta^{*T}\phi + \eta,$$

where

$$\eta = \Delta_u(s)u + \Delta_y(s)y$$

and Δ_u, Δ_y are proper transfer functions analytic in $\Re[s] \geq -\frac{\delta_0}{2}$ for some known $\delta_0 > 0$.

(a) Design a normalizing signal m_s that guarantees $\frac{\eta}{m_s} \in \mathcal{L}_\infty$ when (i) Δ_u, Δ_y are biproper, (ii) Δ_u, Δ_y are strictly proper. In each case specify the upper bound for $\frac{|\eta|}{m_s}$.

(b) Calculate the bound for $\frac{|\eta|}{m_s}$ when (i) $\Delta_u(s) = \frac{e^{-\tau s}-1}{s+2}$, $\Delta_y(s) = \mu\frac{s^2}{(s+1)^2}$, (ii) $\Delta_u(s) = \frac{\mu s}{\mu s+2}$, $\Delta_y(s) = \frac{\mu s}{(\mu s+1)^2}$, where $0 < \mu \ll 1$ and $0 < \tau \ll 1$.

(c) Design and simulate a robust adaptive law to estimate θ^* for the following example:

$$\theta^* = [1, 0.1]^T,$$
$$\phi = [u, y]^T,$$
$$z = \frac{s}{s+5}y,$$
$$\Delta_u(s) = \frac{e^{-\tau s} - 1}{s+2}, \qquad \Delta_y(s) = \frac{\mu(s-1)}{(\mu s+1)^2},$$

where $\tau = 0.01$, $\mu = 0.01$.

Chapter 4
Parameter Identification: Discrete Time

4.1 Introduction

Most of the physical systems to be controlled or identified are continuous-time and are often modeled as such. The implementation of an online estimator or controller, however, is often done using a digital computer, which implies a discrete-time representation or design as shown in Figure 4.1, where S/H denotes the sample and hold device and k is a discrete time that takes values $k = 0, 1, 2, 3, \ldots$, i.e., $t = kT$, where T is the sampling period. The design of the discrete-time online estimator may be done by following two different approaches. In the first approach, we design a continuous-time online parameter estimator based on the continuous-time model of the system, as we did in Chapter 3, and then obtain a discrete-time parameter estimator using a discrete-time approximation of the continuous-time one. In the second approach, we obtain a discrete-time approximation of

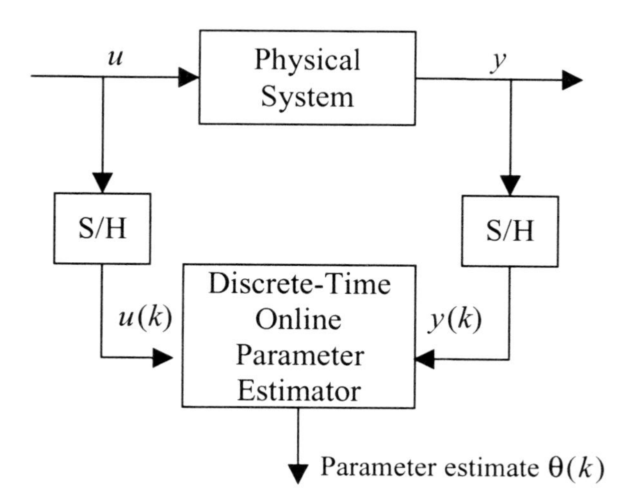

Figure 4.1. *Implementation of online parameter estimator.*

the continuous-time model of the physical system, which we use to develop discrete-time online estimators using discrete-time techniques to be developed in this chapter. In the following sections we cover both approaches and demonstrate that the discrete-time online estimators or adaptive laws have a form and share properties very similar to those of their counterparts in the continuous-time case covered in Chapter 3.

A wide class of discrete-time adaptive laws is of the form

$$\varepsilon(k) = \frac{z(k) - \theta^T(k-1)\phi(k)}{m^2(k)}, \tag{4.1}$$

$$\theta(k) = \theta(k-1) + \Gamma\varepsilon(k)\phi(k), \tag{4.2}$$

where $\theta(k)$ is the estimate of θ^* at $t = kT$, $\phi(k)$ is the regressor, $z(k)$ is the output of the SPM

$$z = \theta^{*T}\phi,$$

$m(k) \geq c > 0$ (where c is a constant)[5] is the normalization signal, and $\Gamma = \Gamma^T > 0$ is the adaptive gain.

Lemma 4.1.1 (key lemma). *If* $\frac{\|\Gamma\| |\phi(k)|^2}{m^2(k)} < 2$, *then* (4.1), (4.2) *has the following properties:*

(i) $\theta(k) \in \ell_\infty$.

(ii) $\varepsilon(k), \varepsilon(k)m(k), \varepsilon(k)\phi(k), |\theta(k) - \theta(k-N)| \in \ell_2 \cap \ell_\infty$.

(iii) $\varepsilon(k), \varepsilon(k)m(k), |\varepsilon(k)\phi(k)|, |\theta(k) - \theta(k-N)| \to 0$ *as* $k \to \infty$, *where* $N \geq 1$ *is any finite integer.*

(iv) *If, in addition,* $\frac{\phi(k)}{m(k)}$ *is persistently exciting (PE), i.e., satisfies*

$$\sum_{i=0}^{l-1} \frac{\phi(k+i)\phi^T(k+i)}{m^2(k+i)} \geq \alpha_0 l I$$

$\forall k$ and some fixed integer $l > 1$ and constant $\alpha_0 > 0$, then $\theta(k) \to \theta^*$ exponentially fast.

Proof. (i)–(iii) From (4.1), (4.2), and $\tilde{\theta}(k) \triangleq \theta(k) - \theta^*$ we obtain the error equation

$$\tilde{\theta}(k) = \left[I - \frac{\Gamma\phi(k)\phi^T(k)}{m^2(k)}\right]\tilde{\theta}(k-1), \tag{4.3}$$

$$\varepsilon(k) = -\frac{\tilde{\theta}^T(k-1)\phi(k)}{m^2(k)}. \tag{4.4}$$

Consider the function

$$V(k) = \frac{1}{2}\tilde{\theta}^T(k)\Gamma^{-1}\tilde{\theta}(k). \tag{4.5}$$

[5] Note that the constant c can be taken to be equal to 1 without loss of generality, as done in the continuous-time case.

Then

$$V(k) - V(k-1) = -\frac{\tilde{\theta}^T(k-1)\phi(k)\phi^T(k)\tilde{\theta}(k-1)}{m^2(k)} + \frac{\tilde{\theta}^T(k-1)\phi(k)\phi^T(k)\Gamma\phi(k)\phi^T(k)\tilde{\theta}(k-1)}{2m^4(k)}.$$

Using $\varepsilon(k)m^2(k) = -\tilde{\theta}^T(k-1)\phi(k)$, we obtain

$$V(k) - V(k-1) = -\varepsilon^2(k)m^2(k)\left[1 - \frac{\phi^T(k)\Gamma\phi(k)}{2m^2(k)}\right].$$

Since $\frac{\phi^T(k)\Gamma\phi(k)}{m^2(k)} \leq \frac{\|\Gamma\||\phi(k)|^2}{m^2(k)} < 2$, it follows that

$$V(k) - V(k-1) \leq -c_0\varepsilon^2(k)m^2(k) \leq 0 \tag{4.6}$$

for some constant $c_0 > 0$. From (4.5), (4.6) we have that $V(k)$ and therefore $\theta(k) \in \ell_\infty$ and $V(k)$ has a limit, i.e., $\lim_{k\to\infty} V(k) = V_\infty$. Consequently, using (4.6) we obtain

$$c_0 \sum_{k=1}^{\infty} \varepsilon^2(k)m^2(k) \leq V(0) - V_\infty < \infty,$$

which implies $\varepsilon(k)m(k) \in \ell_2$ and $\varepsilon(k)m(k) \to 0$ as $k \to \infty$. Since $m(k) \geq c > 0$, we also have that $\varepsilon(k) \in \ell_2$ and $\varepsilon(k) \to 0$ as $k \to \infty$. We have $\varepsilon(k)\phi(k) = \varepsilon(k)m(k)\frac{\phi(k)}{m(k)}$. Since $\frac{\phi(k)}{m(k)}$ is bounded and $\varepsilon(k)m(k) \in \ell_2$, we have that $\varepsilon(k)\phi(k) \in \ell_2$ and $|\varepsilon(k)\phi(k)| \to 0$ as $k \to \infty$. This implies (using (4.2)) that $|\theta(k) - \theta(k-1)| \in \ell_2$ and $|\theta(k) - \theta(k-1)| \to 0$ as $k \to \infty$. Now

$$\theta(k) - \theta(k-N) = \theta(k) - \theta(k-1) + \theta(k-1) - \theta(k-2) + \cdots + \theta(k-N+1) - \theta(k-N)$$

for any finite N. Using the Schwarz inequality, we have

$$|\theta(k) - \theta(k-N)|^2 \leq |\theta(k) - \theta(k-1)|^2 + |\theta(k-1) - \theta(k-2)|^2 + \cdots + |\theta(k-N+1) - \theta(k-N)|^2.$$

Since each term on the right-hand side of the inequality is in ℓ_2 and goes to zero with $k \to \infty$, it follows that $|\theta(k) - \theta(k-N)| \in \ell_2$ and $|\theta(k) - \theta(k-N)| \to 0$ as $k \to \infty$.

(iv) From (4.3), we have

$$\tilde{\theta}(k) = A(k)\tilde{\theta}(k-1),$$

where

$$A(k) = I - \frac{\Gamma\phi(k)\phi^T(k)}{m^2(k)}.$$

Let us consider Theorem A.12.22 in the Appendix and let $N(k) = \frac{\phi}{m}(2 - \frac{\phi^T\Gamma\phi}{m^2})^{1/2}$, $P = \Gamma^{-1}$. We can show that

$$A^T(k)PA(k) - P = -N(k)N^T(k),$$

which according to Theorem A.12.22 implies that the equilibrium $\tilde{\theta}_e(k) = 0$ of (4.3) is stable. Let us now choose

$$K(k) = -\frac{\Gamma\phi(k)}{m(k)\left(2 - \frac{\phi^T(k)\Gamma\phi(k)}{m^2(k)}\right)^{1/2}},$$

which is uniformly bounded due to $\frac{\phi}{m} \in \ell_\infty$ and $\frac{\phi^T(k)\Gamma\phi(k)}{m^2(k)} < 2$. Then

$$\bar{A}(k) = A(k) - K(k)N^T(k) = I$$

and the transition matrix for $\bar{A}(k)$ is $\Phi(k, j) = I$. According to Theorem A.12.22, we can conclude that the equilibrium $\tilde{\theta}_e = 0$ of (4.3) is e.s. if we establish that $(\bar{A}(k), N(k)) = (I, N(k))$ is UCO, i.e., there exist constants $\alpha, \beta, l > 0$ such that

$$\alpha I \leq \sum_{j=0}^{l-1} N(k+j)N^T(k+j) \leq \beta I. \tag{4.7}$$

Since

$$\alpha_1 \frac{\phi(k)\phi^T(k)}{m^2(k)} \leq N(k)N^T(k) \leq \frac{2\phi(k)\phi^T(k)}{m^2(k)},$$

where

$$\alpha_1 = 2 - \max_k \frac{\|\Gamma\||\phi(k)|^2}{m^2(k)} > 0,$$

we have

$$\alpha_1 \sum_{j=0}^{l-1} \frac{\phi(k+j)\phi^T(k+j)}{m^2(k+j)} \leq \sum_{j=0}^{l-1} N(k+j)N^T(k+j) \leq 2\sum_{j=0}^{l-1} \frac{\phi(k+j)\phi^T(k+j)}{m^2(k+j)}.$$

Since $\frac{\phi}{m} \in \ell_\infty$, condition (iv) becomes

$$\bar{\beta} I \geq \sum_{j=0}^{l-1} \frac{\phi(k+j)\phi^T(k+j)}{m^2(k+j)} \geq \alpha_0 l I$$

for some $\bar{\beta} > 0$, which can be used to establish that (4.7) is satisfied with $\alpha = \alpha_1\alpha_0 l$, $\beta = 2\bar{\beta}$. Therefore, the equilibrium $\tilde{\theta}_e = 0$ of (4.3) is e.s., which implies that $\theta(k) \to \theta^*$ as $k \to \infty$ exponentially fast. □

Lemma 4.1.1 is a fundamental one in the sense that it describes the stability properties of a wide class of adaptive laws obtained by discretizing the continuous-time ones or by generating them using a discrete-time parametric model, as we demonstrate in the following sections.

4.2 Discretization of Continuous-Time Adaptive Laws

Let us consider the continuous-time adaptive law

$$\dot{\theta} = \Gamma \varepsilon \phi, \tag{4.8}$$

$$\varepsilon = \frac{z - \theta^T \phi}{m_s^2}, \tag{4.9}$$

based on the SPM

$$z = \theta^{*T} \phi,$$

where $\theta(t)$, the estimate of θ^*, is updated continuously with time. Let us now assume that $z(t)$, $\phi(t)$ are measured at times $t = kT$, where $k = 0, 1, 2, 3, \dots$ and T is the sampling period, i.e.,

$$z(k) = \theta^{*T} \phi(k), \tag{4.10}$$

where $z(k) \stackrel{\Delta}{=} z(kT)$, $\phi(k) = \phi(kT)$. Let us now discretize the continuous-time adaptive law (4.8), (4.9) using the Euler backward approximation method, i.e.,

$$\dot{\theta} \cong \frac{\theta(k) - \theta(k-1)}{T} = \Gamma \varepsilon(k) \phi(k).$$

We obtain

$$\varepsilon(k) = \frac{z(k) - \theta^T(k)\phi(k)}{m_s^2(k)}, \tag{4.11}$$

$$\theta(k) = \theta(k-1) + \Gamma_1 \varepsilon(k) \phi(k), \tag{4.12}$$

where $\Gamma_1 = T\Gamma$. The above adaptive law cannot be implemented as is, because at time $t = kT$ the calculation of $\theta(k)$ requires $\varepsilon(k)$, which in turn depends on $\theta(k)$. If we use (4.12) to substitute for $\theta(k)$ in (4.11), we obtain

$$\varepsilon(k) = \frac{z(k) - \theta^T(k-1)\phi(k) - \phi^T(k)\Gamma_1\phi(k)\varepsilon(k)}{m_s^2(k)}$$

or

$$\varepsilon(k) = \frac{z(k) - \theta^T(k-1)\phi(k)}{m^2(k)}, \tag{4.13}$$

where $m^2(k) = m_s^2(k) + \phi^T(k)\Gamma_1\phi(k)$. In (4.13), $\varepsilon(k)$ depends on $\theta(k-1)$, $z(k)$, $\phi(k)$, $m_s(k)$, which are all available for measurement at instant k and therefore can be generated at time k.

Theorem 4.2.1. *Consider the adaptive law* (4.12), (4.13) *with $m_s(k) > c > 0$ for some constant c. If $\Gamma_1 = \Gamma_1^T > 0$ is designed such that $\lambda_{\max}(\Gamma_1) < 2\lambda_{\min}(\Gamma_1)$, then properties* (i)–(iv) *of the key lemma are applicable.*

Proof. The adaptive law (4.12), (4.13) is of the same form as (4.1), (4.2); therefore, the key lemma is applicable if the condition

$$\frac{\|\Gamma_1\| |\phi(k)|^2}{m^2(k)} < 2$$

is satisfied. We have

$$\frac{\|\Gamma_1\||\phi(k)|^2}{m^2(k)} \leq \frac{\lambda_{\max}(\Gamma_1)|\phi(k)|^2}{m_s^2(k) + |\phi(k)|^2\lambda_{\min}(\Gamma_1)} < 2$$

due to $m_s^2(k) \geq c > 0$ and $\lambda_{\max}(\Gamma_1) < 2\lambda_{\min}(\Gamma_1)$. Therefore, all conditions of the key lemma are applicable and the result follows. □

Remark 4.2.2 We should note that a simple choice for Γ_1 is $\Gamma_1 = \gamma I$, where $\gamma > 0$. Since γ also appears in $m^2(k)$, there is no requirement for an upper bound for γ.

In a similar fashion the other continuous-time adaptive laws of Chapter 3 can be discretized using the Euler backward approximation method or other discretization methods. The challenge is to show that the resulting discrete-time adaptive law has properties similar to those established by the key lemma.

4.3 Discrete-Time Parametric Model

Consider the SISO system described by the discrete-time deterministic autoregressive moving average (DARMA) form

$$y(k) = -\sum_{j=1}^{n} a_j y(k-j) + \sum_{j=0}^{m} b_j u(k-j-d), \tag{4.14}$$

where a_j, $j = 1, \ldots, n$, and b_j, $j = 0, \ldots, m$, are constant parameters, d is the time delay, $m \leq n$, and $k = 0, 1, 2, \ldots$. Using the z transform and setting all initial conditions to zero, we express the system (4.14) in the transfer function form

$$A(z)Y(z) = B(z)U(z), \tag{4.15}$$

where

$$A(z) = z^n + a_1 z^{n-1} + \cdots + a_n,$$
$$B(z) = z^{-d}(b_0 z^n + b_1 z^{n-1} + \cdots + b_m z^{n-m}).$$

If we treat z as the shift operator, i.e.,

$$zy(k) = y(k+1), \qquad z^{-1}y(k) = y(k-1),$$

then (4.15) may also be expressed as

$$A(z)y(k) = B(z)u(k). \tag{4.16}$$

If a_j, $j = 1, \ldots, n$, and b_j, $j = 0, 1, \ldots, m$, are the unknown constant parameters, then we can separate them from the known signals $u(k-j)$, $y(k-j)$, $j = 0, 1, \ldots, n$, by expressing (4.15) in the form of SPM

$$z = \theta^{*T}\phi, \tag{4.17}$$

where

$$z(k) = y(k),$$
$$\phi(k) = [u(k-d), u(k-d-1), \dots, u(k-d-m), -y(k-1), \dots, -y(k-n)]^T,$$
$$\theta^* = [b_0, b_1, \dots, b_m, a_1, a_2, \dots, a_n]^T.$$

If some of the parameters in θ^* are known, then the dimension of θ^*can be reduced. For example, if in (4.15) a_1, a_n, b_m are known, then (4.17) can be rewritten in the form

$$z_r = \theta_r^{*T} \phi_r,$$

where

$$z_r(k) = y(k) + a_1 y(k-1) + a_n y(k-n) - b_m u(k-m-d),$$
$$\phi_r(k) = [u(k-d), u(k-d-1), \dots, u(k-d-m+1), -y(k-2), \dots, -y(k-n+1)]^T,$$
$$\theta_r^* = [b_0, b_1, \dots, b_{m-1}, a_2, \dots, a_{n-1}]^T.$$

In the following sections we design and analyze adaptive laws for estimating θ^* in the discrete-time SPM (4.17). The accuracy of estimation of θ^* depends on the properties of the adaptive law used as well as on the properties of the plant input sequence $u(k)$. If $u(k)$ does not belong to the class of sufficiently rich sequences (to be defined in the next section), then the accurate estimation of θ^* cannot be guaranteed in general.

4.4 Sufficiently Rich Inputs

In Lemma 4.1.1, we have established that if $\frac{\phi(k)}{m(k)}$ is PE, i.e., satisfies

$$\sum_{i=0}^{l-1} \frac{\phi(k+i)\phi^T(k+i)}{m^2(k+i)} \geq \alpha_0 l I \tag{4.18}$$

$\forall k$ and some fixed constants $l > 1$, $\alpha_0 > 0$, then the exponential convergence of $\theta(k)$ to the unknown vector θ^* in the SPM $z = \theta^{*T}\phi$ is guaranteed by the adaptive law (4.1), (4.2). The condition (4.18) is the same as

$$\sum_{j=k}^{k+l-1} \frac{\phi(j)\phi^T(j)}{m^2(j)} \geq \alpha_0 l I$$

for some $l > 1$, $\alpha_0 > 0$, and $\forall k$, which resembles the PE condition in the continuous-time case. Condition (4.18) is also referred to as *strong persistence of excitation* [46] in order to distinguish it from its weaker version

$$\lim_{k\to\infty} \lambda_{\min} \sum_{j=1}^{k} \frac{\phi(j)\phi^T(j)}{m^2(j)} = \infty,$$

which is often referred to as the *weak persistence of excitation* condition [46]. It is easy to see that if the signal $\frac{\phi(k)}{m(k)}$ is strongly PE, then it is weakly PE as well, but the converse is not always true; i.e., there exist discrete-time signals that are weakly PE but not strongly PE.

For the parametric model (4.17) of the LTI plant (4.14), the regressor vector $\phi(k)$ can be expressed as

$$\phi(k) = H(z)u(k), \tag{4.19}$$

where

$$H(z) = \left[z^{-d}, z^{-d-1}, \dots, z^{-d-m}, -z^{-1}\frac{B(z)}{A(z)}, -z^{-2}\frac{B(z)}{A(z)}, \dots, -z^{-n}\frac{B(z)}{A(z)}\right]^T. \tag{4.20}$$

It is clear from (4.19) that the PE property of $\phi(k)$ and subsequently that of $\frac{\phi(k)}{m(k)}$ depends on the properties of $H(z)$ and input sequence $u(k)$.

Definition 4.4.1. *The input sequence $u(k)$ is called* sufficiently rich *of order n if it contains at least $\frac{n}{2}$ distinct frequencies.*

Theorem 4.4.2. *Consider* (4.19) *and assume that all the poles of $H(z)$ are within the unit circle and the complex vectors $H(e^{j\omega_1})$, $H(e^{j\omega_2}), \dots, H(e^{j\omega_{\bar{n}}})$, where $\bar{n} = n + m + 1$ are linearly independent vectors over $\mathcal{C}^n$ $\forall \omega_1, \omega_2, \dots, \omega_{\bar{n}} \in [0, 2\pi)$, where $\omega_i \neq \omega_j$, for $i \neq j$. If $u(k)$ is sufficiently rich of order $\bar{n} = n + m + 1$, then $\phi(k)$ is weakly PE.*

Proof. The proof is presented in the web resource [94]. □

Since all the poles of $H(z)$ are within the unit circle and $u(k)$ is a bounded sequence, it follows that $\phi(k)$ is a vector of bounded sequences. If $m(k)$ is a bounded sequence too, then it follows that if $\phi(k)$ is PE, then in general $\frac{\phi(k)}{m(k)}$ is PE too (provided that $m(k)$ is not designed to cancel elements of $\phi(k)$). We should note that when $\phi(k)$ is bounded, $m(k)$ can be designed to be equal to 1 as long as the conditions of Lemma 4.1.1 are satisfied. The poles of $H(z)$ include those of the plant, which means that the plant has to have stable poles for stable poles in $H(z)$.

Lemma 4.4.3. *If the plant* (4.15) *is stable, i.e., has all poles within the unit circle and has no zero-pole cancellations, then all the conditions of Theorem* 4.4.2 *for $H(z)$ are satisfied; i.e.,*

(i) *all poles of $H(z)$ are within the unit circle;*

(ii) *the complex vectors $H(e^{j\omega_1}), \dots, H(e^{j\omega_{\bar{n}}})$, where $\bar{n} = n + m + 1$, are linearly independent vectors over $\mathcal{C}^n$ $\forall \omega_1, \omega_2, \dots, \omega_{\bar{n}} \in [0, 2\pi)$, where $\omega_i \neq \omega_j$, for $i \neq j$.*

Lemma 4.4.3 implies that if we overparameterize the plant, we cannot guarantee parameter convergence, since overparameterization implies zero-pole cancellations. For example, let us assume that the real plant is of the form

$$y(k) = \frac{b_0(z + b_1)}{(z + a_1)(z + a_0)}u(k) \tag{4.21}$$

of order 2 with four unknown parameters. Let us assume that we made a mistake in the order and, instead of (4.21), we assume that the plant model is of order 3 and of the form

$$y_p(k) = \frac{b_0(z+b_2)(z+b_1)}{(z+a_2)(z+a_1)(z+a_0)}u(k). \tag{4.22}$$

It is clear that for output matching for all $u(k)$, i.e., $y_p(k) = y(k)\ \forall u(k)$, the parameter $b_2 = a_2$, which implies zero-pole cancellation. This implies that $y(k)$ carries no information about a_2, b_2 and no method can be used to estimate a_2, b_2 from the measurements of $y(k)$, $u(k)$. A parameterization of (4.22) in the form of the SPM will lead to the regressor

$$\phi(k) = H(z)u(k),$$

where the vectors $H(e^{j\omega_1}), \ldots, H(e^{j\omega_{\bar{n}}})$ cannot be guaranteed to be independent for any $\bar{n} > 4$ and any set of $\omega_1, \omega_2, \ldots, \omega_{\bar{n}} \in \mathcal{R}$, and therefore $\phi(k)$ cannot be PE no matter how rich we choose $u(k)$ to be.

4.5 Gradient Algorithms

Consider the discrete-time parametric model

$$z(k) = \theta^{*T}\phi(k), \tag{4.23}$$

where $\theta^* \in \mathcal{R}^n$ is the unknown constant vector and $z \in \mathcal{R}$, $\phi \in \mathcal{R}^n$ are known sequences at each instant $k = 0, 1, 2, \ldots$. The estimation error $\varepsilon(k)$ is generated as

$$\begin{aligned} \hat{z}(k) &= \theta^T(k-1)\phi(k), \\ \varepsilon(k) &= \frac{z(k) - \hat{z}(k)}{m^2(k)} = \frac{z(k) - \theta^T(k-1)\phi(k)}{m^2(k)}, \end{aligned} \tag{4.24}$$

where $m^2(k) \geq c > 0$ (for some constant c) is the normalizing signal, designed to guarantee $\frac{\phi}{m} \in \ell_\infty$. The estimation error $\varepsilon(k)$ at time k depends on the most recent estimate of θ^*, which at time k is assumed to be $\theta(k-1)$, as $\theta(k)$ is generated afterwards using $\varepsilon(k)$.

The adaptive law is of the form

$$\theta(k) = \theta(k-1) + \Gamma\varepsilon(k)\phi(k), \quad \theta(0) = \theta_0 \tag{4.25}$$

with $\varepsilon(k)$ given by (4.24). Different choices for $\Gamma = \Gamma^T > 0$ and $m^2(k)$ give rise to different adaptive laws, which are summarized below.

4.5.1 Projection Algorithm

In this case

$$\begin{aligned} \varepsilon(k) &= \frac{z(k) - \theta^T(k-1)\phi(k)}{\phi^T(k)\phi(k)}, \\ \theta(k) &= \theta(k-1) + \varepsilon(k)\phi(k) \end{aligned} \tag{4.26}$$

is developed by solving the minimization problem

$$\text{Minimize } J = \frac{1}{2}|\theta(k) - \theta(k-1)|^2 \quad \text{w.r.t. } \theta(k),$$

subject to the constraint $z(k) = \theta^T(k)\phi(k)$.

Using the Lagrange multiplier method to get rid of the constraint, the minimization problem becomes

$$\text{Minimize } J_\lambda = \frac{1}{2}|\theta(k) - \theta(k-1)|^2 + \lambda[z(k) - \theta^T(k)\phi(k)] \quad \text{w.r.t. } \theta(k) \text{ and } \lambda.$$

The necessary conditions for a minimum are

$$\frac{\partial J_\lambda}{\partial \theta(k)} = \theta(k) - \theta(k-1) - \lambda\phi(k) = 0, \tag{4.27}$$

$$\frac{\partial J_\lambda}{\partial \lambda} = z(k) - \theta^T(k)\phi(k) = 0. \tag{4.28}$$

Substituting for $\theta(k)$ from (4.27) in (4.28) and solving for λ, we have

$$\lambda = \frac{z(k) - \theta^T(k-1)\phi(k)}{\phi^T(k)\phi(k)}.$$

Substituting for λ in (4.27), we obtain (4.26). The name projection algorithm [46] arises from its geometric interpretation, where the estimate $\theta(k)$ is the orthogonal projection of $\theta(k-1)$ to the hypersurface $H = \{\theta | z(k) = \theta^T\phi(k)\}$ (see Problem 11).

One of the problems of the projection algorithm (4.26) is the possibility of division by zero when $\phi^T(k)\phi(k) = 0$. This possibility is eliminated by modifying (4.26) to

$$\begin{aligned} \varepsilon(k) &= \frac{z(k) - \theta^T(k-1)\phi(k)}{c + \phi^T(k)\phi(k)}, \\ \theta(k) &= \theta(k-1) + \alpha\varepsilon(k)\phi(k), \end{aligned} \tag{4.29}$$

where $c > 0$, $\alpha > 0$ are design constants.

Theorem 4.5.1. *If* $c > 0$ *and* $0 < \alpha < 2$, *then the results of the key lemma are applicable for the modified projection algorithm* (4.29).

Proof. Comparing (4.29) with (4.1), (4.2), we have $\Gamma = \alpha I$, $m^2(k) = c + \phi^T(k)\phi(k)$. For $c > 0$ and $0 < \alpha < 2$, we have

$$\frac{\alpha|\phi(k)|^2}{c + |\phi(k)|^2} < 2,$$

and therefore all conditions of the key lemma are satisfied and (4.29) shares the same properties as (4.1), (4.2) described by the key lemma. □

A modified version of the projection algorithm is the so-called *orthogonalized projection algorithm*

$$\varepsilon(k) = \frac{z(k) - \theta^T(k-1)\phi(k)}{\phi^T(k)P(k-1)\phi(k)},$$
$$\theta(k) = \theta(k-1) + P(k-1)\phi(k)\varepsilon(k), \tag{4.30}$$
$$P(k) = P(k-1) - \frac{P(k-1)\phi(k)\phi^T(k)P(k-1)}{\phi^T(k)P(k-1)\phi(k)},$$

with initial conditions $\theta(0)$, $P(0) = I$. In this case, it can be shown (see [46] and the web resource [94]) that $\theta(k)$ will converge to θ^* in m steps, provided that

$$\text{rank}[\phi(1), \ldots, \phi(m)] = n = \text{dimension of } \theta^*.$$

The possible division by zero in (4.30) can be avoided by using $\theta(k) = \theta(k-1)$ and $P(k) = P(k-1)$ whenever $\phi^T(k)P(k-1)\phi(k) = 0$. The above algorithm is a way of sequentially solving a set of linear equations for the unknown vector θ^*. It is referred to as the orthogonalized projection algorithm because $P(k-1)\phi(j) = 0\ \forall j \leq k-1$ and hence $P(k-1)\phi(k)$ is the component of $\phi(k)$, which is orthogonal to $\phi(j)\ \forall j < k$ [46].

4.5.2 Gradient Algorithm Based on Instantaneous Cost

Let us now follow an approach similar to that used in the continuous-time case and consider the minimization of the cost

$$J(\theta) = \frac{(z(k) - \theta^T(k-1)\phi(k))^2}{2m^2(k)}$$

w.r.t. $\theta(k-1)$. Using the steepest-descent method, we have

$$\theta(k) = \theta(k-1) - \Gamma\nabla J,$$

where $\Gamma = \Gamma^T > 0$ and

$$\nabla J = -\frac{z(k) - \theta^T(k-1)\phi(k)}{m^2(k)}\phi(k) = -\varepsilon(k)\phi(k),$$

leading to the gradient algorithm

$$\varepsilon(k) = \frac{z(k) - \theta^T(k-1)\phi(k)}{m^2(k)},$$
$$\theta(k) = \theta(k-1) + \Gamma\varepsilon(k)\phi(k),$$

which is exactly in the form of (4.1), (4.2), the properties of which are established by the key lemma.

4.6 LS Algorithms

A wide class of LS algorithms may be developed by using the parametric model

$$z(k) = \theta^{*T}\phi(k).$$

The estimation model and estimation error are given as

$$\begin{aligned}\hat{z}(k) &= \theta^T(k-1)\phi(k),\\ \varepsilon(k) &= \frac{z(k)-\hat{z}(k)}{m^2(k)} = \frac{z(k)-\theta^T(k-1)\phi(k)}{m^2(k)},\end{aligned} \tag{4.31}$$

where $m(k) > c > 0$ is the normalizing signal designed to bound $|\phi(k)|$ from above. The procedure for designing discrete-time LS algorithms is very similar to the one in continuous time, as demonstrated in the following sections.

4.6.1 Pure LS

We consider the cost function

$$J_N(\theta) = \frac{1}{2}\sum_{k=1}^{N}\frac{(z(k)-\theta^T\phi(k))^2}{m^2(k)} + \frac{1}{2}(\theta-\theta_0)^T P_0^{-1}(\theta-\theta_0), \tag{4.32}$$

where $\theta_0 = \theta(0)$. Letting

$$\begin{aligned}Z_k &\triangleq \left[\frac{z(1)}{m(1)}, \frac{z(2)}{m(2)}, \dots, \frac{z(k)}{m(k)}\right]^T,\\ \Phi_k &\triangleq \left[\frac{\phi(1)}{m(1)}, \frac{\phi(2)}{m(2)}, \dots, \frac{\phi(k)}{m(k)}\right]^T,\end{aligned}$$

the cost function (4.32) can be expressed as

$$J_k(\theta) = \frac{1}{2}(Z_k - \Phi_k\theta)^T(Z_k-\Phi_k\theta) + \frac{1}{2}(\theta-\theta_0)^T P_0^{-1}(\theta-\theta_0). \tag{4.33}$$

Setting

$$\nabla J = -\Phi_k^T(Z_k - \Phi_k\theta) + P_0^{-1}(\theta-\theta_0) = 0,$$

we obtain

$$(\Phi_k^T\Phi_k + P_0^{-1})\theta = P_0^{-1}\theta_0 + \Phi_k^T Z_k. \tag{4.34}$$

Therefore, the estimate of θ^* at time k which minimizes (4.32) is

$$\theta(k) = (\Phi_k^T\Phi_k + P_0^{-1})^{-1}(P_0^{-1}\theta_0 + \Phi_k^T Z_k). \tag{4.35}$$

Let us define

$$P^{-1}(k) \triangleq \Phi_k^T\Phi_k + P_0^{-1}. \tag{4.36}$$

Then

$$P^{-1}(k) - P^{-1}(k-1) = \Phi_k^T \Phi_k - \Phi_{k-1}^T \Phi_{k-1} = \frac{\phi(k)\phi^T(k)}{m^2(k)}.$$

Hence

$$P^{-1}(k) = P^{-1}(k-1) + \frac{\phi(k)\phi^T(k)}{m^2(k)}. \tag{4.37}$$

Using the relationship

$$(A + BC)^{-1} = A^{-1} - A^{-1}B(I + CA^{-1}B)^{-1}CA^{-1} \tag{4.38}$$

and letting $A = P^{-1}(k-1)$, $B = \frac{\phi(k)}{m(k)}$, $C = \frac{\phi^T(k)}{m(k)}$, it follows from (4.37) that

$$P(k) = P(k-1) - \frac{P(k-1)\phi(k)\phi^T(k)P(k-1)}{m^2(k) + \phi^T(k)P(k-1)\phi(k)}. \tag{4.39}$$

Substituting (4.36) into (4.35), we obtain

$$\theta(k) = P(k)\left(P_0^{-1}\theta_0 + \Phi_{k-1}^T Z_{k-1} + \frac{\phi(k)z(k)}{m^2(k)}\right). \tag{4.40}$$

Using (4.34) at $k-1$ and (4.36) in (4.40), we have

$$\theta(k) = P(k)\left(P^{-1}(k-1)\theta(k-1) + \frac{\phi(k)z(k)}{m^2(k)}\right). \tag{4.41}$$

Substituting for $P^{-1}(k-1)$ from (4.37) in (4.41), we obtain

$$\theta(k) = P(k)\left(P^{-1}(k)\theta(k-1) - \frac{\phi(k)\phi^T(k)\theta(k-1)}{m^2(k)} + \frac{\phi(k)z(k)}{m^2(k)}\right),$$

i.e.,

$$\theta(k) = \theta(k-1) + P(k)\phi(k)\varepsilon(k). \tag{4.42}$$

The *pure LS algorithm* is therefore given by

$$\begin{aligned}
P(k) &= P(k-1) - \frac{P(k-1)\phi(k)\phi^T(k)P(k-1)}{m^2(k) + \phi^T(k)P(k-1)\phi(k)}, \quad P(0) = P_0 = P_0^T > 0,\\
\varepsilon(k) &= \frac{z(k) - \theta^T(k-1)\phi(k)}{m^2(k)},\\
\theta(k) &= \theta(k-1) + P(k)\phi(k)\varepsilon(k), \quad \theta(0) = \theta_0,
\end{aligned} \tag{4.43}$$

where $m^2(k) > c > 0$ is the normalizing signal designed to bound $\phi(k)$ from above. Noting that

$$P(k)\phi(k) = P(k-1)\phi(k)\frac{m^2(k)}{m^2(k) + \phi^T(k)P(k-1)\phi(k)},$$

the equation for $\theta(k)$ can be rewritten as

$$\theta(k) = \theta(k-1) + \frac{P(k-1)\phi(k)(z(k) - \theta^T(k-1)\phi(k))}{m^2(k) + \phi^T(k)P(k-1)\phi(k)}. \tag{4.44}$$

Theorem 4.6.1. *The pure LS algorithm given by* (4.43) *has the following properties:*

(i) $\varepsilon(k), \varepsilon(k)m(k), \varepsilon(k)\phi(k), \theta(k), P(k) \in \ell_\infty$.

(ii) $\lim_{k\to\infty}\theta(k) = \bar{\theta}$, *where* $|\bar{\theta} - \theta^*| \le c_1|\theta(0) - \theta^*|$ *and* $c_1 = \frac{\lambda_{\max}(P_0^{-1})}{\lambda_{\min}(P_0^{-1})}$.

(iii) $\varepsilon(k), \varepsilon(k)m(k), \varepsilon(k)\phi(k), |\theta(k) - \theta(k-N)| \in \ell_2$.

(iv) $\varepsilon(k), \varepsilon(k)m(k), |\varepsilon(k)\phi(k)|, |\theta(k) - \theta(k-N)| \to 0$ *as* $k \to \infty$, *where* N *is any positive finite integer.*

(v) *If* $\frac{\phi(k)}{m(k)}$ *is weakly (or strongly) PE, i.e., satisfies* $\lim_{k\to\infty}\lambda_{\min}\{\sum_{j=0}^{k}\frac{\phi(j)\phi^T(j)}{m^2(j)}\} = \infty$ *(or* $\sum_{j=k}^{k+l-1}\frac{\phi(j)\phi^T(j)}{m^2(j)} \ge \alpha_0 lI$ *for some* $\alpha_0 > 0, l > 1$ *and* $\forall k \ge 1$*), then* $\theta(k) \to \theta^*$ *as* $k \to \infty$.

(vi) *Let the SPM be that of the plant given by (4.15). If the plant has stable poles and no zero-pole cancellations and* $u(k)$ *is sufficiently rich of order* $n + m + 1$*, then* $\theta(k) \to \theta^*$ *as* $k \to \infty$.

Proof. (i)–(iv) From (4.43) we have that

$$P(k) - P(k-1) \le 0,$$

i.e.,

$$P(k) \le P(k-1) \le P(k-2) \le P(0),$$

which implies that $P(k) \in \ell_\infty$. Since $P(0) \ge P(k) = P^T(k) \ge 0$ and $P(k) \le P(k-1)$ $\forall k$, it follows that $\lim_{k\to\infty} P(k) = \bar{P}$, where $\bar{P} = \bar{P}^T \ge 0$ is a constant matrix. From (4.43) we have

$$\tilde{\theta}(k) = \left[I - P(k)\frac{\phi(k)\phi^T(k)}{m^2(k)}\right]\tilde{\theta}(k-1).$$

From (4.37), we have

$$I = P(k)P^{-1}(k) = P(k)P^{-1}(k-1) + P(k)\frac{\phi(k)\phi^T(k)}{m^2(k)}.$$

Then

$$\tilde{\theta}(k) = P(k)P^{-1}(k-1)\tilde{\theta}(k-1)$$

and

$$P^{-1}(k)\tilde{\theta}(k) = P^{-1}(k-1)\tilde{\theta}(k-1) = P^{-1}(0)\tilde{\theta}(0),$$

i.e.,

$$\tilde{\theta}(k) = P(k)P^{-1}(0)\tilde{\theta}(0)$$

and

$$\lim_{k\to\infty}\tilde{\theta}(k) = \bar{P}P^{-1}(0)\tilde{\theta}(0) \stackrel{\Delta}{=} \bar{\theta} - \theta^*,$$

i.e.,

$$\lim_{k\to\infty}\theta(k) = \bar{\theta},$$

where $\bar{\theta}$ is some constant vector.

Consider the function

$$V(k) = \tilde{\theta}^T(k)P^{-1}(k)\tilde{\theta}(k).$$

Then

$$V(k) - V(k-1) = \tilde{\theta}^T(k)P^{-1}(k)\tilde{\theta}(k) - \tilde{\theta}^T(k-1)P^{-1}(k-1)\tilde{\theta}(k-1).$$

Using the identity $\tilde{\theta}(k) = P(k)P^{-1}(k-1)\tilde{\theta}(k-1)$ developed above, we obtain

$$V(k) - V(k-1) = \tilde{\theta}^T(k-1)[P^{-1}(k-1)P(k)P^{-1}(k-1) - P^{-1}(k-1)]\tilde{\theta}(k-1).$$

Multiplying both sides of the $P(k)$ equation in (4.43) by $P^{-1}(k-1)$ from left and right, we have

$$P^{-1}(k-1)P(k)P^{-1}(k-1) = P^{-1}(k-1) - \frac{\phi(k)\phi^T(k)}{m^2(k) + \phi^T(k)P(k-1)\phi(k)},$$

which we substitute into the previous equation to obtain

$$\begin{aligned} V(k) - V(k-1) &= -\frac{(\tilde{\theta}^T(k-1)\phi(k))^2}{m^2(k) + \phi^T(k)P(k-1)\phi(k)} \\ &= -\frac{\varepsilon^2(k)m^4(k)}{m^2(k) + \phi^T(k)P(k-1)\phi(k)} \leq 0. \end{aligned} \tag{4.45}$$

Equation (4.45) implies that $V(k)$ is a nonincreasing function, which implies (using the expression for $V(k)$) that

$$\tilde{\theta}^T(k)P^{-1}(k)\tilde{\theta}(k) \leq \tilde{\theta}^T(0)P_0^{-1}\tilde{\theta}(0). \tag{4.46}$$

From (4.37), it follows that

$$\lambda_{\min}(P^{-1}(k)) \geq \lambda_{\min}(P^{-1}(k-1)) \geq \cdots \geq \lambda_{\min}(P_0^{-1}). \tag{4.47}$$

Therefore,

$$\begin{aligned} |\tilde{\theta}(k)|^2\lambda_{\min}(P_0^{-1}) \leq \lambda_{\min}(P^{-1}(k))|\tilde{\theta}(k)|^2 &\leq \tilde{\theta}^T(k)P^{-1}(k)\tilde{\theta}(k) \\ &\leq \tilde{\theta}^T(0)P_0^{-1}\tilde{\theta}(0). \end{aligned}$$

Hence

$$|\tilde{\theta}(k)|^2 \leq \frac{\tilde{\theta}^T(0)P_0^{-1}\tilde{\theta}(0)}{\lambda_{\min}(P_0^{-1})} \leq \frac{\lambda_{\max}(P_0^{-1})}{\lambda_{\min}(P_0^{-1})}|\tilde{\theta}(0)|^2,$$

which establishes (i). Taking summation on both sides of (4.45) from 1 to N, we have

$$V(N) - V(0) = -\sum_{k=1}^{N}\varepsilon^2(k)m^2(k)\frac{m^2(k)}{m^2(k) + \phi^T(k)P(k-1)\phi(k)}$$

$$= -\sum_{k=1}^{N} \frac{\varepsilon^2(k)m^2(k)}{1+\bar{\phi}^T P(k-1)\bar{\phi}(k)},$$

where $\bar{\phi}(k) \stackrel{\Delta}{=} \frac{\phi(k)}{m(k)}$. Since $\lambda_{\max}(P(k-1)) \leq \frac{1}{\lambda_{\min}(P^{-1}(k-1))}$ and $\lambda_{\min}(P^{-1}(k-1)) \geq \lambda_{\min}(P_0^{-1})$, we have $\frac{1}{1+\bar{\phi}^T(k)P(k-1)\bar{\phi}(k)} \geq \frac{1}{1+c_0}$, where $c_0 = \sup_k |\bar{\phi}(k)|^2 \frac{1}{\lambda_{\min}(P_0^{-1})}$ is a finite constant due to $\bar{\phi}(k) \in \ell_\infty$. Hence

$$\frac{1}{1+c_0}\sum_{k=1}^{N} \varepsilon^2(k)m^2(k) \leq V(0) - V(N).$$

Since $\lim_{x\to\infty} V(N) = V_\infty$ is finite, it follows that $\varepsilon(k)m(k) \in \ell_2$ and $\varepsilon(k)m(k) \to 0$ as $k \to \infty$. Since $m(k) \geq c > 0$ we also have $\varepsilon(k) \in \ell_2$ and $\varepsilon(k) \to 0$ as $k \to \infty$. Now from $\varepsilon(k)\phi(k) = \varepsilon(k)m(k)\bar{\phi}(k)$, where $\bar{\phi}(k) \in \ell_\infty$, we have $\varepsilon(k)\phi(k) \in \ell_2$ and $|\varepsilon(k)\phi(k)| \to 0$ as $k \to \infty$. From (4.43) we have

$$\theta(k) - \theta(k-1) = P(k)\varepsilon(k)\phi(k).$$

Since $\lambda_{\max}(P(k)) \leq \lambda_{\max}(P(k-1)) \leq \cdots \leq \lambda_{\max}(P_0)$, we have $|\theta(k)-\theta(k-1)| \leq |\varepsilon(k)\phi(k)|\lambda_{\max}(P_0)$, which implies that $|\theta(k) - \theta(k-1)| \in \ell_2$ due to $\varepsilon(k)\phi(k) \in \ell_2$ and $|\theta(k) - \theta(k-1)| \to 0$ as $k \to \infty$. Now

$$|\theta(k)-\theta(k-N)|^2 \leq |\theta(k)-\theta(k-1)|^2+|\theta(k-1)-\theta(k-2)|^2+\cdots+|\theta(k-N+1)-\theta(k-N)|^2,$$

which implies $|\theta(k) - \theta(k-N)| \in \ell_2$ and $|\theta(k) - \theta(k-N)| \to 0$ as $k \to \infty$ for any finite N.

(v) We have already established that

$$V(k) = \tilde{\theta}^T(k)P^{-1}(k)\tilde{\theta}(k)$$

is a bounded sequence, which means that if $\lambda_{\min}(P^{-1}(k)) \to \infty$ as $k \to \infty$, then for $V(k)$ to be bounded we should have $\tilde{\theta}(k) \to 0$ as $k \to \infty$. Now

$$P^{-1}(k) = P^{-1}(k-1) + \frac{\phi(k)\phi^T(k)}{m^2(k)}$$
$$= P^{-1}(0) + \sum_{j=1}^{k} \frac{\phi(j)\phi^T(j)}{m^2(j)}, \quad k = 1, 2, \ldots.$$

Therefore, if $\lim_{k\to\infty} \lambda_{\min}(\sum_{j=1}^{k} \frac{\phi(j)\phi^T(j)}{m^2(j)}) = \infty$, then $\lambda_{\min}(P^{-1}(k)) \to \infty$ as $k \to \infty$, which in turn implies that $\tilde{\theta}(k) \to 0$ and that $\theta(k) \to \theta^*$ as $k \to \infty$. We will now show that if $\frac{\phi(k)}{m(k)}$ is PE, i.e., satisfies $\sum_{j=k}^{k+l-1} \frac{\phi(j)\phi^T(j)}{m^2(j)} \geq \alpha_0 l I$ for some $\alpha_0 > 0, l > 1$ and $\forall k \geq 0$, then $\lim_{k\to\infty} \lambda_{\min}(\sum_{j=1}^{k} \frac{\phi(j)\phi^T(j)}{m^2(j)}) = \infty$ is satisfied, which, as we showed above, implies $\theta(k) \to \theta^*$ as $k \to \infty$. Define

$$S_\phi(k,l) \stackrel{\Delta}{=} \sum_{j=k}^{k+l-1} \frac{\phi(j)\phi^T(j)}{m^2(j)} \geq \alpha_0 l I$$

and

$$S(nl) \triangleq \sum_{j=1}^{nl} \frac{\phi(j)\phi^T(j)}{m^2(j)}$$

for any integer $n, l \geq 1$. Then

$$S(nl) = \sum_{i=0}^{n-1} S_\phi(il+1, l),$$

which implies that

$$\lambda_{\min}(S(nl)) \geq n\alpha_0 l.$$

Therefore,

$$\lim_{n\to\infty} \lambda_{\min}(S(nl)) = \lim_{N\to\infty} \lambda_{\min}\left(\sum_{j=1}^{N} \frac{\phi(j)\phi^T(j)}{m^2(j)}\right) = \infty,$$

which, as we have shown above, implies $\theta(k) \to \theta^*$ as $k \to \infty$.

(vi) The proof follows by using Theorem 4.4.2. $\square$

4.7 Modified LS Algorithms

Instead of the cost (4.32), we may choose the following cost that involves the weighting parameter $a(k)$:

$$\frac{1}{2}\sum_{k=1}^{N} a(k)\frac{(z(k) - \theta^T\phi(k))^2}{m^2(k)} + \frac{1}{2}(\theta - \theta(0))^T P_0^{-1}(\theta - \theta(0)), \tag{4.48}$$

where P_0 is a positive definite matrix and $a(k)$ is a nonnegative sequence of weighting coefficients. The LS algorithm takes the form

$$\begin{aligned} P(k) &= P(k-1) - \frac{a(k)P(k-1)\phi(k)\phi^T(k)P(k-1)}{m^2(k) + a(k)\phi^T(k)P(k-1)\phi(k)}, \\ \varepsilon(k) &= \frac{z(k) - \theta^T(k-1)\phi(k)}{m^2(k)}, \end{aligned} \tag{4.49}$$

$$\theta(k) = \theta(k-1) + \sqrt{a(k)}P(k)\phi(k)\varepsilon(k). \tag{4.50}$$

As before, using the fact that

$$P(k)\phi(k) = \sqrt{a(k)}\frac{P(k-1)\phi(k)m^2(k)}{m^2(k) + a(k)\phi^T(k)P(k-1)\phi(k)}, \tag{4.51}$$

the equation for $\theta(k)$ may also be written in the form

$$\theta(k) = \theta(k-1) + \frac{a(k)P(k-1)\phi(k)[z(k) - \theta^T(k-1)\phi(k)]}{m^2(k) + a(k)\phi^T(k)P(k-1)\phi(k)}. \tag{4.52}$$

Table 4.1. *Static data weighting factors for various weighted LS algorithms.*

Pure LS	$a(k) = 1$
Constant weighting	$a(k) = \text{constant} > 0$
Robust weighting	$a(k) = \begin{cases} a_1 & \text{if } \phi^T(k)P(k-1)\phi(k) \geq \delta, \\ a_2 & \text{otherwise}, \end{cases} \quad a_1 \gg a_2 > 0,$
Modified weighting	$a(k) = \begin{cases} \frac{\phi^T(k)P(k-1)\phi(k)}{\phi^T(k)\phi(k)} & \text{if } \phi^T(k)\phi(k) \geq \delta > 0, \\ l & \text{otherwise}, \end{cases}$ where l, δ are arbitrary positive constants

Remark 4.7.1 We should note that the derivation of (4.49)–(4.52) follows from that of (4.43), (4.45) if in the cost (4.32) we replace $z(k)$, $\phi(k)$ with $\bar{z}(k) = \sqrt{a(k)}z(k)$, $\bar{\phi}(k) = \sqrt{a(k)}\phi(k)$, respectively.

The LS algorithm described by (4.49), (4.50) or (4.49), (4.52) is referred to as the *weighted LS algorithm*. The standard LS algorithm of section 4.5 is a special case of the weighted LS algorithm for $a(k) = 1$. Different choices for $a(k)$ will lead to different algorithms with possibly different properties. Table 4.1 presents some of the choices for $a(k)$ presented in the literature [41, 46, 88, 111].

A more general class of LS algorithms can be obtained by using the following cost function:

$$S_N(\theta) = \beta(N-1)S_{N-1}(\theta) + \frac{[z(N) - \theta^T\phi(N)]^2}{m^2(N)}, \tag{4.53}$$

where

$$S_N(\theta) = \frac{1}{2}\sum_{k=1}^{N} a(k)\frac{(z(k) - \theta^T\phi(k))^2}{m^2(k)} + \frac{1}{2}(\theta - \theta(0))^T P_0^{-1}(\theta - \theta(0)) \tag{4.54}$$

is the weighted LS cost and $\beta(N) \in (0, 1)$ is a design sequence. In this case the LS algorithm becomes

$$P(k) = \frac{1}{\beta(k)}\left[P(k-1) - \frac{a(k)P(k-1)\phi(k)\phi^T(k)P(k-1)}{m^2(k)\beta(k) + a(k)\phi^T(k)P(k-1)\phi(k)}\right], \tag{4.55}$$

$$P(0) = P_0 = P_0^T > 0,$$

$$\theta(k) = \theta(k-1) + \sqrt{a(k)}P(k)\phi(k)\varepsilon(k). \tag{4.56}$$

Equation (4.56) may be also written in the form

$$\theta(k) = \theta(k-1) + \frac{a(k)P(k-1)\phi(k)[z(k) - \theta^T(k-1)\phi(k)]}{m^2(k)\beta(k) + a(k)\phi^T(k)P(k-1)\phi(k)}. \tag{4.57}$$

The above LS algorithm is referred to as the *LS with dynamic data weighting*. Some choices for $\beta(k)$ that are presented in the literature are in [41, 46, 88, 111]. Two of these choices are listed in Table 4.2

Table 4.2. *Various dynamic data weighting factors.*

Constant forgetting	$\beta(k) = \beta$, $0 < \beta < 1$
Startup forgetting	$\beta(k) = \beta_1 \beta(k-1) + 1 - \beta_1$, $0 < \beta(0), \beta_1 < 1$

Table 4.3. *Covariance resetting/modification techniques.*

Least-size resetting	$P(k) = p_0 I$ when $\mathrm{tr}[P(k)] < p_{\min}$, where $p_0, p_{\min} > 0$ are design constants
Periodic resetting	$P(k) = p_0 I$ when $k = nT$, for $n = 1, 2, \dots$, where $T > 0$ is the period for resetting
Scheduled resetting	$P(k) = p_i I$ when $k = k_I$ for $i = 1, 2, \dots$, where $0 < k_1 < k_2 < \cdots$, $p_i > 0\ \forall i$
Covariance modification	$P(k) = P(k) + Q$ whenever $\mathrm{tr}[P(k)] < p_{\max}$, where $p_{\max} > 0$, $Q = Q^T > 0$ are design terms

The problem of the covariance matrix $P(k)$ becoming small in certain directions exists as in the continuous-time case. Table 4.3 presents some of the modifications suggested in the literature for preventing $P(k)$ from becoming small.

The above modifications give rise to a wide range of algorithms where the user can tune the various design parameters described in the modifications as well as the adaptive gains in order to obtain improved convergence rates and results. Unfortunately, there is no systematic procedure for choosing the various design parameters. Computer simulations such as the use of the Adaptive Control Toolbox may help in tuning the adaptive systems for best performance.

4.8 Parameter Identification Based on DPM

See the web resource [94].

4.9 Parameter Identification Based on B-SPM

See the web resource [94].

4.10 Parameter Projection

In many parameter identification problems, we have some a priori knowledge as to where the unknown parameter θ^* is located in $\mathcal{R}^n$. This knowledge could be of the upper and/or lower bounds for each element of θ^*, upper bound for the norm of θ^*, etc. We apply the gradient projection method described in section A.10 to modify the adaptive laws of the

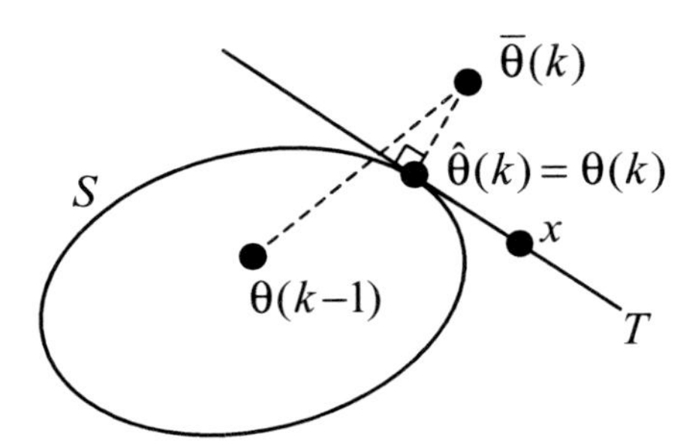

Figure 4.2. *Discrete-time parameter projection.*

previous sections as follows. We replace the gradient algorithm

$$\varepsilon(k) = \frac{z(k) - \theta^T(k-1)\phi(k)}{m^2(k)},$$
$$\theta(k) = \theta(k-1) + \Gamma\varepsilon(k)\phi(k)$$

with

$$\begin{aligned}\varepsilon(k) &= \frac{z(k) - \theta^T(k-1)\phi(k)}{m^2(k)},\\ \bar{\theta}(k) &= \theta(k-1) + \Gamma\varepsilon(k)\phi(k),\\ \theta(k) &= \Pr(\bar{\theta}(k)),\end{aligned} \tag{4.58}$$

where $\Pr(\cdot)$ is the projection operator and the rest of the signals are as defined in (4.1), (4.2). For simplicity, let us take $\Gamma = \alpha I$, where $\alpha > 0$ is a constant scalar. Let

$$S = \{\theta \in \mathcal{R}^n \mid g(\theta) \leq 0\} \tag{4.59}$$

be the convex set with a smooth function $g(\theta)$, where θ should be constrained to belong to. We assume that $\theta^* \in S$. If at instant $k-1$ we have $\theta(k-1) \in S$, it is possible that at instant k, $\bar{\theta}(k)$ is outside S, as shown in Figure 4.2.

In this case, we take $\theta(k)$ as the closest point to $\bar{\theta}(k)$ that satisfies $g(\theta) = 0$, denoted by $\hat{\theta}(k)$. This implies that $\hat{\theta}(k)$ is the orthogonal projection of $\bar{\theta}(k)$ on the boundary of S; i.e., the projection operator $\Pr(\cdot)$ in (4.58) is defined as

$$\theta(k) = \Pr(\bar{\theta}(k)) = \begin{cases}\bar{\theta}(k) & \text{if } \bar{\theta}(k) \in S,\\ \hat{\theta}(k) = \perp\text{proj of } \bar{\theta}(k) \text{ on } S & \text{if } \bar{\theta}(k) \notin S.\end{cases} \tag{4.60}$$

If T is the tangent plane to S at $\hat{\theta}(k)$, then, for any point x on T, we have

$$(x - \hat{\theta}(k))^T(\bar{\theta}(k) - \hat{\theta}(k)) = 0$$

due to orthogonality. Furthermore, for any point y on the left side of T (the same side as S), we have

$$(y - \hat{\theta}(k))^T(\bar{\theta}(k) - \hat{\theta}(k)) \leq 0.$$

Let us take $y = \theta^*$. Then

$$(\hat{\theta}(k) - \theta^*)^T(\bar{\theta}(k) - \hat{\theta}(k)) \geq 0$$

or

$$\tilde{\hat{\theta}}^T(k)(\bar{\theta}(k) - \hat{\theta}(k)) \geq 0, \tag{4.61}$$

which also implies $\tilde{\hat{\theta}}^T(k)\tilde{\hat{\theta}}(k) \leq \tilde{\bar{\theta}}^T(k)\tilde{\bar{\theta}}(k)$, where the notation $\tilde{x} = x - \theta^*$ is used. These properties of the orthogonal projection will be used in the analysis. Before we analyze the effect of projection, let us demonstrate the design of projection algorithms for a number of examples.

Example 4.10.1 Let us assume that we have the a priori information that $|\theta^*| \leq M$ for some known $M > 0$. In this case we can choose

$$S = \{\theta \in \mathcal{R}^n \,|\, \theta^T\theta - M^2 \leq 0\}.$$

Then, whenever $|\bar{\theta}(k)| > M$, the orthogonal projection will produce

$$\theta(k) = \frac{M}{|\bar{\theta}(k)|}\bar{\theta}(k); \tag{4.62}$$

i.e., $\bar{\theta}(k)$ will be projected to the boundary of S so that $|\theta(k)| = \frac{M}{|\bar{\theta}(k)|}|\bar{\theta}(k)| = M$. In this case, since S is a sphere, the vector $\theta(k)$ has the same direction as $\bar{\theta}(k)$, but its magnitude is reduced to $|\theta(k)| = M$. The gradient algorithm with projection in this case may be written as

$$\begin{aligned} \varepsilon(k) &= \frac{z(k) - \theta^T(k-1)\phi(k)}{m^2(k)}, \\ \bar{\theta}(k) &= \theta(k-1) + \alpha\varepsilon(k)\phi(k), \\ \theta(k) &= \bar{\theta}(k)\min\left(1, \frac{M}{|\bar{\theta}(k)|}\right). \end{aligned} \tag{4.63}$$

■

Example 4.10.2 Let us assume that we have the a priori information that $|\theta^* - \theta_0| \leq M$ for some known $M, \theta_0 > 0$. We choose

$$S = \{\theta \in \mathcal{R}^n | (\theta - \theta_0)^T(\theta - \theta_0) - M^2 \leq 0\},$$

and the projection operator becomes

$$\theta(k) = \Pr(\bar{\theta}(k)) = \begin{cases} \bar{\theta}(k) & \text{if } |\bar{\theta}(k) - \theta_0| \leq M, \\ \theta_0 + \frac{M(\bar{\theta}(k) - \theta_0)}{|\bar{\theta}(k) - \theta_0|} & \text{if } |\bar{\theta}(k) - \theta_0| > M, \end{cases} \tag{4.64}$$

which may be written in the compact form

$$\theta(k) = \Pr(\bar{\theta}(k)) = \theta_0 + (\bar{\theta}(k) - \theta_0)\min\left(1, \frac{M}{|\bar{\theta}(k) - \theta_0|}\right).$$

■

Example 4.10.3 Let us assume that for each element θ_i^* of θ^* we have the a priori information that

$$L_i \le \theta_i^* \le U_i \quad \text{for } i = 1, \dots, n$$

for some known numbers L_i, U_i. In this case, the projection set is simply a cross product of n closed sets on the real line, i.e.,

$$S = \{\theta \in \mathcal{R}^n | L_i \le \theta_i \le U_i, i = 1, \dots, n\}.$$

For this set, it is easy to establish that the orthogonal projection on S is given by

$$\theta_i(k) = \Pr(\bar{\theta}_i(k)) = \begin{cases} \bar{\theta}_i(k) & \text{if } L_i \le \bar{\theta}_i(k) \le U_i, \\ L_i & \text{if } \bar{\theta}_i(k) < L_i, \\ U_i & \text{if } \bar{\theta}_i(k) > U_i. \end{cases}$$

∎

Theorem 4.10.4. *Consider the gradient algorithm* (4.58) *with* $\Gamma = \alpha I$, $0 < \alpha < 2$, *and the parameter projection* (4.60) *to the convex set* S *given by* (4.59). *The parameter estimates* $\theta(k) \in S$ $\forall k$, *provided that* $\theta^*, \theta(0) \in S$. *Furthermore,* (4.58)–(4.60) *guarantees all the properties of the corresponding gradient algorithm without parameter projection.*

Proof. If $\theta(0) \in S$, then (4.60) guarantees that $\theta(k) \in S$ $\forall k$. The gradient algorithm may be written as

$$\theta(k) = \theta(k-1) + \alpha\varepsilon(k)\phi(k) + g(k),$$

where

$$g(k) = \begin{cases} 0 & \text{if } \bar{\theta}(k) \in S, \\ \hat{\theta}(k) - \bar{\theta}(k) & \text{if } \bar{\theta}(k) \notin S, \end{cases}$$

$\bar{\theta}(k) = \theta(k-1) + \alpha\varepsilon(k)\phi(k)$, and $\hat{\theta}(k)$ is the orthogonal projection of $\bar{\theta}(k)$ to S. Let us consider

$$V(k) = \frac{\tilde{\theta}^T(k)\tilde{\theta}(k)}{2}.$$

We have

$$\begin{aligned} \Delta V(k) &\triangleq V(k) - V(k-1) \\ &= \frac{1}{2}(\tilde{\theta}(k-1) + \alpha\varepsilon(k)\phi(k) + g(k))^T(\tilde{\theta}(k-1) + \alpha\varepsilon(k)\phi(k) + g(k)) \\ &\quad - \frac{1}{2}\tilde{\theta}^T(k-1)\tilde{\theta}(k-1). \end{aligned}$$

We should, therefore, obtain the same expression as in the unconstrained case plus the terms due to g, i.e.,

$$\Delta V(k) = -\alpha\varepsilon^2(k)m^2(k)\left(1 - \frac{\alpha}{2}\frac{\phi^T(k)\phi(k)}{m^2(k)}\right) + \frac{1}{2}g^T(k)g(k) + g^T(k)(\bar{\theta}(k) - \theta^*).$$

Since $\frac{\phi^T(k)\phi(k)}{m^2(k)} \le 1$ and $0 < \alpha < 2$, we have

$$\Delta V(k) \le -c_0\varepsilon^2(k)m^2(k) + \frac{1}{2}g^T(k)g(k) + g^T(k)(\bar{\theta}(k) - \theta^*)$$

for some $c_0 > 0$. If we establish that $\frac{1}{2}g^T(k)g(k) + g^T(k)(\bar{\theta}(k) - \theta^*) \leq 0$, then the effect of parameter projection could only make $\Delta V(k)$ more negative and

$$\Delta V(k) \leq -c_0\varepsilon^2(k)m^2(k),$$

which is the same inequality we have in the case of unconstrained parameter estimation. For the case $\bar{\theta}(k) \notin S$, let us drop the argument k for clarity of presentation and write

$$g^T g + 2g^T(\bar{\theta} - \theta^*) = (\hat{\theta} - \bar{\theta})^T(\hat{\theta} - \bar{\theta}) + 2(\hat{\theta} - \bar{\theta})^T(\bar{\theta} - \theta^*).$$

After some manipulation of the right-hand side, we obtain

$$g^T g + 2g^T(\bar{\theta} - \theta^*) = -2(\hat{\theta} - \theta^*)^T(\bar{\theta} - \hat{\theta}) - |\hat{\theta} - \bar{\theta}|^2.$$

Using the projection property (4.61), we have that $(\hat{\theta} - \theta^*)^T(\bar{\theta} - \hat{\theta}) \geq 0$ and therefore

$$g^T g + 2g^T(\bar{\theta} - \theta^*) \leq 0 \quad \forall k.$$

Hence the algorithm (4.60) has the same properties as that without projection, i.e., $g = 0$. □

For a general adaptation gain Γ as well as in the case of LS where the adaptive gain is $P(k)$, a transformation of coordinates has to be performed in order to guarantee the properties of Theorem 4.10.4. In this case, we use the transformation

$$\bar{\theta}'(k) = P^{-1/2}\bar{\theta}(k),$$

where

$$P^{-1} = (P^{-1/2})^T P^{-1/2}$$

and P is equal to Γ in the case of the gradient algorithm and to $P(k)$ in the case of LS.

The set S is also transformed to another set S' such that if $\bar{\theta}(k) \in S$, then $\bar{\theta}'(k) \in S'$. The parameter projection in this case becomes

$$\begin{aligned} \bar{\theta}(k) &= \theta(k-1) + P\varepsilon(k)\phi(k), \\ \bar{\theta}'(k) &= P^{-1/2}\bar{\theta}(k), \\ \theta'(k) &= \begin{cases} \bar{\theta}'(k) & \text{if } \bar{\theta}'(k) \in S', \\ \hat{\theta}'(k) = \perp\text{proj of } \bar{\theta}'(k) \text{ on } S' & \text{if } \bar{\theta}'(k) \notin S', \end{cases} \\ \theta(k) &= P^{1/2}\theta'(k), \end{aligned} \tag{4.65}$$

where $P = \Gamma = \Gamma^T > 0$ is a constant in the case of the gradient algorithm and $P = P(k)$ is the covariance matrix in the case of the LS algorithm.

The problem with this more general parameter projection algorithm is computational, since the matrix P has to be inverted to obtain P^{-1} (in the case of LS, $P^{-1}(k)$ may be generated recursively) and decomposed into $(P^{-1/2})^T P^{-1/2}$. In the case of LS, this decomposition has to be done at each instant k.

The stability analysis of (4.65) can be found in [46, 88, 112].

4.11 Robust Parameter Identification

In the presence of modeling errors such as unmodeled dynamics, bounded disturbances, small unknown time delays, etc., the SPM

$$z = \theta^{*T}\phi,$$

where z, ϕ are known signals, is no longer a valid model in general, no matter how we manipulate the signals and parameterize the system. In the presence of modeling errors, the SPM is usually of the form

$$z = \theta^{*T}\phi + \eta,$$

where η is the modeling error term, unknown in general but hopefully small compared to the elements of the regressor ϕ.

As in the continuous-time case, the stability of an adaptive law (such as those developed in previous sections) designed for $\eta = 0$ but applied to the SPM with $\eta \neq 0$ cannot be guaranteed no matter how small $\eta \neq 0$ is, unless ϕ is PE [41, 88, 111]. The robustness of the discrete-time adaptive laws developed in the previous sections can be established using two different approaches depending on the situation we encounter.

In approach I, if z, ϕ, η are bounded and the regressor ϕ is guaranteed to be PE, then the effect of the modeling error η is a possible nonzero parameter estimation error at steady state whose size depends on the size of η. The problem we have to address in this case is how to guarantee that ϕ is PE. As in the continuous-time case, if the elements of θ^* are the parameters of an LTI plant, then ϕ can be guaranteed to be PE by choosing u to be dominantly rich.

In approach II, which is of interest in adaptive control, z, ϕ, or η cannot be assumed to be bounded and ϕ cannot be guaranteed to be PE. In this case, the discrete-time adaptive laws have to be modified to counteract possible instabilities. The two modifications used in this case are (i) getting rid of the pure delay action (the equivalent of pure integral action in the continuous-time case) of the adaptive law and (ii) using dynamic normalization to bound the modeling errors from above.

4.11.1 Dominantly Rich Excitation

Let us consider the plant (4.14) but with modeling errors

$$y(k) = -\sum_{j=1}^{n} a_j y(k-j) + \sum_{j=0}^{m} b_j u(k-j-d) + \eta(k), \tag{4.66}$$

where $\eta(k)$ could be of the general form

$$\eta(k) = \mu(\Delta_1(z)y(k) + \Delta_2(z)u(k)) + d_0(k), \tag{4.67}$$

$\Delta_1(z)$, $\Delta_2(z)$ are proper transfer functions with poles in $|z| < 1$, μ is a small parameter that rates the size of the dynamic uncertainty, and $d_0(k)$ is a bounded disturbance.

Following the procedure in section 4.3, we obtain the SPM

$$z(k) = \theta^{*T}\phi(k) + \eta(k), \tag{4.68}$$

where

$$z(k) = y(k),$$
$$\phi(k) = [u(k-d), \ldots, u(k-d-m), -y(k-1), \ldots, -y(k-n)]^T,$$
$$\theta^* = [b_0, b_1, \ldots, b_m, a_1, a_2, \ldots, a_n]^T.$$

Let us consider the gradient algorithm

$$\begin{aligned} \varepsilon(k) &= \frac{z(k) - \theta^T(k-1)\phi(k)}{m^2(k)}, \\ \theta(k) &= \theta(k-1) + \Gamma\varepsilon(k)\phi(k), \quad \theta(0) = \theta_0. \end{aligned} \tag{4.69}$$

We consider the case where the plant is stable and the plant input $u(k)$ is bounded. In this case, $\phi(k)$, $z(k)$ are bounded and the normalizing signal could be designed to satisfy the conditions of Lemma 4.1.1. One choice for $m^2(k)$ is $m^2(k) = 1 + \phi^T(k)\phi(k)$. Using $\tilde{\theta}(k) = \theta(k) - \theta^*$, we express (4.69) as

$$\tilde{\theta}(k) = A(k)\tilde{\theta}(k-1) + \Gamma\frac{\phi(k)\eta(k)}{m^2(k)}, \tag{4.70}$$

where

$$A(k) = I - \frac{\Gamma\phi(k)\phi^T(k)}{m^2(k)}.$$

If ϕ is PE, then the homogenous part of (4.70), i.e.,

$$\tilde{\theta}(k) = A(k)\tilde{\theta}(k-1),$$

is e.s. as established by Lemma 4.1.1, provided that $\|\Gamma\| < 2$. This implies that the bounded input $\frac{\Gamma\phi(k)\eta(k)}{m^2(k)}$ will produce a bounded state $\tilde{\theta}(k)$. In this case, we can show that

$$|\tilde{\theta}(k)| \le c_0\eta_0 + \varepsilon_k \quad \forall k,$$

where c_0 is a constant that depends on the rate of convergence, η_0 is an upper bound of $\frac{|\eta(k)|}{m(k)} \le \eta_0$, and ε_k is a term exponentially decaying to zero.

It is clear that the size of the parameter error at steady state is of the order of the modeling error term η_0. The smaller $\eta(k)$ is, the closer $\theta(k)$ will remain to θ^* at steady state. The problem is how to choose the input $u(k)$ so that $\phi(k)$ is PE despite the presence of $\eta(k)$. In this case, we can express

$$\phi(k) = H_0(z)u(k) + \mu H_1(z)u(k) + B_1(z)d_0(k), \tag{4.71}$$

where $H_0(z)$, $\mu H_1(z)$, $B_1(z)$ are transfer matrices with proper elements and poles in $|z| < 1$. It is clear that the modeling error term $\mu H_1(z)u(k) + B_1(z)d_0(k)$ could destroy the PE property of $\phi(k)$ if $u(k)$ is chosen to be just sufficiently rich of order $n+m+1$, as done in the case of $\eta(k) = 0$.

Lemma 4.11.1. *Let $u(k)$ be dominantly rich; i.e., it is sufficiently rich of order $n+m+1$ and achieves its richness with frequencies ω_i, $i = 1, 2, \ldots, N$, where $N \ge \frac{n+m+1}{2}$, $|\omega_i| < O(\frac{1}{\mu})$, $|\omega_i - \omega_j| > O(\mu)$ for $i \ne j$, $|u(k)| > O(\mu) + O(d_0)$, and $H_0(z)$, $H_1(z)$ satisfy the following conditions:*

(a) *The vectors* $H_0(e^{j\omega_1}), \ldots, H_0(e^{j\omega_{\bar{n}}})$, *where* $\bar{n} = n + m + 1$, *are linearly independent for all possible* $\omega_1, \ldots, \omega_{\bar{n}} \in [0, 2\pi)$ *and* $\omega_i \neq \omega_j$, $i \neq j$.

(b) $|\det(\bar{H})| > O(\mu)$, *where* $\bar{H} \triangleq [H_0(e^{j\omega_1}), \ldots, H_0(e^{j\omega_{\bar{n}}})]$ *for* $|\omega_i| < O(\frac{1}{\mu})$.

(c) $|H_1(e^{j\omega})| \leq c$ *for some constant* c *independent of* μ *and* $\forall \omega \in \mathcal{R}$.

Then there exists $\mu^* > 0$ *such that* $\forall \mu \in [0, \mu^*)$, ϕ *is PE of order* $n + m + 1$ *with level of excitation* $\alpha_0 > O(\mu)$.

Proof. The proof is similar to that of Lemma 3.11.6 in the continuous-time case and is omitted. □

Conditions (a), (b) are satisfied if there are no zero-pole cancellations in the modeled part of the plant transfer function $\frac{B(z)}{A(z)}$ and zeros are not within $O(\mu)$ distance from poles.

Lemma 4.11.1 guarantees that if the plant is stable and the input is chosen to be dominantly rich, i.e., rich but with frequencies away from the range of the modeling error, then the parameter error at steady state is of the order of the modeling error.

4.11.2 Robustness Modifications

In adaptive control, we cannot assume that $z(k)$, $\phi(k)$, and $\eta(k)$ in the SPM (4.68) are bounded a priori. The adaptive law has to guarantee bounded estimates even when $z(k)$, $\phi(k)$, $\eta(k)$ cannot be assumed to be bounded a priori. Furthermore, $\phi(k)$ cannot be assumed to be PE since such property can no longer be guaranteed via the choice of $u(k)$, which in adaptive control is a result of feedback, and may not be rich in frequencies. Consequently, the adaptive laws of the previous sections have to be modified to be robust with respect to modeling errors.

Let us consider the SPM (4.68), i.e.,

$$z(k) = \theta^{*T}\phi(k) + \eta(k),$$

where $\eta(k)$ is the modeling error. The robustness of the adaptive laws developed for $\eta(k) = 0$ can be established using two modifications:

(i) Design the normalizing signal $m(k)$ to bound $\eta(k)$ from above in addition to $\phi(k)$.

(ii) Modify the adaptive law using the equivalent of σ-modification, dead zone, projection, etc.

We demonstrate the design of robust adaptive laws for the SPM

$$z(k) = \theta^{*T}\phi(k) + \eta(k),$$
$$\eta(k) = \mu\Delta_1(z)u(k) + \mu\Delta_2(z)y(k)$$

and the gradient algorithm (4.69). The modeling error perturbations $\mu\Delta_1(z)$, $\mu\Delta_2(z)$ are assumed to be analytic in $|z| \geq \sqrt{p_0}$ for some $1 > p_0 \geq 0$, which implies that $\mu\Delta_1(zp_0^{1/2})$,

$\mu\Delta_2(zp_0^{1/2})$ are analytic in $|z| \geq 1$ or that the magnitude of each pole of $\mu\Delta_1(z)$, $\mu\Delta_2(z)$ is less than $\sqrt{p_0}$.

We can establish using Lemma A.12.33 that

$$\frac{|\eta(k)|}{m(k)} \leq c\mu, \qquad \frac{|\phi(k)|}{m(k)} \leq c$$

for some generic constant $c > 0$ if $m(k)$ is designed as

$$\begin{aligned} m^2(k) &= 1 + \phi^T(k)\phi(k) + m_s(k), \\ m_s(k) &= \delta_0 m_s(k-1) + |u(k)|^2 + |y(k)|^2 \end{aligned} \tag{4.72}$$

for some δ_0, $1 > \delta_0 > p_0$.

Fixed σ-Modification [50]

In this case, the gradient algorithm is modified as

$$\begin{aligned} \theta(k) &= (1-\sigma)\theta(k-1) + \Gamma\varepsilon(k)\phi(k), \\ \varepsilon(k) &= \frac{z(k) - \theta^T(k-1)\phi(k)}{m^2(k)}, \end{aligned} \tag{4.73}$$

where $0 < \sigma < 1$ is a small fixed constant and $m^2(k)$ is given by (4.72).

Theorem 4.11.2. *The gradient algorithm* (4.73) *with the fixed σ-modification, where σ, Γ are chosen to satisfy* $0 < \sigma < 1$, $1 - \sigma - \frac{\lambda_{\max}(\Gamma)}{2} > 0$, *guarantees the following:*

(i) $\theta(k), \varepsilon(k), \varepsilon(k)m(k) \in \ell_\infty$.

(ii) $\varepsilon(k), \varepsilon(k)m(k), |\theta(k) - \theta(k-1)| \in \mathcal{S}(\sigma + \eta_0^2)$.[6]

Proof. Defining $\tilde{\theta}(k) = \theta(k) - \theta^*$, we express (4.73) as

$$\begin{aligned} \tilde{\theta}(k) &= \alpha_0\tilde{\theta}(k-1) + \Gamma\varepsilon(k)\phi(k) + \sigma\theta^*, \\ \varepsilon(k) &= -\frac{\tilde{\theta}^T(k-1)\phi(k)}{m^2(k)} + \frac{\eta(k)}{m^2(k)}, \end{aligned}$$

where $\alpha_0 = 1 - \sigma$. We choose

$$V(k) = \frac{\tilde{\theta}^T(k)\Gamma^{-1}\tilde{\theta}(k)}{2}.$$

Then

$$\Delta V(k) = V(k) - V(k-1) = \left(\frac{\alpha_0^2 - 1}{2}\right)\tilde{\theta}^T(k-1)\Gamma^{-1}\tilde{\theta}(k-1) + \alpha_0\tilde{\theta}^T(k-1)\varepsilon(k)\phi(k)$$

[6]*Definition*: The sequence vector $x(k) \in S(\mu)$ if it satisfies $\sum_k^{k+N} x^T(k)x(k) \leq c_0\mu N + c_1$ for all positive numbers k, a given constant $\mu \geq 0$, and some positive number N, where $c_0, c_1 \geq 0$. We say that $x(k)$ is μ-small in the mean square sense if $x(k) \in X(\mu)$.

$$+ \alpha_0\sigma\tilde{\theta}^T(k-1)\Gamma^{-1}\theta^* + \frac{1}{2}\varepsilon^2(k)\phi^T(k)\Gamma\phi(k) + \sigma\varepsilon(k)\phi^T(k)\theta^*$$
$$+ \frac{1}{2}\sigma^2\theta^{*T}\Gamma^{-1}\theta^*.$$

Substituting for $\alpha_0 = 1-\sigma \leq 1$ and $\tilde{\theta}^T(k-1)\phi(k) = -\varepsilon(k)m^2(k)+\eta(k)$, we obtain

$$\Delta V(k) \leq -\frac{\sigma(2-\sigma)}{2}\tilde{\theta}^T(k-1)\Gamma^{-1}\tilde{\theta}(k-1) - \alpha_0\varepsilon^2(k)m^2(k) + \alpha_0\varepsilon(k)\eta(k)$$
$$+ \frac{1}{2}\varepsilon^2(k)m^2(k)\lambda_{\max}(\Gamma) + \sigma|\theta^*||\tilde{\theta}(k-1)|\|\Gamma^{-1}\| + \sigma|\varepsilon(k)m(k)||\theta^*|$$
$$+ \frac{\sigma^2}{2}|\theta^*|^2\|\Gamma^{-1}\|.$$

Let us denote all positive nonzero constants by the generic symbol c and use

$$\varepsilon(k)\eta(k) \leq |\varepsilon(k)m(k)|\eta_0,$$

where η_0 is the upper bound of $\frac{|\eta(k)|}{m(k)}$, to obtain

$$\Delta V(k) \leq -\frac{\sigma(2-\sigma)}{2}\tilde{\theta}^T(k-1)\Gamma^{-1}\tilde{\theta}(k-1) - \varepsilon^2(k)m^2(k)\left[1-\sigma-\frac{\lambda_{\max}(\Gamma)}{2}\right]$$
$$+ c(1-\sigma)\sigma|\tilde{\theta}(k-1)| + c(\sigma+\eta_0)|\varepsilon(k)m(k)| + \sigma^2 c.$$

Using the condition $\alpha_1 = 1-\sigma-\frac{\lambda_{\max}(\Gamma)}{2} > c > 0$ and the inequality

$$-ax^2 + bx \leq -\frac{a}{2}x^2 + \frac{b^2}{2a},$$

we have

$$-\frac{\sigma(2-\sigma)}{2}\tilde{\theta}^T(k-1)\Gamma^{-1}\tilde{\theta}(k-1) + c(1-\sigma)\sigma|\tilde{\theta}(k-1)|$$
$$\leq -\frac{\sigma(2-\sigma)}{4}\tilde{\theta}^T(k-1)\Gamma^{-1}\tilde{\theta}(k-1) + \frac{(c(1-\sigma))^2\sigma}{(2-\sigma)}$$
$$\leq -\frac{\sigma}{4}\tilde{\theta}^T(k-1)\Gamma^{-1}\tilde{\theta}(k-1) + c\sigma$$

and

$$-\alpha_1\varepsilon^2(k)m^2(k) + c(\sigma+\eta_0)|\varepsilon(k)m(k)| \leq -\frac{\alpha_1}{2}\varepsilon^2(k)m^2(k) + \frac{c(\sigma+\eta_0)^2}{\alpha_1}$$
$$\leq -\frac{\alpha_1}{2}\varepsilon^2(k)m^2(k) + c\sigma + c\eta_0^2,$$

which yield

$$\Delta V(k) \leq -\frac{\sigma}{4}\tilde{\theta}^T(k-1)\Gamma^{-1}\tilde{\theta}(k-1) - \frac{\alpha_1}{2}\varepsilon^2(k)m^2(k) + c\sigma + c\eta_0^2$$
$$= -\frac{\sigma}{2}V(k-1) - \frac{\alpha_1}{2}\varepsilon^2(k)m^2(k) + c\sigma + c\eta_0^2, \tag{4.74}$$

i.e.,

$$V(k) \le \left(1 - \frac{\sigma}{2}\right) V(k-1) + c\sigma + c\eta_0^2.$$

It follows from (4.74) that, for $V(k-1) > 2c\frac{\sigma+\eta_0^2}{\sigma} \stackrel{\Delta}{=} c_0$, we have $V(k) < V(k-1)$. It is also clear that $V(k) \le c_0$ for $V(k-1) \le c_0$. This implies that for any $k = 0, 1, \dots$, either $V(k+1) < V(k)$ or $V(j) \le c_0 \; \forall j > k$. Therefore, $V(k) \le \max(V(0), c_0) \; \forall k \ge 0$, which implies that $V(k)$ and therefore $\theta(k)$ are bounded from above. From $\theta(k) \in \ell_\infty$, it follows that $\varepsilon(k), \varepsilon(k)m(k) \in \ell_\infty$. Writing (4.74) as

$$V(k) - V(k-1) \le -\frac{\alpha_1}{2}\varepsilon^2(k)m^2(k) + c(\sigma + \eta_0^2)$$

and taking summation on both sides, we have

$$\frac{1}{N}\sum_{k=0}^{N} \varepsilon^2(k)m^2(k) \le \frac{c}{N}(V(0) - V(N)) + c(\sigma + \eta_0^2),$$

which implies that $\varepsilon(k)m(k) \in \mathcal{S}(\sigma + \eta_0^2)$. Similarly, we can establish that

$$\varepsilon(k), |\theta(k) - \theta(k-1)| \in \mathcal{S}(\sigma + \eta_0^2). \qquad \square$$

It is clear that the fixed σ-modification acts as a disturbance that remains even when the modeling error is set equal to zero. Therefore, robustness is achieved at the expense of destroying the ideal properties of the algorithm obtained when $\eta_0 = 0$.

Switching σ-Modification [113]

As in the continuous-time case, the drawbacks of the fixed σ-modification can be removed by switching it off whenever the parameter estimates are within certain acceptable bounds. If, for example, we know that $|\theta^*| \le \bar{M}_0$ for some known constant $\bar{M}_0 > 0$, we modify (4.73) to obtain

$$\theta(k) = (1 - \sigma_s)\theta(k-1) + \Gamma\varepsilon(k)\phi(k), \tag{4.75}$$

where

$$\sigma_s = \begin{cases} 0 & \text{if } |\theta(k-1)| \le M_0, \\ \sigma & \text{if } |\theta(k-1)| > M_0 \end{cases}$$

and $M_0 > 0$ is a design constant dependent on $\bar{M}_0$.

Theorem 4.11.3. *The gradient algorithm* (4.75) *with switching σ-modification, where M_0, σ, Γ are designed to satisfy*

- $0 < \sigma < 2$,
- $\lambda_{\max}(\Gamma) < 1$,
- $\lambda_{\min}(\Gamma^{-1}) > \frac{2+\sigma}{2-\sigma}$,

- $M_0 \geq \frac{2(1+\lambda_{\max}(\Gamma^{-1}))\bar{M}_0}{a_0}$, $a_0 = \frac{[(2-\sigma)\lambda_{\min}(\Gamma^{-1})-(2+\sigma)]}{2}$,

guarantees the following:

(i) $\theta(k), \varepsilon(k), \varepsilon(k)m(k) \in \ell_\infty$.

(ii) $\varepsilon(k), \varepsilon(k)m(k), |\theta(k) - \theta(k-1)| \in \mathcal{S}(\eta_0^2)$.

Proof. We express (4.75) in terms of the parameter error $\tilde{\theta}(k) = \theta(k) - \theta^*$,

$$\tilde{\theta}(k) = \tilde{\theta}(k-1) + \Gamma\varepsilon(k)\phi(k) - \sigma_s\theta(k-1),$$
$$\varepsilon(k) = -\frac{\tilde{\theta}^T(k-1)\phi(k)}{m^2(k)} + \frac{\eta(k)}{m^2(k)},$$

and consider

$$V(k) = \frac{\tilde{\theta}^T(k)\Gamma^{-1}\tilde{\theta}(k)}{2}.$$

We can establish as in the proof of Theorem 4.11.2 that

$$\begin{aligned}\Delta V(k) \triangleq V(k) - V(k-1) &= \varepsilon^2(k)\frac{\phi^T(k)\Gamma\phi(k)}{2} - \varepsilon^2(k)m^2(k) + \varepsilon(k)\eta(k)\\ &\quad + \frac{\sigma_s^2}{2}\theta^T(k-1)\Gamma^{-1}\theta(k-1) - \sigma_s\tilde{\theta}^T(k-1)\Gamma^{-1}\theta(k-1) - \sigma_s\varepsilon(k)\theta^T(k-1)\phi(k)\\ &\leq -\frac{\varepsilon^2(k)m^2(k)}{2}\left(1 - \frac{\phi^T(k)\Gamma\phi(k)}{m^2(k)}\right) - \frac{\varepsilon^2(k)m^2(k)}{2} + |\varepsilon(k)m(k)|\eta_0\\ &\quad + \frac{\sigma_s^2}{2}\theta^T(k-1)\Gamma^{-1}\theta(k-1) - \sigma_s\theta^T(k-1)\Gamma^{-1}(\theta(k-1) - \theta^*)\\ &\quad + \sigma_s\theta^T(k-1)\frac{\phi(k)\phi^T(k)}{m^2(k)}(\theta(k-1) - \theta^*) - \sigma_s\frac{\eta(k)\phi^T(k)}{m^2(k)}\theta(k-1),\end{aligned}$$

where η_0 is the upper bound for $\frac{|\eta(k)|}{m(k)}$. The identity $\theta^T\phi = \phi^T\theta$ is used to obtain the last two terms of the above inequality. Using the condition that $\lambda_{\max}(\Gamma) < 1$, $\frac{|\phi(k)|}{m(k)} \leq 1, \sigma_s \leq \sigma < 1$, and the inequality $-\frac{a^2}{2} + ab \leq \frac{b^2}{2}$, we have

$$\begin{aligned}\Delta V(k) &\leq -\frac{1-\lambda_{\max}(\Gamma)}{2}\varepsilon^2(k)m^2(k) + \frac{\eta_0^2}{2} - \sigma_s\left(1 - \frac{\sigma_s}{2}\right)\theta^T(k-1)\Gamma^{-1}\theta(k-1)\\ &\quad + \sigma_s\theta^T(k-1)\Gamma^{-1}\theta^* + \sigma_s|\theta(k-1)|^2 + \sigma_s|\theta^*||\theta(k-1)| + \sigma_s\eta_0|\theta(k-1)|\\ &\leq -\frac{1-\lambda_{\max}(\Gamma)}{2}\varepsilon^2(k)m^2(k) + \eta_0^2 - \sigma_s a_0|\theta(k-1)|^2\\ &\quad + \sigma_s(1+\lambda_{\max}(\Gamma^{-1}))|\theta^*||\theta(k-1)| + \sigma_s\eta_0|\theta(k-1)|\\ &\leq -\frac{1-\lambda_{\max}(\Gamma)}{2}\varepsilon^2(k)m^2(k) - \frac{\sigma_s}{2}a_0|\theta(k-1)|^2 + \left(1 + \frac{\sigma}{2a_0}\right)\eta_0^2\\ &\quad + \sigma_s(1+\lambda_{\max}(\Gamma^{-1}))|\theta^*||\theta(k-1)|,\end{aligned}$$

where $a_0 = \left(1 - \frac{\sigma}{2}\right)\lambda_{\min}(\Gamma^{-1}) - 1 - \frac{\sigma}{2} \leq \left(1 - \frac{\sigma_s}{2}\right)\lambda_{\min}(\Gamma^{-1}) - 1 - \frac{\sigma_s}{2}$ satisfies $a_0 \geq c > 0$ due to the second condition of Theorem 4.11.3. The last term in the final equation above satisfies

$$\begin{aligned}
&-\sigma_s|\theta(k-1)|\left(\frac{a_0}{2}|\theta(k-1)| - (1 + \lambda_{\max}(\Gamma^{-1}))|\theta^*|\right) \\
&\leq -\frac{a_0}{2}\sigma_s|\theta(k-1)|\left(|\theta(k-1)| - \frac{2}{a_0}(1 + \lambda_{\max}(\Gamma^{-1}))|\theta^*|\right) \\
&\leq -\frac{a_0}{2}\sigma_s|\theta(k-1)|\left(|\theta(k-1)| - M_0 + M_0 - \frac{2}{a_0}(1 + \lambda_{\max}(\Gamma^{-1}))\bar{M}_0\right) \\
&\leq -\frac{a_0}{2}\sigma_s|\theta(k-1)|(|\theta(k-1)| - M_0) \leq 0,
\end{aligned}$$

where the last two inequalities are obtained by using the bound $\bar{M}_0 \geq |\theta^*|$ and the properties of M_0, σ_s. Therefore,

$$\Delta V(k) \leq -\frac{1 - \lambda_{\max}(\Gamma)}{2}\varepsilon^2(k)m^2(k) - \sigma_s c|\theta(k-1)|(|\theta(k-1)| - M_0) + c\eta_0^2.$$

Since $\Delta V(k) < 0$ for large $|\theta(k-1)|$ and consequently large $|\tilde{\theta}(k-1)|$, it follows that $V(k)$ and therefore $\theta(k-1), \tilde{\theta}(k-1) \in \ell_\infty$. The rest of the proof follows by using the fact that $-\sigma_s|\theta(k-1)|(|\theta(k-1)| - M_0) \leq 0$. □

It follows from the proof of Theorem 4.11.3 that when $\eta_0 = 0$, i.e., the parametric model is free of modeling errors, the switching σ-modification does not affect the ideal properties of the gradient algorithm. The only drawback is that an upper bound for $|\theta^*|$ should be known a priori in order to design M_0. This is not the case in the fixed σ-modification, where no bounds for $|\theta^*|$ are required.

4.11.3 Parameter Projection

The possible parameter drift due to modeling errors can be eliminated by using parameter projection to force the parameter estimates to lie inside a compact set in the parameter space and therefore guarantee to be bounded at each instant of time.

Let the compact set be defined as

$$S = \{\theta \in \mathcal{R}^n | \theta^T\theta \leq M_0^2\}, \tag{4.76}$$

where M_0 is chosen large enough for $\theta^* \in S$. The gradient algorithm with projection becomes

$$\begin{aligned}
\bar{\theta}(k) &= \theta(k-1) + \alpha\varepsilon(k)\phi(k), \quad \theta(0) \in S, \\
\theta(k) &= \begin{cases} \bar{\theta}(k) & \text{if } |\bar{\theta}(k)| \leq M_0, \\ \frac{M_0}{|\bar{\theta}(k)|}\bar{\theta}(k) & \text{otherwise}, \end{cases} \\
\varepsilon(k) &= \frac{z(k) - \theta^T(k-1)\phi(k)}{m^2(k)}.
\end{aligned} \tag{4.77}$$

Theorem 4.11.4. *The gradient algorithm* (4.77) *with* $M_0 \geq |\theta^*|$, $0 < \alpha < 2$ *guarantees the following:*

(i) $\varepsilon(k), \varepsilon(k)m(k), \theta(k) \in \ell_\infty$.

(ii) $\varepsilon(k), \varepsilon(k)m(k), |\theta(k) - \theta(k-1)| \in \mathcal{S}(\eta_0^2)$.

Proof. The parameter equations are of the form

$$\tilde{\theta}(k) = \begin{cases} \tilde{\theta}(k-1) + \alpha\varepsilon(k)\phi(k) & \text{if } |\bar{\theta}(k)| \le M_0, \\ \tilde{\theta}(k-1) + \alpha\varepsilon(k)\phi(k) - \left(1 - \frac{M_0}{|\bar{\theta}(k)|}\right)\bar{\theta}(k) & \text{if } |\bar{\theta}(k)| > M_0, \end{cases}$$

$$\varepsilon(k) = -\frac{\tilde{\theta}^T(k-1)\phi(k)}{m^2(k)} + \frac{\eta(k)}{m^2(k)}.$$

Choosing

$$V(k) = \frac{\tilde{\theta}^T(k)\tilde{\theta}(k)}{2},$$

we have

$$\begin{aligned}\Delta V(k) &\triangleq V(k) - V(k-1) \\ &= \frac{1}{2}(\tilde{\theta}(k-1) + \alpha\varepsilon(k)\phi(k) + g(k))^T(\tilde{\theta}(k-1) + \alpha\varepsilon(k)\phi(k) + g(k)) \\ &\quad - \frac{\tilde{\theta}^T(k-1)\tilde{\theta}(k-1)}{2},\end{aligned}$$

where

$$g(k) = \begin{cases} 0 & \text{if } |\bar{\theta}(k)| \le M_0, \\ -\left(1 - \frac{M_0}{|\bar{\theta}(k)|}\right)\bar{\theta}(k) & \text{if } |\bar{\theta}(k)| > M_0. \end{cases}$$

For clarity of presentation, let us drop the argument k, $k-1$ and rewrite the above equation as

$$\Delta V(k) = \frac{\alpha^2}{2}\varepsilon^2\phi^T\phi + \alpha\varepsilon\phi^T\tilde{\theta} + \frac{g^Tg}{2} + g^T(\bar{\theta} - \theta^*).$$

Let us examine the term g, which is due to projection. For $|\bar{\theta}| \le M_0$ we have $g = 0$, and for $|\bar{\theta}| > M_0$ we have

$$\begin{aligned}\frac{g^Tg}{2} + g^T(\bar{\theta} - \theta^*) &= \frac{1}{2}\left(1 - \frac{M_0}{|\bar{\theta}|}\right)^2 \bar{\theta}^T\bar{\theta} - \left(1 - \frac{M_0}{|\bar{\theta}|}\right)\bar{\theta}^T(\bar{\theta} - \theta^*) \\ &= \frac{(|\bar{\theta}| - M_0)^2}{2} - (|\bar{\theta}| - M_0)\left(|\bar{\theta}| - \frac{\bar{\theta}^T\theta^*}{|\bar{\theta}|}\right) \\ &\le (|\bar{\theta}| - M_0)\left(\frac{|\bar{\theta}| - M_0}{2} - (|\bar{\theta}| - |\theta^*|)\right) \\ &\le -(|\bar{\theta}| - M_0)(M_0 - |\theta^*|) \le 0.\end{aligned}$$

Therefore, $\frac{g^Tg}{2} + g^T(\bar{\theta} - \theta^*) \le 0 \ \forall k$. Hence

$$\Delta V(k) \le \frac{\alpha^2}{2}\varepsilon^2(k)\phi^T(k)\phi(k) + \alpha\varepsilon(k)\phi^T(k)\tilde{\theta}(k-1) \quad \forall k.$$

Substituting $\phi(k)^T\tilde{\theta}(k-1) = -\varepsilon(k)m^2(k) + \eta(k)$, we obtain

$$\begin{aligned}
\Delta V(k) &\leq -\alpha\varepsilon^2(k)m^2(k)\left(1 - \alpha\frac{\phi^T(k)\phi(k)}{2m^2(k)}\right) + \alpha\varepsilon(k)\eta(k) \\
&\leq -\alpha_0\varepsilon^2(k)m^2(k) + \alpha|\varepsilon(k)|\eta_0 \\
&\leq -\frac{\alpha_0}{2}\varepsilon^2(k)m^2(k) + \frac{\alpha_0}{2}\eta_0^2,
\end{aligned}$$

where $\alpha_0 = 1 - \frac{\alpha}{2} > 0$. Since the projection guarantees that $|\theta(k)| < M_0 \ \forall k$, we can establish that $\theta(k), \varepsilon(k), \varepsilon(k)m(k) \in \ell_\infty$ and, from the expression of $\Delta V(k)$ above, that $\varepsilon(k)m(k) \in \mathcal{S}(\eta_0^2)$. Using these properties, we can also establish that $\varepsilon(k)$, $|\theta(k) - \theta(k-1)| \in \mathcal{S}(\eta_0^2)$. □

The parameter projection algorithm given by (4.77) has properties very similar to those of the switching σ-modification, which is often referred to as a soft projection.

Other modifications include the dead zone, ε-modification, etc., as shown in the continuous-time case. The purpose of these modifications is to prevent parameter drift by modifying the pure delay in the adaptive law. The switching σ-modification and parameter projection are the two modifications that guarantee robustness without destroying the ideal properties of the adaptive laws. The dead zone, fixed-σ, and ε-modifications guarantee robustness at the expense of destroying the ideal properties of the adaptive law in the absence of modeling errors [46, 56, 86, 88, 91].

4.12 Case Study: Online Parameter Estimation of Traffic Flow Characteristics

Highway traffic is often modeled as a fluid, characterized by the flow rate, q, in number of vehicles/unit of time; density, ρ, in number of vehicles/unit of length; and speed, v, in unit of length/unit of time. These variables are related in a nonlinear fashion, and many attempts have been made to describe these relationships mathematically. Based on empirical studies it is widely accepted by most traffic experts that the speed, density, and flow are related by the nonlinear functions

$$v(k) = v_f \exp\left[-\frac{1}{a}\left(\frac{\rho(k)}{\rho_c}\right)^a\right], \tag{4.78}$$

$$q(k) = \rho(k)v_f \exp\left[-\frac{1}{a}\left(\frac{\rho(k)}{\rho_c}\right)^a\right], \tag{4.79}$$

where $v(k)$, $q(k)$, $\rho(k)$ are the speed, flow rate, and density measurements at each instant k, respectively; v_f, ρ_c, and a are model parameters with v_f the free flow speed, ρ_c the critical density, and a an exponent. Equation (4.78) gives the relationship between speed and density, and (4.79) the corresponding relationship between flow rate and density. The plot of q versus k described by (4.78) is known as the fundamental diagram and is shown in Figure 4.3.

The free flow speed $v_f = \tan\theta$ is the slope of the curve at zero density and zero flow, and denotes the speed when the traffic is very light. When the density is less than ρ_d, traffic

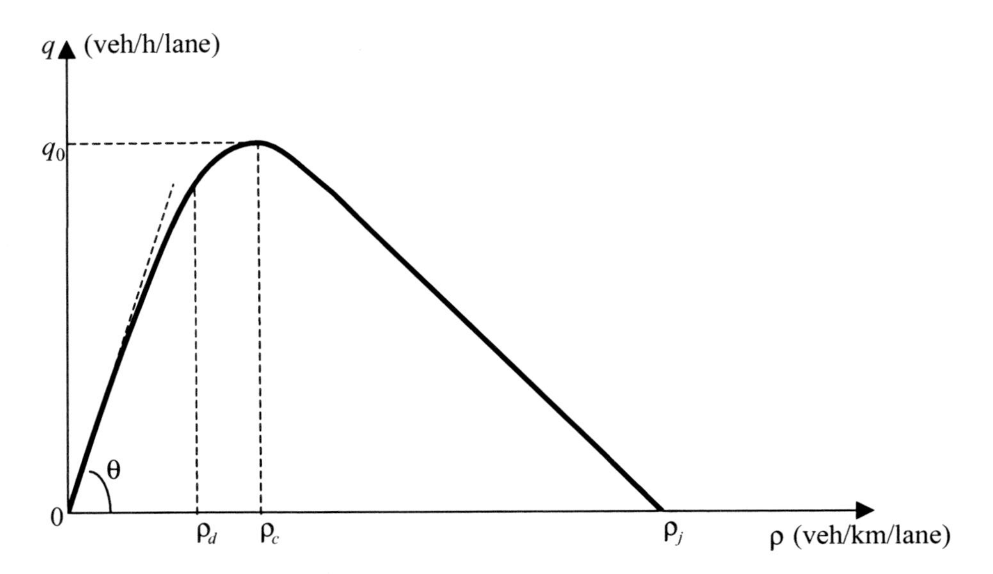

Figure 4.3. *The fundamental traffic flow diagram.*

is considered to be light, and when between ρ_d and ρ_c is considered to be dense. When the density is above the critical density ρ_c, where the traffic flow is maximum, traffic is congested, and at density ρ_j the traffic is at standstill.

Since the characteristics of traffic flow vary from one region to another and may depend on the geometry of the highway as well as on the time of day, volume of traffic, and other effects, the parameters of the model v_f, ρ_c, and a are not fixed and have to be estimated online. The equations (4.78), (4.79) are nonlinear in the unknown parameters v_f, ρ_c, and a, and the results of this chapter do not apply unless we manage to express the unknown parameters in the form of a parametric model where the unknown parameters appear linearly or in the special bilinear form. However, based on empirical data we know a priori that the parameter a takes values between 0.9–1.1 approximately. Using this information, we proceed as follows.

Using the natural logarithm on both sides of (4.78), we have

$$\ln v(k) = \ln v_f - \frac{\rho(k)^a}{a\rho_c^a}, \tag{4.80}$$

which we express as

$$z(k) = \theta^{*T}\phi(k), \tag{4.81}$$

where

$$z(k) = \ln v(k), \qquad \theta^* = \left[\ln v_f, \frac{1}{\rho_c^a}\right]^T, \qquad \phi(k) = \left[1, -\frac{\rho(k)^a}{a}\right]^T. \tag{4.82}$$

Even though we obtain a parametric model in the form of SPM, we cannot use it since $\phi(k)$ depends on the unknown parameter a. The approach we follow in this case is to assume different values from the range 0.9–1.1 for a and run parallel estimators corresponding to each value of a. Then we choose the estimator that gives us the smallest prediction error over a given interval. We demonstrate the approach for three different values of a, 0.95, 1.0, and 1.05. For each value of a, we compute the accumulated prediction error $\varepsilon_a(N)$ over N samples

$$\varepsilon_a(N) = \sum_{k=0}^{N} \left[\hat{v}_f(k) \exp\left(-\frac{1}{a} \left(\frac{\rho(k)}{\hat{\rho}_c(k)} \right)^a \right) - v(k) \right],$$

where $\hat{v}_f(k)$, $\hat{\rho}_c(k)$ is the online estimate of v_f, ρ_c, respectively, generated by the respective adaptive law based on the assumed value of a.

For each parametric model corresponding to the chosen value of a, we use the following gradient algorithm to generate the estimates $\hat{v}_f(k)$, $\hat{\rho}_c(k)$:

$$\varepsilon(k) = \frac{z(k) - \theta^T(k-1)\phi(k)}{m^2(k)}, \quad m^2(k) = 1 + \phi^T(k)\phi(k),$$

$$\bar{\theta}_i(k) = \theta_i(k-1) + \gamma_i \varepsilon(k)\phi_i(k),$$

$$\theta_i(k) = \Pr(\bar{\theta}_i(k)) = \begin{cases} \bar{\theta}_i(k) & \text{if } L_i \le \bar{\theta}_i(k) \le U_i, \\ L_i & \text{if } \bar{\theta}_i(k) < L_i, \\ U_i & \text{if } \bar{\theta}_i(k) > U_i, \end{cases}$$

where

$$v_f(0) = 130, \quad \theta_1(0) = \ln v_f(0) = 4.8675, \quad \rho_c(0) = 25,$$

$$\theta_2(0) = \left(\frac{1}{\rho_c(0)} \right)^a = \left(\frac{1}{25} \right)^a, \quad 100 \le v_f \le 200, \quad L_1 = 4.6052,$$

$$U_1 = 5.2983, \quad 10 \le \rho_c \le 50, \quad L_2 = \left(\frac{1}{50} \right)^a,$$

$$U_2 = \left(\frac{1}{10} \right)^a, \quad \gamma_1 = \gamma_2 = 1000.$$

Therefore, the estimated free flow speed and critical density are

$$\hat{v}_f(k) = \exp(\hat{\theta}_1(k)), \qquad \hat{\rho}_c(k) = (\hat{\theta}_2(k))^{-\frac{1}{a}}.$$

We compute the accumulated estimation error ε_a for $a_1 = 0.95$, $a_2 = 1.00$, $a_3 = 1.05$ by running a parallel estimator for each value of a. We choose the value of a that gives the smallest ε_a at each instant N and take the corresponding values of $\hat{v}_f(k)$, $\hat{\rho}_c(k)$ as the estimates. Real-time traffic data from the I-80 in northern California generated by the Berkeley Highway Laboratory (BHL) are used to test the proposed algorithm. Figure 4.4 shows the evolution of the estimated parameters and their convergence to constant values. Figure 4.5 shows the fundamental diagram based on the measured data and estimated parameters. The results could be improved further by increasing the number of parallel estimators with additional values of the parameter a from the assumed range.

See the web resource [94] for examples using the Adaptive Control Toolbox.

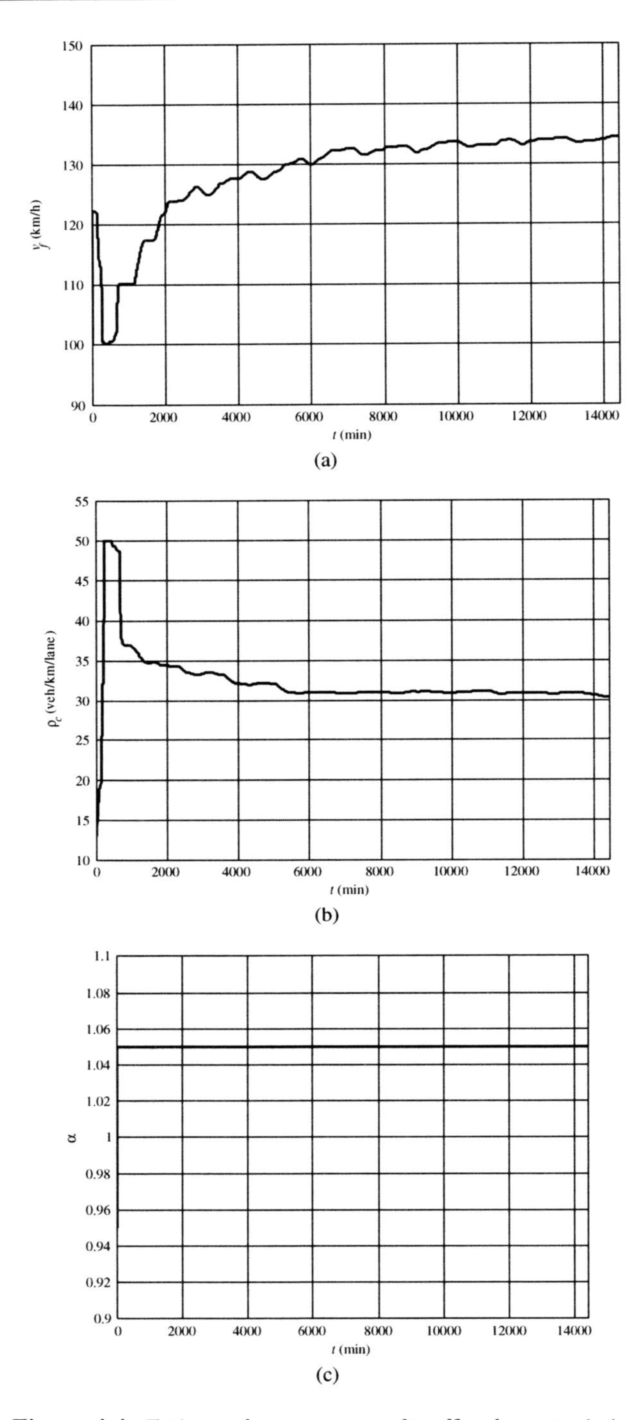

Figure 4.4. *Estimated parameters of traffic characteristics.*

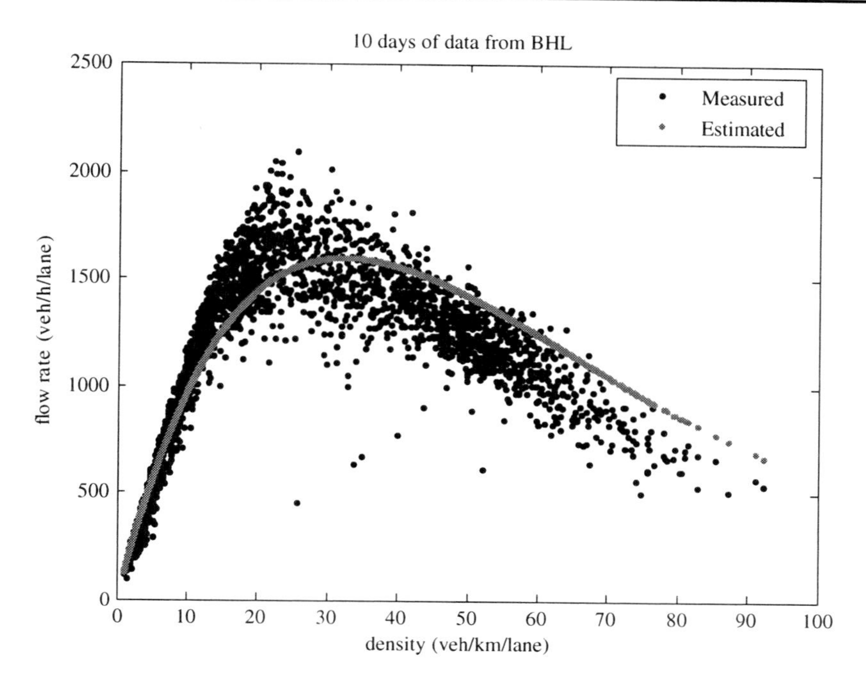

Figure 4.5. *Fundamental diagram based on measured data and estimated parameters.*

Problems

1. Consider the continuous-time gradient algorithm with integral cost function

$$\dot{\theta} = -\Gamma(R\theta + Q),$$
$$\dot{R} = -\beta R + \frac{\phi\phi^T}{m_s^2}, \quad R(0) = 0,$$
$$\dot{Q} = -\beta Q - \frac{z\phi}{m_s^2}, \quad Q(0) = 0,$$

where $\theta, \phi, Q \in \mathcal{R}^n$, $R, \Gamma \in \mathcal{R}^{n \times n}$, and $\Gamma = \Gamma^T > 0$ is a constant gain matrix. Use the Euler backward approximation method to obtain a discrete-time version of this gradient algorithm.

2. Consider the continuous-time plant

$$y_p = \frac{k_p}{s^2 + a_1 s + a_0} u_p,$$

where k_p, a_0, a_1 are unknown positive constant parameters. Design a discrete-time algorithm to estimate k_p, a_0, a_1 online from the samples of u_p, y_p or samples of their filtered values.

3. Consider the discrete-time DARMA system

$$y(k) = a_1 y(k-1) + a_2 y(k-2) + b_0 u(k-3),$$

where a_1, a_2, b_0 are the unknown constant parameters.

   (a) Express the unknown parameters in terms of the SPM $z = \theta^{*T}\phi$.

   (b) Find conditions for a_1, a_2, b_0, and $u(k)$ which guarantee that the regressor $\phi(k)$ is PE.

4. Consider the discrete-time system

$$y(k) = \frac{b_0(z+b_1)}{(z+a_0)(z-0.5)}u(k).$$

   (a) Express the system in the SPM form assuming that b_0, b_1, a_0 are unknown constants.

   (b) Show that if $b_1 = a_0$, then the regressor vector obtained in (a) cannot be PE for any input sequence $u(k)$.

5. Consider the projection algorithm (4.26) applied to the SPM

$$z(k) = \theta^{*T}\phi(k).$$

   (a) Show that the division by zero in the projection algorithm can be avoided by modifying the SPM to
$$z(k) = \bar{\theta}^{*T}\bar{\phi}(k),$$
where $\bar{\theta}^* = [\theta^{*T}, 0]^T$, $\bar{\phi}(k) = [\phi^T(k), 1]^T$, and then applying the algorithm to the modified SPM.

   (b) Comment on the advantages of the approach in (a) versus that of adding a constant c to the denominator in (4.26) to obtain (4.29), or using $\theta(k) = \theta(k-1)$ when $\phi^T(k)\phi(k) \cong 0$.

6. Consider the DARMA system

$$y(k+2) = a_1 y(k) + a_2 y(k-1) + b_1 u(k+1) + 0.2u(k-1),$$

where a_1, a_2, b_1 are the unknown constant parameters to be estimated online.

   (a) Express the unknown parameters in an SPM.

   (b) Use a projection algorithm with the modification suggested in Problem 5(a) and without the modification to estimate a_1, a_2, b_1.

   (c) Simulate the algorithms in (b) for $a_1 = 0.7$, $a_2 = -0.18$, $b_1 = 1$, and an input $u(k)$ of your choice. Demonstrate that when $u(k)$ is not sufficiently rich, the parameters do not converge to their true values.

7. Repeat Problem 6 for the gradient algorithm of section 4.5.2.

8. Repeat Problem 6 for the orthogonalized projection algorithm (4.30).

9. Repeat Problem 6 for the LS algorithm (4.43).

10. Prove the relationship

$$(A + BC)^{-1} = A^{-1} - A^{-1}B(I + CA^{-1}B)^{-1}CA^{-1},$$

where A, B, C are matrices of appropriate dimensions.

11. Show that in the projection algorithm (4.26), $\theta(k)$ is the orthogonal projection of $\theta(k-1)$ to the surface $H = \{\theta | z(k) = \theta^T \phi(k)\}$. (*Hint*: Show that $(x - \theta(k))^T(\theta(k-1) - \theta(k)) = 0$ for any x that satisfies $z(k) = x^T\phi(k)$.)

12. Consider the system

$$y(k) = a_0 y(k-1) + u(k),$$

where $0 < a_0 < 1$ is the unknown parameter.

(a) Use the weighted LS to estimate a_0.

(b) Use simulations to examine the effect of the four choices of $\alpha(k)$ given in Table 4.1 on the convergence of the estimate of a_0 in the case of $a_0 = 0.99$ and $u(k)$ of your choice.

13. (a) Use the weighted LS algorithm with dynamic data weighting (4.55), (4.56) to estimate the parameter a_0 of the system

$$y(k) = a_0 y(k-1) + u(k).$$

(b) Use simulations to examine the effect of the modifications given in Table 4.2 on the parameter convergence in the case of $a_0 = 0.99$ and $u(k)$ of your choice.

14. Show that in the parameter projection algorithm (4.63), the projected parameter vector $\hat{\theta}(k) = \frac{M}{|\bar{\theta}(k)|}\bar{\theta}(k)$ is the orthogonal projection of $\bar{\theta}(k)$ to the surface $\theta^T\theta = M^2$ when $|\bar{\theta}(k)| > M$.

15. Show that in the parameter projection algorithm (4.64), the projected parameter vector $\hat{\theta}(k) = \theta_0 + \frac{M(\bar{\theta}(k)-\theta_0)}{|\bar{\theta}(k)-\theta_0|}$ is the orthogonal projection of $\bar{\theta}(k)$ to the surface $(\bar{\theta}(k) - \theta_0)^T(\bar{\theta}(k) - \theta_0) = M^2$ when $|\bar{\theta}(k) - \theta_0| > M$.

16. Consider the parameter error equation (4.70), where $\frac{\phi(k)}{m(k)}$ is PE and $\frac{|\phi|}{m} \le 1$, $\frac{|\eta|}{m} \le \eta_0$ for some constant $\eta_0 > 0$. Show that

$$|\tilde{\theta}(k)| \le c_1\eta_0 + \varepsilon_t,$$

where $c_1 > 0$ is some constant that depends on the rate of convergence and ε_t is exponentially decaying to zero.

17. Establish (4.71).

18. Consider the plant

$$y(k) = \frac{z + a_0}{(z + a_0 + \varepsilon)(z + b_0)}u(k),$$

where $0 < \varepsilon \ll 1$, $0 < a_0 < 0.8$, $0 < b_0 < 0.9$.

(a) Model the plant as a first-order dominant part with an additive uncertainty.

(b) Design a robust adaptive law to estimate the unknown parameters of the dominant part.

(c) Simulate the scheme you developed in (b) for $a_0 = 0.5$, $b_0 = 0.85$, $\varepsilon = 0, 0.08, 0.09$, and values of $u(k)$ for your choice.

19. Consider the following gradient algorithm with dead zone for the SPM

$$z(k) = \theta^{*T}\phi(k) + \eta(k)$$

given by

$$\varepsilon(k) = \frac{z(k) - \theta^T(k-1)\phi(k)}{m^2(k)},$$
$$\theta(k) = \begin{cases} \theta(k-1) & \text{if } |\varepsilon(k)m(k)| < \bar{\eta}_0, \\ \theta(k-1) + \Gamma\varepsilon(k)\phi(k) & \text{if } |\varepsilon(k)m(k)| \geq \bar{\eta}_0, \end{cases}$$

where $\bar{\eta}_0 > \frac{2\eta_0}{\alpha}$, $\alpha = 2 - \lambda_{\max}(\Gamma)$, $\frac{|\eta(k)|}{m(k)} \leq \eta_0$. Show that the adaptive law above with dead zone guarantees that $\varepsilon(k)$, $\varepsilon(k)\phi(k)$, $\theta(k) \in \ell_\infty$ and $\varepsilon(k), \varepsilon(k)\phi(k), |\theta(k) - \theta(k-1)| \in \mathcal{S}(\bar{\eta}_0^2 + \eta_0^2)$, provided $0 < \lambda_{\max}(\Gamma) < 2$ and $\frac{|\phi|}{m} \leq 1$.

20. Consider the gradient algorithm

$$\varepsilon(k) = \frac{z(k) - \theta^T(k-1)\phi(k)}{m^2(k)},$$
$$\theta(k) = \theta(k-1) + \Gamma\varepsilon(k)\phi(k),$$

where $z(k) = \theta^{*T}\phi(k) + \eta(k)$ and $\Gamma = \text{diag}\{\gamma_1, \gamma_2, \ldots, \gamma_n\}$, $m^2(k)$ is chosen so that $\frac{|\phi|}{m} \leq 1$, and $\frac{|\eta(k)|}{m(k)} \leq \eta_0$ for some $\eta_0 \geq 0$. Let us assume that we have the a priori information $L_i \leq \theta_i^* \leq U_i$, $i = 1, 2, \ldots, n$, where L_i, U_i are known constants.

(a) Design a robust adaptive law that guarantees $L_i \leq \theta_i(k) \leq U_i$ $\forall k$ when $\eta(k) = 0$ $\forall k$.

(b) Analyze the adaptive law in (a) for robustness when $\eta(k) \neq 0$.

Chapter 5
Continuous-Time Model Reference Adaptive Control

5.1 Introduction

In this chapter, we design and analyze a wide class of adaptive control schemes based on *model reference control* (*MRC*) referred to as *model reference adaptive control* (*MRAC*). In MRC, the desired plant behavior is described by a reference model which is simply an LTI system with a transfer function $W_m(s)$ and is driven by a reference input. The control law $C(s, \theta_c^*)$ is then developed so that the closed-loop plant has a transfer function equal to $W_m(s)$. This transfer function matching guarantees that the plant will behave like the reference model for any reference input signal.

Figure 5.1 shows the basic structure of MRC. The plant transfer function is $G_p(s, \theta_p^*)$, where θ_p^* is a vector with all the coefficients of G_p. The controller transfer function is $C(s, \theta_c^*)$, where θ_c^* is a vector with the coefficients of $C(s)$.

The transfer function $C(s, \theta_c^*)$, and therefore θ_c^*, is designed so that the closed-loop transfer function of the plant from the reference input r to y_p is equal to $W_m(s)$; i.e.,

$$\frac{y_p(s)}{r(s)} = W_m(s) = \frac{y_m(s)}{r(s)}. \tag{5.1}$$

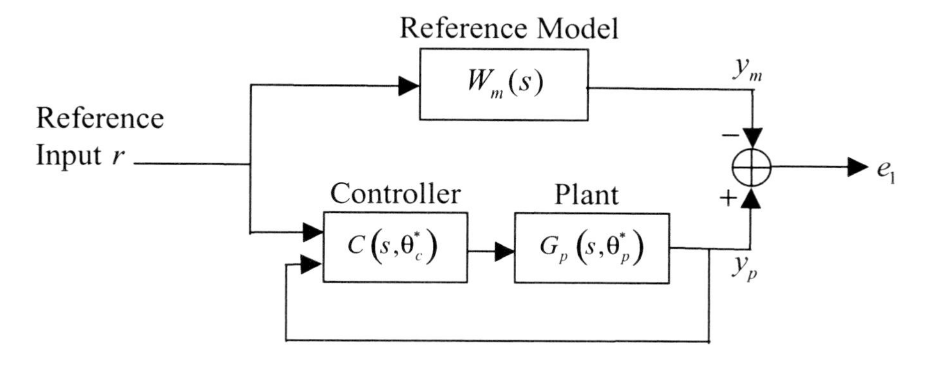

Figure 5.1. *Model reference control.*

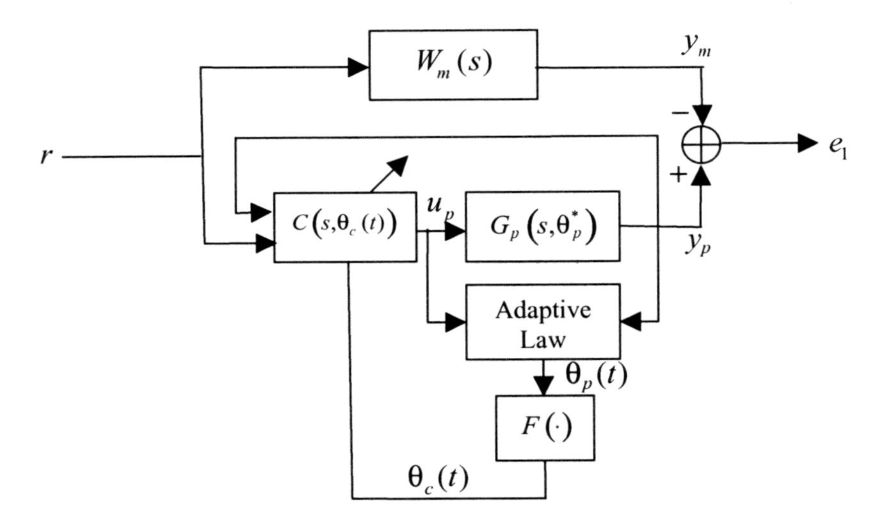

Figure 5.2. *Indirect MRAC.*

For this transfer matching to be possible, $G_p(s)$, $W_m(s)$ have to satisfy certain assumptions. These assumptions enable the calculation of the controller parameter vector θ_c^* as

$$\theta_c^* = F(\theta_p^*), \tag{5.2}$$

where F is a function of the plant parameters θ_p^*, to satisfy the matching equation (5.1). This transfer function matching guarantees that the tracking error $e_1 = y_p - y_m$ converges to zero for any given reference input signal r.

If the plant parameter vector θ_p^* is known, then the controller parameters θ_c^* can be calculated using (5.2), and the controller $C(s, \theta_c^*)$ can be implemented. We are considering the case where θ_p^* is unknown. In this case the use of the certainty equivalence (CE) approach, where the unknown parameters are replaced with their estimates, leads to the adaptive control scheme referred to as *indirect MRAC*, shown in Figure 5.2.

The unknown plant parameter vector θ_p^* is estimated at each time t, denoted by $\theta_p(t)$, using an adaptive law of the form studied in the previous chapters. The plant parameter estimate $\theta_p(t)$ at each time t is then used to calculate the controller parameter vector $\theta_c(t) = F(\theta_p(t))$ used in the controller $C(s, \theta_c)$. This class of MRAC is called *indirect MRAC* because the controller parameters are not updated directly but calculated at each time t using the estimated plant parameters.

Another way of designing MRAC schemes is to parameterize the plant transfer function in terms of the desired controller parameter vector θ_c^*. This is possible in the MRC case because, as we will show later in this chapter, the structure of the MRC law is such that we can use (5.2) to write

$$\theta_p^* = F^{-1}(\theta_c^*),$$

where F^{-1} is the inverse of the mapping $F(.)$, and then express $G_p(s, \theta_p^*) = G_p(s, F^{-1}(\theta_c^*)) = \bar{G}_p(s, \theta_c^*)$. The adaptive law for estimating θ_c^* online can now be developed by using

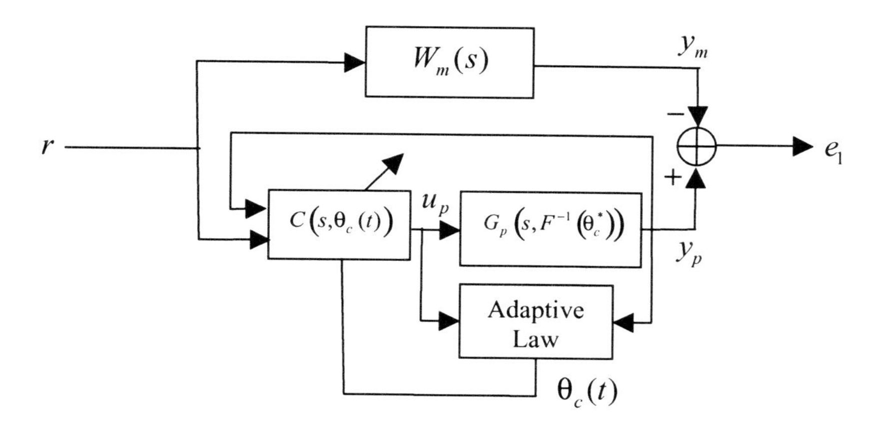

Figure 5.3. *Direct MRAC.*

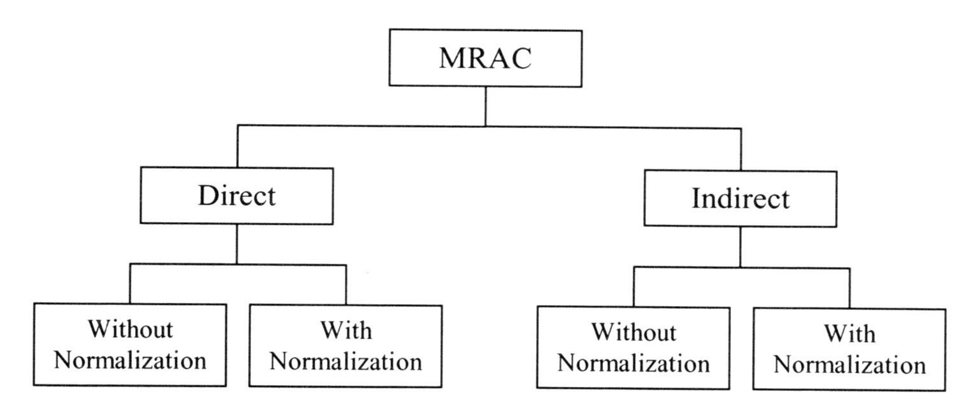

Figure 5.4. *Classes of MRAC.*

$y_p = \bar{G}_p(s, \theta_c^*)u_p$ to obtain a parametric model of the form considered in Chapter 3 with θ_c^* as the unknown parameter vector. The MRAC can then be developed using the CE approach, as shown in Figure 5.3. In this case, the controller parameter $\theta_c(t)$ is updated directly without any intermediate calculations, and for this reason the scheme is called *direct MRAC*.

The division of MRAC to indirect and direct is, in general, unique to MRC structures where a parametric model for the plant in the form of SPM, B-SPM, DPM, or B-DPM can be obtained in terms of the unknown controller parameter vector θ_c^*. For control structures other than MRC, we will see that this is not always possible, and the indirect approach is often the only approach we can use. In addition to direct and indirect MRAC classification, the adaptive law for estimating the unknown parameters could employ the normalizing signal discussed in Chapter 3 or could involve no normalization. This leads to the various classes of MRAC shown in Figure 5.4.

In this chapter, we start with simple examples to demonstrate the design and analysis of adaptive control schemes and then proceed to the general MRAC case.

5.2 Simple MRAC Schemes

In this section, we use several examples to illustrate the design and analysis of simple MRAC schemes where the estimation error is simply the regulation or tracking error. This approach dominated the literature of adaptive control for continuous-time plants with relative degree $n^* = 1$ because of the simplicity of design and stability analysis [41, 50, 86, 87].

5.2.1 Scalar Example: Adaptive Regulation

Consider the scalar plant

$$\dot{x} = ax + u, \quad x(0) = x_0, \tag{5.3}$$

where a is a constant but unknown. The control objective is to determine a bounded function $u = f(t, x)$ such that the state $x(t)$ is bounded and converges to zero as $t \to \infty$ for any given initial condition x_0. Let $-a_m$ be the desired closed-loop pole, where $a_m > 0$ is chosen by the designer. If the plant parameter a is known, the control law

$$u = -k^* x, \quad k^* = a + a_m, \tag{5.4}$$

could be used to meet the control objective; i.e., with (5.4), the closed-loop plant is

$$\dot{x} = -a_m x,$$

whose equilibrium $x_e = 0$ is e.s. in the large. Because a is unknown, k^* cannot be calculated and, therefore, (5.4) cannot be implemented. A possible procedure to follow in the unknown parameter case is to use the same control law as given in (5.4) but with k^* replaced by its estimate $k(t)$; i.e., we use

$$u = -k(t)x \tag{5.5}$$

and search for an adaptive law to update $k(t)$ continuously with time.

The adaptive law for generating $k(t)$ is developed by viewing the problem as an online identification problem for k^*. This is accomplished by first obtaining an appropriate parameterization for the plant (5.3) in terms of the unknown k^* and then using a similar approach, as in Chapter 3, to estimate k^* online. We illustrate this procedure below.

We add and subtract the desired control input $-k^* x$ in the plant equation to obtain

$$\dot{x} = ax - k^* x + k^* x + u.$$

Because $a - k^* = -a_m$, we have $\dot{x} = -a_m x + k^* x + u$, and the following two parametric models may be generated:

$$\begin{aligned} \text{SSPM:} \quad & \dot{x} = -a_m x + k^* x + u, \\ \text{DPM:} \quad & z = \frac{1}{s + a_m} k^* x, \quad z = x - \frac{1}{s + a_m} u. \end{aligned} \tag{5.6}$$

Equation (5.6) shows two parameterizations of the plant equation (5.3) in terms of the unknown controller parameter k^*. The first equation in (5.6) is in the form of SSPM, whereas the second is in the form of DPM. Since x, u are measured and $a_m > 0$ is known, a wide class of adaptive laws may be generated by using either the parametric model given in (5.6) or other parametric models generated using (5.6).

Let us consider the SSPM. The estimation model is given by

$$\dot{\hat{x}} = -a_m\hat{x} + kx + u, \quad \hat{x}(0) = 0.$$

By substituting $u = -k(t)x$ in the estimation model, we obtain that $\hat{x}(t) = 0 \; \forall t \geq 0$ and the estimation error $\varepsilon = x - \hat{x} = x \; \forall t \geq 0$. Therefore, the estimation error equation is obtained by simply substituting for $u = -k(t)x$ in the SSPM to obtain

$$\dot{x} = -a_m x - \tilde{k}x, \quad \tilde{k} = k - k^*. \tag{5.7}$$

Equation (5.7) relates the parameter error $\tilde{k}$ to the regulation error x or estimation error $\varepsilon = x$ and motivates the use of the Lyapunov-like function

$$V(x, \tilde{k}) = \frac{x^2}{2} + \frac{\tilde{k}^2}{2\gamma}$$

to generate the adaptive law for k. Taking the time derivative of V, we obtain

$$\dot{V} = -a_m x^2 - \tilde{k}x^2 + \frac{\tilde{k}\dot{\tilde{k}}}{\gamma},$$

where we choose $\dot{\tilde{k}}$ to make $\dot{V}$ negative. It is clear that for

$$\dot{\tilde{k}} = \dot{k} = \gamma x^2, \quad k(0) = k_0 > 0, \tag{5.8}$$

we have

$$\dot{V} = -a_m x^2 \leq 0.$$

In (5.8), we use the fact that $\dot{\tilde{k}} = \dot{k} - \dot{k}^* = \dot{k}$ due to the fact that k^* is a constant. Since V is a positive definite function and $\dot{V} \leq 0$, we have $V \in \mathcal{L}_\infty$, which implies that $x, \tilde{k} \in \mathcal{L}_\infty$ and therefore all signals in the closed-loop plant are bounded. Furthermore, from $V > 0$ and $\dot{V} \leq 0$, we have that $\lim_{t\to\infty} V(t) = V_\infty < V(0)$. Therefore,

$$\lim_{t\to\infty} \int_0^t a_m x^2(\tau)d\tau = V(0) - V_\infty,$$

which implies that $x \in \mathcal{L}_2$. From $x \in \mathcal{L}_2 \cap \mathcal{L}_\infty$ and $\dot{x} \in \mathcal{L}_\infty$, which follows from (5.7), we have that $x(t) \to 0$ as $t \to \infty$ (see Barbalat's lemma in the Appendix). From $x(t) \to 0$ and the boundedness of k, we establish that $\dot{k}(t) \to 0$, $u(t) \to 0$ as $t \to \infty$.

We have shown that the combination of the control law (5.5) with the adaptive law (5.8) meets the control objective in the sense that it guarantees boundedness for all signals and forces the plant state to converge to zero. It is worth mentioning that, as in the simple parameter identification examples considered in Chapter 3, we cannot establish that $k(t)$

converges to k^*, i.e., that the pole of the closed-loop plant converges to the desired one given by $-a_m$. The lack of parameter convergence is less crucial in adaptive control than in parameter identification because in most cases the control objective can be achieved without requiring the parameters to converge to their true values.

The simplicity of this scalar example allows us to solve for x explicitly and study the properties of $k(t)$, $x(t)$ as they evolve with time. We can verify that

$$\begin{aligned} x(t) &= \frac{2ce^{-ct}}{c + k_0 - a + (c - k_0 + a)e^{-2ct}} x(0), \\ k(t) &= a + \frac{c[(c + k_0 - a)e^{2ct} - (c - k_0 + a)]}{(c + k_0 - a)e^{2ct} + (c - k_0 + a)}, \end{aligned} \tag{5.9}$$

where $c^2 = \gamma x_0^2 + (k_0 - a)^2$ and $x_0 = x(0)$, satisfy the differential equations (5.7) and (5.8) of the closed-loop plant. Equation (5.9) can be used to investigate the effects of initial conditions and adaptive gain γ on the transient and asymptotic behavior of $x(t)$, $k(t)$. We have $\lim_{t\to\infty} k(t) = a + c$ if $c > 0$ and $\lim_{t\to\infty} k(t) = a - c$ if $c < 0$; i.e.,

$$\lim_{t\to\infty} k(t) = k_\infty = a + \sqrt{\gamma x_0^2 + (k_0 - a)^2}. \tag{5.10}$$

Therefore, for $x_0 \neq 0$, $k(t)$ converges to a stabilizing gain whose value depends on γ and the initial conditions x_0, k_0. It is clear from (5.10) that the value of k_∞ is independent of whether k_0 is a destabilizing gain ($0 < k_0 < a$) or a stabilizing one ($k_0 > a$) as long as $(k_0 - a)^2$ is the same. The use of different initial condition k_0, however, will affect the transient behavior, as is obvious from (5.9). In the limit as $t \to \infty$, the closed-loop pole converges to $-(k_\infty - a) = -\sqrt{\gamma x_0^2 + (k_0 - a)^2}$, which depends on the initial conditions and may be different from $-a_m$. Because the control objective is to achieve signal boundedness and regulation of the state $x(t)$ to zero, the convergence of $k(t)$ to k^* is not crucial.

The adaptive control scheme developed and analyzed above is given by the following equations:

$$u = -k(t)x, \quad \dot{k} = \gamma x^2, \quad k(0) = k_0, \tag{5.11}$$

where x is the measured state of the plant and, for this example, is identical to the estimation error.

The design parameters in (5.11) are the initial parameter k_0 and the adaptive gain $\gamma > 0$. For signal boundedness and asymptotic regulation of x to zero, our analysis allows k_0, γ to be arbitrary. It is clear, however, from (5.9) that their values affect the transient performance of the closed-loop plant as well as the steady-state value of the closed-loop pole. For a given k_0, $x_0 \neq 0$, large γ leads to a larger value of c in (5.9) and, therefore, to a faster convergence of $x(t)$ to zero. Large γ, however, may make the differential equation for k "stiff" (i.e., $\dot{k}$ large), which will require a very small step size or sampling period to implement it on a digital computer and may pose computational problems for some differential equation solver algorithms.

5.2.2 Scalar Example: Direct MRAC without Normalization

Consider the following first-order plant:

$$\dot{x} = ax + bu, \tag{5.12}$$

where a, b are unknown parameters but the sign of b is known. The control objective is to choose an appropriate control law u such that all signals in the closed-loop plant are bounded and x tracks the state x_m of the reference model given by

$$\dot{x}_m = -a_m x_m + b_m r$$

or

$$x_m = \frac{b_m}{s + a_m} r \tag{5.13}$$

for any bounded piecewise continuous signal $r(t)$, where $a_m > 0$, b_m are known and $x_m(t)$, $r(t)$ are measured at each time t. It is assumed that a_m, b_m, and r are chosen so that x_m represents the desired state response of the plant.

For x to track x_m for *any* reference input signal $r(t)$, the control law should be chosen so that the closed-loop plant transfer function from the input r to output x is equal to that of the reference model. We propose the control law

$$u = -k^* x + l^* r, \tag{5.14}$$

where k^*, l^* are calculated so that the closed-loop transfer function from r to x is equal to that of the reference model, i.e.,

$$\frac{x(s)}{r(s)} = \frac{bl^*}{s - a + bk^*} = \frac{b_m}{s + a_m} = \frac{x_m(s)}{r(s)}. \tag{5.15}$$

Equation (5.15) is satisfied if we choose

$$l^* = \frac{b_m}{b}, \qquad k^* = \frac{a_m + a}{b}, \tag{5.16}$$

provided of course that $b \neq 0$, i.e., the plant (5.12) is controllable. The control law (5.14), (5.16) guarantees that the transfer function of the closed-loop plant, i.e., $\frac{x(s)}{r(s)}$, is equal to that of the reference model. Such a transfer function matching guarantees that $x(t) = x_m(t)$ $\forall t \geq 0$ when $x(0) = x_m(0)$ or $|x(t) - x_m(t)| \to 0$ exponentially fast when $x(0) \neq x_m(0)$ for any bounded reference signal $r(t)$.

When the plant parameters a, b are unknown, (5.14) cannot be implemented. Therefore, instead of (5.14), we propose the control law

$$u = -k(t)x + l(t)r, \tag{5.17}$$

where $k(t)$, $l(t)$ are the estimates of k^*, l^*, respectively, at time t, and search for an adaptive law to generate $k(t)$, $l(t)$ online.

As in the example in section 5.2.1, we can view the problem as an online identification problem of the unknown constants k^*, l^*. We start with the plant equation (5.12) which we express in terms of k^*, l^* by adding and subtracting the desired input term $-bk^*x + bl^*r = -(a_m + a)x + b_m r$ to obtain

$$\dot{x} = -a_m x + b_m r + b(k^* x - l^* r + u),$$

i.e.,

$$x = \frac{b_m}{s + a_m} r + \frac{b}{s + a_m}(k^* x - l^* r + u). \tag{5.18}$$

Since $x_m = \frac{b_m}{s+a_m} r$ is a known bounded signal, we express (5.18) in terms of the tracking error defined as $e \triangleq x - x_m$, i.e.,

$$\text{B-DPM:} \quad e = \frac{b}{s + a_m}(k^* x - l^* r + u)$$

or

$$\text{B-SSPM:} \quad \dot{e} = -a_m e + b(k^* x - l^* r + u). \tag{5.19}$$

The B-DPM can also be expressed in the form of the B-SPM by filtering through with $\frac{1}{s+a_m}$. A wide class of adaptive laws for estimating k^*, l^* can be generated using any of the parametric models in (5.19) and combined with (5.17) to form adaptive control schemes.

A simple adaptive control scheme can be developed by using the B-SSPM in (5.19). The estimation model based on the B-SSPM is given by

$$\dot{\hat{e}} = -a_m \hat{e} + \hat{b}(kx - lr + u), \quad \hat{e}(0) = 0,$$

where $\hat{b}$ is the estimate of b. However, by substituting the control law $u = -k(t)x + l(t)r$, we obtain that $\hat{e}(t) = 0 \; \forall t \geq 0$, which means that there is no need to form the estimation model, and in this case the estimation error is simply the tracking error e. By substituting the control law in the B-SSPM model, we obtain

$$\dot{e} = -a_m e + b(-\tilde{k}x + \tilde{l}r), \tag{5.20}$$

where $\tilde{k} = k - k^*$, $\tilde{l} = l - l^*$ are the parameter errors. The error equation (5.20) relates the parameter errors $\tilde{k}, \tilde{l}$ to the tracking error e and motivates the following Lyapunov-like function for designing the adaptive laws to generate the parameter estimates k, l:

$$V(e, \tilde{k}, \tilde{l}) = \frac{e^2}{2} + \frac{\tilde{k}^2}{2\gamma_1}|b| + \frac{\tilde{l}^2}{2\gamma_2}|b|, \tag{5.21}$$

where $\gamma_1, \gamma_2 > 0$. The time derivative $\dot{V}$ is given by

$$\dot{V} = -a_m e^2 - b\tilde{k}ex + b\tilde{l}er + \frac{|b|\tilde{k}}{\gamma_1}\dot{\tilde{k}} + \frac{|b|\tilde{l}}{\gamma_2}\dot{\tilde{l}}. \tag{5.22}$$

Because $|b| = b\,\text{sgn}(b)$, the indefinite terms in (5.22) disappear if we choose

$$\dot{k} = \gamma_1 \, ex\text{sgn}(b), \qquad \dot{l} = -\gamma_2 \, er\text{sgn}(b), \tag{5.23}$$

which leads to

$$\dot{V} = -a_m e^2. \tag{5.24}$$

Treating $x_m(t), r(t)$ in (5.20) as bounded functions of time, it follows from (5.21), (5.24) that V is a Lyapunov function for the differential equations described by (5.20) and (5.23), and the equilibrium $e_e = 0$, $\tilde{k}_e = 0$, $\tilde{l}_e = 0$ is u.s. Furthermore, $e, \tilde{k}, \tilde{l} \in \mathcal{L}_\infty$. Because $e = x - x_m$ and $x_m \in \mathcal{L}_\infty$, we also have $x \in \mathcal{L}_\infty$ and $u \in \mathcal{L}_\infty$; therefore, all signals in the closed loop are bounded. Now from (5.24) we can establish that $e \in \mathcal{L}_2$ and from (5.20) that $\dot{e} \in \mathcal{L}_\infty$, which implies that $e(t) \to 0$ as $t \to \infty$.

We have established that the control law (5.17) together with the adaptive law (5.23) guarantees boundedness for all signals in the closed-loop system. In addition, the plant state $x(t)$ tracks the state of the reference model $x_m(t)$ asymptotically with time for any reference input signal $r(t)$, which is bounded and piecewise continuous. These results do not imply that $k(t) \to k^*$ and $l(t) \to l^*$ as $t \to \infty$; i.e., the transfer function of the closed-loop plant may not approach that of the reference model as $t \to \infty$. To achieve such a result, the reference input $r(t)$ has to be sufficiently rich of order 2. For example, $r(t) = \sin \omega t$ for some $\omega \neq 0$ guarantees the exponential convergence of $x(t)$ to $x_m(t)$ and of $k(t), l(t)$ to k^*, l^*, respectively. In general, a sufficiently rich reference input $r(t)$ is not desirable, especially in cases where the control objective involves tracking of signals that are not rich in frequencies.

5.2.3 Scalar Example: Indirect MRAC without Normalization

Let us consider the same plant and control objective as in section 5.2.2. As we showed above, when a, b are known, the control law

$$u = -k^* x + l^* r \tag{5.25}$$

with

$$k^* = \frac{a_m + a}{b}, \qquad l^* = \frac{b_m}{b} \tag{5.26}$$

could be used to meet the control objective. In the unknown parameter case, we propose

$$u = -k(t)x + l(t)r, \tag{5.27}$$

where $k(t), l(t)$ are the online estimates of k^*, l^* at time t, respectively. In direct adaptive control, $k(t), l(t)$ are generated directly by an adaptive law. In indirect adaptive control, we follow a different approach. We calculate $k(t), l(t)$ by using the relationship (5.26) and the estimates $\hat{a}$, $\hat{b}$ of the unknown parameters a, b as

$$k(t) = \frac{a_m + \hat{a}(t)}{\hat{b}(t)}, \qquad l(t) = \frac{b_m}{\hat{b}(t)}, \tag{5.28}$$

where $\hat{a}, \hat{b}$ are generated by an adaptive law that we design.

The adaptive law for generating $\hat{a}, \hat{b}$ is obtained by following the same procedure as in the identification examples of Chapter 3; i.e., we rewrite (5.12) in the form of the SSPM

$$\dot{x} = -a_m x + (a + a_m)x + bu.$$

The estimation model based on the above SSPM is given by

$$\dot{\hat{x}} = -a_m \hat{x} + (\hat{a} + a_m)x + \hat{b}u, \tag{5.29}$$

where $\hat{a}, \hat{b}$ are the estimates of a, b at time t, respectively. Note that in the estimation model we treat the signals x, u in the term $(a + a_m)x + bu$ as input signals which are available for measurement; therefore, we do not use the estimate of x. This structure of estimation model is known as the series-parallel model, as it uses both x, $\hat{x}$. Using (5.27), (5.28) in (5.29), we obtain

$$\dot{\hat{x}} = -a_m \hat{x} + b_m r.$$

Since we are designing the estimation model (5.29) and the reference model (5.13), we can choose $\hat{x}(0) = x_m(0)$, which implies that $\hat{x}(t) = x_m(t)\ \forall t \geq 0$. Therefore, the estimation error $\varepsilon = x - \hat{x} = x - x_m = e$ is simply the tracking error, which can be shown to satisfy

$$\dot{e} = -a_m e - \tilde{a}x - \tilde{b}u, \tag{5.30}$$

where $\tilde{a} \triangleq \hat{a} - a$, $\tilde{b} \triangleq \hat{b} - b$ are the parameter errors, by subtracting (5.29) from the SSPM of the plant. Equation (5.30) relates the parameter errors $\tilde{a}$, $\tilde{b}$ to the tracking error e and motivates the choice of

$$V = \frac{1}{2}\left(e^2 + \frac{\tilde{a}^2}{\gamma_1} + \frac{\tilde{b}^2}{\gamma_2}\right) \tag{5.31}$$

for some $\gamma_1, \gamma_2 > 0$, as a potential Lyapunov function candidate for (5.30). The time derivative of V along any trajectory of (5.30) is given by

$$\dot{V} = -a_m e^2 - \tilde{a}xe - \tilde{b}ue + \frac{\tilde{a}\dot{\tilde{a}}}{\gamma_1} + \frac{\tilde{b}\dot{\tilde{b}}}{\gamma_2}. \tag{5.32}$$

Hence, for

$$\dot{\tilde{a}} = \dot{\hat{a}} = \gamma_1 ex, \quad \dot{\tilde{b}} = \dot{\hat{b}} = \gamma_2 eu, \tag{5.33}$$

we have

$$\dot{V} = -a_m e^2 \leq 0,$$

which implies that $e, \hat{a}, \hat{b} \in \mathcal{L}_\infty$ and that $e \in \mathcal{L}_2$ by following the usual arguments. Furthermore, $x_m, e \in \mathcal{L}_\infty$ imply that $x \in \mathcal{L}_\infty$. The boundedness of u, however, cannot be established unless we show that $k(t), l(t)$ are bounded. The boundedness of $\frac{1}{\hat{b}}$ and therefore of $k(t), l(t)$ cannot be guaranteed by the adaptive law (5.33) because (5.33) may generate estimates $\hat{b}(t)$ arbitrarily close or even equal to zero. The requirement that $\hat{b}(t)$ be bounded away from zero is a controllability condition for the estimated plant. Since the control law (5.27) is based on the estimated plant at each time t rather than the actual unknown, this controllability condition arises naturally. One method for preventing $\hat{b}(t)$ from going through zero is to modify the adaptive law for $\hat{b}(t)$ so that adaptation takes place in a closed subset of $\mathcal{R}$ which does not include the zero element. Such a modification is achieved using the following a priori knowledge:

The $\text{sgn}(b)$ and a lower bound $b_0 > 0$ for $|b|$ are known.

Applying the projection method with the constraint $\hat{b}\,\text{sgn}(b) \geq b_0$ to the adaptive law (5.33), we obtain

$$\dot{\hat{a}} = \gamma_1 ex, \dot{\hat{b}} = \begin{cases} \gamma_2 eu & \text{if } |\hat{b}| > b_0 \\ & \text{or if } |\hat{b}| = b_0 \text{ and } eu\,\text{sgn}(b) \geq 0, \\ 0 & \text{otherwise,} \end{cases} \tag{5.34}$$

where $\hat{b}(0)$ is chosen so that $\hat{b}(0)\,\mathrm{sgn}(b) > b_0$. A simple explanation of the projection is as follows: As an example assume that we know a priori that $b \geq 1$. We choose the initial condition $\hat{b}(0) > 1$. If $\hat{b}(t) > 1$, we adapt as usual. If $\hat{b}(t) = 1$ and $\dot{\hat{b}}(t) < 0$ at some time t, $\hat{b}(t)$ is decreasing and will soon become less than 1, in which case we set $\dot{\hat{b}}(t) = 0$. If, on the other hand, $\hat{b}(t) = 1$ and $\dot{\hat{b}}(t) \geq 0$, then $\hat{b}(t)$ is nondecreasing and, therefore, it will not become less than 1, in which case we adapt as usual. This way $\hat{b}(t)$ is guaranteed not to go below 1 at any time t.

It follows from (5.34) that if $\hat{b}(0)\,\mathrm{sgn}(b) \geq b_0$, then whenever $\hat{b}(t)\,\mathrm{sgn}(b) = |\hat{b}(t)|$ becomes equal to b_0 we have $\dot{\hat{b}}\hat{b} \geq 0$, which implies that $|\hat{b}(t)| \geq b_0 \; \forall t \geq 0$. Furthermore, the time derivative of (5.31) along the trajectory of (5.30), (5.34) satisfies

$$\dot{V} = \begin{cases} -a_m e^2 & \text{if } |\hat{b}| > b_0 \text{ or if } |\hat{b}| = b_0 \text{ and } eu\,\mathrm{sgn}(b) \geq 0, \\ -a_m e^2 - \tilde{b}eu & \text{if } |\hat{b}| = b_0 \text{ and } eu\,\mathrm{sgn}(b) < 0. \end{cases}$$

Now for $|\hat{b}| = b_0$ we have $(\hat{b}-b)\,\mathrm{sgn}(b) < 0$. Therefore, for $|\hat{b}| = b_0$ and $eu\,\mathrm{sgn}(b) < 0$, we have

$$\tilde{b}eu = (\hat{b} - b)eu = (\hat{b} - b)\,\mathrm{sgn}(b)(eu\,\mathrm{sgn}(b)) > 0,$$

which implies that the effect of the extra term $-\tilde{b}eu \leq 0$ due to the projection is to make $\dot{V}$ more negative. Therefore,

$$\dot{V} \leq -a_m e^2 \leq 0 \quad \forall t \geq 0,$$

which implies that $V, \hat{a}, \hat{b} \in \mathcal{L}_\infty$. Using the usual arguments, we have $e \in \mathcal{L}_2$ and $\dot{e} \in \mathcal{L}_\infty$, which imply that $e(t) = x(t) - x_m(t) \to 0$ as $t \to \infty$ and therefore that $\dot{\hat{a}}(t), \dot{\hat{b}}(t) \to 0$ as $t \to \infty$. As in the direct case it can be shown that if the reference input signal $r(t)$ is sufficiently rich of order 2, then $\tilde{b}, \tilde{a}$ and therefore $\tilde{k}, \tilde{l}$ converge to zero exponentially fast.

In the above indirect MRAC scheme, the adaptive laws are driven by the tracking error, which is equal to the estimation error without normalization.

5.2.4 Scalar Example: Direct MRAC with Normalization

Let us consider the tracking problem defined in section 5.2.2 for the first-order plant

$$\dot{x} = ax + bu, \tag{5.35}$$

where a, b are unknown (with $b \neq 0$). The control law

$$u = -k^* x + l^* r, \tag{5.36}$$

where

$$k^* = \frac{a_m + a}{b}, \qquad l^* = \frac{b_m}{b}, \tag{5.37}$$

guarantees that all signals in the closed-loop plant are bounded and the plant state x converges exponentially to the state x_m of the reference model

$$x_m = \frac{b_m}{s + a_m} r. \tag{5.38}$$

Because a, b are unknown, we replace (5.36) with

$$u = -k(t)x + l(t)r, \tag{5.39}$$

where $k(t), l(t)$ are the online estimates of k^*, l^*, respectively. We design the adaptive laws for updating $k(t), l(t)$ by first developing appropriate parametric models for k^*, l^* of the form studied in Chapter 3. As in section 5.2.2, if we add and subtract the desired input term $-bk^*x + bl^*r$ in the plant equation (5.35) and use (5.37) to eliminate the unknown a, we obtain

$$\dot{x} = -a_m x + b_m r + b(u + k^* x - l^* r),$$

which together with (5.38) and the definition of $e = x - x_m$ gives

$$e = \frac{b}{s + a_m}[u + k^* x - l^* r]. \tag{5.40}$$

Equation (5.40) can also be rewritten as

$$e = b(\theta^{*T}\phi + u_f), \tag{5.41}$$

where $\theta^* = [k^*, l^*]$, $\phi = \frac{1}{s+a_m}[x, -r]^T$, $u_f = \frac{1}{s+a_m}u$, which is in the form of the B-SPM. Using the techniques in section 3.9, the adaptive law for estimating θ^*, b is given by

$$\begin{aligned} \dot{\theta} &= \Gamma\varepsilon\phi\,\mathrm{sgn}(b), \quad \theta(t) = [k(t), l(t)]^T, \\ \dot{\hat{b}} &= \gamma\varepsilon\xi, \quad \xi = \theta^T\phi + u_f, \\ \varepsilon &= \frac{e - \hat{b}(\theta^T\phi + u_f)}{m_s^2} = \frac{e - \hat{b}\xi}{m_s^2}, \end{aligned} \tag{5.42}$$

where the normalizing signal m_s is chosen as $m_s^2 = 1 + \phi^T\phi + u_f^2$.

It has been shown in section 3.9 that the above adaptive law guarantees that

(i) $\theta, \hat{b} \in \mathcal{L}_\infty$,

(ii) $\varepsilon, \varepsilon m_s, \dot{\theta}, \dot{\hat{b}} \in \mathcal{L}_2 \cap \mathcal{L}_\infty$,

independent of the boundedness of ϕ, e, u.

In the following steps we use properties (i)–(ii) of the adaptive law to first establish signal boundedness and then convergence of the tracking error e to zero.

Step 1. Express the plant output x and input u in terms of the parameter errors $\tilde{k}, \tilde{l}$
From (5.38)–(5.40) we have

$$x = x_m - \frac{b}{s + a_m}(\tilde{k}x - \tilde{l}r) = \frac{1}{s + a_m}(b_m r + b\tilde{l}r - b\tilde{k}x), \tag{5.43}$$

where $\tilde{k} = k - k^*$, $\tilde{l} = l - l^*$, and from (5.35), (5.43) we have

$$u = \frac{(s - a)}{b}x = \frac{s - a}{b(s + a_m)}(b_m r + b\tilde{l}r - b\tilde{k}x). \tag{5.44}$$

For simplicity, let us denote the norm $\|(\cdot)_t\|_{2\delta} = (\int_0^t e^{-\delta(t-\tau)}(\cdot)^T(\cdot)d\tau)^{1/2}$ by $\|\cdot\|$. Again for the sake of clarity and ease of exposition, let us also denote any positive finite constant whose actual value does not affect stability with the same symbol c. Using the properties of the $\mathcal{L}_{2\delta}$ norm in (5.43), (5.44) and the fact that $r, \tilde{l} \in \mathcal{L}_\infty$, we have

$$\|x\| \le c + c\|\tilde{k}x\|, \quad \|u\| \le c + c\|\tilde{k}x\|$$

for any $\delta \in [0, 2a_m)$, which imply that the fictitious normalizing signal defined as

$$m_f^2 \triangleq 1 + \|x\|^2 + \|u\|^2$$

satisfies

$$m_f^2 \le c + c\|\tilde{k}x\|^2. \tag{5.45}$$

We verify, using the boundedness of $r, \tilde{l}, \tilde{k}$, that $\frac{\phi_1}{m_f}, \frac{\dot{x}}{m_f}, \frac{m_s}{m_f} \in \mathcal{L}_\infty$ as follows: From the definition of ϕ_1, we have $|\phi_1(t)| \le c\|x\| \le cm_f$. Similarly, from (5.43) and the boundedness of $r, \tilde{l}, \tilde{k}$, we have

$$|x(t)| \le c + c\|x\| \le c + cm_f.$$

Because $\dot{x} = -a_m x + b_m r + b\tilde{l}r - b\tilde{k}x$, it follows that $|\dot{x}| \le c + cm_f$. Next, let us consider the signal $m_s^2 = 1 + \phi_1^2 + \phi_2^2 + u_f^2$. Because $|u_f| \le c\|u\| \le cm_f$ and $\frac{\phi_1}{m_f}, \phi_2 \in \mathcal{L}_\infty$, it follows that $m_s \le cm_f$.

Step 2. Use the swapping lemma and properties of the $\mathcal{L}_{2\delta}$ norm to upper bound $\|\tilde{k}x\|$ with terms that are guaranteed by the adaptive law to have finite $\mathcal{L}_2$ gains We start with the identity

$$\tilde{k}x = \left(1 - \frac{\alpha_0}{s + \alpha_0}\right)\tilde{k}x + \frac{\alpha_0}{s + \alpha_0}\tilde{k}x = \frac{1}{s + \alpha_0}(\dot{\tilde{k}}x + \tilde{k}\dot{x}) + \frac{\alpha_0}{s + \alpha_0}\tilde{k}x, \tag{5.46}$$

where $\alpha_0 > 0$ is an arbitrary constant. From (5.43) we also have that

$$\tilde{k}x = -\frac{(s + a_m)}{b}e + \tilde{l}r,$$

where $e = x - x_m$, which we substitute in the second term of the right-hand side of (5.46) to obtain

$$\tilde{k}x = \frac{1}{s + \alpha_0}(\dot{\tilde{k}}x + \tilde{k}\dot{x}) - \frac{\alpha_0}{b}\frac{(s + a_m)}{(s + \alpha_0)}e + \frac{\alpha_0}{s + \alpha_0}\tilde{l}r.$$

Because $\tilde{k}, \tilde{l}, r \in \mathcal{L}_\infty$, we have

$$\|\tilde{k}x\| \le \frac{c}{\alpha_0}\|\dot{\tilde{k}}x\| + \frac{c}{\alpha_0}\|\dot{x}\| + \alpha_0 c\|e\| + c \tag{5.47}$$

for any $0 < \delta < 2a_m < \alpha_0$.

The gain of the first two terms on the right-hand side of (5.47) can be reduced by choosing α_0 large. So the only term that needs further examination is $\alpha_0 c\|e\|$. The tracking error e, however, is related to the normalized estimation error ε through the equation

$$e = \varepsilon m_s^2 + \hat{b}\xi$$

that follows from (5.42). Because $\varepsilon, \varepsilon m_s \in \mathcal{L}_\infty \cap \mathcal{L}_2$ and $\hat{b} \in \mathcal{L}_\infty$, the signal we need to concentrate on is ξ, which is given by

$$\xi = k\phi_1 + l\phi_2 + \frac{1}{s+a_m}u.$$

We consider the equation

$$\frac{1}{s+a_m}u = \frac{1}{s+a_m}(-kx + lr) = -k\phi_1 - l\phi_2 + \frac{1}{s+a_m}(\dot{k}\phi_1 + \dot{l}\varphi_2)$$

obtained by using Swapping Lemma 1 (see Lemma A.11.1 in the Appendix) or the equation

$$(s+a_m)(k\phi_1 + l\phi_2) = kx - lr + (\dot{k}\phi_1 + \dot{l}\phi_2) = -u + \dot{k}\phi_1 + \dot{l}\phi_2,$$

where s is treated as the differential operator.

Using any one of the above equations to substitute for $k_1\phi_1 + l\phi_2$ in the equation for ξ, we obtain

$$\xi = \frac{1}{s+a_m}(\dot{k}\phi_1 + \dot{l}\phi_2).$$

Hence

$$e = \varepsilon m_s^2 + \hat{b}\frac{1}{s+a_m}(\dot{k}\phi_1 + \dot{l}\phi_2). \tag{5.48}$$

Since $\varepsilon m_s, \hat{b}, \dot{l}, \phi_2, r \in \mathcal{L}_\infty$, it follows from (5.48) that

$$\|e\| \le c + \|\varepsilon m_s^2\| + c\|\dot{k}\phi_1\|, \tag{5.49}$$

and therefore (5.47) and (5.49) imply that

$$\|\tilde{k}x\| \le c + c\alpha_0 + \frac{c}{\alpha_0}\|\dot{\tilde{k}}x\| + \frac{c}{\alpha_0}\|\dot{x}\| + \alpha_0 c\|\varepsilon m_s^2\| + \alpha_0 c\|\dot{k}\phi_1\|. \tag{5.50}$$

Step 3. Use the B–G lemma to establish boundedness Using (5.50) and the normalizing properties of m_f, we can write (5.45) in the form

$$\begin{aligned} m_f^2 &\le c + c\alpha_0^2 + \frac{c}{\alpha_0^2}\|\dot{k}m_f\|^2 + \frac{c}{\alpha_0^2}\|m_f\|^2 + \alpha_0^2 c\|\varepsilon m_s m_f\|^2 + \alpha_0^2 c\|\dot{k}m_f\|^2 \\ &\le c + c\alpha_0^2 + c\alpha_0^2\|\tilde{g}m_f\|^2 + \frac{c}{\alpha_0^2}\|m_f\|^2, \end{aligned} \tag{5.51}$$

where $\tilde{g}^2 \triangleq \frac{1}{\alpha_0^4}|\dot{k}|^2 + |\varepsilon^2 m_s^2| + |\dot{k}|$ and $\tilde{g} \in \mathcal{L}_2$ due to properties of the adaptive law. The inequality (5.51) implies that

$$m_f^2(t) \le c + c\alpha_0^2 + c\int_0^t e^{-\delta(t-\tau)}\left(\alpha_0^2\tilde{g}^2(\tau) + \frac{1}{\alpha_0^2}\right)m_f^2(\tau)d\tau\,\|m_f\|^2.$$

We apply B–G Lemma 3 (see Lemma A.6.3) with $c_0 = 0$, $c_1 = c + c\alpha_0^2$, $c_2 = c$, $k(\tau) = \alpha_0^2 \tilde{g}^2(\tau) + \frac{1}{\alpha_0^2}$, $\alpha = \delta$, and $y(t) = m_f^2(t)$ to obtain

$$m_f^2(t) \leq (c + c\alpha_0^2)e^{-\delta t}\Phi(t, 0) + (c + c\alpha_0^2)\delta \int_0^t e^{-\delta(t-\tau)}\Phi(t, \tau)d\tau,$$

where $\Phi(t, \tau) = e^{\frac{c}{\alpha_0^2}(t-\tau)} e^{c\alpha_0^2 \int_\tau^t \tilde{g}^2(\sigma)d\sigma}$. Since $\tilde{g} \in \mathcal{L}_2$, we have $\int_\tau^t \tilde{g}^2(\sigma)d\sigma \leq c$ and therefore

$$\Phi(t, \tau) \leq ce^{\frac{c}{\alpha_0^2}(t-\tau)}.$$

Hence

$$m_f^2(t) \leq (c + c\alpha_0^2)e^{-\beta t} + (c + c\alpha_0^2)\delta \int_0^t e^{-\beta(t-\tau)}d\tau, \tag{5.52}$$

where $\beta = \delta - \frac{c}{\alpha_0^2}$. Since α_0 is arbitrary, it can be chosen so that $\beta > 0$ for any given $\delta > 0$. Therefore, we can establish, using (5.52), that

$$m_f^2(t) \leq c + ce^{-\beta t},$$

i.e., $m_f \in \mathcal{L}_\infty$. From $m_f \in \mathcal{L}_\infty$ we have $x, \dot{x}, m_s, \phi_1 \in \mathcal{L}_\infty$, which imply that u and all signals in the closed-loop plant are bounded.

Step 4. Establish convergence of the tracking error to zero We show the convergence of the tracking error to zero by using (5.48). Since $\phi_1, \phi_2, \hat{b}, m_s \in \mathcal{L}_\infty$, we can conclude from $\varepsilon m_s, \dot{k}, \dot{l} \in \mathcal{L}_2$ that $e \in \mathcal{L}_2$. Since $e \in \mathcal{L}_2 \cap \mathcal{L}_\infty$ and $\dot{e} = \dot{x} - \dot{x}_m \in \mathcal{L}_\infty$, we have $e(t) \to 0$ as $t \to \infty$.

The above example demonstrates that even though the design approach for the adaptive controller is rather straightforward the use of normalization makes the estimation error different from the tracking error. Since boundedness of the estimation error guaranteed by the adaptive law does not directly imply boundedness for the tracking error as was the case with the adaptive control examples without normalization, additional steps are used to establish that. These steps are rather complicated and long even for simple examples. However, once understood, the analysis for the case of higher-order plants is very similar.

5.2.5 Scalar Example: Indirect MRAC with Normalization

Let us consider the same plant

$$\dot{x} = ax + bu.$$

We can establish that the parameters a, b satisfy the SPM

$$z = \theta^{*T}\phi,$$

where

$$z = \frac{s}{s+\lambda}x, \qquad \theta^* = [a, b]^T, \qquad \phi = \frac{1}{s+\lambda}[x, u]^T,$$

$\lambda > 0$ is a design constant. A wide class of adaptive laws may be designed using this SPM. Let us consider the gradient algorithm

$$\dot{\theta} = \Gamma \varepsilon \phi, \quad \varepsilon = \frac{z - \theta^T \phi}{m_s^2}, \tag{5.53}$$

where $\Gamma = \Gamma^T > 0$ and m_s is chosen so that $\frac{\phi}{m_s} \in \mathcal{L}_\infty$. As shown in Chapter 3, the above adaptive law guarantees that

(i) $\theta, \varepsilon, \varepsilon m_s \in \mathcal{L}_\infty$,

(ii) $\dot{\theta}, \varepsilon, \varepsilon m_s \in \mathcal{L}_2$,

independent of the boundedness of ϕ, x, u.

The indirect MRAC scheme is formed as

$$u(t) = -k(t)x(t) + l(t)r(t), \quad k(t) = \frac{a_m + \hat{a}(t)}{\hat{b}(t)}, \quad l(t) = \frac{b_m}{\hat{b}(t)}, \tag{5.54}$$

where $\hat{a}, \hat{b}$ are the elements of $\theta = [\hat{a}, \hat{b}]^T$. Due to division by $\hat{b}$, the gradient algorithm has to guarantee that $\hat{b}$ does not become equal to zero. Therefore, instead of (5.53), we use

$$\dot{\hat{a}} = \gamma_1 \varepsilon \phi_1, \quad \dot{\hat{b}} = \begin{cases} \gamma_2 \varepsilon \phi_2 & \text{if } |\hat{b}| < b_0 \text{ or if } |\hat{b}| = b_0 \text{ and } \varepsilon \phi_2 \operatorname{sgn}(b) \geq 0, \\ 0 & \text{otherwise,} \end{cases} \quad \hat{b}(0)\operatorname{sgn}(b) > 0, \quad \varepsilon = \frac{z - \hat{a}\phi_1 - \hat{b}\phi_2}{m_s^2}, \tag{5.55}$$

where ϕ_1, ϕ_2 are the elements of $\phi = [\phi_1, \phi_2]^T$, b is assumed to satisfy $|b| \geq b_0$ for some known $b_0 > 0$, and the sign of b is known.

As shown in Chapter 3, the above adaptive law guarantees that

(i) $\hat{b}(t)\operatorname{sgn}(b) > 0 \; \forall t$;

(ii) $\hat{a}, \hat{b}, \varepsilon, \varepsilon m_s \in \mathcal{L}_\infty$;

(iii) $\varepsilon, \varepsilon m_s, \dot{\hat{a}}, \dot{\hat{b}} \in \mathcal{L}_2$.

The analysis steps are the same as in the direct MRAC scheme with normalization.

Step 1. Express the plant output x and input u in terms of the parameter errors $\tilde{k}, \tilde{l}$
The equations are the same as in Step 1 of the direct MRAC with normalization, i.e.,

$$x = \frac{1}{s + a_m}(b_m r + b\tilde{l}r - b\tilde{k}x), \quad u = \frac{s - a}{b(s + a_m)}(b_m r + b\tilde{l}r - b\tilde{k}x), \tag{5.56}$$

and the fictitious normalizing signal $m_f^2 \triangleq 1 + \|x\|^2 + \|u\|^2$ satisfies

$$m_f^2 \leq c + c\|\tilde{l}r - \tilde{k}x\|^2. \tag{5.57}$$

Furthermore, as in section 5.2.4, m_f bounds from above ϕ_1, ϕ_2, m_s, $\dot{x}$, x, u for $0 \leq \delta < 2\min(a_m, \lambda)$.

Step 2. Obtain an upper bound for $\|\tilde{l}r - \tilde{k}x\|^2$ with terms guaranteed by the adaptive law to have finite $\mathcal{L}_2$ gains From the adaptive law (5.55) we have that

$$-\varepsilon m_s^2 = \tilde{a}\phi_1 + \tilde{b}\phi_2,$$

where $\tilde{a} = \hat{a} - a$, $\tilde{b} = \hat{b} - b$, $\phi_1 = \frac{1}{s+\lambda}x$, $\phi_2 = \frac{1}{s+\lambda}u$. The adaptive law guarantees that $\varepsilon m_s, \dot{\tilde{a}}, \dot{\tilde{b}} \in \mathcal{L}_2$. Therefore, we need to use these properties and obtain a bound for $\|\tilde{l}r - \tilde{k}x\|^2$ which depends on εm_s, $\dot{\tilde{a}}$, $\dot{\tilde{b}}$ in a way that allows the use of the B–G lemma to establish boundedness for m_f the same way as in section 5.2.4.

From (5.56), we have that

$$r = \frac{s + a_m}{b_m}x - \frac{b}{b_m}(\tilde{l}r - \tilde{k}x).$$

Since

$$1 + \frac{b}{b_m}\tilde{l} = 1 + \frac{b}{b_m}\left(\frac{b_m}{\hat{b}} - \frac{b_m}{b}\right) = \frac{b}{\hat{b}},$$

we have

$$r = \frac{\hat{b}}{b}\left(\frac{s + a_m}{b_m}x + \frac{b}{b_m}\tilde{k}x\right)$$

and hence

$$\begin{aligned} -\tilde{k}x + \tilde{l}r &= -\tilde{k}x + \tilde{l}\frac{\hat{b}}{b}\left(\frac{s + a_m}{b_m}x + \frac{b}{b_m}\tilde{k}x\right) \\ &= -\tilde{k}x + \frac{\tilde{l}\hat{b}}{bb_m}((a + a_m)x + bu + b\tilde{k}x). \end{aligned}$$

Substituting for $\tilde{k} = \frac{\hat{a}+a_m}{\hat{b}} - \frac{a+a_m}{b}$, $\tilde{l} = \frac{b_m}{\hat{b}} - \frac{b_m}{b}$, we show that

$$-\tilde{k}x + \tilde{l}r = -\frac{1}{b}(\tilde{a}x + \tilde{b}u). \tag{5.58}$$

Let us now consider the identity

$$\begin{aligned} -\tilde{k}x + \tilde{l}r &= \left(1 - \frac{\alpha_0}{s + \alpha_0}\right)(-\tilde{k}x + \tilde{l}r) + \frac{\alpha_0}{s + \alpha_0}(-\tilde{k}x + \tilde{l}r) \\ &= \frac{s}{s + \alpha_0}(-\tilde{k}x + \tilde{l}r) + \frac{\alpha_0}{s + \alpha_0}(-\tilde{k}x + \tilde{l}r) \\ &= \frac{1}{s + \alpha_0}(-\dot{\tilde{k}}x + \dot{\tilde{l}}r - \tilde{k}\dot{x} + \tilde{l}\dot{r}) + \frac{\alpha_0}{s + \alpha_0}(-\tilde{k}x + \tilde{l}r) \end{aligned} \tag{5.59}$$

for any arbitrary constant $\alpha_0 > 0$. Let us also consider the equation

$$-(s+\lambda)(\varepsilon m_s^2) = (s+\lambda)(\tilde{a}\phi_1 + \tilde{b}\phi_2) = \dot{\tilde{a}}\phi_1 + \dot{\tilde{b}}\phi_2 + \tilde{a}x + \tilde{b}u.$$

Hence

$$\tilde{a}x + \tilde{b}u = -(s+\lambda)(\varepsilon m_s^2) - \dot{\tilde{a}}\phi_1 - \dot{\tilde{b}}\phi_2. \tag{5.60}$$

Using (5.58) and (5.60), we have

$$\begin{aligned}\frac{\alpha_0}{s+\alpha_0}(-\tilde{k}x + \tilde{l}r) &= -\frac{\alpha_0}{b(s+\alpha_0)}(\tilde{a}x + \tilde{b}u) \\ &= \frac{\alpha_0(s+\lambda)}{b(s+\alpha_0)}(\varepsilon m_s^2) + \frac{\alpha_0}{b(s+\alpha_0)}(\dot{\tilde{a}}\phi_1 + \dot{\tilde{b}}\phi_2).\end{aligned} \tag{5.61}$$

Substituting (5.61) into (5.59), we have

$$-\tilde{k}x + \tilde{l}r = \frac{1}{s+\alpha_0}(-\dot{\tilde{k}}x + \dot{\tilde{l}}r - \tilde{k}\dot{x} + \tilde{l}\dot{r}) + \frac{\alpha_0(s+\lambda)}{b(s+\alpha_0)}(\varepsilon m_s^2) + \frac{\alpha_0}{b(s+\alpha_0)}(\dot{\tilde{a}}\phi_1 + \dot{\tilde{b}}\phi_2).$$

Since $\|\frac{1}{s+\alpha_0}\|_{\infty\delta} \le \frac{c}{\alpha_0}$, $\|\frac{\alpha_0}{s+\alpha_0}\|_{\infty\delta} \le c$, $\tilde{k}, \tilde{l} \in \mathcal{L}_\infty$, and $\dot{x}$, $\dot{r}$, m_s, ϕ_1, ϕ_2 are bounded from above by m_f, we have that

$$\|-\tilde{k}x + \tilde{l}r\| \le \frac{c}{\alpha_0}\|g_1 m_f\| + \frac{c}{\alpha_0}\|m_f\| + c\|g_2 m_f\| + c\alpha_0\|g_3 m_f\|, \tag{5.62}$$

where $g_1 = c(|\dot{\tilde{k}}| + |\dot{\tilde{l}}|) \in \mathcal{L}_2$, $g_2 = c(|\dot{\tilde{a}}| + |\dot{\tilde{b}}|) \in \mathcal{L}_2$, $g_3 = c|\varepsilon m_s| \in \mathcal{L}_2$.

Step 3. Use the B–G lemma to establish boundedness Using (5.62) in (5.57), we have

$$m_f^2 \le c + \frac{c}{\alpha_0^2}\|g_1 m_f\|^2 + \frac{c}{\alpha_0^2}\|m_f\|^2 + c\|g_2 m_f\|^2 + c\alpha_0^2\|g_3 m_f\|^2$$

or

$$m_f^2(t) \le c + c\int_0^t e^{-\delta(t-\tau)}\left(\frac{1}{\alpha_0^2} + \frac{1}{\alpha_0^2}g_1^2(\tau) + g_2^2(\tau) + \alpha_0^2 g_3^2(\tau)\right)m_f^2(\tau)d\tau.$$

Since α_0 is arbitrary and can be made large and $g_1, g_2, g_3 \in \mathcal{L}_2$, the boundedness of m_f follows by applying B–G Lemma 3 in exactly the same way as in Step 3 of section 5.2.4. Since m_f bounds all signals from above, it follows that all signals are bounded.

Step 4. Establish the convergence of the tracking error to zero From (5.56), we have

$$e = x - x_m = \frac{b}{s+a_m}(-\tilde{k}x + \tilde{l}r),$$

which together with (5.58), (5.60) implies that

$$e = -\frac{1}{s+a_m}(\tilde{a}x + \tilde{b}u) = \frac{s+\lambda}{s+a_m}(\varepsilon m_s^2) + \frac{1}{s+a_m}(\dot{\tilde{a}}\phi_1 + \dot{\tilde{b}}\phi_2).$$

Since $m_s, \phi_1, \phi_2 \in \mathcal{L}_\infty$ and $\varepsilon m_s, \dot{\tilde{a}}, \dot{\tilde{b}} \in \mathcal{L}_2$, it follows that $\varepsilon m_s^2, \dot{\tilde{a}}\phi_1 + \dot{\tilde{b}}\phi_2 \in \mathcal{L}_2$, which in turn implies that $e \in \mathcal{L}_2$. Since

$$\dot{e} = \frac{bs}{s + a_m}(-\tilde{k}x + \tilde{l}r)$$

implies that $\dot{e} \in \mathcal{L}_\infty$ due to $\tilde{k}x, \tilde{l}r \in \mathcal{L}_\infty$, by applying Lemma A.4.7, we can conclude from $e \in \mathcal{L}_2 \cap \mathcal{L}_\infty$ and $\dot{e} \in \mathcal{L}_\infty$ that $e(t) \to 0$ as $t \to \infty$.

5.2.6 Vector Case: Full-State Measurement

Let us now consider the nth-order plant

$$\dot{x} = Ax + Bu, \quad x \in \mathcal{R}^n, \tag{5.63}$$

where $A \in \mathcal{R}^{n\times n}$, $B \in \mathcal{R}^{n\times q}$ are unknown constant matrices and (A, B) is controllable. The control objective is to choose the input vector $u \in \mathcal{R}^q$ such that all signals in the closed-loop plant are bounded and the plant state x follows the state $x_m \in \mathcal{R}^n$ of a reference model. The reference model is described by

$$\dot{x}_m = A_m x_m + B_m r, \tag{5.64}$$

where $A_m \in \mathcal{R}^{n\times n}$ is a stable matrix, $B_m \in \mathcal{R}^{n\times q}$, and $r \in \mathcal{R}^q$ is a bounded reference input vector. The reference model and input r are chosen so that $x_m(t)$ represents a desired trajectory that x has to follow.

If the matrices A, B were known, we could apply the control law

$$u = -K^*x + L^*r \tag{5.65}$$

and obtain the closed-loop plant

$$\dot{x} = (A - BK^*)x + BL^*r. \tag{5.66}$$

Hence, if $K^* \in \mathcal{R}^{q\times n}$ and $L^* \in \mathcal{R}^{q\times q}$ are chosen to satisfy the algebraic equations

$$A - BK^* = A_m, \qquad BL^* = B_m, \tag{5.67}$$

then the transfer matrix of the closed-loop plant is the same as that of the reference model and $x(t) \to x_m(t)$ exponentially fast for any bounded reference input signal $r(t)$. We should note that in general, no K^*, L^* may exist to satisfy the matching condition (5.67) for the given matrices A, B, A_m, B_m, indicating that the control law (5.65) may not have enough structural flexibility to meet the control objective. In some cases, if the structure of A, B is known, A_m, B_m may be designed so that (5.67) has a solution for K^*, L^*.

Let us assume that K^*, L^* in (5.67) exist, i.e., that there is sufficient structural flexibility to meet the control objective, and propose the control law

$$u = -K(t)x + L(t)r, \tag{5.68}$$

where $K(t)$, $L(t)$ are the estimates of K^*, L^*, respectively, to be generated by an appropriate adaptive law.

By adding and subtracting the desired input term, namely, $-B(K^*x - L^*r)$ in the plant equation, and using (5.67), we obtain

$$\dot{x} = A_m x + B_m r + B(K^*x - L^*r + u), \tag{5.69}$$

which is the extension of the scalar equation (5.18) in section 5.2.2 to the vector case. Defining the tracking error $e = x - x_m$ and subtracting (5.64) from (5.69), we obtain the tracking error equation

$$\dot{e} = A_m e + B(K^*x - L^*r + u),$$

which is in the form of the B-SSPM. The estimation model is given by

$$\dot{\hat{e}} = A_m \hat{e} + B(Kx - Lr + u), \quad \hat{e}(0) = 0.$$

Due to the control law $u = -K(t)x + L(t)r$, the signal $\hat{e}(t) = 0 \; \forall t \geq 0$ and the estimation error $\varepsilon = e - \hat{e} = e$. Therefore, the estimation model is not needed, and the tracking error e is used as the estimation error.

Following the same procedure as in section 5.2.2, we can show that the tracking error $e = x - x_m$ and parameter error $\tilde{K} \triangleq K - K^*$, $\tilde{L} \triangleq L - L^*$ satisfy the equation

$$\dot{e} = A_m e + B(-\tilde{K}x + \tilde{L}r), \tag{5.70}$$

which also depends on the unknown matrix B. In the scalar case we manage to get away with the unknown B by assuming that its sign is known. An extension of the scalar assumption of section 5.2.2 to the vector case is as follows: Let us assume that L^* is either positive definite or negative definite and $\Gamma^{-1} = L^* \operatorname{sgn}(l)$, where $l = 1$ if L^* is positive definite and $l = -1$ if L^* is negative definite. Then $B = B_m L^{*-1}$ and (5.70) becomes

$$\dot{e} = A_m e + B_m L^{*-1}(-\tilde{K}x + \tilde{L}r).$$

We propose the following Lyapunov function candidate:

$$V(e, \tilde{K}, \tilde{L}) = e^T P e + \operatorname{tr}(\tilde{K}^T \Gamma \tilde{K} + \tilde{L}^T \Gamma \tilde{L}),$$

where $P = P^T > 0$ satisfies the Lyapunov equation

$$P A_m + A_m^T P = -Q$$

for some $Q = Q^T > 0$. Then

$$\dot{V} = -e^T Q e + 2e^T P B_m L^{*-1}(-\tilde{K}x + \tilde{L}r) + 2\operatorname{tr}(\tilde{K}^T \Gamma \dot{\tilde{K}} + \tilde{L}^T \Gamma \dot{\tilde{L}}).$$

Now

$$e^T P B_m L^{*-1} e K x = \operatorname{tr}(x^T \tilde{K}^T \Gamma B_m^T P e) \operatorname{sgn}(l) = \operatorname{tr}(\tilde{K}^T \Gamma B_m^T P e x^T) \operatorname{sgn}(l)$$

and

$$e^T P B_m L^{*-1} \tilde{L} r = \operatorname{tr}(\tilde{L}^T \Gamma B_m^T P e r^T) \operatorname{sgn}(l).$$

Therefore, for

$$\dot{\tilde{K}} = \dot{K} = B_m^T P e x^T \operatorname{sgn}(l), \qquad \dot{\tilde{L}} = \dot{L} = -B_m^T P e r^T \operatorname{sgn}(l), \tag{5.71}$$

we have

$$\dot{V} = -e^T Q e \leq 0.$$

From the properties of V, $\dot{V}$ we establish, as in the scalar case, that $K(t)$, $L(t)$, $e(t)$ are bounded and that $e(t) \to 0$ as $t \to \infty$. The adaptive control scheme developed is given by (5.68) and (5.71). The matrix $B_m^T P$ acts as an adaptive gain matrix, where P is obtained by solving the Lyapunov equation $PA_m + A_m^T P = -Q$ for some arbitrary $Q = Q^T > 0$. Different choices of Q will not affect boundedness and the asymptotic behavior of the scheme, but they will affect the transient response. The assumption that the unknown L^* in the matching equation $BL^* = B_m$ is either positive or negative definite imposes an additional restriction on the structure and elements of B, B_m. Since B is unknown, this assumption may not be realistic in some applications.

The above MRAC scheme belongs to the class of direct MRAC without normalization in the adaptive law. The same methodology can be extended to the other classes of MRAC shown in Figure 5.4.

The MRAC schemes designed in this section are based on the assumption that the plant state is available for measurement. In the following sections, we relax this assumption and develop different classes of MRAC schemes shown in Figure 5.4 for a general class of plants.

5.3 MRC for SISO Plants

In the general case, the design of the control law is not as straightforward as it appears to be in the case of the examples of section 5.2. For this reason, we use this section to formulate the MRC problem for a general class of LTI SISO plants and solve it for the case where the plant parameters are known exactly. The significance of the existence of a control law that solves the MRC problem is twofold: First, it demonstrates that given a set of assumptions about the plant and reference model, there is enough structural flexibility to meet the control objective; second, it provides the form of the control law that is to be combined with an adaptive law to form MRAC schemes in the case of unknown plant parameters to be treated in the sections to follow.

5.3.1 Problem Statement

Consider the SISO LTI plant described by the vector differential equation

$$\begin{aligned} \dot{x}_p &= A_p x_p + B_p u_p, \quad x_p(0) = x_0, \\ y_p &= C_p^T x_p, \end{aligned} \tag{5.72}$$

where $x_p \in \mathcal{R}^n$; $y_p, u_p \in \mathcal{R}$; and A_p, B_p, C_p have the appropriate dimensions. The transfer function of the plant is given by

$$y_p = G_p(s) u_p \tag{5.73}$$

with $G_p(s)$ expressed in the form

$$G_p(s) = k_p \frac{Z_p(s)}{R_p(s)}, \tag{5.74}$$

where Z_p, R_p are monic polynomials and k_p is a constant referred to as the high-frequency gain. The reference model, selected by the designer to describe the desired characteristics of the plant, is described by the differential equation

$$\begin{aligned} \dot{x}_m &= A_m x_m + B_m r, \quad x_m(0) = x_{m0}, \\ y_m &= C_m^T x_m, \end{aligned} \tag{5.75}$$

where $x_m \in \mathcal{R}^{p_m}$ for some integer p_m; $y_m, r \in \mathcal{R}$; and r is the reference input which is assumed to be a uniformly bounded piecewise continuous function of time. The transfer function of the reference model given by

$$y_m = W_m(s)r$$

is expressed in the same form as (5.74), i.e.,

$$W_m(s) = k_m \frac{Z_m(s)}{R_m(s)}, \tag{5.76}$$

where $Z_m(s)$, $R_m(s)$ are monic polynomials and k_m is a constant.

The MRC objective is to determine the plant input u_p so that all signals are bounded and the plant output y_p tracks the reference model output y_m as close as possible for *any* given reference input $r(t)$ of the class defined above. We refer to the problem of finding the desired u_p to meet the control objective as the *MRC problem*.

In order to meet the MRC objective with a control law that is free of differentiators and uses only measurable signals, we assume that the plant and reference models satisfy the following assumptions.

Plant assumptions:

P1. $Z_p(s)$ is a monic Hurwitz polynomial.

P2. An upper bound n of the degree n_p of $R_p(s)$ is known.

P3. The relative degree $n^* = n_p - m_p$ of $G_p(s)$ is known, where m_p is the degree of $Z_p(s)$.

P4. The sign of the high-frequency gain k_p is known.

Reference model assumptions:

M1. $Z_m(s)$, $R_m(s)$ are monic Hurwitz polynomials of degree q_m, p_m, respectively, where $p_m \leq n$.

M2. The relative degree $n_m^* = p_m - q_m$ of $W_m(s)$ is the same as that of $G_p(s)$, i.e., $n_m^* = n^*$.

Remark 5.3.1 Assumption P1 requires that the plant transfer function $G_p(s)$ be minimum phase. We make no assumptions, however, about the location of the poles of $G_p(s)$; i.e., the plant is allowed to have unstable poles. We allow the plant to be uncontrollable or unobservable; i.e., we allow common zeros and poles in the plant transfer function. Since, by assumption P1, all the plant zeros are in $\mathcal{C}^-$, any zero-pole cancellation can occur only in $\mathcal{C}^-$, which implies that the plant (5.72) is both stabilizable and detectable.

The minimum-phase assumption (P1) is a consequence of the control objective which is met by designing an MRC control law that cancels the zeros of the plant and replaces them with those of the reference model in an effort to force the closed-loop plant transfer function from r to y_p to be equal to $W_m(s)$. For stability, such cancellations should occur in $\mathcal{C}^-$, which implies that $Z_p(s)$ should satisfy assumption P1.

5.3.2 MRC Schemes: Known Plant Parameters

In addition to assumptions P1–P4 and M1, M2, let us also assume that the coefficients of $G_p(s)$ are known exactly. Because the plant is LTI and known, the design of the MRC scheme is achieved using linear system theory. The MRC objective is met if u_p is chosen so that the closed-loop transfer function from r to y_p has stable poles and is equal to $W_m(s)$, the transfer function of the reference model. Such a transfer function matching guarantees that for any reference input signal $r(t)$, the plant output y_p converges to y_m exponentially fast.

Let us consider the feedback control law

$$u_p = \theta_1^{*T}\frac{\alpha(s)}{\Lambda(s)}u_p + \theta_2^{*T}\frac{\alpha(s)}{\Lambda(s)}y_p + \theta_3^* y_p + c_0^* r \tag{5.77}$$

shown in Figure 5.5, where

$$\alpha(s) \triangleq \alpha_{n-2}(s) = [s^{n-2}, s^{n-3}, \dots, s, 1]^T \quad \text{for } n \geq 2,$$
$$\alpha(s) \triangleq 0 \quad \text{for } n = 1.$$

$c_0^*, \theta_3^* \in \mathcal{R}$; $\theta_1^*, \theta_2^* \in \mathcal{R}^{n-1}$ are constant parameters to be designed; and $\Lambda(s)$ is an arbitrary monic Hurwitz polynomial of degree $n-1$ that contains $Z_m(s)$ as a factor, i.e.,

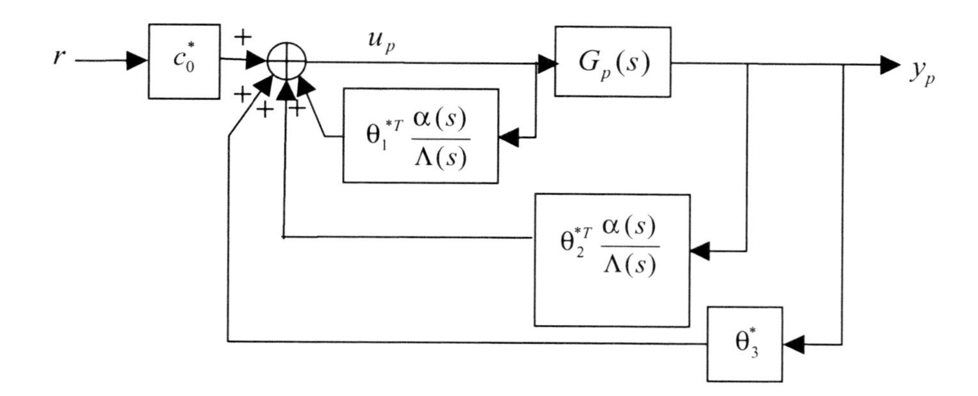

Figure 5.5. *Structure of the MRC scheme* (5.77).

$$\Lambda(s) = \Lambda_0(s) Z_m(s),$$

which implies that $\Lambda_0(s)$ is monic, Hurwitz, and of degree $n_0 = n - 1 - q_m$. The controller parameter vector

$$\theta^* = [\theta_1^{*T}, \theta_2^{*T}, \theta_3^*, c_0^*]^T \in \mathcal{R}^{2n}$$

is to be chosen so that the transfer function from r to y_p is equal to $W_m(s)$.

The I/O properties of the closed-loop plant shown in Figure 5.5 are described by the transfer function equation

$$y_p = G_c(s) r, \tag{5.78}$$

where

$$G_c(s) = \frac{c_0^* k_p Z_p(s) \Lambda^2(s)}{\Lambda(s)[(\Lambda(s) - \theta_1^{*T}\alpha(s))R_p(s) - k_p Z_p(s)(\theta_2^{*T}\alpha(s) + \theta_3^* \Lambda(s))]}. \tag{5.79}$$

We can now meet the control objective if we select the controller parameters θ_1^*, θ_2^*, θ_3^*, c_0^* so that the closed-loop poles are stable and the closed-loop transfer function $G_c(s) = W_m(s)$, i.e.,

$$\frac{c_0^* k_p Z_p \Lambda^2}{\Lambda[(\Lambda - \theta_1^{*T}\alpha)R_p - k_p Z_p(\theta_2^{*T}\alpha + \theta_3^* \Lambda)]} = k_m \frac{Z_m}{R_m}, \tag{5.80}$$

is satisfied for all $s \in \mathcal{C}$. Choosing

$$c_0^* = \frac{k_m}{k_p} \tag{5.81}$$

and using $\Lambda(s) = \Lambda_0(s) Z_m(s)$, the matching equation (5.80) becomes

$$(\Lambda - \theta_1^{*T}\alpha)R_p - k_p Z_p(\theta_2^{*T}\alpha + \theta_3^* \Lambda) = Z_p \Lambda_0 R_m \tag{5.82}$$

or

$$\theta_1^{*T}\alpha(s)R_p(s) + k_p(\theta_2^{*T}\alpha(s) + \theta_3^* \Lambda(s))Z_p(s) = \Lambda(s)R_p(s) - Z_p(s)\Lambda_0(s)R_m(s). \tag{5.83}$$

Equating the coefficients of the powers of s on both sides of (5.83), we can express (5.82) in terms of the algebraic equation

$$S\bar{\theta}^* = p, \tag{5.84}$$

where $\bar{\theta}^* = [\theta_1^{*T}, \theta_2^{*T}, \theta_3^*]^T$; S is an $(n + n_p - 1) \times (2n - 1)$ matrix that depends on the coefficients of R_p, $k_p Z_p$, and Λ; and p is an $n + n_p - 1$ vector with the coefficients of $\Lambda R_p - Z_p \Lambda_0 R_m$. The existence of $\bar{\theta}^*$ to satisfy (5.84) and, therefore, (5.83) will very much depend on the properties of the matrix S. For example, if $n > n_p$, more than one $\bar{\theta}^*$ will satisfy (5.84), whereas if $n = n_p$ and S is nonsingular, (5.84) will have only one solution.

Remark 5.3.2 For the design of the control input (5.77), we assume that $n \geq n_p$. Since the plant is known exactly, there is no need to assume an upper bound for the degree of the plant; i.e., since n_p is known, n can be set equal to n_p. We use $n \geq n_p$ on purpose in order to use the result in the unknown plant parameter case treated in sections 5.4 and 5.5, where only the upper bound n for n_p is known.

Remark 5.3.3 Instead of using (5.84), one can solve (5.82) for θ_1^*, θ_2^*, θ_3^* as follows: Dividing both sides of (5.82) by $R_p(s)$, we obtain

$$\Lambda - \theta_1^{*T}\alpha - k_p \frac{Z_p}{R_p}(\theta_2^{*T}\alpha + \theta_3^*\Lambda) = Z_p\left(Q + k_p \frac{\Delta^*}{R_p}\right),$$

where $Q(s)$ (of degree $n - 1 - m_p$) is the quotient and $k_p\Delta^*$ (of degree at most $n_p - 1$) is the remainder of $\frac{\Lambda_0 R_m}{R_p}$, respectively. Then the solution for θ_i^*, $i = 1, 2, 3$, can be found by inspection, i.e.,

$$\theta_1^{*T}\alpha(s) = \Lambda(s) - Z_p(s)Q(s), \tag{5.85}$$

$$\theta_2^{*T}\alpha(s) + \theta_3^*\Lambda(s) = \frac{Q(s)R_p(s) - \Lambda_0(s)R_m(s)}{k_p}, \tag{5.86}$$

where the equality in the second equation is obtained by substituting for $\Delta^*(s)$ using the identity $\frac{\Lambda_0 R_m}{R_p} = Q + \frac{k_p\Delta^*}{R_p}$. The parameters θ_i^*, $i = 1, 2, 3$, can now be obtained directly by equating the coefficients of the powers of s on both sides of (5.85), (5.86).

Equations (5.85) and (5.86) indicate that in general the controller parameters θ_i^*, $i = 1, 2, 3$, are nonlinear functions of the coefficients of the plant polynomials $Z_p(s)$, $R_p(s)$ due to the dependence of $Q(s)$ on the coefficients of $R_p(s)$. When $n = n_p$ and $n^* = 1$, however, $Q(s) = 1$, and the θ_i^*'s are linear functions of the coefficients of $Z_p(s)$, $R_p(s)$.

Lemma 5.3.4. *Let the degrees of R_p, Z_p, Λ, Λ_0, and R_m be as specified in* (5.77). *Then the solution $\bar{\theta}^*$ of* (5.83) *or* (5.84) *always exists. In addition, if R_p, Z_p are coprime and $n = n_p$, then the solution $\bar{\theta}^*$ is unique.*[7]

Remark 5.3.5 It is clear from (5.81), (5.82) that the control law (5.77) places the poles of the closed-loop plant at the roots of the polynomial $Z_p(s)\Lambda_0(s)R_m(s)$ and changes the high-frequency gain from k_p to k_m by using the feedforward gain c_0^*. Therefore, the MRC scheme can be viewed as a special case of a general pole placement scheme where the desired closed-loop characteristic equation is given by

$$Z_p(s)\Lambda_0(s)R_m(s) = 0.$$

The transfer function matching (5.80) is achieved by canceling the zeros of the plant, i.e., $Z_p(s)$, and replacing them by those of the reference model, i.e., by designing $\Lambda = \Lambda_0 Z_m$. Such a cancellation is made possible by assuming that $Z_p(s)$ is Hurwitz and by designing Λ_0, Z_m to have stable zeros.

A state-space realization of the control law (5.77) is

$$\begin{aligned} \dot{\omega}_1 &= F\omega_1 + gu_p, \quad \omega_1(0) = 0, \\ \dot{\omega}_2 &= F\omega_2 + gy_p, \quad \omega_2(0) = 0, \\ u_p &= \theta^{*T}\omega, \end{aligned} \tag{5.87}$$

[7] For the proof, see the web resource [94].

where $\omega_1, \omega_2 \in \mathcal{R}^{n-1}$,

$$\theta^* = [\theta_1^{*T}, \theta_2^{*T}, \theta_3^*, c_0^*]^T, \qquad \omega = [\omega_1^T, \omega_2^T, y_p, r]^T,$$

$$F = \begin{bmatrix} -\lambda_{n-2} & -\lambda_{n-3} & -\lambda_{n-4} & \cdots & -\lambda_0 \\ 1 & 0 & 0 & \cdots & 0 \\ 0 & 1 & 0 & \cdots & 0 \\ \vdots & \vdots & \ddots & \ddots & \vdots \\ 0 & 0 & \cdots & 1 & 0 \end{bmatrix}, \qquad g = \begin{bmatrix} 1 \\ 0 \\ \vdots \\ 0 \end{bmatrix}, \tag{5.88}$$

λ_i are the coefficients of

$$\Lambda(s) = s^{n-1} + \lambda_{n-2}s^{n-2} + \cdots + \lambda_1 s + \lambda_0 = \det(sI - F),$$

and (F, g) is the state-space realization of $\frac{\alpha(s)}{\Lambda(s)}$, i.e., $(sI - F)^{-1}g = \frac{\alpha(s)}{\Lambda(s)}$.

We obtain the state-space representation of the overall closed-loop plant by augmenting the state x_p of the plant (5.72) with the states ω_1, ω_2 of the controller (5.87), i.e.,

$$\begin{aligned} \dot{Y}_c &= A_c Y_c + B_c c_0^* r, \quad Y_c(0) = Y_0, \\ y_p &= C_c^T Y_c, \end{aligned} \tag{5.89}$$

where

$$\begin{aligned} Y_c &= [x_p^T, \omega_1^T, \omega_2^T]^T \in \mathcal{R}^{n_p+2n-2}, \\ A_c &= \begin{bmatrix} A_p + B_p\theta_3^* C_p^T & B_p\theta_1^{*T} & B_p\theta_2^{*T} \\ g\theta_3^* C_p^T & F + g\theta_1^{*T} & g\theta_2^{*T} \\ gC_p^T & 0 & F \end{bmatrix}, \qquad B_c = \begin{bmatrix} B_p \\ g \\ 0 \end{bmatrix}, \\ C_c^T &= [C_p^T, 0, 0], \end{aligned} \tag{5.90}$$

and Y_0 is the vector with initial conditions. We have already established that the transfer function from r to y_p is given by

$$\frac{y_p(s)}{r(s)} = \frac{c_0^* k_p Z_p \Lambda^2}{\Lambda[(\Lambda - \theta_1^{*T}\alpha)R_p - k_p Z_p(\theta_2^{*T}\alpha + \theta_3^*\Lambda)]} = W_m(s),$$

which implies that

$$C_c^T(sI - A_c)^{-1}B_c c_0^* = \frac{c_0^* k_p Z_p \Lambda^2}{\Lambda[(\Lambda - \theta_1^{*T}\alpha)R_p - k_p Z_p(\theta_2^{*T}\alpha + \theta_3^*\Lambda)]} = W_m(s),$$

and therefore

$$\det(sI - A_c) = \Lambda[(\Lambda - \theta_1^{*T}\alpha)R_p - k_p Z_p(\theta_2^{*T}\alpha + \theta_3^*\Lambda)] = \Lambda Z_p \Lambda_0 R_m,$$

where the last equality is obtained by using the matching equation (5.82). It is clear that the eigenvalues of A_c are equal to the roots of the polynomials Λ, Z_p, and R_m; therefore, A_c is a stable matrix. The stability of A_c and the boundedness of r imply that the state vector Y_c in (5.89) is bounded.

Since $C_c^T(sI - A_c)^{-1}B_c c_0^* = W_m(s)$, the reference model may be realized by the triple $(A_c, B_c c_0^*, C_c)$ and described by the nonminimal state-space representation

$$\begin{aligned} \dot{Y}_m &= A_c Y_m + B_c c_0^* r, \quad Y_m(0) = Y_{m0}, \\ y_m &= C_c^T Y_m, \end{aligned} \tag{5.91}$$

where $Y_m \in \mathcal{R}^{n_p+2n-2}$. Letting $e = Y_c - Y_m$ be the state error and $e_1 = y_p - y_m$ the output tracking error, it follows from (5.89) and (5.91) that

$$\dot{e} = A_c e, \qquad e_1 = C_c^T e;$$

i.e., the tracking error e_1 satisfies

$$e_1 = C_c^T e^{A_c t}(Y_c(0) - Y_m(0)).$$

Since A_c is a stable matrix, $e_1(t)$ converges exponentially to zero. The rate of convergence depends on the location of the eigenvalues of A_c, which are equal to the roots of $\Lambda(s)\Lambda_0(s)R_m(s)Z_p(s) = 0$. We can affect the rate of convergence by designing $\Lambda(s)\Lambda_0(s)R_m(s)$ to have fast zeros, but we are limited by the dependence of A_c on the zeros of $Z_p(s)$, which are fixed by the given plant.

Example 5.3.6 Let us consider the second-order plant

$$y_p = \frac{-3(s+4)}{s^2 - 3s + 2} u_p = \frac{-3(s+4)}{(s-1)(s-2)} u_p$$

and the reference model

$$y_m = \frac{1}{s+1} r.$$

The order of the plant is $n_p = 2$. Its relative degree $n^* = 1$ is equal to that of the reference model. We choose the polynomial $\Lambda(s)$ as

$$\Lambda(s) = s + 2 = \Lambda_0(s)$$

and the control input

$$u_p = \theta_1^* \frac{1}{s+2} u_p + \theta_2^* \frac{1}{s+2} y_p + \theta_3^* y_p + c_0^* r,$$

which gives the closed-loop transfer function

$$\frac{y_p(s)}{r(s)} = \frac{-3c_0^*(s+4)(s+2)}{(s+2-\theta_1^*)(s-1)(s-2) + 3(s+4)(\theta_2^* + \theta_3^*(s+2))} = G_c(s).$$

Forcing $G_c(s) = \frac{1}{s+1}$, we have $c_0^* = -\frac{1}{3}$, and the matching equation (5.83) becomes

$$\theta_1^*(s-1)(s-2) - 3(\theta_2^* + \theta_3^*(s+2))(s+4) = (s+2)(s-1)(s-2) - (s+4)(s+2)(s+1),$$

i.e.,

$$(\theta_1^* - 3\theta_3^*)s^2 + (-3\theta_1^* - 3\theta_2^* - 18\theta_3^*)s + 2\theta_1^* - 12\theta_2^* - 24\theta_3^* = -8s^2 - 18s - 4.$$

Equating the powers of s, we have

$$\begin{aligned} \theta_1^* - 3\theta_3^* &= -8, \\ \theta_1^* + \theta_2^* + 6\theta_3^* &= 6, \\ \theta_1^* - 6\theta_2^* - 12\theta_3^* &= -2, \end{aligned}$$

i.e.,

$$\begin{bmatrix} 1 & 0 & -3 \\ 1 & 1 & 6 \\ 1 & -6 & -12 \end{bmatrix} \begin{bmatrix} \theta_1^* \\ \theta_2^* \\ \theta_3^* \end{bmatrix} = \begin{bmatrix} -8 \\ 6 \\ -2 \end{bmatrix},$$

which gives

$$\begin{bmatrix} \theta_1^* \\ \theta_2^* \\ \theta_3^* \end{bmatrix} = \begin{bmatrix} -2 \\ -4 \\ 2 \end{bmatrix}.$$

The control input is therefore given by

$$u_p = -2\frac{1}{s+2}u_p - 4\frac{1}{s+2}y_p + 2y_p - \left(\frac{1}{3}\right)r$$

and is implemented as follows:

$$\begin{aligned} \dot{\omega}_1 &= -2\omega_1 + u_p, \\ \dot{\omega}_2 &= -2\omega_2 + y_p, \\ u_p &= -2\omega_1 - 4\omega_2 + 2y_p - \left(\frac{1}{3}\right)r. \end{aligned}$$

■

The control law (5.87) is the backbone of the corresponding MRAC schemes to be designed in the subsequent sections to handle the case of unknown plant parameters. It is natural to expect that assumptions P1–P4 that are used to meet the MRC objective in the case of known plant parameters will also be needed in the case of unknown plant parameters. Since MRAC has been one of the most popular adaptive control schemes, newcomers to adaptive control often misunderstand adaptive control as a method applicable only to minimum-phase plants with known relative degree, etc. As we will show in Chapter 6, if we change the MRC objective to a pole placement one, assumptions P1, P2, P4 are no longer needed.

In the following sections we develop different classes of MRAC schemes and analyze their properties. As we will demonstrate, the design and analysis approach is very similar to that used for the simple MRAC examples presented in section 5.2.

5.4 Direct MRAC with Unnormalized Adaptive Laws

In this section, we extend the scalar example presented in section 5.2.2 to the general class of plants (5.72) where only the output is available for measurement. The complexity of the schemes increases with the relative degree n^* of the plant. The simplest cases are the ones where $n^* = 1$ and 2. Because of their simplicity, they are still quite popular in the literature of continuous-time MRAC and are presented in separate sections.

5.4.1 Relative Degree $n^* = 1$

Let us assume that the relative degree of the plant

$$y_p = G_p(s)u_p = k_p \frac{Z_p(s)}{R_p(s)} u_p \tag{5.92}$$

is $n^* = 1$. The reference model

$$y_m = W_m(s)r$$

is chosen to have the same relative degree, and both $G_p(s)$ and $W_m(s)$ satisfy assumptions P1–P4 and M1 and M2, respectively. In addition $W_m(s)$ is designed to be SPR. The design of the MRAC law to meet the control objective defined in section 5.3.1 proceeds as follows.

We have shown in section 5.3.2 that the control law

$$\begin{aligned} \dot{\omega}_1 &= F\omega_1 + gu_p, \quad \omega_1(0) = 0, \\ \dot{\omega}_2 &= F\omega_2 + gy_p, \quad \omega_2(0) = 0, \\ u_p &= \theta^{*T}\omega, \end{aligned} \tag{5.93}$$

where $\omega = [\omega_1^T, \omega_2^T, y_p, r]^T$ and $\theta^* = [\theta_1^{*T}, \theta_2^{*T}, \theta_3^*, c_0^*]^T$, calculated from the matching equation (5.81) and (5.82) meets the MRC objective defined in section 5.3.1. Because the parameters of the plant are unknown, the desired controller parameter vector θ^* cannot be calculated from the matching equation, and therefore (5.93) cannot be implemented. Following the certainty equivalence (CE) approach, instead of (5.93) we use

$$\begin{aligned} \dot{\omega}_1 &= F\omega_1 + gu_p, \quad \omega_1(0) = 0, \\ \dot{\omega}_2 &= F\omega_2 + gy_p, \quad \omega_2(0) = 0, \\ u_p &= \theta^{T}\omega, \end{aligned} \tag{5.94}$$

where $\theta(t)$ is the online estimate of the unknown parameter vector θ^* to be generated by an adaptive law. We derive such an adaptive law by following a procedure similar to that used in the case of the example of section 5.2.2. We first obtain a composite state-space representation of the plant and controller, i.e.,

$$\begin{aligned} \dot{Y}_c &= A_0 Y_c + B_c u_p, \quad Y_c(0) = Y_0, \\ y_p &= C_c^T Y_c, \\ u_p &= \theta^T \omega, \end{aligned}$$

where $Y_c = [x_p^T, \omega_1^T, \omega_2^T]^T$,

$$A_0 = \begin{bmatrix} A_p & 0 & 0 \\ 0 & F & 0 \\ gC_p^T & 0 & F \end{bmatrix}, \qquad B_c = \begin{bmatrix} B_p \\ g \\ 0 \end{bmatrix},$$
$$C_c^T = [C_p^T, 0, 0],$$

and then add and subtract the desired input $B_c\theta^{*T}\omega$ to obtain

$$\dot{Y}_c = A_0 Y_c + B_c\theta^{*T}\omega + B_c(u_p - \theta^{*T}\omega).$$

If we now absorb the term $B_c\theta^{*T}\omega$ into the homogeneous part of the above equation, we end up with the representation

$$\begin{aligned} \dot{Y}_c &= A_c Y_c + B_c c_0^* r + B_c(u_p - \theta^{*T}\omega), \quad Y_c(0) = Y_0, \\ y_p &= C_c^T Y_c, \end{aligned} \tag{5.95}$$

where A_c is as defined in (5.90). Equation (5.95) is the same as the closed-loop equation (5.89) in the known parameter case except for the additional input term $B_c(u_p - \theta^{*T}\omega)$ that depends on the choice of the input u_p. It serves as the parameterization of the plant equation in terms of the desired controller parameter vector θ^*. Letting $e = Y_c - Y_m$ and $e_1 = y_p - y_m$, where Y_m is the state of the nonminimal representation of the reference model given by (5.91), we obtain the tracking error equation

$$\begin{aligned} \dot{e} &= A_c e + B_c(u_p - \theta^{*T}\omega), \quad e(0) = e_0, \\ e_1 &= C_c^T e. \end{aligned} \tag{5.96}$$

Since

$$C_c^T(sI - A_c)^{-1} B_c c_0^* = W_m(s),$$

we have

$$e_1 = W_m(s)\rho^*(u_p - \theta^{*T}\omega), \tag{5.97}$$

where $\rho^* = \frac{1}{c_0^*}$, which is in the form of the bilinear parametric model discussed in Chapters 2 and 3. We can now use (5.97) to generate a wide class of adaptive laws for estimating θ^* by using the results of Chapter 3. We should note that (5.96) and (5.97) hold for any relative degree and will also be used in later sections.

We should also note that (5.96) is similar in form to the B-SSPM, except that the state e is not available for measurement and the matrix A_c, even though stable, is unknown. Despite these differences, we can still form the estimation model as

$$\hat{e}_1 = W_m(s)\rho(u_p - \theta^T\omega) = C_c^T(sI - A_c)^{-1} B_c c_0^* \rho(u_p - \theta^T\omega).$$

Substituting for the control law $u_p = \theta^T\omega$, we have that $\hat{e}_1(t) = 0 \; \forall t \geq 0$ and the estimation error $\varepsilon = e_1 - \hat{e}_1 = e_1$. Therefore, the estimation error equation is the same as the tracking error equation (5.96). Substituting for the control law in (5.96), we obtain the error equation

$$\begin{aligned} \dot{e} &= A_c e + \bar{B}_c \rho^* \tilde{\theta}^T \omega, \quad e(0) = e_0, \\ e_1 &= C_c^T e, \end{aligned} \tag{5.98}$$

where $\bar{B}_c = B_c c_0^*$ or

$$e_1 = W_m(s)(\rho^* \tilde{\theta}^T \omega),$$

which relates the parameter error $\tilde{\theta} = \theta(t) - \theta^*$ to the tracking error e_1. Because $W_m(s) = C_c^T(sI - A_c)^{-1} B_c c_0^*$ is SPR and A_c is stable, (5.98) motivates the Lyapunov function

$$V(\tilde{\theta}, e) = \frac{e^T P_c e}{2} + \frac{\tilde{\theta}^T \Gamma^{-1} \tilde{\theta}}{2} |\rho^*|, \tag{5.99}$$

where $\Gamma = \Gamma^T > 0$ and $P_c = P_c^T > 0$ satisfies the algebraic equations

$$P_c A_c + A_c^T P_c = -qq^T - \nu_c L_c,$$
$$P_c \bar{B}_c = C_c,$$

where q is a vector, $L_c = L_c^T > 0$, and $\nu_c > 0$ is a small constant, that are implied by the MKY lemma (see Lemma A.9.8 in the Appendix). The time derivative $\dot{V}$ of V along the solution of (5.99) is given by

$$\dot{V} = -\frac{e^T qq^T e}{2} - \frac{\nu_c}{2} e^T L_c e + e^T P_c \bar{B}_c \rho^* \tilde{\theta}^T \omega + \tilde{\theta}^T \Gamma^{-1} \dot{\tilde{\theta}} |\rho^*|.$$

Since $e^T P_c \bar{B}_c = e^T C_c = e_1$ and $\rho^* = |\rho^*| \operatorname{sgn}(\rho^*)$, we can make $\dot{V} \leq 0$ by choosing

$$\dot{\tilde{\theta}} = \dot{\theta} = -\Gamma e_1 \omega \operatorname{sgn}(\rho^*), \tag{5.100}$$

which leads to

$$\dot{V} = -\frac{e^T qq^T e}{2} - \frac{\nu_c}{2} e^T L_c e. \tag{5.101}$$

Equations (5.99) and (5.101) imply that V and, therefore, $e, \tilde{\theta} \in \mathcal{L}_\infty$. Because $e = Y_c - Y_m$ and $Y_m \in \mathcal{L}_\infty$, we have $Y_c \in \mathcal{L}_\infty$, which implies that $x_p, y_p, \omega_1, \omega_2 \in \mathcal{L}_\infty$. Because $u_p = \theta^T \omega$ and $\theta, \omega \in \mathcal{L}_\infty$ we also have $u_p \in \mathcal{L}_\infty$. Therefore, all the signals in the closed-loop plant are bounded. It remains to show that the tracking error $e_1 = y_p - y_m$ goes to zero as $t \to \infty$.

From (5.99) and (5.101) we establish that e and therefore $e_1 \in \mathcal{L}_2$. Furthermore, using $\theta, \omega, e \in \mathcal{L}_\infty$ in (5.98), we have that $\dot{e}, \dot{e}_1 \in \mathcal{L}_\infty$. Hence $e_1, \dot{e}_1 \in \mathcal{L}_\infty$ and $e_1 \in \mathcal{L}_2$, which (see the Appendix) imply that $e_1(t) \to 0$ as $t \to \infty$.

The MRAC scheme is summarized by the equations

$$\begin{aligned}
\dot{\omega}_1 &= F\omega_1 + g u_p, \\
\dot{\omega}_2 &= F\omega_2 + g y_p, \\
u_p &= \theta^T \omega, \\
\dot{\theta} &= -\Gamma e_1 \omega \operatorname{sgn}\left(\frac{k_p}{k_m}\right).
\end{aligned} \tag{5.102}$$

Its stability properties are summarized by the following theorem.

Theorem 5.4.1. *The MRAC scheme* (5.102) *has the following properties:*

(i) *All signals in the closed-loop plant are bounded, and the tracking error e_1 converges to zero asymptotically with time for any reference input $r \in \mathcal{L}_\infty$.*

(ii) *If r is sufficiently rich of order $2n$, $\dot{r} \in \mathcal{L}_\infty$, and $Z_p(s)$, $R_p(s)$ are relatively coprime, then the parameter error $|\tilde{\theta}| = |\theta - \theta^*|$ and the tracking error e_1 converge to zero exponentially fast.*

Proof. Part (i) has already been completed above. For the proof of (ii), see the web resource [94]. □

Example 5.4.2 Let us consider the second-order plant

$$y_p = \frac{k_p(s + b_0)}{(s^2 + a_1 s + a_0)} u_p,$$

where $k_p > 0$, $b_0 > 0$, and k_p, b_0, a_1, a_0 are unknown constants. The desired performance of the plant is specified by the reference model

$$y_m = \frac{3}{s+3} r.$$

The control law is designed as

$$\begin{aligned} \dot{\omega}_1 &= -2\omega_1 + u_p, \quad \omega_1(0) = 0, \\ \dot{\omega}_2 &= -2\omega_2 + y_p, \quad \omega_2(0) = 0, \\ u_p &= \theta_1\omega_1 + \theta_2\omega_2 + \theta_3 y_p + c_0 r \end{aligned}$$

by choosing $F = -2$, $g = 1$, and $\Lambda(s) = s + 2$. The adaptive law is given by

$$\dot{\theta} = -\Gamma e_1 \omega, \quad \theta(0) = \theta_0,$$

where $e_1 = y_p - y_m$, $\theta = [\theta_1, \theta_2, \theta_3, c_0]^T$, and $\omega = [\omega_1, \omega_2, y_p, r]^T$. We can choose $\Gamma = \text{diag}\{\gamma_i\}$ for some $\gamma_i > 0$ and obtain the decoupled adaptive law

$$\dot{\theta}_i = -\gamma_i e_1 \omega_i, \quad i = 1, \ldots, 4,$$

where $\theta_4 = c_0$, $\omega_3 = y_p$, $\omega_4 = r$; or we can choose Γ to be any positive definite matrix.

For parameter convergence, we choose r to be sufficiently rich of order 4. As an example, we select $r = A_1 \sin \omega_1 t + A_2 \sin \omega_2 t$ for some nonzero constants $A_1, A_2, \omega_1, \omega_2$ with $\omega_1 \neq \omega_2$. We should emphasize that we may not always have the luxury of choosing r to be sufficiently rich. For example, if the control objective requires $r =$ constant in order for y_p to follow a constant set point at steady state, then the use of a sufficiently rich input r of order 4 will destroy the desired tracking properties of the closed-loop plant. ■

5.4.2 Relative Degree $n^* = 2$

Let us again consider the parameterization of the plant in terms of θ^*, developed in the previous section, i.e.,

$$\begin{aligned} \dot{e} &= A_c e + B_c(u_p - \theta^{*T}\omega), \\ e_1 &= C_c^T e, \end{aligned} \tag{5.103}$$

or

$$e_1 = W_m(s)\rho^*(u_p - \theta^{*T}\omega). \tag{5.104}$$

In the relative degree $n^* = 1$ case, we are able to design $W_m(s)$ to be SPR, which together with the control law $u_p = \theta^T\omega$ enables us to obtain an error equation that motivates an appropriate Lyapunov function. With $n^* = 2$, $W_m(s)$ can no longer be designed to be

SPR,[8] and therefore the procedure of section 5.4.1 fails to apply here. Instead, let us use the identity $(s+p_0)(s+p_0)^{-1}=1$ for some $p_0>0$ to rewrite (5.103), (5.104) as

$$\begin{aligned}\dot{e} &= A_c e + \bar{B}_c (s+p_0)\rho^*(u_f - \theta^{*T}\phi), \quad e(0)=e_0,\\ e_1 &= C_c^T e,\end{aligned} \tag{5.105}$$

i.e.,

$$e_1 = W_m(s)(s+p_0)\rho^*(u_f - \theta^{*T}\phi), \tag{5.106}$$

where $\bar{B}_c = B_c c_0^*$,

$$u_f = \frac{1}{s+p_0}u_p, \quad \phi = \frac{1}{s+p_0}\omega,$$

and $W_m(s)$, $p_0>0$ are chosen so that $W_m(s)(s+p_0)$ is SPR. If we choose u_p so that

$$u_f = \theta^T\phi, \tag{5.107}$$

then the estimation error is the same as the tracking error, and the estimation or tracking error equation takes the form

$$\begin{aligned}\dot{e} &= A_c e + \bar{B}_c (s+p_0)\rho^*\tilde{\theta}^T\phi, \quad e(0)=e_0,\\ e_1 &= C_c^T e,\end{aligned} \tag{5.108}$$

or, in the transfer function form,

$$e_1 = W_m(s)(s+p_0)\rho^*\tilde{\theta}^T\phi,$$

which can be transformed into the desired form by using the transformation

$$\bar{e} = e - \bar{B}_c\rho^*\tilde{\theta}^T\phi, \tag{5.109}$$

i.e.,

$$\begin{aligned}\dot{\bar{e}} &= A_c\bar{e} + B_1\rho^*\tilde{\theta}^T\phi, \quad \bar{e}(0)=\bar{e}_0,\\ e_1 &= C_c^T\bar{e},\end{aligned} \tag{5.110}$$

where $B_1 = A_c\bar{B}_c + \bar{B}_c p_0$ and $C_c^T\bar{B}_c = C_p^T B_p c_0^* = 0$ due to $n^*=2$. With (5.110), we can proceed as in the case of $n^*=1$ and develop an adaptive law for θ. Equation (5.110) is developed based on the assumption that we can choose u_p so that $u_f = \frac{1}{s+p_0}u_p = \theta^T\phi$. We have

$$u_p = (s+p_0)u_f = (s+p_0)\theta^T\phi,$$

which implies that

$$u_p = \theta^T\omega + \dot{\theta}^T\phi. \tag{5.111}$$

Since $\dot{\theta}$ is made available by the adaptive law, the control law (5.111) can be implemented without the use of differentiators. Let us now go back to the error equation (5.110). Since

$$C_c^T(sI-A_c)^{-1}B_1 = C_c^T(sI-A_c)^{-1}\bar{B}_c(s+p_0) = W_m(s)(s+p_0)$$

[8] A necessary condition for $W_m(s)$ to be SPR is that $n^*=0$ or 1.

is SPR, (5.110) is of the same form as (5.98), and the adaptive law for generating θ is designed by considering

$$V(\tilde{\theta},\bar{e}) = \frac{\bar{e}^T P_c \bar{e}}{2} + \frac{\tilde{\theta}^T \Gamma^{-1} \tilde{\theta}}{2} |\rho^*|,$$

where $P_c = P_c^T > 0$ satisfies the MKY lemma (Lemma A.9.8). As in the case of $n^* = 1$, for

$$\dot{\tilde{\theta}} = \dot{\theta} = -\Gamma e_1 \phi \,\text{sgn}(\rho^*), \tag{5.112}$$

the time derivative $\dot{V}$ of V along the solution of (5.110), (5.112) is given by

$$\dot{V} = -\frac{\bar{e}^T q q^T \bar{e}}{2} - \frac{\nu_c}{2} \bar{e}^T L_c \bar{e} \le 0,$$

which implies that $\bar{e},\tilde{\theta},e_1 \in \mathcal{L}_\infty$ and $\bar{e},e_1 \in \mathcal{L}_2$. Because $e_1 = y_p - y_m$, we also have $y_p \in \mathcal{L}_\infty$. The signal vector ϕ is expressed as

$$\phi = \frac{1}{s+p_0} \begin{bmatrix} (sI-F)^{-1} g G_p^{-1}(s) y_p \\ (sI-F)^{-1} g y_p \\ y_p \\ r \end{bmatrix} \tag{5.113}$$

by using $u_p = G_p^{-1}(s) y_p$. We can observe that each element of ϕ is the output of a proper stable transfer function whose input is y_p or r. Since $y_p, r \in \mathcal{L}_\infty$, we have $\phi \in \mathcal{L}_\infty$. Now $\bar{e},\theta,\phi \in \mathcal{L}_\infty$ imply (from (5.109)) that e and, therefore, $Y_c \in \mathcal{L}_\infty$. Because $\omega,\phi,e_1 \in \mathcal{L}_\infty$, we have $\dot{\theta} \in \mathcal{L}_\infty$ and $u_p \in \mathcal{L}_\infty$ and, therefore, all signals in the closed-loop plant are bounded. From (5.110), we also have that $\dot{\bar{e}} \in \mathcal{L}_\infty$, i.e., $\dot{e}_1 \in \mathcal{L}_\infty$, which, together with $e_1 \in \mathcal{L}_\infty \cap \mathcal{L}_2$, implies that $e_1(t) \to 0$ as $t \to \infty$.

The equations of the overall MRAC scheme are

$$\begin{aligned} \dot{\omega}_1 &= F\omega_1 + g u_p, & \omega_1(0) &= 0, \\ \dot{\omega}_2 &= F\omega_2 + g y_p, & \omega_2(0) &= 0, \\ \dot{\phi} &= -p_0 \phi + \omega, & \phi(0) &= 0, \\ u_p &= \theta^T \omega + \dot{\theta}^T \phi, & \omega &= [\omega_1^T, \omega_2^T, y_p, r]^T, \\ \dot{\theta} &= -\Gamma e_1 \phi \,\text{sgn}\left(\frac{k_p}{k_m}\right), \end{aligned} \tag{5.114}$$

where $\omega_1, \omega_2 \in \mathcal{R}^{n-1}$, $(s+p_0)W_m(s)$ is SPR, and F, G are as defined before.

Theorem 5.4.3. *The MRAC scheme* (5.114) *guarantees that*

(i) *All signals in the closed-loop plant are bounded, and the tracking error e_1 converges to zero asymptotically.*

(ii) *If R_p, Z_p are coprime and r is sufficiently rich of order $2n$, then the parameter error $|\tilde{\theta}| = |\theta - \theta^*|$ and the tracking error e_1 converge to zero exponentially fast.*

Proof. Part (i) has already been completed above. For the proof of (ii), see the web resource [94]. □

Example 5.4.4 Let us consider the second-order plant

$$y_p = \frac{k_p}{s^2 + a_1 s + a_0} u_p,$$

where $k_p > 0$ and a_1, a_0 are constants. The reference model is chosen as

$$y_m = \frac{9}{(s+3)^2} r.$$

Using (5.114), the control law is designed as

$$\begin{aligned}\dot{\omega}_1 &= -\omega_1 + u_p,\\ \dot{\omega}_2 &= -\omega_2 + y_p,\\ \dot{\phi} &= -\phi + \omega,\\ u_p &= \theta^T \omega - \phi^T \Gamma \phi e_1,\end{aligned}$$

where $\omega = [\omega_1, \omega_2, y_p, r]^T$, $e_1 = y_p - y_m$, $p_0 = 1$, $\Lambda(s) = s + 1$, and $\frac{9(s+1)}{(s+3)^2}$ is SPR. The adaptive law is given by

$$\dot{\theta} = -\Gamma e_1 \phi,$$

where $\Gamma = \Gamma^T > 0$ is any positive definite matrix and $\theta = [\theta_1, \theta_2, \theta_3, c_0]^T$. For parameter convergence, the reference input r is chosen as

$$r = A_1 \sin \omega_1 t + A_2 \sin \omega_2 t$$

for some $A_1, A_2 \neq 0$ and $\omega_1 \neq \omega_2$. ■

Remark 5.4.5 The control law (5.111) is a modification of the CE control law $u_p = \theta^T \omega$ and is motivated from stability considerations. The additional term $\dot{\theta}^T \phi = -\phi^T \Gamma \phi e_1 \operatorname{sgn}(\rho^*)$ is a nonlinear one that disappears asymptotically with time; i.e., $u_p = \theta^T \omega + \dot{\theta}^T \phi$ converges to the CE control law $u_p = \theta^T \omega$ as $t \to \infty$. The number and complexity of the additional terms in the CE control law increase with the relative degree n^*.

In a similar manner, MRAC schemes can be developed for $n^* = 3$ and above. For higher relative degrees, the complexity of the control law increases as it becomes a function of higher-order nonlinearities. The reader is referred to the web resource [94] for more details on the design and analysis of MRAC with unnormalized adaptive laws for plants with $n^* \geq 3$.

5.4.3 Relative Degree Greater than 2

When the relative degree is greater than 2 the procedure is very similar, but the control law becomes more complex. In the web resource [94], we present the relative degree-3 case to illustrate the approach and complexity.

5.5 Direct MRAC with Normalized Adaptive Laws

In this section we present and analyze a class of MRAC schemes that dominated the literature of adaptive control due to the simplicity of their design as well as their robustness properties in the presence of modeling errors. Their design is based on the CE approach that combines a control law, motivated from the known parameter case, with an adaptive law generated using the techniques of Chapter 3. The adaptive law is driven by the normalized estimation error and is based on an appropriate parameterization of the plant that involves the unknown desired controller parameters.

We have already demonstrated this class of MRAC for a simple example in section 5.2.4. The design and analysis of direct MRAC with normalized adaptive laws for the SISO plant (5.72) is very similar to that of the example in section 5.2.4.

Let us use the MRC law (5.87) and CE to propose the following control law for the case of unknown parameters,

$$u_p = \theta_1^T(t)\frac{\alpha(s)}{\Lambda(s)}u_p + \theta_2^T(t)\frac{\alpha(s)}{\Lambda(s)}y_p + \theta_3(t)y_p + c_0(t)r, \tag{5.115}$$

whose state-space realization is given by

$$\begin{aligned} \dot{\omega}_1 &= F\omega_1 + gu_p, \quad \omega_1(0) = 0, \\ \dot{\omega}_2 &= F\omega_2 + gy_p, \quad \omega_2(0) = 0, \\ u_p &= \theta^T\omega, \end{aligned} \tag{5.116}$$

where $\theta = [\theta_1^T, \theta_2^T, \theta_3, c_0]^T$ and $\omega = [\omega_1^T, \omega_2^T, y_p, r]^T$, and search for an adaptive law to generate $\theta(t)$.

In section 5.4.1, we developed the DPM

$$e_1 = W_m(s)\rho^*(u_p - \theta^{*T}\omega), \tag{5.117}$$

where $\rho^* = \frac{1}{c_0^*} = \frac{k_p}{k_m}$, $\theta^* = [\theta_1^{*T}, \theta_2^{*T}, \theta_3^*, c_0^*]^T$, by adding and subtracting the desired control input $\theta^{*T}\omega$ in the overall representation of the plant and controller states (see (5.97)).

We can express (5.117) in the form of the B-SPM as

$$e_1 = \rho^*(\theta^{*T}\phi + u_f), \tag{5.118}$$

where $\phi = -W_m(s)\omega$, $u_f = W_m(s)u_p$. We can relate it to the B-SPM in Chapter 2 by defining $z = e_1$, $z_0 = u_f$. Either (5.117) or (5.118) may be used to design an adaptive law for generating $\theta(t)$ in (5.116).

As a demonstration, we consider the parametric model (5.118). Using the results of Chapter 3, we write

$$\begin{aligned} \dot{\theta} &= \Gamma\varepsilon\phi\,\mathrm{sgn}\left(\frac{k_p}{k_m}\right), & \dot{\rho} &= \gamma\varepsilon\xi, \\ \varepsilon &= \frac{e_1 - \rho\xi}{m_s^2}, & m_s^2 &= 1 + \phi^T\phi + u_f^2, \\ \xi &= \theta^T\phi + u_f, & \phi &= -W_m(s)\omega, \\ u_f &= W_m(s)u_p. \end{aligned} \tag{5.119}$$

As shown in Chapter 3, the above adaptive law guarantees that (i) $\theta, \rho \in \mathcal{L}_\infty$, (ii) ε, εm_s, $\varepsilon\phi$, $\dot{\theta}$, $\dot{\rho} \in \mathcal{L}_2 \cap \mathcal{L}_\infty$ independent of the boundedness of ϕ, ρ, u_p, y_p. The MRAC scheme is described by (5.116), (5.119). Its stability properties are given by the following theorem.

Theorem 5.5.1. *The MRAC scheme* (5.116), (5.119) *applied to the plant* (5.72) *has the following properties:*

(i) *All signals are uniformly bounded.*

(ii) *The tracking error $e_1 = y_p - y_m$ converges to zero as $t \to \infty$.*

(iii) *If the reference input signal r is sufficiently rich of order $2n$, $\dot{r} \in \mathcal{L}_\infty$, and R_p, Z_p are coprime, the tracking error e_1 and parameter error $\tilde{\theta} = \theta - \theta^*$ converge to zero exponentially fast.*[9]

Below we present an outline of the proof whose procedure is very similar to that for the example presented in section 5.2.4.

Proof. The proof is completed by using the following steps.

Step 1. Express the plant input and output in terms of the adaptation error $\tilde{\theta}^T\omega$ The transfer function between the input $r + \frac{1}{c_0^*}\tilde{\theta}^T\omega$ and the plant output y_p after cancellation of all the common zeros and poles is given by

$$y_p = W_m(s)\left(r + \frac{1}{c_0^*}\tilde{\theta}^T\omega\right). \tag{5.120}$$

Because $y_p = G_p(s)u_p$ and $G_p^{-1}(s)$ has stable poles, we have

$$u_p = G_p^{-1}(s)W_m(s)\left(r + \frac{1}{c_0^*}\tilde{\theta}^T\omega\right), \tag{5.121}$$

where $G_p^{-1}(s)W_m(s)$ is biproper. We now define the fictitious normalizing signal m_f as

$$m_f^2 \triangleq 1 + \|u_p\|^2 + \|y_p\|^2, \tag{5.122}$$

where $\|\cdot\|$ denotes the $\mathcal{L}_{2\delta}$ norm for some $\delta > 0$. Using the properties of the $\mathcal{L}_{2\delta}$ norm, it follows that

$$m_f \le c + c\|\tilde{\theta}^T\omega\|, \tag{5.123}$$

where c is used to denote any finite constant and $\delta > 0$ is such that $W_m(s - \frac{\delta}{2})$, $G_p^{-1}(s - \frac{\delta}{2})$ have stable poles. Furthermore, for $\theta \in \mathcal{L}_\infty$ (guaranteed by the adaptive law), the signal m_f bounds most of the signals and their derivatives from above.

[9] A detailed proof of Theorem 5.5.1 is given in the web resource [94].

Step 2. Use the swapping lemmas and properties of the $\mathcal{L}_{2\delta}$ norm to upper bound $\|\tilde{\theta}^T\omega\|$ with terms that are guaranteed by the adaptive law to have finite $\mathcal{L}_2$ gains This is the most complicated step, and it involves the use of Swapping Lemmas 1 and 2 to obtain the inequality

$$\|\tilde{\theta}^T\omega\| \leq \frac{c}{\alpha_0}m_f + c\alpha_0^{n^*}\|\tilde{g}m_f\|, \tag{5.124}$$

where $\tilde{g}^2 \triangleq \varepsilon^2 m_s^2 + |\dot{\theta}|^2$, $\tilde{g}$ is guaranteed by the adaptive law to belong to $\mathcal{L}_2$, and $\alpha_0 > 0$ is an arbitrary constant to be chosen.

Step 3. Use the B–G lemma to establish boundedness From (5.123) and (5.124), it follows that

$$m_f^2 \leq c + \frac{c}{\alpha_0^2}m_f^2 + c\alpha_0^{2n^*}\|\tilde{g}m_f\|^2 \tag{5.125}$$

or

$$m_f^2 \leq c + c\int_0^t \alpha_0^{2n^*}\tilde{g}^2(\tau)m_f^2(\tau)d\tau$$

for any $\alpha_0 > \alpha_0^*$ and some $\alpha_0^* > 0$ for which $1 - \frac{c}{\alpha_0^{*2}} > 0$. Applying the B–G lemma and using $\tilde{g} \in \mathcal{L}_2$, the boundedness of m_f follows. Using $m_f \in \mathcal{L}_\infty$, we establish the boundedness of all the signals in the closed-loop plant.

Step 4. Show that the tracking error converges to zero The convergence of e_1 to zero is established by showing that $e_1 \in \mathcal{L}_2$ and $\dot{e}_1 \in \mathcal{L}_\infty$, and using Lemma A.4.7.

Step 5. Establish that the parameter error converges to zero The convergence of the estimated parameters to their true values is established by first showing that the signal vector ϕ, which drives the adaptive law under consideration, can be expressed as

$$\phi = H(s)r + \bar{\phi},$$

where $H(s)$ is a stable transfer matrix and $\bar{\phi} \in \mathcal{L}_2$. If r is sufficiently rich of order $2n$ and Z_p, R_p are coprime, then it can be shown that $\phi_m = H(s)r$ is PE, which implies that ϕ is PE due to $\bar{\phi} \in \mathcal{L}_2$. The PE property of ϕ guarantees that $\tilde{\theta}$ and e_1 converge to zero as shown in Chapter 3. □

5.6 Indirect MRAC

As demonstrated in sections 5.2.3, 5.2.5 using a scalar example, in indirect MRAC the plant parameters, namely k_p and the coefficients of $Z_p(s)$, $R_p(s)$, are estimated online, and their estimates at each time t are used to calculate the controller parameters. The online adaptive laws could be designed with or without normalization as in the direct MRAC case. The indirect MRAC without normalization is easier to analyze, but it involves a more complex control law whose complexity increases with the relative degree of the plant. In contrast, the design of indirect MRAC with normalization is strictly based on the CE principle and is easy to follow; however, the stability analysis is more complicated.

5.6.1 Indirect MRAC with Unnormalized Adaptive Laws

As in the direct MRAC case with unnormalized adaptive laws, the complexity of the control law increases with the value of the relative degree n^* of the plant. In this section we demonstrate the case for $n^* = 1$. The same methodology is applicable to the case of $n^* \geq 2$ at the expense of additional algebraic manipulations.

We propose the same control law

$$\begin{aligned} \dot{\omega}_1 &= F\omega_1 + gu_p, \quad \omega_1(0) = 0, \\ \dot{\omega}_2 &= F\omega_2 + gy_p, \quad \omega_2(0) = 0, \\ u_p &= \theta^T\omega, \qquad \omega = [\omega_1^T, \omega_2^T, y_p, r]^T \end{aligned} \tag{5.126}$$

as in the direct MRAC case, where $\theta(t)$ is calculated using the estimate of k_p and the estimates of the coefficients of the plant polynomials

$$\begin{aligned} R_p(s) &= s^n + a_{n-1}s^{n-1} + a_p^T\alpha_{n-2}(s), \\ Z_p(s) &= s^{n-1} + b_p^T\alpha_{n-2}(s), \end{aligned}$$

where $a_p, b_p \in \mathcal{R}^{n-1}$, $\alpha_{n-2}(s) = [s^{n-2}, s^{n-3}, \dots, s, 1]^T$, and $\theta_p^* = [k_p, b_p^T, a_{n-1}, a_p^T]^T$ is the vector with plant parameters. Using the mapping that relates the desired controller parameters $\theta^* = [\theta_1^{*T}, \theta_2^{*T}, \theta_3^*, c_0^*]^T$ and θ_p^* specified by the matching equations (5.81), (5.85), and (5.86) with $Q(s) = 1$ due to $n^* = 1$, we obtain the algebraic equations

$$\begin{aligned} c_0^* &= \frac{k_m}{k_p}, \\ \theta_1^* &= \lambda - b_p, \\ \theta_2^* &= \frac{a_p - a_{n-1}\lambda + r_{n-1}\lambda - \nu}{k_p}, \\ \theta_3^* &= \frac{a_{n-1} - r_{n-1}}{k_p}, \end{aligned} \tag{5.127}$$

where λ, r_{n-1}, ν are the coefficients of

$$\begin{aligned} \Lambda(s) &= s^{n-1} + \lambda^T\alpha_{n-2}(s), \\ \Lambda_0(s)R_m(s) &= s^n + r_{n-1}s^{n-1} + \nu^T\alpha_{n-2}(s). \end{aligned}$$

If we let $\hat{k}_p(t), \hat{b}_p(t), \hat{a}_p(t), \hat{a}_{n-1}(t)$ be the estimates of k_p, b_p, a_p, a_{n-1}, respectively, at each time t, then $\theta(t) = [\theta_1^T(t), \theta_2^T(t), \theta_3(t), c_0(t)]^T$ can be calculated as

$$\begin{aligned} c_0(t) &= \frac{k_m}{\hat{k}_p(t)}, \\ \theta_1(t) &= \lambda - \hat{b}_p(t), \\ \theta_2(t) &= \frac{\hat{a}_p(t) - \hat{a}_{n-1}(t)\lambda + r_{n-1}\lambda - \nu}{\hat{k}_p(t)}, \\ \theta_3(t) &= \frac{\hat{a}_{n-1}(t) - r_{n-1}}{\hat{k}_p(t)}. \end{aligned} \tag{5.128}$$

The adaptive law for generating $\hat{k}_p$, $\hat{b}_p$, $\hat{a}_p$, $\hat{a}_{n-1}$ is constructed by considering the parametric model (5.117), i.e.,

$$e_1 = W_m(s)\rho^*(u_p - \theta^{*T}\omega),$$

where $\rho^* = \frac{1}{c_0^*} = \frac{k_p}{k_m}$. As in the direct case, since $n^* = 1$ we can choose $W_m(s)$ to be SPR. Let us rewrite the above equation as

$$e_1 = W_m(s)\frac{1}{k_m}(k_p u_p - k_p\theta^{*T}\omega - \hat{k}_p u_p + \hat{k}_p\theta^T\omega)$$

by using the identity $\hat{k}_p u_p = \hat{k}_p\theta^T\omega$. We use (5.127) and (5.128) to substitute for $k_p\theta^*$, $\hat{k}_p\theta$ and obtain

$$e_1 = W_m(s)\frac{1}{k_m}(\tilde{k}_p(\lambda^T\omega_1 - u_p) + \tilde{a}_p^T\omega_2 + k_p b_p^T\omega_1 - \hat{k}_p\hat{b}_p^T\omega_1 + \tilde{a}_{n-1}(y_p - \lambda^T\omega_2)),$$

where $\tilde{k}_p \triangleq \hat{k}_p - k_p$, $\tilde{a}_p \triangleq \hat{a}_p - a_p$, $\tilde{a}_{n-1} \triangleq \hat{a}_{n-1} - a_{n-1}$ are the parameter errors. Using the identity $k_p b_p^T\omega_1 - \hat{k}_p\hat{b}_p^T\omega_1 = -\tilde{k}_p\hat{b}_p^T\omega_1 - k_p\tilde{b}_p^T\omega_1$, where $\tilde{b}_p \triangleq \hat{b}_p - b_p$, we obtain

$$e_1 = W_m(s)\frac{1}{k_m}(\tilde{k}_p\xi_1 + \tilde{a}_{n-1}\xi_2 + \tilde{a}_p^T\omega_2 - k_p\tilde{b}_p^T\omega_1), \tag{5.129}$$

where $\xi_1 \triangleq \lambda^T\omega_1 - u_p - \hat{b}_p^T\omega_1$, $\xi_2 \triangleq y_p - \lambda^T\omega_2$. A minimal state-space representation of (5.129) is

$$\begin{aligned}
\dot{e} &= A_c e + B_c(\tilde{k}_p\xi_1 + \tilde{a}_{n-1}\xi_2 + \tilde{a}_p^T\omega_2 - k_p\tilde{b}_p^T\omega_1),\\
e_1 &= C_c^T e,
\end{aligned}$$

where $C_c^T(sI - A_c)^{-1}B_c = \frac{1}{k_m}W_m(s)$. Using the Lyapunov-like function

$$V = \frac{e^T P_c e}{2} + \frac{\tilde{k}_p^2}{2\gamma_p} + \frac{\tilde{a}_{n-1}^2}{2\gamma_1} + |k_p|\frac{\tilde{b}_p^T\Gamma_1^{-1}\tilde{b}_p}{2} + \frac{\tilde{a}_p^T\Gamma_2^{-1}\tilde{a}_p}{2},$$

where $P_c = P_c^T > 0$ satisfies the LKY lemma (Lemma A.9.7), $\gamma_1, \gamma_p > 0$, and $\Gamma_i = \Gamma_i^T > 0$, $i = 1, 2$, it follows that for the adaptive laws

$$\begin{aligned}
\dot{\hat{a}}_{n-1} &= -\gamma_1 e_1\xi_2,\\
\dot{\hat{a}}_p &= -\Gamma_2 e_1\omega_2,\\
\dot{\hat{b}}_p &= \Gamma_1 e_1\omega_1 \operatorname{sgn}(k_p),\\
\dot{\hat{k}}_p &= \begin{cases} -\gamma_p e_1\xi_1 & \text{if } |\hat{k}_p| > k_0 \text{ or if } |\hat{k}_p| = k_0 \text{ and } e_1\xi_1 \operatorname{sgn}(k_p) \le 0,\\ 0 & \text{otherwise,} \end{cases}
\end{aligned}$$

where $\hat{k}_p(0)\operatorname{sgn}(k_p) \ge k_0 > 0$ and k_0 is a known lower bound for $|k_p|$, we can establish that

$$\dot{V} \le -\nu_c\frac{e^T L_c e}{2},$$

where $\nu_c > 0$, $L_c = L_c^T > 0$. Following the usual stability arguments, we show that all signals are bounded and $e_1 \to 0$ as $t \to \infty$.

For $n^* \geq 2$, the control law $u_p = \theta^T \omega$ has to be modified as in the direct case by adding some nonlinear terms to $\theta^T \omega$ whose complexity increases with increasing n^*. We leave this part as an exercise for the ambitious reader.

5.6.2 Indirect MRAC with Normalized Adaptive Law

As in the direct MRAC case, the design of indirect MRAC with normalized adaptive laws is conceptually simple. The simplicity arises from the fact that the control and adaptive laws are designed independently and are combined using the CE approach. The adaptive law is developed by first expressing the plant in the form of a linear parametric model as shown in Chapter 3. Starting with the plant equation (5.73), which we express in the form

$$y_p = \frac{b_m s^m + b_{m-1} s^{m-1} + \cdots + b_0}{s^n + a_{n-1} s^{n-1} + \cdots + a_0} u_p,$$

where $b_m = k_p$ is the high-frequency gain, we obtain the following plant parametric model:

$$z = \theta_p^{*T} \phi, \tag{5.130}$$

where

$$z = \frac{s^n}{\Lambda_p(s)} y_p, \quad \phi = \left[\frac{\alpha_m^T(s)}{\Lambda_p(s)} u_p, -\frac{\alpha_{n-1}^T(s)}{\Lambda_p(s)} y_p \right]^T,$$
$$\theta_p^* = [b_m, \ldots, b_0, a_{n-1}, \ldots, a_0]^T,$$

where $\alpha_i^T(s) \triangleq [s^i, s^{i-1}, \ldots, s, 1]$ and $\Lambda_p(s) = s^n + \lambda_p^T \alpha_{n-1}(s)$ with $\lambda_p = [\lambda_{n-1}, \ldots, \lambda_0]^T$ is a Hurwitz polynomial. The parametric model (5.130) may be used to generate a wide class of adaptive laws. Using the estimate $\theta_p(t)$ of θ_p^*, the MRAC law may be formed as follows: The controller parameter vectors $\theta_1(t), \theta_2(t), \theta_3(t), c_0(t)$ in the control law

$$u_p = \theta_1^T \frac{\alpha(s)}{\Lambda(s)} u_p + \theta_2^T \frac{\alpha(s)}{\Lambda(s)} y_p + \theta_3 y_p + c_0 r = \theta^T \omega, \tag{5.131}$$

where $\omega = [\frac{\alpha^T(s)}{\Lambda(s)} u_p, \frac{\alpha^T(s)}{\Lambda(s)} y_p, y_p, r]^T$, $\alpha(s) = \alpha_{n-2}(s)$, and $\theta = [\theta_1^T, \theta_2^T, \theta_3, c_0]^T$ is calculated using the mapping $\theta(t) = f(\theta_p(t))$. The mapping $f(\cdot)$ is obtained by using the matching equations (5.81), (5.85), (5.86), i.e.,

$$\begin{aligned} c_0^* &= \frac{k_m}{k_p}, \\ \theta_1^{*T} \alpha(s) &= \Lambda(s) - Z_p(s) Q(s), \\ \theta_2^{*T} \alpha(s) + \theta_3^* \Lambda(s) &= \frac{Q(s) R_p(s) - \Lambda_0(s) R_m(s)}{k_p}, \end{aligned} \tag{5.132}$$

where $Q(s)$ is the quotient of $\frac{\Lambda_0 R_m}{R_p}$ and $\Lambda(s) = \Lambda_0(s) Z_m(s)$. That is, if $\hat{R}_p(s,t)$, $\hat{\bar{Z}}_p(s,t)$ are the estimated values of the polynomials $R_p(s)$, $\bar{Z}_p(s) \triangleq k_p Z_p(s)$, respectively, at each time t, then $c_0, \theta_1, \theta_2, \theta_3$ are obtained as solutions to the polynomial equations

$$\begin{aligned} c_0 &= \frac{k_m}{\hat{k}_p}, \\ \theta_1^T \alpha(s) &= \Lambda(s) - \frac{1}{\hat{k}_p} \hat{\bar{Z}}_p(s,t).\hat{Q}(s,t), \\ \theta_2^T \alpha(s) + \theta_3 \Lambda(s) &= \frac{1}{\hat{k}_p} (\hat{Q}(s,t).\hat{R}_p(s,t) - \Lambda_0(s) R_m(s)), \end{aligned} \tag{5.133}$$

provided $\hat{k}_p \neq 0$, where $\hat{Q}(s,t)$ is the quotient of $\frac{\Lambda_0(s) R_m(s)}{\hat{R}_p(s,t)}$. Here $A(s,t) \cdot B(s,t)$ denotes the frozen time product of two polynomials. In other words, the s does not differentiate the time-varying coefficients of the polynomials during multiplication. The polynomials $\hat{R}_p(s,t)$, $\hat{\bar{Z}}_p(s,t)$ are evaluated from the estimate

$$\theta_p = [\hat{b}_m, \ldots, \hat{b}_0, \hat{a}_{n-1}, \ldots, \hat{a}_0]^T$$

of θ_p^*, i.e.,

$$\begin{aligned} \hat{R}_p(s,t) &= s^n + \hat{a}_{n-1} s^{n-1} + \cdots + \hat{a}_0, \\ \hat{\bar{Z}}_p(s,t) &= \hat{b}_m s^m + \hat{b}_{m-1} s^{m-1} + \cdots + \hat{b}_0, \\ \hat{k}_p &= \hat{b}_m. \end{aligned}$$

As in section 5.6.2, the estimate $\hat{b}_m = \hat{k}_p$ should be constrained from going through zero by using projection.

The equations of the indirect MRAC scheme are described by (5.131), (5.133), where θ is generated by any adaptive law based on the parametric model (5.130). As an example we consider the gradient algorithm

$$\begin{aligned} \dot{\theta}_p &= \Pr(\Gamma \varepsilon \phi), \\ \varepsilon &= \frac{z - \theta_p^T \phi}{m_s^2}, \quad m_s^2 = 1 + \phi^T \phi, \end{aligned} \tag{5.134}$$

where $\Pr(\cdot)$ guarantees that the first element of θ_p, namely $\hat{b}_m = \hat{k}_p$, satisfies $\hat{b}_m \operatorname{sgn}(k_p) \geq k_0 > 0 \; \forall t \geq 0$, provided $\hat{b}_m(0) \operatorname{sgn}(k_p) \geq k_0 > 0$. The $\Pr(\cdot)$ operator does not affect the other elements of θ_p.

Theorem 5.6.1. *The indirect MRAC scheme described by* (5.131)–(5.134) *guarantees that all signals are bounded and the tracking error* $e_1 = y_p - y_m$ *converges to zero as* $t \to \infty$.

Proof. The proof follows the same steps as in the case of the example in section 5.2.5 and it is presented in the web resource [94]. □

5.7 Robust MRAC

In this section we consider MRAC schemes that are designed for a simplified model of the plant but are applied to a higher-order plant. We assume that the plant is of the form

$$y_p = G_0(s)(1 + \Delta_m(s))(u_p + d_u), \tag{5.135}$$

where $G_0(s)$ is the modeled part of the plant, $\Delta_m(s)$ is an unknown multiplicative perturbation with stable poles, and d_u is a bounded input disturbance. We assume that the overall plant transfer function and G_0 are strictly proper. This implies that $G_0\Delta_m$ is also strictly proper. We design the MRAC scheme assuming that the plant model is of the form

$$y_p = G_0(s)u_p, \quad G_0(s) = k_p \frac{Z_p(s)}{R_p(s)}, \tag{5.136}$$

where $G_0(s)$ satisfies assumptions P1–P4 in section 5.3.1, but we implement it on the plant (5.135). The effect of the dynamic uncertainty $\Delta_m(s)$ and disturbance d_u on the stability and performance of the MRAC scheme is analyzed in the next sections.

We first treat the case where the parameters of $G_0(s)$ are known exactly. In this case no adaptation is needed, and therefore the overall closed-loop plant can be studied using linear system theory.

5.7.1 MRC: Known Plant Parameters

Let us consider the MRC law

$$u_p = \theta_1^{*T} \frac{\alpha(s)}{\Lambda(s)} u_p + \theta_2^{*T} \frac{\alpha(s)}{\Lambda(s)} y_p + \theta_3^* y_p + c_0^* r \tag{5.137}$$

developed in section 5.3.2 and shown to meet the MRC objective for the plant model (5.136). Let us now apply (5.137) to the actual plant (5.135) and analyze its properties with respect to the multiplicative uncertainty $\Delta_m(s)$ and input disturbance d_u. The closed-loop plant is represented as shown in Figure 5.6, where

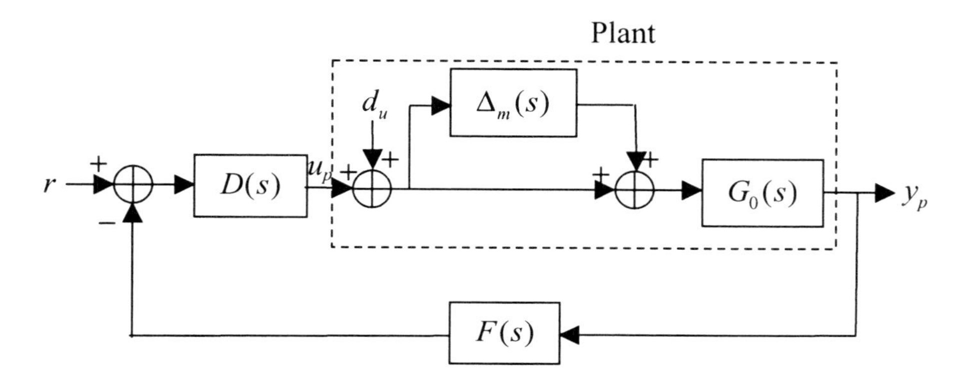

Figure 5.6. *Closed-loop MRC scheme.*

$$D(s) = \frac{\Lambda(s)c_0^*}{\Lambda(s) - \theta_1^{*T}\alpha(s)}, \qquad F(s) = -\frac{\theta_2^{*T}\alpha(s) + \theta_3^*\Lambda(s)}{c_0^*\Lambda(s)}.$$

The stability of the closed-loop system shown in Figure 5.6 depends on the location of the roots of the characteristic equation

$$1 + FDG = 0,$$

where $G = G_0(1 + \Delta_m)$. It can be shown that

$$y_p = \frac{DG}{1 + FDG} r + \frac{G}{1 + FDG} d_u \tag{5.138}$$

and D, F, G_0 satisfy the matching equation

$$\frac{DG_0}{1 + FDG_0} = W_m. \tag{5.139}$$

The tracking error $e_1 = y_p - y_m$, $y_m = W_m(s)r$ satisfies the equation

$$\begin{aligned} e_1 &= \left[\frac{DG}{1 + FDG} - \frac{DG_0}{1 + FDG_0}\right] r + \frac{G}{1 + FDG} d_u, \\ &= \frac{W_m \Delta_m}{(1 + FDG)} r + \frac{G}{1 + FDG} d_u. \end{aligned} \tag{5.140}$$

It is clear that for $\Delta_m = 0$, $d_u = 0$ the tracking error is equal to zero, provided of course that the roots of $1 + FDG_0 = 0$ are in $\Re[s] < 0$. Therefore, given that stability is satisfied, the tracking performance will very much depend on the size of Δ_m, d_u and the factors that multiply these terms.

The following theorem describes the stability and performance properties of the control law (5.137) designed to meet the MRC objective for the simplified plant (5.136) but applied to the actual plant (5.135) with modeling error.

Theorem 5.7.1. *If*

$$\left\| \frac{\theta_2^{*T}\alpha(s) + \theta_3^*\Lambda(s)}{\Lambda(s)} \frac{k_p}{k_m} W_m(s)\Delta_m(s) \right\|_\infty < 1, \tag{5.141}$$

then the closed-loop plant (5.135), (5.137) *is stable in the sense that all the closed-loop poles are in* $\Re[s] < 0$. *Furthermore, the tracking error satisfies*

$$\lim_{t\to\infty} \sup_{\tau \ge t} |e_1(\tau)| \le \Delta r_0 + c d_0, \tag{5.142}$$

where r_0, d_0 *are upper bounds for* $|r(t)|$, $|d_u(t)|$, *respectively;* $c \ge 0$ *is a finite constant; and*

$$\Delta \triangleq \left\| \frac{W_m(s)(\Lambda(s) - C_1^*(s))R_p(s)}{Z_p(s)(k_m\Lambda(s) - k_p W_m(s) D_1^*(s)\Delta_m(s))} W_m(s)\Delta_m(s) \right\|_{2\delta}$$

with $C_1^*(s) = \theta_1^{*T}\alpha(s)$, $D_1^*(s) = \theta_2^{*T}\alpha(s) + \theta_3^*\Lambda(s)$, *and* $\delta \in [0, \delta^*]$ *for some* $\delta^* > 0$.

Proof. The roots of $1 + FDG = 0$ are the same as the roots of

$$(\Lambda - C_1^*)R_p - D_1^* k_p Z_p(1 + \Delta_m) = 0. \tag{5.143}$$

Using the matching equation (5.139) or (5.82), i.e., $(\Lambda - C_1^*)R_p - k_p D_1^* Z_p = Z_p \Lambda_0 R_m$, the characteristic equation (5.143) becomes

$$Z_p(\Lambda_0 R_m - k_p D_1^* \Delta_m) = 0,$$

where Λ_0 is a factor of $\Lambda = \Lambda_0 Z_m$. Since Z_p is Hurwitz, we examine the roots of

$$\Lambda_0 R_m - k_p D_1^* \Delta_m = 0,$$

which can also be written as

$$1 - \frac{D_1^* k_p}{\Lambda_0 R_m}\Delta_m = 1 - \frac{k_p D_1^* W_m \Delta_m}{k_m \Lambda} = 0.$$

Because the poles of $\frac{D_1^* W_m \Delta_m}{\Lambda}$ are stable, it follows from the Nyquist criterion that for all Δ_m satisfying

$$\left\| \frac{k_p}{k_m} \frac{D_1^*(s) W_m(s) \Delta_m(s)}{\Lambda(s)} \right\|_\infty < 1, \tag{5.144}$$

the roots of (5.143) are stable. Hence, (5.138), (5.144) imply that for $r, d_u \in \mathcal{L}_\infty$ we have $y_p \in \mathcal{L}_\infty$. From $u_p = D(r - F y_p)$, we also have $u_p \in \mathcal{L}_\infty$.

Since $F(s) = -\frac{D_1^*(s)}{c_0^* \Lambda(s)}$ has stable poles and $r, d_u, u_p, y_p \in \mathcal{L}_\infty$, it follows from Figure 5.6 that all signals in the closed-loop scheme are bounded. Substituting for F, D, G_0 in the tracking error equation (5.140) and using the matching equation, we have

$$e_1 = \frac{W_m(\Lambda - C_1^*)R_p}{Z_p(k_m \Lambda - k_p W_m D_1^* \Delta_m)} W_m \Delta_m r + \frac{k_p(\Lambda - C_1^*) W_m (1 + \Delta_m)}{k_m \Lambda - k_p W_m D_1^* \Delta_m} d_u. \tag{5.145}$$

Due to $G(s)$, $G_0(s)$ being strictly proper and the fact that $W_m(s)$ has the same relative degree as $G_0(s)$, it follows that $W_m(s)\Delta_m(s)$ is strictly proper, which implies that the transfer function $\frac{e_1(s)}{r(s)}$ is strictly proper and $\frac{e_1(s)}{d_u(s)}$ is proper. Furthermore, both $\frac{e_1(s)}{r(s)}$, $\frac{e_1(s)}{d_u(s)}$ have stable poles (due to (5.141)), which implies that there exists a constant $\delta^* > 0$ such that $\frac{e_1(s)}{r(s)}$, $\frac{e_1(s)}{d_u(s)}$ are analytic in $\Re[s] \geq -\frac{\delta^*}{2}$. Using the properties of the $\mathcal{L}_{2\delta}$ norm, Lemma A.5.9, (5.145), and the fact that $r, d_u \in \mathcal{L}_\infty$, the bound for the tracking error given by (5.142) follows. $\square$

Remark 5.7.2 The expression for the tracking error given by (5.140) suggests that by increasing the loop gain FDG_0, we may be able to improve the tracking performance. Because F, D, G_0 are constrained by the matching equation, any changes in the loop gain must be performed under the constraint of the matching equation (5.82).

Example 5.7.3 Let us consider the plant

$$y = \frac{b}{s + a}(1 + \Delta_m(s))u = G(s)u, \tag{5.146}$$

where $G(s)$ is the overall strictly proper transfer function of the plant and Δ_m is a multiplicative perturbation. The control objective is to choose u such that all signals in the closed-loop plant are bounded and y tracks as closely as possible the output y_m of the reference model

$$y_m = \frac{b_m}{s + a_m} r,$$

where $a_m, b_m > 0$. The plant (5.146) can be modeled as

$$y = \frac{b}{s + a} u. \tag{5.147}$$

The MRC law based on (5.147) given by

$$u = \theta^* y + c_0 r,$$

where $\theta^* = \frac{a - a_m}{b}$, $c_0 = \frac{b_m}{b}$, meets the control objective for the plant model. Let us now implement the same control law on the actual plant (5.146). The closed-loop plant is given by

$$y = \frac{b_m(1 + \Delta_m(s))}{s + a_m - b\theta^* \Delta_m(s)} r,$$

whose characteristic equation is

$$s + a_m - b\theta^* \Delta_m(s) = 0$$

or

$$1 - \frac{b\theta^* \Delta_m(s)}{s + a_m} = 0.$$

Since $\frac{b\theta^* \Delta_m(s)}{s + a_m}$ is strictly proper with stable poles, it follows from the Nyquist criterion that a sufficient condition for the closed-loop system to be stable is that $\Delta_m(s)$ satisfies

$$\left\| \frac{b\theta^* \Delta_m(s)}{s + a_m} \right\|_\infty = \left\| \frac{(a - a_m)\Delta_m(s)}{s + a_m} \right\|_\infty < 1. \tag{5.148}$$

The tracking error $e_1 = y - y_m$ satisfies

$$e_1 = \frac{b_m(s + a_m + b\theta^*)}{(s + a_m)(s + a_m - b\theta^* \Delta_m(s))} \Delta_m(s) r.$$

Because $r \in \mathcal{L}_\infty$ and the transfer function $\frac{e_1(s)}{r(s)}$ has stable poles for all Δ_m satisfying (5.148), we have that

$$\lim_{t \to \infty} \sup_{\tau \geq t} |e_1(\tau)| \leq \Delta r_0,$$

where $|r(t)| \leq r_0$ and

$$\Delta = \left\| \frac{b_m}{s + a_m} \frac{(s + a_m + \theta^*)}{(s + a_m - \theta^* \Delta_m(s))} \Delta_m(s) \right\|_{2\delta}$$

for any $\delta \in [0, \delta^*]$ and some $\delta^* > 0$. Therefore, the smaller the term $\| \frac{\Delta_m(s)}{s+a_m} \|_\infty$ is, the better the stability margin and tracking performance will be. Let us consider the case where

$$\Delta_m(s) = -\frac{2\mu s}{1+\mu s}$$

and $\mu > 0$ is small, which arises from the parameterization of the nonminimum-phase plant

$$y = \frac{b(1-\mu s)}{(s+a)(1+\mu s)} u = \frac{b}{s+a}\left(1 - \frac{2\mu s}{1+\mu s}\right) u. \tag{5.149}$$

For this $\Delta_m(s)$, condition (5.148) becomes

$$\left\| \frac{(a-a_m)}{(s+a_m)} \frac{2\mu s}{(1+\mu s)} \right\|_\infty < 1,$$

which is satisfied, provided that

$$\mu < \frac{1}{2|a-a_m|}.$$

Similarly, it can be shown that $\Delta = O(\mu)$; i.e., the faster the unmodeled pole and zero in (5.149) are, the better the tracking performance. As $\mu \to 0$, $\Delta \to 0$ and therefore $\lim_{t\to\infty} \sup_{\tau \ge t} |e_1(\tau)| \to 0$. ∎

The above analysis shows that MRC control law is robust with respect to a multiplicative uncertainty $\Delta_m(s)$ and input disturbance in the sense that stability is preserved if the dynamic uncertainty is "small" and the input disturbance is bounded. When the MRC control law is combined with an adaptive law to form MRAC, the resulting closed-loop system is nonlinear, and the effect of the modeling error and disturbance cannot be analyzed using linear system theory for LTI systems. We have already shown in Chapter 3 that the adaptive laws without robust modifications can cause parameter drift in the parameter identification case. Parameter drift may cause high gain feedback in the adaptive control case causing several signals in the closed loop to go unbounded. In the following sections we combine the robust adaptive laws developed in Chapter 3 with the MRC control law to form robust MRAC schemes.

5.7.2 Robust Direct MRAC

In this section, we first use an example to illustrate the design and stability analysis of a robust MRAC scheme with a normalized adaptive law. We then extend the results to a general SISO plant with unmodeled dynamics and bounded disturbances.

Example 5.7.4 Consider the SISO plant

$$y = \frac{1}{s-a}(1+\Delta_m(s))u \tag{5.150}$$

with a strictly proper transfer function, where a is unknown and $\Delta_m(s)$ is a multiplicative plant uncertainty. Let us consider the adaptive control law

$$u = -\theta y, \tag{5.151}$$

$$\begin{aligned} \dot{\theta} &= \gamma\varepsilon\phi, \quad \gamma > 0, \\ \varepsilon &= \frac{z - \hat{z}}{m^2}, \quad m_s^2 = 1 + \phi^2, \\ \phi &= \frac{1}{s + a_m} y, \quad \hat{z} = \theta\phi, \quad z = y - \frac{1}{s + a_m} u, \end{aligned} \tag{5.152}$$

where $-a_m$ is the desired closed-loop pole and θ is the estimate of $\theta^* = a + a_m$ designed for the plant model

$$y = \frac{1}{s - a} u$$

and applied to the plant (5.150) with $\Delta_m(s) \neq 0$. The plant uncertainty $\Delta_m(s)$ introduces a disturbance term in the adaptive law that may easily cause θ to drift to infinity and certain signals to become unbounded no matter how small $\Delta_m(s)$ is. The adaptive control law (5.151), (5.152) is, therefore, not robust with respect to the plant uncertainty $\Delta_m(s)$. Several examples of nonrobust behavior are presented in [41, 50–52, 56]. The adaptive control scheme (5.151), (5.152), however, can be made robust if we replace the adaptive law (5.152) with a robust one developed by following the procedure of Chapter 3, as follows.

We first express the desired controller parameter $\theta^* = a + a_m$ in the form of a linear parametric model by rewriting (5.150) as

$$z = \theta^*\phi + \eta,$$

where z, ϕ are as defined in (5.152) and

$$\eta = \frac{1}{s + a_m} \Delta_m(s) u$$

is the modeling error term. If we now assume that a bound for the stability margin of the poles of $\Delta_m(s)$ is known, that is, $\Delta_m(s)$ is analytic in $\Re[s] \geq -\frac{\delta_0}{2}$ for some known constant $\delta_0 > 0$, then we can verify that the signal m_s generated as

$$m_s^2 = 1 + n_d, \quad \dot{n}_d = -\delta_0 n_d + u^2 + y^2, \quad n_d(0) = 0, \quad \delta_0 < 2a_m, \tag{5.153}$$

guarantees that $\frac{\eta}{m_s}, \frac{\phi}{m_s} \in \mathcal{L}_\infty$ and therefore qualifies to be used as a normalizing signal. Hence, we can combine normalization with any modification, such as leakage, dead zone, or projection, to form a robust adaptive law. Let us consider the switching σ-modification, i.e.,

$$\begin{aligned} \dot{\theta} &= \gamma\varepsilon\phi - \sigma_s\gamma\theta, \\ \varepsilon &= \frac{z - \theta\phi}{m_s^2}, \end{aligned} \tag{5.154}$$

where σ_s is as defined in Chapter 3. According to the results in Chapter 3, the robust adaptive law given by (5.153), (5.154) guarantees that $\varepsilon, \varepsilon m_s, \theta, \dot{\theta} \in \mathcal{L}_\infty$ and $\varepsilon, \varepsilon m_s, \dot{\theta} \in \mathcal{S}(\frac{\eta^2}{m_s^2})$ or $\mathcal{S}(\eta_0^2)$, where η_0^2 is an upper bound for $\frac{\eta^2}{m_s^2}$. Using the properties of the $\mathcal{L}_{2\delta}$ norm, we have

$$|\eta(t)| \leq \Delta_2 \|u_t\|_{2\delta_0}, \quad \Delta_2 \triangleq \left\| \frac{1}{s + a_m} \Delta_m(s) \right\|_{2\delta_0}.$$

Since $m_s^2 = 1 + \|u_t\|_{2\delta_0}^2 + \|y_t\|_{2\delta_0}^2$, it follows that $\frac{|\eta|}{m_s} \leq \Delta_2$. Therefore, $\varepsilon, \varepsilon m_s, \dot{\theta} \in \mathcal{S}(\Delta_2^2)$.

We analyze the stability properties of the MRAC scheme described by (5.151), (5.153), and (5.154) when applied to the plant (5.150) with $\Delta_m(s) \neq 0$ as follows.

We start by writing the closed-loop plant equation as

$$y = \frac{1}{s + a_m}(-\tilde{\theta} y + \Delta_m(s)u) = -\frac{1}{s + a_m}\tilde{\theta} y + \eta, \quad u = -\theta y. \tag{5.155}$$

Using Swapping Lemma 1 and noting that $\varepsilon m_s^2 = -\tilde{\theta}\phi + \eta$, we obtain from (5.155) that

$$\begin{aligned} y &= -\tilde{\theta}\phi + \frac{1}{s + a_m}(\dot{\tilde{\theta}}\phi) + \eta \\ &= \varepsilon m_s^2 + \frac{1}{s + a_m}\dot{\tilde{\theta}}\phi. \end{aligned} \tag{5.156}$$

Using the properties of the $\mathcal{L}_{2\delta}$ norm $\|(\cdot)_t\|_{2\delta}$, which for simplicity we denote by $\|\cdot\|$, it follows from (5.156) that

$$\|y\| \leq \|\varepsilon m_s^2\| + \left\| \frac{1}{s + a_m} \right\|_{\infty\delta} \|\dot{\tilde{\theta}}\phi\|$$

for any $0 < \delta \leq \delta_0$. Because $u = -\theta y$ and $\theta \in \mathcal{L}_\infty$, it follows that

$$\|u\| \leq c\|y\| \leq c\|\varepsilon m_s^2\| + c\|\dot{\tilde{\theta}}\phi\|,$$

where $c \geq 0$ is used to denote any finite constant. Therefore, the fictitious normalizing signal m_f satisfies

$$m_f^2 \stackrel{\Delta}{=} 1 + \|u\|^2 + \|y\|^2 \leq 1 + c\|\varepsilon m_s^2\|^2 + c\|\dot{\tilde{\theta}}\phi\|^2. \tag{5.157}$$

We can establish that m_f guarantees that $\frac{m_s}{m_f}, \frac{\phi}{m_f}, \frac{\eta}{m_f} \in \mathcal{L}_\infty$ by using the properties of the $\mathcal{L}_{2\delta}$ norm, which implies that (5.157) can be written as

$$m_f^2 \leq 1 + c\|\varepsilon m_s m_f\|^2 + c\|\dot{\tilde{\theta}} m_f\|^2 \leq 1 + c\|\tilde{g} m_f\|^2,$$

where $\tilde{g}^2 \stackrel{\Delta}{=} \varepsilon^2 m_s^2 + \dot{\tilde{\theta}}^2$ or

$$m_f^2 \leq 1 + c\int_0^t e^{-\delta(t-\tau)}\tilde{g}^2(\tau)m_f^2(\tau)d\tau. \tag{5.158}$$

Applying B–G Lemma 3 to (5.158), we obtain

$$m_f^2(t) \leq \Phi(t, 0) + \delta\int_0^t \Phi(t, \tau)d\tau,$$

where

$$\Phi(t, \tau) = e^{-\delta(t-\tau)}e^{c\int_\tau^t \tilde{g}^2(s)ds}.$$

Since the robust adaptive law guarantees that $\varepsilon m_s, \dot{\tilde{\theta}} \in \mathcal{S}(\Delta_2^2)$, we have $\tilde{g} \in \mathcal{S}(\Delta_2^2)$, i.e.,

$$c\int_\tau^t \tilde{g}^2(s)ds \le c\Delta_2^2(t-\tau)+c$$

and

$$\Phi(t,\tau) \le e^{-(\delta - c\Delta_2^2)(t-\tau)}.$$

Hence, for

$$c\Delta_2^2 < \delta,$$

$\Phi(t,\tau)$ is bounded from above by an exponential decaying to zero, which implies that $m_f \in \mathcal{L}_\infty$. Because of the normalizing properties of m_f, we have $\phi, y, u \in \mathcal{L}_\infty$, which implies that all signals are bounded. The condition $c\Delta_2^2 < \delta$ implies that the multiplicative plant uncertainty $\Delta_m(s)$ should satisfy

$$\left\| \frac{\Delta_m(s)}{s+a_m} \right\|_{2\delta_0}^2 < \frac{\delta}{c},$$

where c can be calculated by keeping track of all the constants. It can be shown that the constant c depends on $\|\frac{1}{s+a_m}\|_{\infty\delta}$ and the upper bound for $|\theta(t)|$. Because $0 < \delta \le \delta_0$ is arbitrary, we can choose it to be equal to δ_0. The bound for $\sup_t |\theta(t)|$ can be calculated from the Lyapunov-like function used to analyze the adaptive law. Such a bound, however, may be conservative. If, instead of the switching σ, we use projection, then the bound for $|\theta(t)|$ is known a priori, and the calculation of the constant c is easier [114, 115].

The effect of the unmodeled dynamics on the regulation error y is analyzed as follows: From (5.156) and $m_s, \phi \in \mathcal{L}_\infty$, we have

$$\int_t^{t+T} y^2 d\tau \le \int_t^{t+T} \varepsilon^2 m_s^2 d\tau + c\int_t^{t+T} \dot{\theta}^2 d\tau \quad \forall t \ge 0$$

for some constant $c \ge 0$ and any $T > 0$. Using the mean square sense (m.s.s.) property of $\varepsilon m_s, \dot{\theta}$ guaranteed by the adaptive law, we have

$$\frac{1}{T}\int_t^{t+T} y^2 d\tau \le c\Delta_2^2 + \frac{c}{T};$$

therefore, $y \in \mathcal{S}(\Delta_2^2)$, i.e., the regulation error is of the order of the modeling error in m.s.s. The m.s.s. bound for y^2 does not imply that at steady state y^2 is of the order of the modeling error characterized by Δ_2. A phenomenon known as bursting [105, 116], where y^2 assumes large values over short intervals of time, cannot be excluded by the m.s.s. bound.

The stability and robustness analysis for the MRAC scheme presented above is rather straightforward due to the simplicity of the plant. It cannot be directly extended to the general case without additional steps. A more elaborate and yet more systematic method that extends to the general case is presented below by the following steps.

Step 1. Express the plant input u and output y in terms of the parameter error $\tilde{\theta}$
We have

$$y = -\frac{1}{s+a_m}\tilde{\theta}y + \eta,$$
$$u = (s-a)y - \Delta_m(s)u = -\frac{s-a}{s+a_m}\tilde{\theta}y + \eta_u, \tag{5.159}$$

where $\eta_u = -\frac{a+a_m}{s+a_m}\Delta_m(s)u$. Using the $\mathcal{L}_{2\delta}$ norm for some $\delta \in (0, 2a_m)$, we have

$$\|y\| \leq c\|\tilde{\theta}y\| + \|\eta\|, \qquad \|u\| \leq c\|\tilde{\theta}y\| + \|\eta_u\|,$$

and therefore the fictitious normalizing signal m_f satisfies

$$m_f^2 = 1 + \|u\|^2 + \|y\|^2 \leq 1 + c(\|\tilde{\theta}y\|^2 + \|\eta\|^2 + \|\eta_u\|^2). \tag{5.160}$$

Step 2. Use the swapping lemmas and properties of the $\mathcal{L}_{2\delta}$ norm to upper bound $\|\tilde{\theta}y\|$ with terms that are guaranteed by the adaptive law to have small in m.s.s. gains
We use Swapping Lemma 2 (given in the Appendix) to write the identity

$$\tilde{\theta}y = \left(1 - \frac{\alpha_0}{s+\alpha_0}\right)\tilde{\theta}y + \frac{\alpha_0}{s+\alpha_0}\tilde{\theta}y = \frac{1}{s+\alpha_0}(\dot{\tilde{\theta}}y + \tilde{\theta}\dot{y}) + \frac{\alpha_0}{s+\alpha_0}\tilde{\theta}y, \tag{5.161}$$

where $\alpha_0 > 0$ is an arbitrary constant. Now, from the equation for y in (5.159) we obtain

$$\tilde{\theta}y = (s+a_m)(\eta - y),$$

which we substitute into the last term in (5.161) to obtain

$$\tilde{\theta}y = \frac{1}{s+\alpha_0}(\dot{\tilde{\theta}}y + \tilde{\theta}\dot{y}) + \alpha_0\frac{s+a_m}{s+\alpha_0}(\eta - y).$$

Therefore, by choosing δ, α_0 to satisfy $\alpha_0 > a_m > \frac{\delta}{2} > 0$, we obtain

$$\|\tilde{\theta}y\| \leq \left\|\frac{1}{s+\alpha_0}\right\|_{\infty\delta}(\|\dot{\tilde{\theta}}y\| + \|\tilde{\theta}\dot{y}\|) + \alpha_0\left\|\frac{s+a_m}{s+\alpha_0}\right\|_{\infty\delta}(\|\eta\| + \|y\|).$$

Hence

$$\|\tilde{\theta}y\| \leq \frac{2}{\alpha_0}(\|\dot{\tilde{\theta}}y\| + \|\tilde{\theta}\dot{y}\|) + \alpha_0 c(\|\eta\| + \|y\|),$$

where $c = \|\frac{s+a_m}{s+\alpha_0}\|_{\infty\delta}$. Using (5.156), it follows that

$$\|y\| \leq \epsilon m_s^2 + c\|\dot{\tilde{\theta}}\phi\|.$$

Therefore,

$$\|\tilde{\theta}y\| \leq \frac{2}{\alpha_0}(\|\dot{\tilde{\theta}}y\| + \|\tilde{\theta}\dot{y}\|) + \alpha_0 c(\|\epsilon m_s^2\| + \|\dot{\tilde{\theta}}\phi\| + \|\eta\|). \tag{5.162}$$

The gain of the first term in the right-hand side of (5.162) can be made small by choosing large α_0. The m.s.s. gain of the second and third terms is guaranteed by the adaptive law to be of the order of the modeling error denoted by the bound Δ_2, i.e., $\epsilon m_s, \dot{\tilde{\theta}} \in \mathcal{S}(\Delta_2^2)$. The last term also has a gain which is of the order of the modeling error. This implies that the gain of $\|\tilde{\theta}y\|$ is small, provided that Δ_2 is small and α_0 is chosen to be large.

Step 3. Use the B–G lemma to establish boundedness The normalizing properties of m_f and $\theta \in \mathcal{L}_\infty$ guarantee that $\frac{y}{m_f}, \frac{\phi}{m_f}, \frac{m_s}{m_f} \in \mathcal{L}_\infty$. Because

$$\|\dot{y}\| \leq |a|\|y\| + \|\Delta_m(s)\|_{\infty\delta}\|u\| + \|u\|,$$

it follows that $\frac{\dot{y}}{m_f} \in \mathcal{L}_\infty$. Due to the fact that $\Delta_m(s)$ is proper and analytic in $\Re[s] \geq -\frac{\delta_0}{2}$, $\|\Delta_m(s)\|_{\infty\delta}$ is a finite number, provided $0 < \delta \leq \delta_0$. Furthermore, $\frac{\|\eta\|}{m_f} \leq \Delta_\infty$, where

$$\Delta_\infty = \left\| \frac{\Delta_m(s)}{s + a_m} \right\|_{\infty\delta}, \tag{5.163}$$

and therefore (5.162) may be written in the form

$$\|\tilde{\theta} y\| \leq \frac{c}{\alpha_0}(\|\dot{\tilde{\theta}} m_f\| + m_f) + \alpha_0 c(\|\varepsilon m_s m_f\| + \|\dot{\tilde{\theta}} m_f\| + \Delta_\infty m_f). \tag{5.164}$$

Using (5.164) and $\frac{\|\eta\|}{m_f} \leq c\Delta_\infty$, $\frac{\|\eta_u\|}{m_f} \leq c\Delta_\infty$ in (5.160), we obtain

$$m_f^2 \leq 1 + c\left(\frac{1}{\alpha_0^2} + \alpha_0^2 \Delta_\infty^2\right) m_f^2 + c\|\tilde{g} m_f\|^2,$$

where $\tilde{g}^2 = \frac{|\dot{\theta}|^2}{\alpha_0^2} + \alpha_0^2 |\varepsilon m_s|^2 + \alpha_0^2 |\dot{\tilde{\theta}}|^2$ and $\alpha_0 \geq 1$. For $c(\frac{1}{\alpha_0^2} + \alpha_0^2 \Delta_\infty^2) < 1$, we have

$$m_f^2 \leq c + c\|\tilde{g} m_f\|^2 = c + c\int_0^t e^{-\delta(t-\tau)} \tilde{g}^2(\tau) m_f^2(\tau) d\tau.$$

Applying B–G Lemma 3, we obtain

$$m_f^2 \leq ce^{-\delta t} e^{c\int_0^t \tilde{g}^2(\tau)d\tau} + c\delta \int_0^t e^{-\delta(t-s)} e^{c\int_s^t \tilde{g}^2(\tau)d\tau} ds.$$

Because $\dot{\theta}, \varepsilon m_s \in \mathcal{S}(\Delta_2^2)$, it follows that

$$c\int_s^t \tilde{g}^2(\tau) d\tau \leq c\Delta_2^2 \left(\frac{1}{\alpha_0^2} + \alpha_0^2\right)(t - s) + c.$$

Hence, for

$$c\Delta_2^2 \left(\frac{1}{\alpha_0^2} + \alpha_0^2\right) < \delta \tag{5.165}$$

we have

$$e^{-\delta t} e^{c\int_0^t \tilde{g}^2(\tau)d\tau} \leq e^{-\bar{\alpha} t},$$

where $\bar{\alpha} = \delta - c\Delta_2^2(\frac{1}{\alpha_0^2} + \alpha_0^2)$, which implies that m_f is bounded. The boundedness of m_f implies that all the other signals are bounded too. The constant δ in (5.165) may be replaced by δ_0 since no restriction on δ is imposed except that $\delta \in (0, \delta_0]$. The constant $c > 0$ may be determined by following the calculations in each of the steps and is left as an exercise for the reader.

Step 4. Obtain a bound for the regulation error y The regulation error, i.e., y, is expressed in terms of signals that are guaranteed by the adaptive law to be of the order of the modeling error in m.s.s. This is achieved by using Swapping Lemma 1 for the error equation (5.155) and the equation $\varepsilon m_s^2 = -\tilde{\theta}\phi + \eta$ to obtain (5.155), which as shown before implies that $y \in \mathcal{S}(\Delta_2^2)$. That is, the regulation error is of the order of the modeling error in m.s.s.

The conditions that $\Delta_m(s)$ has to satisfy for robust stability are summarized as follows:

$$c\left(\frac{1}{\alpha_0^2} + \alpha_0^2\Delta_\infty^2\right) < 1, \qquad c\Delta_2^2\left(\frac{1}{\alpha_0^2} + \alpha_0^2\right) < \delta_0,$$

where

$$\Delta_\infty = \left\| \frac{\Delta_m(s)}{s + a_m} \right\|_{\infty\delta_0}, \qquad \Delta_2 = \left\| \frac{\Delta_m(s)}{s + a_m} \right\|_{2\delta_0}.$$

The constant $\delta_0 > 0$ is such that $\Delta_m(s)$ is analytic in $\Re[s] \geq -\frac{\delta_0}{2}$ and c denotes finite constants that can be calculated. The constant $\alpha_0 > \max(1, \frac{\delta_0}{2})$ is arbitrary and can be chosen to satisfy the above inequalities for small Δ_2, Δ_∞.

Let us now simulate the above robust MRAC scheme summarized by the equations

$$\begin{aligned}
u &= -\theta y, \\
\dot{\theta} &= \gamma\varepsilon\phi - \sigma_s\gamma\theta, \quad \varepsilon = \frac{z - \theta\phi}{m_s^2}, \\
\phi &= \frac{1}{s + a_m}y, \quad z = y - \frac{1}{s + a_m}u, \\
m_s^2 &= 1 + n_d, \quad \dot{n}_d = -\delta_0 n_d + u^2 + y^2, \quad n_d(0) = 0,
\end{aligned}$$

and applied to the plant

$$y = \frac{1}{s - a}(1 + \Delta_m(s))u,$$

where for simulation purposes we assume that $a = a_m = 2$, $\Delta_m(s) = -\frac{\mu s}{1+\mu s}$ with $\mu \geq 0$, $\gamma = 10$, $\delta_0 = 0.5$, and σ_s is the switching-σ with parameters $M_0 = 5$ and $\sigma_0 = 0.5$. Note that the plant is nonminimum-phase for $\mu > 0$. Figure 5.7 shows the response of $y(t)$ for different values of μ, which characterizes the size of the perturbation $\Delta_m(s)$. For small μ, we have boundedness and good regulation performance. As μ increases, stability deteriorates, and for $\mu > 0.5$ the plant becomes unstable. ∎

General case Let us now consider the SISO plant given by

$$y_p = G_0(s)(1 + \Delta_m(s))(u_p + d_u), \tag{5.166}$$

where

$$G_0(s) = k_p\frac{Z_p(s)}{R_p(s)} \tag{5.167}$$

is the transfer function of the modeled part of the plant. The high-frequency gain k_p and the polynomials $Z_p(s)$, $R_p(s)$ satisfy assumptions P1–P4 given in section 5.3, and the overall transfer function of the plant is strictly proper. The multiplicative uncertainty $\Delta_m(s)$ satisfies the following assumptions:

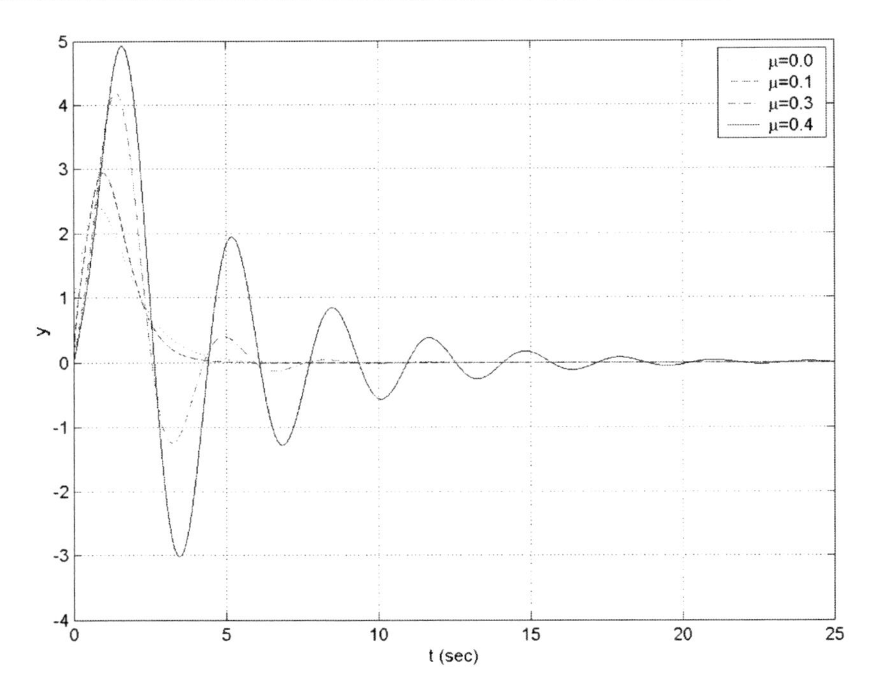

Figure 5.7. *Simulation results of the MRAC scheme of Example 5.7.4 for different values of μ.*

S1. $\Delta_m(s)$ is analytic in $\Re[s] \geq -\frac{\delta_0}{2}$ for some known $\delta_0 > 0$.

S2. There exists a strictly proper transfer function $W(s)$ analytic in $\Re[s] \geq -\frac{\delta_0}{2}$ and such that $W(s)\Delta_m(s)$ is strictly proper.

Assumptions S1 and S2 imply that Δ_∞, Δ_2 defined as

$$\Delta_\infty \triangleq \|W(s)\Delta_m(s)\|_{\infty\delta_0}, \qquad \Delta_2 \triangleq \|W(s)\Delta_m(s)\|_{2\delta_0}$$

are finite constants which for robustness purposes will be required to satisfy certain upper bounds. We should note that the strict properness of the overall plant transfer function and of $G_0(s)$ imply that $G_0(s)\Delta_m(s)$ is a strictly proper transfer function.

The control objective is to choose u_p so that all signals in the closed-loop plant are bounded and the output y_p tracks, as closely as possible, the output of the reference model y_m given by

$$y_m = W_m(s)r = k_m \frac{Z_m(s)}{R_m(s)} r$$

for any bounded reference signal $r(t)$. The transfer function $W_m(s)$ of the reference model satisfies assumptions M1 and M2 given in section 5.3.

The design of the control input u_p is based on the plant model with $\Delta_m(s) \equiv 0$ and $d_u \equiv 0$. The control objective, however, has to be achieved for the plant with $\Delta_m(s) \neq 0$ and $d_u \neq 0$. We start with the control law developed in section 5.5 for the plant model with $\Delta_m(s) \equiv 0$, $d_u \equiv 0$, i.e.,

$$u_p = \theta^T \omega, \tag{5.168}$$

where $\theta = [\theta_1^T, \theta_2^T, \theta_3, c_0]^T$, $\omega = [\omega_1^T, \omega_2^T, y_p, r]^T$. The parameter vector θ is to be generated online by an adaptive law. The signal vectors ω_1, ω_2 are generated, as in section 5.5, by filtering the plant input u_p and output y_p. The control law (5.168) will be robust with respect to the plant uncertainties $\Delta_m(s)$, d_u if we use robust adaptive laws from Chapter 3 to update the controller parameters.

The derivation of the robust adaptive laws is achieved by first developing the appropriate parametric models for the desired controller parameter vector θ^* and then choosing the appropriate robust adaptive law by employing the results of Chapter 3 as follows.

It can be shown that the parametric model for θ^* is given by

$$e_1 = \rho^*(\theta^{*T}\phi + u_f + \eta + d),$$

where

$$\begin{aligned}
\phi &= -W_m(s)\omega, \; u_f = W_m(s)u_p, \\
\eta &= W_m(s)\frac{\Lambda(s) - \theta_1^{*T}\alpha(s)}{\Lambda(s)}\Delta_m(s)u_p, \\
d &= W_m(s)\frac{\Lambda(s) - \theta_1^{*T}\alpha(s)}{\Lambda(s)}(\Delta_m(s)+1)d_u,
\end{aligned}$$

and can be used to develop a wide class of robust adaptive laws. Let us use the gradient algorithm to obtain

$$\begin{aligned}
\dot{\theta} &= \Gamma\varepsilon\phi \operatorname{sgn}\left(\frac{k_p}{k_m}\right) - \sigma_s\Gamma\theta, \\
\dot{\rho} &= \gamma\varepsilon\xi - \sigma_s\gamma\rho, \\
\varepsilon &= \frac{e_1 - \rho\xi}{m_s^2}, \quad \xi = \theta^T\phi + u_f, \\
m_s^2 &= 1 + n_d, \quad \dot{n}_d = -\delta_0 n_d + |u_p|^2 + |y_p|^2, \quad n_d(0) = 0,
\end{aligned} \tag{5.169}$$

where σ_s is the switching σ-modification and $\Gamma = \Gamma^T > 0$, $\gamma > 0$. The constant $\delta_0 > 0$ is chosen so that $W_m(s)$, $\frac{1}{\Lambda(s)}$, $\Delta_m(s)$ are analytic in $\Re[s] \geq -\frac{\delta_0}{2}$. This implies that $\frac{\phi}{m_s}$, $\frac{\eta}{m_s}$, $\frac{u_f}{m_s} \in \mathcal{L}_\infty$.

The adaptive law (5.169) guarantees that (i) $\theta, \rho \in \mathcal{L}_\infty$, (ii) $\varepsilon, \varepsilon m_s, \varepsilon\phi, \dot{\theta}, \dot{\rho} \in \mathcal{S}(\bar{\Delta}_2^2)$, where $\bar{\Delta}_2$ is the upper bound of $\frac{\eta}{m_s} + \frac{d}{m_s}$. The control law (5.168) together with the robust adaptive law (5.169) form the robust direct MRAC scheme whose properties are described by the following theorem.

Theorem 5.7.5. *Consider the MRAC scheme* (5.168), (5.169) *designed for the plant model* $y_p = G_0(s)u_p$ *but applied to the plant* (5.166) *with nonzero plant uncertainties* $\Delta_m(s)$ *and* *bounded input disturbance d_u. If*

$$c\left(\frac{1}{\alpha_0^2}+\alpha_0^{2n^*}\Delta_\infty^2\right)<1, \qquad c\left(\frac{1}{\alpha_0^2}+\alpha_0^{2n^*}\right)\Delta_2^2\le\frac{\delta}{2}, \tag{5.170}$$

where $\Delta_\infty = \|W_m(s)\Delta_m(s)\|_{\infty\delta_0}$, $\Delta_2 = \|\frac{\Lambda(s)-\theta_1^{*T}\alpha(s)}{\Lambda(s)}W_m(s)\Delta_m(s)\|_{2\delta_0}$, $\delta \in (0,\delta_0)$, *is such that* $G_0^{-1}(s)$ *is analytic in* $\Re[s]\ge -\frac{\delta}{2}$, $a_0 > \max[1, \frac{\delta_0}{2}]$ *is an arbitrary constant, and* $c\ge 0$ *denotes finite constants that can be calculated, then all the signals in the closed-loop plant are bounded and the tracking error* e_1 *satisfies*

$$\frac{1}{T}\int_t^{t+T} e_1^2 d\tau \le c(\Delta^2+d_0^2)+\frac{c}{T} \quad \forall t\ge 0$$

for any $T>0$, *where* d_0 *is an upper bound for* $|d_u|$ *and* $\Delta^2 = \frac{1}{\alpha_0^2}+\Delta_\infty^2+\Delta_2^2$.

If, in addition, the reference signal r *is dominantly rich of order* $2n$ *and* Z_p, R_p *are coprime, then the parameter error* $\tilde\theta$ *and tracking error* e_1 *converge exponentially to the residual set*

$$S=\left\{\tilde\theta\in\mathcal{R}^{2n}, e_1\in\mathcal{R} \mid |\tilde\theta|+|e_1|\le c(\Delta+d_0)\right\}.$$

Proof. A summary of the main steps of the proof is given below. For a detailed proof the reader is referred to the web resource [94].

Step 1. Express the plant input and output in terms of the parameter error term $\tilde\theta^T\omega$
In the presence of unmodeled dynamics and bounded disturbances, the plant input u_p and output y_p satisfy

$$\begin{aligned} y_p &= W_m\left(r+\frac{1}{c_0^*}\tilde\theta^T\omega\right)+\eta_y,\\ u_p &= G_0^{-1}W_m\left(r+\frac{1}{c_0^*}\tilde\theta^T\omega\right)+\eta_u, \end{aligned}\tag{5.171}$$

where

$$\eta_y=\frac{\Lambda-C_1^*}{c_0^*\Lambda}W_m(\Delta_m(u_p+d_u)+d_u), \qquad \eta_u=\frac{D_1^*}{c_0^*\Lambda}W_m(\Delta_m(u_p+d_u)+d_u),$$

and $C_1^*(s)=\theta_1^{*T}\alpha(s)$, $D_1^*=\theta_2^{*T}\alpha(s)+\theta_3^*\Lambda(s)$. Using the properties of the $\mathcal{L}_{2\delta}$ norm $\|(\cdot)_t\|_{2\delta}$, which for simplicity we denote as $\|\cdot\|$, and the stability of W_m, G_0^{-1}, it follows that there exists $\delta\in(0,\delta_0]$ such that

$$\begin{aligned} \|y_p\| &\le c+c\|\tilde\theta^T\omega\|+\|\eta_y\|,\\ \|u_p\| &\le c+c\|\tilde\theta^T\omega\|+\|\eta_u\|. \end{aligned}$$

The constant $\delta>0$ is such that $G_0^{-1}(s)$ is analytic in $\Re[s]\ge-\frac{\delta}{2}$. Therefore, the fictitious normalizing signal $m_f^2 \triangleq 1+\|u_p\|^2+\|y_p\|^2$ satisfies

$$m_f^2\le c+c\|\tilde\theta^T\omega\|^2+c\|\eta_y\|^2+c\|\eta_u\|^2. \tag{5.172}$$

Step 2. Use the swapping lemmas and properties of the $\mathcal{L}_{2\delta}$ norm to bound $\|\tilde{\theta}^T\omega\|$ from above with terms that are guaranteed by the robust adaptive laws to have small in m.s.s. gains In this step, we use Swapping Lemmas 1 and 2 and the properties of the $\mathcal{L}_{2\delta}$ norm to obtain the expression

$$\|\tilde{\theta}^T\omega\| \le c\|gm_f\| + c\left(\frac{1}{\alpha_0} + \alpha_0^{n^*}\Delta_\infty\right)m_f + cd_0, \tag{5.173}$$

where $\alpha_0 > \max(1, \frac{\delta_0}{2})$ is arbitrary and $g \in \mathcal{S}(\Delta_2^2 + \frac{d_0^2}{m_s^2})$.

Step 3. Use the B–G lemma to establish boundedness Using (5.173) in (5.172), we obtain

$$m_f^2 \le c + c\|gm_f\|^2 + c\left(\frac{1}{\alpha_0^2} + \alpha_0^{2n^*}\Delta_\infty^2\right)m_f^2 + cd_0^2.$$

We choose α_0 large enough so that for small Δ_∞,

$$c\left(\frac{1}{\alpha_0^2} + \alpha_0^{2n^*}\Delta_\infty^2\right) < 1.$$

We then have $m_f^2 \le c + c\|gm_f\|^2$ for some constant $c \ge 0$, which implies that

$$m_f^2 \le c + c\int_0^t e^{-\delta(t-\tau)}g^2(\tau)m_f^2(\tau)d\tau.$$

Applying B–G Lemma 3, we obtain

$$m_f^2 \le ce^{-\delta t}e^{c\int_0^t g^2(\tau)d\tau} + c\delta\int_0^t e^{-\delta(t-s)}e^{c\int_s^t g^2(\tau)d\tau}ds.$$

Since

$$\int_s^t g^2(\tau)d\tau \le c\Delta_2^2(t-s) + c\int_s^t \frac{d_0^2}{m_s^2}d\tau + c \quad \forall t \ge s \ge 0,$$

it follows that for $c\Delta_2^2 < \frac{\delta}{2}$ we have

$$m_f^2 \le ce^{-\frac{\delta}{2}t}e^{c\int_0^t \frac{d_0^2}{m_s^2}d\tau} + c\delta\int_0^t e^{-\frac{\delta}{2}(t-s)}e^{c\int_s^t \frac{d_0^2}{m_s^2}d\tau}ds.$$

If $d_u = d_0 = 0$, i.e., the input disturbance were zero, the boundedness of m_f would follow directly from the above inequality. In the case of $d_0 \ne 0$, the boundedness of m_f would follow directly if the normalizing signal m_s were chosen as $m_s^2 = 1 + n_d + \beta_0$, where β_0 is chosen large enough so that $c\frac{d_0^2}{\beta_0} < \frac{\delta}{2}$. This unnecessarily high normalization is used to simplify the analysis.

The boundedness of m_f can be established without having to modify m_s as follows: Consider an arbitrary interval $[t_1, t)$ for any $t_1 \geq 0$. We can establish, by following a similar procedure as in Steps 1 and 2, the inequality

$$\begin{aligned} m_s^2(t) &\leq c(1 + m_s^2(t_1))e^{-\frac{\delta}{2}(t-t_1)}e^{c\int_{t_1}^t \frac{d_0^2}{m_s^2}d\tau} \\ &+ c\delta \int_{t_1}^t e^{-\frac{\delta}{2}(t-s)}e^{c\int_s^t \frac{d_0^2}{m_s^2(\tau)}d\tau} ds \quad \forall t \geq t_1 \geq 0. \end{aligned} \tag{5.174}$$

We assume that $m_s^2(t)$ goes unbounded. Then for any given large number $\bar{\alpha} > 0$, there exist constants $t_2 > t_1 > 0$ such that $m_s^2(t_1) = \bar{\alpha}$, $m_s^2(t_2) > f_1(\bar{\alpha})$, where $f_1(\bar{\alpha})$ is any static function satisfying $f_1(\bar{\alpha}) > \bar{\alpha}$. Using the fact that m_s^2 cannot grow or decay faster than an exponential, we can choose f_1 properly so that $m_s^2(t) \geq \bar{\alpha}$ $\forall t \in [t_1, t_2]$ for some $t_1 \geq \bar{\alpha}$, where $t_2 - t_1 > \bar{\alpha}$. Choosing $\bar{\alpha}$ large enough so that $\frac{d_0^2}{\bar{\alpha}} < \frac{\delta}{2}$, it follows from (5.174) that

$$m_s^2(t_2) \leq c(1 + \bar{\alpha})e^{-\frac{\delta}{2}\bar{\alpha}}e^{cd_0^2} + c.$$

We can now choose $\bar{\alpha}$ large enough so that $m_s^2(t_2) < \bar{\alpha}$, which contradicts the hypothesis that $m_s^2(t_2) > \bar{\alpha}$, and therefore $m_s \in \mathcal{L}_\infty$. Because m_s bounds u_p, y_p, ω from above, we conclude that all signals are bounded.

Step 4. Establish bounds for the tracking error e_1 Bounds for e_1 in m.s.s. are established by relating e_1 with the signals that are guaranteed by the adaptive law to be of the order of the modeling error in m.s.s.

Step 5. Establish convergence of the estimated parameter and tracking error to residual sets Parameter convergence is established by expressing the parameter and tracking error equations as a linear system whose homogeneous part is e.s. and whose input is bounded. □

Remark 5.7.6 (robustness without dynamic normalization) The results of Theorem 5.7.5 are based on the use of a dynamic normalizing signal $m_s^2 = 1 + n_d$, which bounds both the signal vector ϕ and modeling error term η from above. The question is whether the signal m_s is necessary for the results of the theorem to hold. In [114, 117], it was shown that if m_s is chosen as $m_s^2 = 1 + \phi^T\phi$, i.e., the same normalization used in the ideal case, then the projection modification or switching σ-modification alone is sufficient to obtain the same qualitative results as those of Theorem 5.7.5. The proof of these results is based on arguments over intervals of time, an approach that was also used in some of the original results on robustness with respect to bounded disturbances [41]. In simulations, adaptive laws using dynamic normalization often lead to better transient behavior than those using static normalization.

Remark 5.7.7 (calculation of robustness bounds) The calculation of the constants c, δ, Δ_i, Δ_∞ is tedious but possible, as shown in [114, 115]. These constants depend on the size of various transfer functions, namely their $\mathcal{H}_{\infty\delta}$, $\mathcal{H}_{2\delta}$ bounds and stability margins, and on the bounds for the estimated parameters. The bounds for the estimated parameters can be calculated from the Lyapunov-like functions which are used to analyze the adaptive laws. In the case of projection, the bounds for the estimated parameters are known a priori. Because the constants c, Δ_i, Δ_∞, δ depend on unknown transfer functions and parameters such as $G_0(s)$, $G_0^{-1}(s)$, θ^*, the conditions for robust stability are quite difficult to check for a given plant. The importance of the robustness bounds is therefore more qualitative than quantitative.

Remark 5.7.8 In the case of MRAC with adaptive laws without normalization, robustness is established by modifying the adaptive law using simple leakage or projection or dead zone. In this case, however, global boundedness results do not exist. Instead, robustness and signal boundedness are established, provided all initial conditions lie inside a finite convex set whose size increases as the modeling error goes to zero. Details of this analysis can be found in [50, 51, 56].

5.8 Case Study: Adaptive Cruise Control Design

The adaptive cruise control (ACC) system is an extension of the conventional cruise control (CC) system and allows automatic vehicle following in the longitudinal direction. The driver sets the desired spacing between his/her vehicle and the leading vehicle in the same lane and the ACC system takes over the brake and gas pedals and maintains the desired spacing. The driver is responsible for maintaining the vehicle in the center of the lane by operating the steering wheel as well as taking care of emergencies that go beyond the designed capabilities of the ACC system.

A forward looking sensor mounted in the front of the vehicle measures the distance from the front of the vehicle to the rear of the leading vehicle in the same lane as well as the relative speed between the two vehicles. As shown in Figure 5.8, the ACC system regulates the following vehicle's speed v towards the leading vehicle's speed v_l and maintains the intervehicle distance x_r close to the desired spacing s_d. The control objective in the vehicle following mode is expressed as

$$v_r \to 0, \quad \delta \to 0 \quad \text{as } t \to \infty, \tag{5.175}$$

where $v_r = v_l - v$ is the speed error or relative speed and $\delta = x_r - s_d$ is the separation or spacing error. The desired intervehicle spacing is given by

$$s_d = s_0 + hv, \tag{5.176}$$

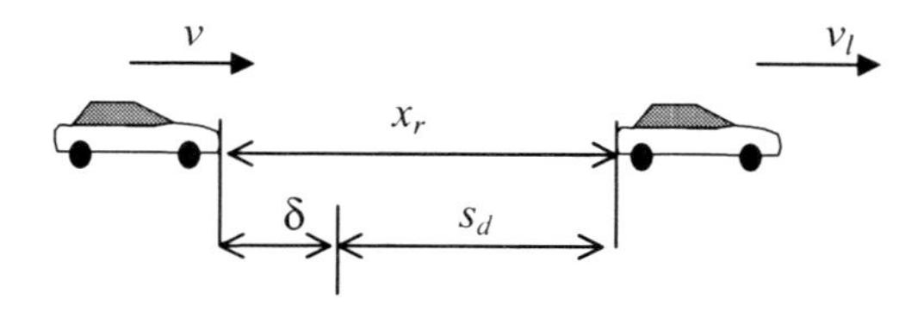

Figure 5.8. *Vehicle following.*

where s_0 is a fixed safety intervehicle spacing so that the vehicles are not touching each other at zero speed and h is the so-called constant time headway. The desired spacing is therefore proportional to the speed; at higher speeds the spacing should be higher in order to have enough time to stop without colliding with the lead vehicle. The time headway represents the time it takes for the vehicle with constant speed v to cover the intervehicle spacing $s_d - s_0$. In addition to (5.175), the control objective has to satisfy the following constraints:

C1. $a_{\min} \leq \dot{v} \leq a_{\max}$, where $a_{\min}$ and $a_{\max}$ are specified.

C2. The absolute value of jerk defined as $|\ddot{v}|$ should be small.

The above constraints are the result of driving comfort and safety concerns and are established using human factor considerations [118]. Constraint C1 restricts the ACC vehicle from generating high acceleration/deceleration. Constraint C2 is related to driver comfort.

In order to design the ACC system we need a model of the longitudinal vehicle dynamics. The full-order longitudinal vehicle model developed based on physical and empirical laws and validated with experimental data is a complex nonlinear system with lookup tables. It is difficult, if not impossible, to design a controller for such a complex nonlinear system. Even if possible, the design of such a control system will most likely be very complex and difficult to understand and implement. On the other hand, the human driver can drive the vehicle without having in mind such a complex nonlinear system. In fact, the vehicle appears to the driver as a first-order system or even as a constant-gain system, i.e., more speed, more gas. The driver's inputs are a low-frequency input, which means high-frequency modes and crucial nonlinearities do not get excited and therefore are not experienced by the driver. Since the driver will be replaced with the computer system one would argue that the generation of higher frequencies for faster driving functions may be desirable. While this argument may be true during emergencies, it raises many human factor questions. With the ACC system on, the driver is like a passenger responsible only for lateral control. The generation of vehicle actions which are outside the frequency range for which the driver feels comfortable may create human factor problems. Based on these human factor considerations the ACC system should be designed to generate low-frequency inputs, and therefore the excitation of high-frequency modes and nonlinearities is avoided. For these reasons it was determined using actual experiments [118] that the simplified first-order model

$$\dot{v} = -av + bu + d, \tag{5.177}$$

where v is the longitudinal speed, u is the throttle/brake command, d is the modeling uncertainty, and a, b are positive constant parameters which vary slowly with speed, is a good approximation of the longitudinal dynamics for the speed to throttle subsystem within the frequency range of interest. We have to emphasize that even though the control design is based on the simplified model (5.177) it has to be tested using the full-order validated model before being applied to the actual vehicle.

In the control design and analysis, we assume that d, $\dot{d}v_l$, and $\dot{v}_l$ are all bounded. We consider the MRAC approach where the throttle/brake command u is designed to force the vehicle speed to follow the output of the reference model

$$v_m = \frac{a_m}{s + a_m}(v_l + k\delta), \tag{5.178}$$

where a_m and k are positive design constants and $v_l + k\delta$ is considered to be the reference input. The term $v_l + k\delta$ represents the desired speed the following vehicle has to reach in order to match the speed of the lead vehicle and reduce the spacing error to zero. Consequently the reference input depends on the value of the speed of the leading vehicle as well as the size of the spacing error.

Let's first assume that a, b, and d in (5.177) are known and consider the following control law:

$$u = k_1^* v_r + k_2^* \delta + k_3^*, \tag{5.179}$$

where

$$k_1^* = \frac{a_m - a}{b}, \qquad k_2^* = \frac{a_m k}{b}, \qquad k_3^* = \frac{a v_l - d}{b}.$$

Using (5.179) in (5.177), the closed-loop transfer function of the vehicle longitudinal dynamics with $v_l + k\delta$ as the input and v as the output is the same as the transfer function of the reference model. It is easy to verify that for any positive constants a_m and k, the closed-loop system is stable and the control objective in (5.175) is achieved when v_l is a constant. Since a, b, and d are unknown and change with vehicle speed, we use the control law

$$u = k_1 v_r + k_2 \delta + k_3, \tag{5.180}$$

where k_i is the estimate of k_i^* to be generated by an adaptive law so that the closed-loop stability is guaranteed and the control objectives are met. We can establish that the tracking error $e = v - v_m$ satisfies

$$e = \frac{b}{s + a_m}(k_1^* v_r + k_2^* \delta + k_3^* + u),$$

which is in the form of the B-DPM. Substituting for u from (5.180), we obtain the error equation

$$e = \frac{b}{s + a_m}(\tilde{k}_1 v_r + \tilde{k}_2 \delta + \tilde{k}_3),$$

where $\tilde{k}_i = k_i - k_i^*$ $(i = 1, 2, 3)$, which implies

$$\dot{e} = -a_m e + b(\tilde{k}_1 v_r + \tilde{k}_2 \delta + \tilde{k}_3).$$

The adaptive law is derived by using the Lyapunov-like function

$$V = \frac{e^2}{2} + \sum_{i=1}^{3} \frac{b}{2\gamma_i} \tilde{k}_i^2,$$

where $\gamma_i > 0$ and b, even though unknown, is always positive. Hence

$$\dot{V} = -a_m e^2 + be(\tilde{k}_1 v_r + \tilde{k}_2 \delta + \tilde{k}_3) + \sum_{i=1}^{3} \frac{b}{\gamma_i} \tilde{k}_i \dot{\tilde{k}}_i.$$

We choose the adaptive laws as

$$\begin{aligned} \dot{k}_1 &= \Pr\{-\gamma_1 e v_r\}, \\ \dot{k}_2 &= \Pr\{-\gamma_2 e \delta\}, \\ \dot{k}_3 &= \Pr\{-\gamma_3 e\}, \end{aligned} \tag{5.181}$$

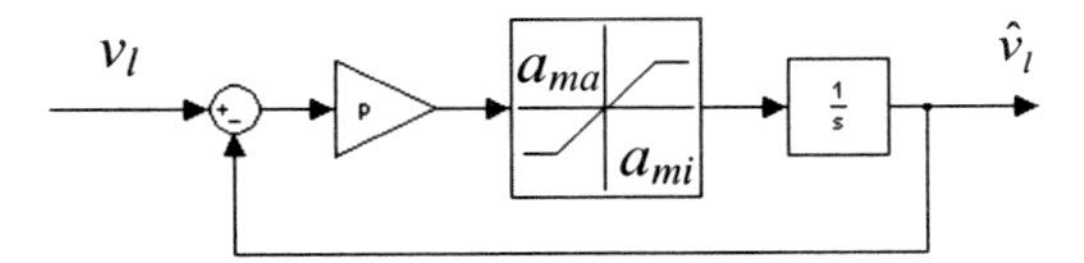

Figure 5.9. *Nonlinear filter used to smooth v_l.*

where $\Pr\{\cdot\}$ is the projection operator which keeps k_i within the intervals $[k_{il}, k_{iu}]$, and k_{il} and k_{iu} are chosen such that $k_i^* \in [k_{il}, k_{iu}]$ $(i = 1, 2, 3)$. Hence

$$\dot{V} = -a_m e^2 - \frac{b}{\gamma_i} \tilde{k}_i \dot{k}_3^*,$$

where $\dot{k}_3^* = \frac{a\dot{v}_l - \dot{d}}{b}$ is a bounded disturbance term. Since the estimated parameters are guaranteed to be bounded by forcing them via projection to remain inside bounded sets, the expression for $\dot{V}$ implies that $e \in \mathcal{L}_\infty$, which in turn implies that all the other signals in the closed-loop system are bounded. In addition, if v_l and d are constants, we can further show that $e \in \mathcal{L}_2$ and e, δ converge to zero with time.

The control law (5.180), (5.181) does not guarantee that the human factor constraints C1, C2 will be satisfied. For example, if the leading vehicle speeds up with high acceleration, the following vehicle in an effort to match its speed and maintain the desired spacing has to generate similar acceleration, violating constraint C1 and possibly C2. In an effort to prevent such phenomena the lead vehicle speed is filtered using the nonlinear filter shown in Figure 5.9. The actual speed v_l is then replaced by the filtered speed $\hat{v}_l$ in the above adaptive control law.

The above adaptive control design is designed for the throttle subsystem. For the brake subsystem another controller is developed. Since the throttle and brake cannot be applied at the same time, a switching logic with hysteresis and the appropriate logic that dictates the switching between brake and throttle need to be developed. Details for the full design can be found in [118]. We have to indicate that during normal driving conditions the throttle subsystem is the one which is mostly active, since small decelerations can also be handled by the throttle subsystem using engine torque, i.e., reducing the throttle angle. Figure 5.10 shows the vehicle speed trajectories and intervehicle spacing error for several vehicle maneuvers that begin with zero speed followed by acceleration/deceleration. The results are obtained by applying the adaptive control scheme developed above based on the simplified model to the full-order nonlinear vehicle model. The following vehicle tracks the speed of the lead vehicle well except when the lead vehicle accelerates with high acceleration at time $t = 50$ sec. Since the following vehicle cannot use high acceleration it lags behind, creating a large spacing error. Then it closes in by accelerating, albeit using a lower acceleration than that of the lead vehicle but higher speed.

Figure 5.11 shows experimental results of vehicle following where the lead vehicle accelerates with high acceleration. During these vehicle maneuvers the parameters of the simplified model vary with speed, but since the adaptive control scheme does not rely on the knowledge of these parameters, performance and stability are not affected.

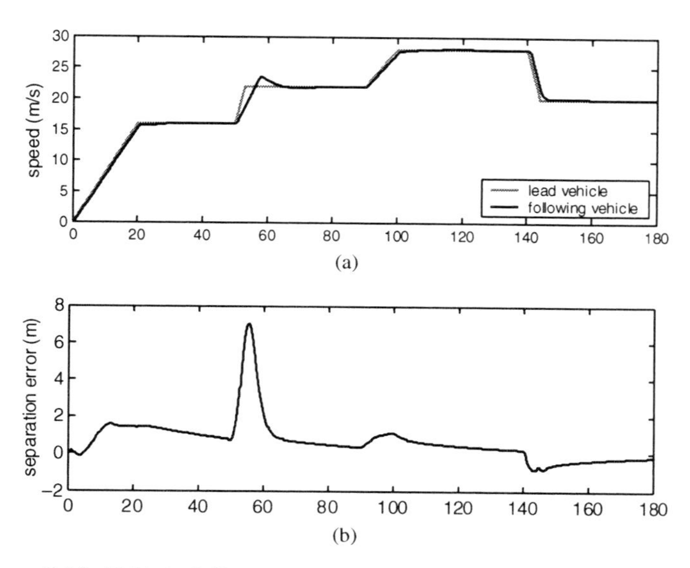

Figure 5.10. *Vehicle following:* (a) *speed tracking,* (b) *separation error.*

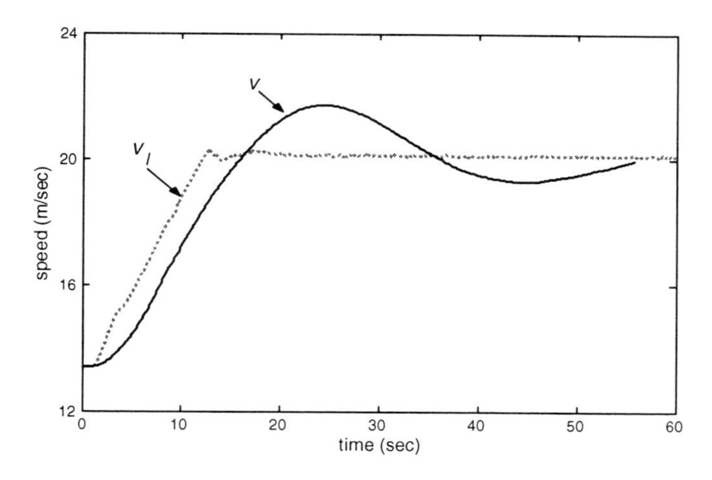

Figure 5.11. *Experimental results of vehicle following using a lead vehicle with high acceleration.*

5.9 Case Study: Adaptive Attitude Control of a Spacecraft

We consider the control problem associated with the descending of a spacecraft onto a landing site such as Mars. The attitude of the spacecraft needs to be controlled in order to avoid tipping or crash. Due to the consumption of fuel during the terminal landing, the moments of inertia I_x, I_y, and I_z are changing with time in an uncertain manner. In order to handle this parametric uncertainty we consider an adaptive control design.

In Figure 5.12, X_B, Y_B, and Z_B are the body frame axes of the spacecraft; X, Y, and

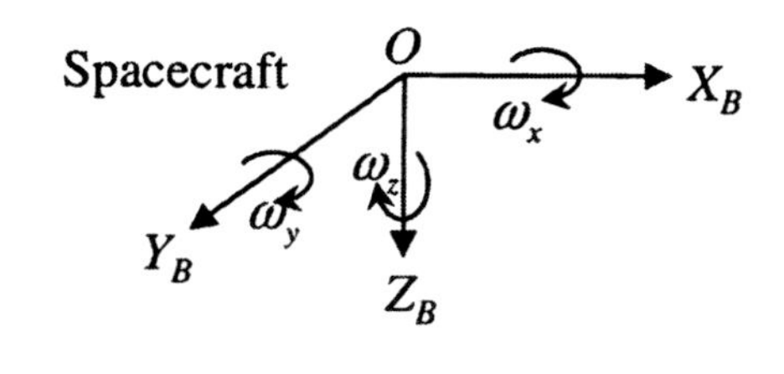

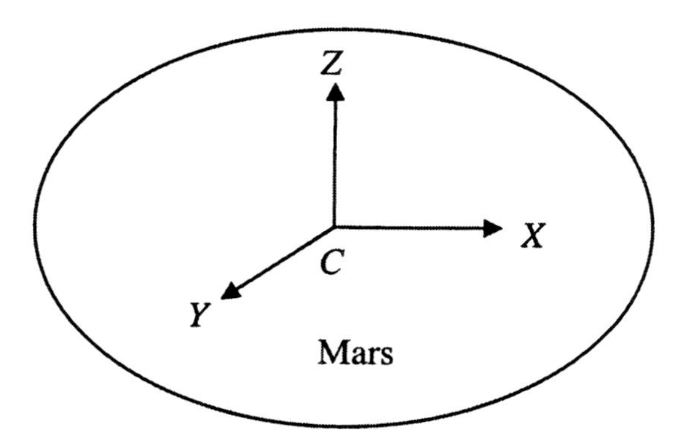

Figure 5.12. *Body frames of spacecraft and inertial reference frame axes of Mars.*

Z are the inertial reference frame axes of Mars; and O and C are the centers of mass of the spacecraft and Mars, respectively.

The dynamics of the spacecraft are described by the following equations:

$$\begin{aligned} T_x &= I_x\dot{\omega}_x + \omega_y\omega_z(I_z - I_y), \\ T_y &= I_y\dot{\omega}_y + \omega_x\omega_z(I_x - I_z), \\ T_z &= I_z\dot{\omega}_z + \omega_x\omega_y(I_y - I_x), \end{aligned} \tag{5.182}$$

where T_x, T_y, and T_z are the input torques; I_x, I_y, and I_z are the moments of inertia;[10] and ω_x, ω_y, and ω_z are the angular velocities with respect to the inertial frame X, Y, and Z axes, as shown in Figure 5.12. Define the coordinates

$$[q_1, q_2, q_3, q_4] = \left[\hat{n}_1 \sin\left(\frac{\varphi}{2}\right), \hat{n}_2 \sin\left(\frac{\varphi}{2}\right), \hat{n}_3 \sin\left(\frac{\varphi}{2}\right), \cos\left(\frac{\varphi}{2}\right)\right], \tag{5.183}$$

where $\hat{n}_1, \hat{n}_2, \hat{n}_3$ are the unit vectors along the axis of rotation and φ is the angle of rotation with $0 \leq \varphi < \pi$; q_1, q_2, q_3, q_4 are the *quaternion*[11] angles of rotation. The attitude

[10] These represent the mass properties of the spacecraft in the space.

[11] A quaternion q contains four Euler symmetric parameters q_1, q_2, q_3, and q_4, $q \cong q_1 i + q_2 j + q_3 k + q_4$, where $i^2 = j^2 = k^2 = -1$ are orthogonal unit vectors. The norm of a quaternion is $q_1^2 + q_2^2 + q_3^2 + q_4^2 = 1$. A quaternion is defined in a four-dimensional space with one real dimension q_4 and three imaginary dimensions q_1, q_2, and q_3. A quaternion represents a rotation about a unit vector through an angle φ, where $0 \leq \varphi < \pi$.

spacecraft dynamics are described as

$$\begin{aligned}\dot{q}_1 &= \frac{1}{2}(q_2\omega_z - q_3\omega_y + q_4\omega_x),\\ \dot{q}_2 &= \frac{1}{2}(-q_1\omega_z + q_3\omega_x + q_4\omega_y),\\ \dot{q}_3 &= \frac{1}{2}(q_1\omega_y - q_2\omega_x + q_4\omega_z),\\ \dot{q}_4 &= \frac{1}{2}(-q_1\omega_x - q_2\omega_y - q_3\omega_z).\end{aligned} \tag{5.184}$$

By assuming a small angle of rotation, i.e., $\sin(\frac{\varphi}{2}) = \frac{\varphi}{2}$, $\cos(\frac{\varphi}{2}) = 1$, we have

$$[q_1, q_2, q_3, q_4] = \left[\hat{n}_1\frac{\varphi}{2}, \hat{n}_2\frac{\varphi}{2}, \hat{n}_3\frac{\varphi}{2}, 1\right]$$

and

$$\begin{aligned}\dot{q}_1 &= \frac{1}{2}(q_2\omega_z - q_3\omega_y + q_4\omega_x) = \frac{1}{2}\left(\hat{n}_2\frac{\varphi}{2}\omega_z - \hat{n}_3\frac{\varphi}{2}\omega_y + \omega_x\right) = \frac{1}{2}\omega_x,\\ \dot{q}_2 &= \frac{1}{2}(-q_1\omega_z + q_3\omega_x + q_4\omega_y) = \frac{1}{2}\left(-\hat{n}_1\frac{\varphi}{2}\omega_z + \hat{n}_3\frac{\varphi}{2}\omega_x + \omega_y\right) = \frac{1}{2}\omega_y,\\ \dot{q}_3 &= \frac{1}{2}(q_1\omega_y - q_2\omega_x + q_4\omega_z) = \frac{1}{2}\left(\hat{n}_1\frac{\varphi}{2}\omega_y - \hat{n}_2\frac{\varphi}{2}\omega_y - \hat{n}_2\frac{\varphi}{2}\omega_x + \omega_z\right) = \frac{1}{2}\omega_z,\\ \dot{q}_4 &= \frac{1}{2}(-q_1\omega_x - q_2\omega_y - q_3\omega_z) = \frac{1}{2}\left(-\hat{n}_1\frac{\varphi}{2}\omega_x - \hat{n}_2\frac{\varphi}{2}\omega_y - \hat{n}_3\frac{\varphi}{2}\omega_z\right) = 0,\end{aligned}$$

i.e.,

$$\begin{bmatrix}\dot{q}_1\\ \dot{q}_2\\ \dot{q}_3\\ \dot{q}_4\end{bmatrix} = \frac{1}{2}\begin{bmatrix}\omega_x\\ \omega_y\\ \omega_z\\ 0\end{bmatrix}, \qquad \begin{bmatrix}\ddot{q}_1\\ \ddot{q}_2\\ \ddot{q}_3\\ \ddot{q}_4\end{bmatrix} = \frac{1}{2}\begin{bmatrix}\dot{\omega}_x\\ \dot{\omega}_y\\ \dot{\omega}_z\\ 0\end{bmatrix}.$$

Using (5.182) and the equation above, we obtain

$$T_x = 2I_x\ddot{q}_1, \qquad T_y = 2I_y\ddot{q}_2, \qquad T_z = 2I_z\ddot{q}_3, \tag{5.185}$$

i.e.,

$$q_i = \frac{1}{2I_j s^2}T_j, \quad i = 1, 2, 3, \quad j = x, y, z.$$

Let us assume that the performance requirements of the attitude control problem are to reach the desired attitude with a settling time less than 0.6 second and a percentage overshoot less than 5%. These performance requirements are used to choose the reference model as a second-order system with two complex poles corresponding to a damping ratio $\zeta = 0.707$ and natural frequency $\omega_n = 10$ rad/sec, i.e.,

$$\frac{q_m}{q_r} = \frac{\omega_n^2}{s^2 + 2\zeta\omega_n s + \omega_n^2} = \frac{100}{s^2 + 14.14s + 100} = W_m(s), \tag{5.186}$$

where q_r and q_m are the reference input and output, respectively. The control objective is to choose the input torques T_x, T_y, T_z so that each q_i, $i = 1, 2, 3$, follows the output of the reference model q_m for any unknown moment of inertia I_x, I_y, and I_z.

The form of the MRC law is

$$\begin{aligned} T_x &= \theta_{1x}^* \frac{1}{s+\lambda} T_x + \theta_{2x}^* \frac{1}{s+\lambda} q_1 + \theta_{3x}^* q_1 + c_{0x}^* q_r, \\ T_y &= \theta_{1y}^* \frac{1}{s+\lambda} T_y + \theta_{2y}^* \frac{1}{s+\lambda} q_2 + \theta_{3y}^* q_3 + c_{0y}^* q_r, \\ T_z &= \theta_{1z}^* \frac{1}{s+\lambda} T_z + \theta_{2z}^* \frac{1}{s+\lambda} q_3 + \theta_{3z}^* q_3 + c_{0z}^* q_r, \end{aligned} \tag{5.187}$$

where the controller parameters θ_{ij}^*, c_{0j}^*, $i = 1, 2, 3$, $j = x, y, z$, are such that the closed-loop transfer function of the system dynamics in each axis is equal to that of the reference model. It can be shown that this matching is achieved if

$$\begin{aligned} \theta_{1j}^* &= -14.14, & \theta_{2j}^* &= 28.28 I_j, \\ \theta_{3j}^* &= -228.28 I_j, & c_{0j}^* &= 200 I_j \end{aligned}$$

for $j = x, y, z$. Substituting these desired controller parameters into the MRC (5.187), we can express the overall MRC law in the compact form

$$T = -14.14 \frac{1}{s+1} T + I_n \left(28.28 \frac{1}{s+1} q - 228.28 q + 200 I q_r \right),$$

where

$$I_n = \mathrm{diag}(I_x, I_y, I_z), \qquad T = [T_y, T_y, T_z]^T, \qquad q = [q_1, q_2, q_3]^T,$$

I is the identity 3×3 matrix, and the design constant λ is taken to be equal to 1. The relationship between the desired controller parameters and unknown plant parameters shows that if we use the direct MRAC approach, we will have to estimate 12 parameters, which is the number of the unknown controller parameters, whereas if we use the indirect MRAC approach, we estimate only three parameters, namely the unknown inertias. For this reason the indirect MRAC approach is more desirable.

Using the CE approach, the control law is given as

$$T = -14.14 \frac{1}{s+1} T + \hat{I}_n \left(28.28 \frac{1}{s+1} q - 228.28 q + 200 I q_r \right), \tag{5.188}$$

where $\hat{I}_n$ is the estimate of I_n to be generated by an online adaptive law as follows.

We consider the plant equations

$$T_x = 2 I_x q_1 s^2, \qquad T_y = 2 I_y q_2 s^2, \qquad T_z = 2 I_z q_3 s^2.$$

Filtering each side of the equations with the filter $\frac{1}{(s+1)^2}$, we obtain

$$\frac{1}{(s+1)^2} T_x = I_x \frac{2s^2}{(s+1)^2} q_1,$$

$$\frac{1}{(s+1)^2}T_y = I_y\frac{2s^2}{(s+1)^2}q_2,$$
$$\frac{1}{(s+1)^2}T_z = I_z\frac{2s^2}{(s+1)^2}q_3.$$

Letting

$$\phi_1 = \frac{2s^2}{(s+1)^2}q_1, \qquad \phi_2 = \frac{2s^2}{(s+1)^2}q_2, \qquad \phi_3 = \frac{2s^2}{(s+1)^2}q_3,$$
$$z_1 = \frac{1}{(s+1)^2}T_x, \qquad z_2 = \frac{1}{(s+1)^2}T_y, \qquad z_3 = \frac{1}{(s+1)^2}T_z,$$

we obtain the SPMs

$$z_1 = I_x\phi_1, \qquad z_2 = I_y\phi_2, \qquad z_3 = I_z\phi_3.$$

Using the a priori information that $150 > I_x > 0$, $100 > I_y > 0$, $79 > I_z > 0$, we design the following adaptive laws:

$$\dot{\hat{I}}_x = \begin{cases} \gamma_1\varepsilon_1\phi_1 & \text{if } 150 > \hat{I}_x > 0 \\ & \text{or if } \hat{I}_x = 0 \text{ and } \varepsilon_1\phi_1 \geq 0 \\ & \text{or if } \hat{I}_x = 150 \text{ and } \varepsilon_1\phi_1 \leq 0, \\ 0 & \text{otherwise,} \end{cases}$$
$$\dot{\hat{I}}_y = \begin{cases} \gamma_2\varepsilon_2\phi_2 & \text{if } 100 > \hat{I}_y > 0 \\ & \text{or if } \hat{I}_y = 0 \text{ and } \varepsilon_2\phi_2 \geq 0 \\ & \text{or if } \hat{I}_y = 100 \text{ and } \varepsilon_2\phi_2 \leq 0, \\ 0 & \text{otherwise,} \end{cases} \tag{5.189}$$
$$\dot{\hat{I}}_z = \begin{cases} \gamma_3\varepsilon_3\phi_3 & \text{if } 79 > \hat{I}_z > 0 \\ & \text{or if } \hat{I}_z = 0 \text{ and } \varepsilon_3\phi_3 \geq 0 \\ & \text{or if } \hat{I}_z = 79 \text{ and } \varepsilon_3\phi \leq 0, \\ 0 & \text{otherwise,} \end{cases}$$

where

$$\varepsilon_1 = \frac{z_1 - \hat{I}_x\phi_1}{m_{s1}^2}, \qquad \varepsilon_2 = \frac{z_2 - \hat{I}_y\phi_2}{m_{s2}^2}, \qquad \varepsilon_3 = \frac{z_3 - \hat{I}_z\phi_3}{m_{s3}^2},$$
$$m_{s1}^2 = 1 + \phi_1^2, \qquad m_{s2}^2 = 1 + \phi_2^2, \qquad m_{s3}^2 = 1 + \phi_3^2,$$

and $\gamma_1 = \gamma_2 = \gamma_3 = 10$ are the adaptive gains. Then

$$\hat{I}_n = \text{diag}(\hat{I}_x, \hat{I}_y, \hat{I}_z).$$

The adaptive laws (5.189) together with the control law (5.188) form the indirect MRAC scheme. In this case due to the simplicity of the plant, the plant unknown parameters

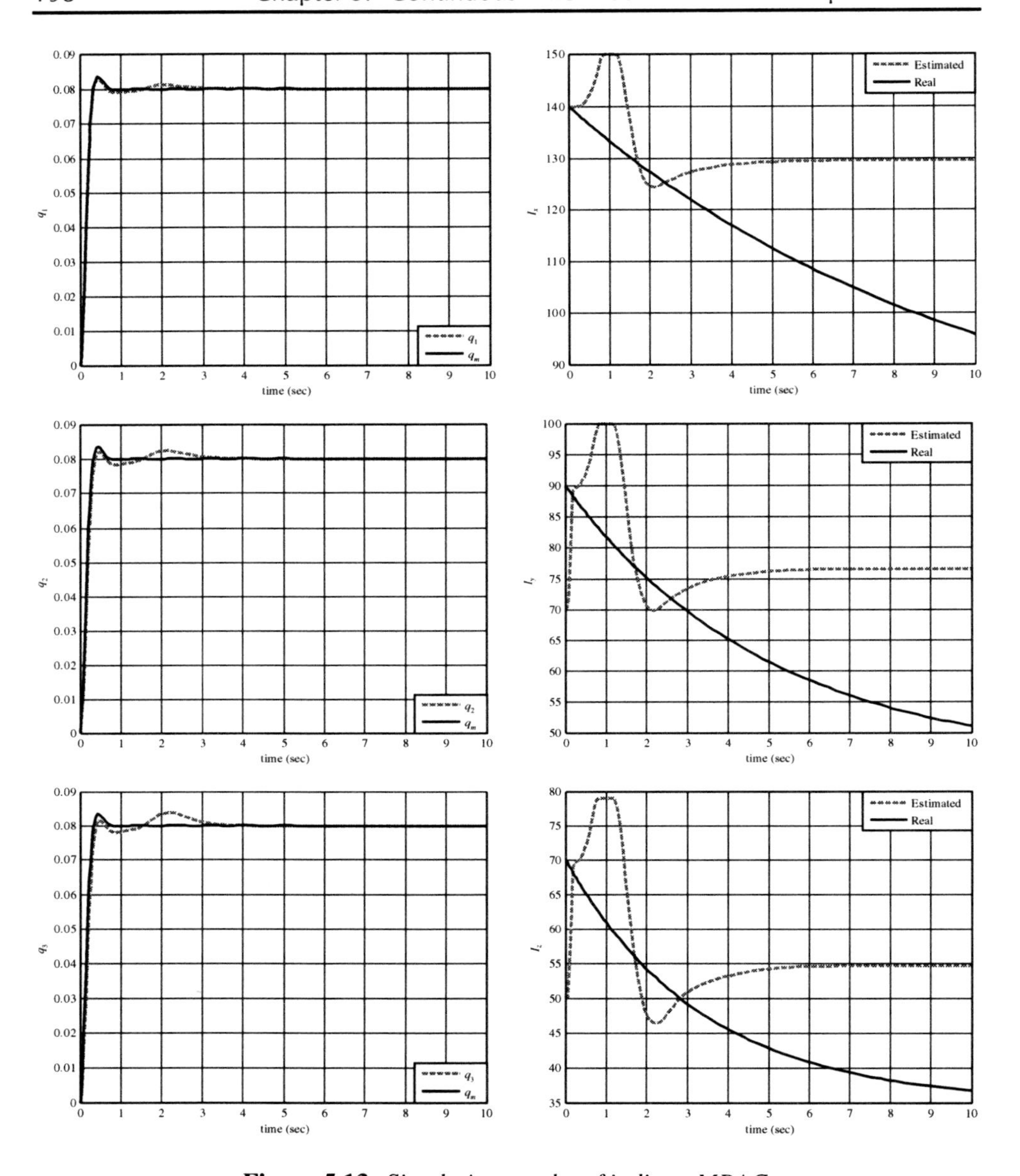

Figure 5.13. *Simulation results of indirect MRAC.*

are incorporated into the control law by performing all calculations a priori. Consequently no calculations need to be performed at each time t. If we followed the direct MRAC approach, we would have to estimate 12 parameters. In this case the indirect approach is the most appropriate one, as we limit the number of estimated parameters to 3. Figure 5.13 shows the tracking of the output q_m of the reference model by the three coordinates q_1, q_2, q_3, as well as the way the inertias change due to fuel reduction and the corresponding estimates generated by the adaptive law. It is clear that the control objective is met despite the fact that the estimated parameters converge to the wrong values due to lack of persistence of excitation.

See the web resource [94] for examples using the Adaptive Control Toolbox.

Problems

1. Consider the first-order plant

$$\dot{x} = ax + bu, \quad x(0) = x_0,$$

where a and b are unknown constants with $b > 0$. Design and analyze a direct MRAC scheme that can stabilize the plant and regulate x towards zero for any given initial condition x_0. Simulate your scheme for different initial conditions, design parameters and $a = -1, 2$ and $b = 5$. Comment on your results.

2. Consider the first-order plant

$$y_p = \frac{b}{s-1} u_p,$$

where $b > 0$ is the only unknown parameter. Design and analyze a direct MRAC scheme that can stabilize the plant and force y_p to follow the output y_m of the reference model

$$y_m = \frac{2}{s+2} r$$

for any bounded and continuous reference signal r. Simulate your scheme for different initial conditions and the choices for $r = 5$, $\sin 2t$, $\frac{1}{1+t}$, and $b = 12$. Comment on your results.

3. The dynamics of a throttle to speed subsystem of a vehicle may be represented by the first-order system

$$V = \frac{b}{s+a}\theta + d,$$

where V is the vehicle speed, θ is the throttle angle, and d is a constant load disturbance. The parameters $b > 0, a$ are unknown constants whose values depend on the operating state of the vehicle that is defined by the gear state, steady-state velocity, drag, etc. Design a cruise control (CC) system by choosing the throttle angle θ so that V follows a desired velocity V_m generated by the reference model

$$V_m = \frac{0.5}{s+0.5} V_s,$$

where V_s is the desired velocity set by the driver.

   (a) Assume that a, b, and d are known exactly. Design an MRC law that meets the control objective.

(b) Design and analyze a direct MRAC scheme to be used in the case of unknown a, b, and d (with $b > 0$).

(c) Design an indirect MRAC scheme to be used in the case of unknown a, b, and d (with $b > 0$).

(d) Simulate and compare your schemes in (b) and (c) by assuming $V_s = 55$ and using the following values for a, b, and d: (i) $a = 0.5$, $b = 1.5$, $d = 10$; (ii) $a = 0.5 + \frac{0.04}{1+V}$, $b = 1.5$, $d = 0.2 + \sin 0.02t$.

4. Consider the SISO plant

$$y_p = k_p \frac{Z_p(s)}{R_p(s)} u_p,$$

where k_p, $Z_p(s)$, and $R_p(s)$ are known. Design an MRC scheme for y_p to track y_m generated by the reference model

$$y_m = k_m \frac{Z_m(s)}{R_m(s)} r,$$

where $R_m(s)$ and $Z_m(s)$ are Hurwitz and $R_m(s)$ has the same degree as $R_p(s)$. Examine stability when $Z_p(s)$ is Hurwitz and when it is not. Comment on your results.

5. Consider the second-order plant

$$y_p = \frac{b_1 s + b_0}{s^2 + a_1 s + a_0} u_p,$$

where a_0, a_1, b_0, and b_1 are constants with $b_0, b_1 > 0$. The reference model is given by

$$y_m = \frac{4}{s+5} r.$$

(a) Assume that a_0, a_1, b_0, and b_1 are known. Design an MRC law that guarantees closed-loop stability and meets the control objective $y_p \to y_m$ as $t \to \infty$ for any bounded reference signal r.

(b) Repeat (a) when a_0, a_1, b_0, and b_1 are unknown and $b_0, b_1 > 0$.

(c) If in (b) $a_0 = -1$, $a_1 = 0$, $b_1 = 1$, and $b_0 > 0$ are known, indicate the simplification that results in the control law.

6. Show that the MRC law given by (5.87) meets the MRC objective for the plant (5.72) for any given nonzero initial conditions.

7. Consider the mass–spring–dashpot system of Figure 2.1 described by (2.5) with unknown constant M, f, and k, and the reference model

$$y_m = \frac{1}{s^2 + \sqrt{2}s + 1} r.$$

(a) Design a direct MRAC law with unnormalized adaptive law so that all signals in the closed-loop system are bounded and the displacement of the mass, x, converges to y_m as $t \to \infty$ for any bounded reference signal r.

(b) Repeat (a) for an indirect MRAC with normalized adaptive law.

(c) Simulate (a) and (b) for $M = 10$ kg, $f = 1$ N $\cdot$ s/m, $k = 9$ N/m using $r = 10$ and $r = 2\sin(3t) + 5\sin(t)$.

8. Consider the plant

$$y_p = \frac{b}{s+a} u_p,$$

where $b \geq 1$ and a are unknown constants. The reference model is given by

$$y_m = \frac{10}{s+10} r.$$

Do the following:

(a) Design a direct MRAC law based on the gradient algorithm.

(b) Repeat (a) for an LS algorithm.

(c) Simulate your design in (a) and (b). For simulations, use $b = 3$, $a = -5$, and r a signal of your choice.

9. Derive an indirect MRAC scheme using unnormalized adaptive laws for a plant with $n^* = 2$. Show boundedness of all signals and convergence of the tracking error to zero.

10. Consider the SISO plant

$$y_p = k_p \frac{Z_p(s)}{R_p(s)} u_p,$$

where $Z_p(s)$ and $R_p(s)$ are monic, $Z_p(s)$ is Hurwitz, and the relative degree $n^* = 1$. The order n of $R_p(s)$ is unknown. Show that the adaptive control law

$$u_p = -\theta y_p \operatorname{sgn}(k_p), \quad \dot{\theta} = y_p^2$$

guarantees signal boundedness and convergence of y_p to zero for any finite n. (*Hint*: The plant may be represented as

$$\begin{aligned} \dot{x}_1 &= A_{11} x_1 + A_{12} y_p, \quad x_1 \in R^{n-1}, \\ \dot{y}_p &= A_{21} x_1 + a_0 y_p + k_p u_p, \end{aligned}$$

where A_{11} is stable.)

11. Consider the MRAC problem of section 5.4.1. Show that the nonzero initial condition appears in the error equation as

$$e_1 = W_m(s)\rho^*(u_p - \theta^{*T}\omega) + C_c^T(sI - A_c)^{-1}e(0).$$

Show that the same stability results as in the case of $e(0) = 0$ can be established when $e(0) \neq 0$ by using the new Lyapunov-like function

$$V = \frac{e^T P_c e}{2} + \frac{\tilde{\theta}^T \Gamma^{-1} \tilde{\theta}}{2} |\rho^*| + \beta e_0^T P_0 e_0,$$

where e_0 is the zero-input response, i.e.,

$$\dot{e}_0 = A_c e_0, \quad e_0(0) = e(0),$$

$P_0 > 0$ satisfies $A_c^T P_0 + P_0 A_c = -I$, and $\beta > 0$ is an arbitrary positive constant.

12. Consider the SISO plant

$$y_p = e^{-\tau s} \frac{p_2}{(s+p_1)(s+p_2)} u_p,$$

where $0 < \tau \ll 1$, $0.5 \le p_1 \le 1$, and $100 \le p_2 \le 1000$.

   (a) Choose the dominant part of the plant and express the unmodeled part as a multiplicative perturbation.

   (b) Design an MRC law for the dominant part of the plant using the reference model

$$y_m = \frac{10}{s+10} r,$$

   assuming that $p_1 = 0.8$ and other parameters are unknown except for their bounds.

   (c) Show that the MRC meets the control objective for the dominant part of the plant.

   (d) Find a bound for the delay τ so that the MRC law designed for the dominant part guarantees stability and small tracking error when applied to the actual plant with nonzero delay.

13. Repeat Problem 12bc when all plant parameters are unknown by considering a robust MRAC approach.

14. Consider the speed control problem described in Problem 3. Suppose the full-order system dynamics are described by

$$V = \frac{b}{s+a}(1+\Delta_m(s))\theta + d,$$

where d is a bounded disturbance and $\Delta_m(s)$ represents the unmodeled dynamics, and the reference model is

$$V_m = \frac{0.5}{s+0.5} V_s$$

as described in Problem 3.

   (a) Design a robust MRAC scheme with and without normalization.

   (b) Simulate and compare your schemes for $a = 0.05(2+\sin 0.02t)$, $b = 1.5$, $d = 10(1+\sin 0.05t)$, and $\Delta_m(s) = -\frac{2\mu s}{\mu s+1}$ for $\mu = 0, 0.2, 0.5, 1$. Comment on your simulation results.

15. Consider the plant

$$y_p = \frac{b}{s+a} u_p + \mu \Delta_a(s) u_p,$$

where $\mu > 0$ is a small parameter, a and $b > 0$ are unknown parameters, and $\Delta_a(s)$ is a strictly proper stable unknown transfer function perturbation independent of μ. The control objective is to choose u_p so that all signals are bounded and y_p tracks, as closely as possible, the output y_m of the reference model

$$y_m = \frac{10}{s+10}r$$

for any bounded reference input r.

(a) Design a robust MRAC to meet the control objective.

(b) Develop bounds for robust stability.

(c) Develop a bound for the tracking error $e_1 = y - y_m$.

(d) Repeat (a), (b), and (c) for the plant

$$y_p = \frac{be^{-\tau s}}{s+a}u_p,$$

where $\tau > 0$ is a small constant.

16. Establish convergence of the estimated parameter and tracking errors to residual sets in the proof for Theorem 5.7.5 when the reference input is dominantly rich and the plant is stable with no zero-pole cancellations.
17. Consider the time-varying plant

$$\dot{x} = a(t)x + b(t)u,$$

where $b(t) > c > 0$ for some unknown constant c and $|\dot{a}| \leq \mu$, $|\dot{b}| \leq \mu$ for some $0 \leq \mu \ll 1$, i.e., the parameters are slowly varying with time. The plant is required to follow the LTI model reference

$$y_m = \frac{5}{s+5}r.$$

(a) Design and analyze an MRAC with unnormalized adaptive law to meet the control objective.

(b) Repeat (a) with an MRAC using a normalized adaptive law.

(c) For both (a) and (b), find a bound on μ for stability.

18. Consider the plant

$$\dot{x} = ax + u + d,$$

where d is a bounded disturbance and a is an unknown parameter.

(a) Show that no linear controller can stabilize the system if a is unknown.

(b) For $d = 0$, establish that the adaptive control law

$$u = -kx, \quad \dot{k} = x^2,$$

guarantees the boundedness of all signals and convergence of x to zero.

(c) Verify that for $k(0) = 5$, $x(0) = 1$, $a = 1$, and

$$d(t) = (1+t)^{-1/5}[5 - (1+t)^{-1/5} - 0.4(1+t)^{-6/5}],$$

the adaptive control law in (b) leads to k that drifts to infinity with time, by solving the associated differential equations. (*Hint*: Consider $x(t) = 5(1+t)^{1/5}$ and find k to satisfy the equations.)

(d) Modify the adaptive control law for robustness and use it to establish boundedness. Obtain a bound for x.

(e) Use simulations to demonstrate the results in (b)–(d).

19. Consider the nonminimum-phase plant

$$y = \frac{1}{s-a}\left[1 - \frac{2\mu s}{1+\mu s}\right],$$

where $0 \leq \mu \ll 1$.

(a) Obtain a reduced-order plant.

(b) Verify that control law $u = -kx$, $\dot{k} = x^2$ guarantees boundedness and convergence of y to zero when applied to the reduced-order plant in (a).

(c) Show that the adaptive law in (b) will lead to instability if $k(0) > \frac{1}{\mu} - a$ when applied to the full-order plant.

(d) Robustify the adaptive control law in (b) to establish boundedness and convergence of x to a small residual set by modifying the adaptive law with normalization and without normalization. Note that in the absence of normalization you can establish only semiglobal stability.

(e) Use simulations to demonstrate your results in (b)–(d).

20. Consider the second-order plant

$$\begin{aligned} \dot{x} &= -x + bz - u, \\ \mu\dot{z} &= -z + 2u, \\ y &= x, \end{aligned}$$

where $0 \leq \mu \ll 1$ and b is unknown and satisfies $2b > 1$. The plant output is required to track the output x_m of the reference model

$$\dot{x}_m = -x_m + r.$$

(a) Obtain a reduced-order model for the plant.

(b) Show that the adaptive control law

$$u = lr, \quad \dot{l} = -\gamma e r, \quad e = x - x_m$$

meets the control objective for the reduced-order plant.

(c) Show that for $\gamma r^2 > \frac{1+\mu}{\mu(2b+\mu)}$ the control law in (b) will lead to instability if applied to the full-order plant.

(d) Robustify the adaptive control law in (b) to guarantee stability when applied to the actual plant.

(e) Use simulations to demonstrate your results in (b)–(d).

21. Consider the plant

$$\dot{x} = -x + bu,$$

where b is the only unknown. The plant output is required to track the output x_m of the reference model

$$\dot{x}_m = -x_m + r.$$

Since the sign of b is unknown the techniques we developed in the chapter do not apply. The following adaptive control law is suggested to meet the control objective:

$$u = lr, \quad \dot{l} = -N(w)\varepsilon x_m, \qquad \dot{\hat{b}} = N(w)\varepsilon\xi,$$

$$N(w) = w^2 \cos w, \quad w = w_0 + \frac{\hat{b}^2}{2},$$

$$\dot{w}_o = \varepsilon^2 m^2, \quad w_0(0) = 0,$$

$$\varepsilon = \frac{x - x_m - N(w)\hat{b}\xi}{m^2}, \qquad \xi = -lx_m + \frac{1}{s+1}u,$$

$$m^2 = 1 + \left(\frac{1}{s+1}u\right)^2.$$

Show that the above adaptive control meets the control objective. (The term $N(w)$ is known as the Nussbaum gain.)

Chapter 6
Continuous-Time Adaptive Pole Placement Control

6.1 Introduction

The desired properties of the plant to be controlled are often expressed in terms of desired pole locations to be placed by the controller. If the plant satisfies the properties of controllability and observability, then a controller always exists to place the poles in the desired locations. While this is possible for poles, it is not the case for zeros. One way to change the zeros of the plant is to cancel them via feedback and replace them with the desired ones via feed forward. This was the case with MRC covered in Chapter 5, where some of the desired poles of the closed-loop plant were assigned to be equal to the zeros of the plant in order to facilitate the zero-pole cancellation. For stability, unstable zeros cannot be canceled with unstable poles; therefore, one of the restrictive assumptions in MRC and hence in MRAC is that the plant has to be minimum-phase; i.e., all zeros have to be stable. The assumption of stable plant zeros is rather restrictive in many applications. For example, the approximation of time delays often encountered in chemical and other industrial processes leads to plant models with unstable zeros.

In this chapter, we relax the assumption of minimum phase by considering control schemes that change the poles of the plant and do not involve unstable zero-pole cancellations. These schemes are referred to as *pole placement* schemes and are applicable to both minimum- and nonminimum-phase LTI plants. The combination of a pole placement control law with a parameter estimator or an adaptive law leads to an *adaptive pole placement control* (*APPC*) scheme that can be used to control a wide class of LTI plants with unknown parameters.

The APPC schemes may be divided into two classes: the *indirect APPC* schemes, where the adaptive law generates online estimates of the coefficients of the plant transfer function which are then used to calculate the parameters of the pole placement control law by solving a certain algebraic equation; and the *direct APPC*, where the parameters of the pole placement control law are generated directly by an adaptive law without any intermediate calculations that involve estimates of the plant parameters. The direct APPC schemes are restricted to scalar plants and to special classes of plants where the desired parameters of the pole placement controller can be expressed in the form of the linear or bilinear parametric models. Efforts to develop direct APPC schemes for a general class of LTI plants led to APPC schemes where both the controller and plant parameters are estimated online simultaneously [4, 119], leading to rather complex adaptive control schemes. The indirect APPC schemes, on the other hand, are easy to design and are applicable to a wide class of LTI plants that are not required to be minimum-phase or stable. The main drawback of indirect APPC is the possible loss of stabilizability of the estimated plant based on which the calculation of the controller parameters is performed. This drawback can be eliminated by modifying the indirect APPC schemes at the expense of adding more complexity. Because of its flexibility in choosing the controller design methodology (state feedback, compensator design, linear quadratic, etc.) and adaptive law, indirect APPC is the most general class of adaptive control schemes. This class also includes indirect MRAC as a special case where some of the poles of the plant are assigned to be equal to the zeros of the plant to facilitate the required zero-pole cancellation for transfer function matching. Indirect APPC schemes have also been known as self-tuning regulators in the literature of adaptive control to distinguish them from direct MRAC schemes.

The use of adaptive laws with normalization and without normalization, leading to different classes of adaptive control schemes in the case of MRAC, does not extend to APPC. In APPC, with the exception of scalar examples, the adaptive laws involve normalization in order to guarantee closed-loop stability.

6.2 Simple APPC Schemes: Without Normalization

In this section we use several examples to illustrate the design and analysis of simple APPC schemes. The important features and characteristics of these schemes are used to motivate and understand the more complicated ones to be introduced in the sections to follow. The simplicity of these examples is enhanced further using adaptive laws without normalization and a single Lyapunov function to establish stability for the overall closed-loop system.

6.2.1 Scalar Example: Adaptive Regulation

Consider the scalar plant

$$\dot{y} = ay + bu, \tag{6.1}$$

where a and b are unknown constants, and the sign of b is known. The control objective is to choose u so that the closed-loop pole is placed at $-a_m$, where $a_m > 0$ is a given constant, y and u are bounded, and $y(t)$ converges to zero as $t \to \infty$.

If a and b were known and $b \neq 0$, then the control law

$$u = -k^* y, \quad k^* = \frac{a + a_m}{b} \tag{6.2}$$

would lead to the closed-loop plant

$$\dot{y} = -a_m y; \tag{6.3}$$

i.e., the control law described by (6.2) changes the pole of the plant from a to $-a_m$ but preserves the zero structure. This is in contrast to MRC, where the zeros of the plant are canceled and replaced with new ones. It is clear from (6.3) that the pole placement law (6.2) meets the control objective exactly. The condition that $b \neq 0$ implies that the plant has to be controllable. It follows from (6.2) that as $b \to 0$, i.e., as the plant loses controllability, the stability gain $k^* \to \infty$, leading to unbounded control input.

We consider the case where a and b are unknown. As in the MRAC case, we use the CE approach to form APPC schemes as follows: We use the same control law as in (6.2) but replace the unknown controller parameter k^* with its online estimate k. The estimate k may be generated in two different ways: the direct one, where k is generated by an adaptive law, and the indirect one, where k is calculated from

$$k = \frac{\hat{a} + a_m}{\hat{b}}, \tag{6.4}$$

provided $\hat{b} \neq 0$, where $\hat{a}$ and $\hat{b}$ are the online estimates of a and b, respectively. We consider each design approach separately.

Direct adaptive regulation In this case the time-varying gain k in the control law

$$u = -ky \tag{6.5}$$

is updated directly by an adaptive law. The adaptive law is developed as follows: We add and subtract the term $-a_m y$ in the plant equation to obtain

$$\dot{y} = -a_m y + (a + a_m)y + bu.$$

Substituting for $a + a_m = bk^*$, we obtain

$$\dot{y} = -a_m y + b(k^* y + u),$$

which is in the form of the B-SSPM or B-DPM. The estimation model is given by

$$\dot{\hat{y}} = -a_m \hat{y} + \hat{b}(ky + u), \quad y(0) = 0.$$

If we substitute for the control law $u = -ky$, we have $\hat{y}(t) = 0 \, \forall t \geq 0$, which implies that the estimation error $\varepsilon = y - \hat{y} = y$ and therefore there is no need to implement the estimation model. Substituting for the control law in the B-SSPM of the plant, we obtain

$$\dot{y} = -a_m y - b\tilde{k}y, \tag{6.6}$$

where $\tilde{k} \triangleq k - k^*$ is the parameter error. Equation (6.6) relates the parameter error term $b\tilde{k}y$ with the regulation error y (which is equal to the estimation error) and motivates the Lyapunov function

$$V = \frac{y^2}{2} + \frac{\tilde{k}^2|b|}{2\gamma},$$

whose time derivative is

$$\dot{V} = -a_m y^2 - b\tilde{k}y^2 + \frac{|b|}{\gamma}\tilde{k}\dot{\tilde{k}} = -a_m y^2 - b\tilde{k}y^2 + b\tilde{k}\frac{\dot{\tilde{k}}}{\gamma}\,\mathrm{sgn}(b).$$

By noting that $\dot{k}^* = 0$ and therefore $\dot{k} = \dot{\tilde{k}}$, the choice of the adaptive law

$$\dot{k} = \gamma y^2 \operatorname{sgn}(b) \tag{6.7}$$

leads to

$$\dot{V} = -a_m y^2 \leq 0,$$

which implies that $y, \tilde{k}, k \in \mathcal{L}_\infty$ and $y \in \mathcal{L}_2$. From (6.6) and $y, \tilde{k}, k \in \mathcal{L}_\infty$, we have $\dot{y}, u \in \mathcal{L}_\infty$; therefore, from $y \in \mathcal{L}_2$, $\dot{y} \in \mathcal{L}_\infty$ it follows using Barbalat's lemma that $y(t) \to 0$ as $t \to \infty$. In summary, the direct APPC scheme (6.5), (6.7) guarantees signal boundedness and regulation of the plant state $y(t)$ to zero. The scheme, however, does not guarantee that the closed-loop pole of the plant is placed at $-a_m$ even asymptotically with time. To achieve such a pole placement result, we need to show that $\hat{k}(t) \to \frac{a+a_m}{b}$ as $t \to \infty$. For parameter convergence, however, y is required to be persistently exciting (PE), which is in conflict with the objective of regulating y to zero. The conflict between PI and regulation or control is well known in adaptive control and cannot be avoided in general.

Indirect adaptive regulation In this approach, the gain $k(t)$ in the control law

$$u = -k(t)y \tag{6.8}$$

is calculated by using the algebraic equation

$$k(t) = \frac{\hat{a}(t) + a_m}{\hat{b}(t)} \tag{6.9}$$

for $\hat{b} \neq 0$, where $\hat{a}$ and $\hat{b}$ are the online estimates of the plant parameters a and b, respectively. The adaptive laws for generating $\hat{a}$ and $\hat{b}$ are constructed using the techniques of Chapter 3 as follows.

We obtain the SSPM by adding and subtracting $a_m y$ in the plant equation

$$\dot{y} = -a_m y + (a + a_m)y + bu. \tag{6.10}$$

The estimation model is formed as

$$\dot{\hat{y}} = -a_m \hat{y} + (\hat{a} + a_m)y + \hat{b}u, \quad \hat{y}(0) = 0.$$

As in the previous example, if we substitute for the control law in the estimation model, we obtain that $\hat{y}(t) = 0 \; \forall t \geq 0$, which implies that the estimation error is simply the plant output or regulation error y. Using (6.8), (6.9), we have

$$(a + a_m)y + bu = (a - \hat{a} + \hat{b}k)y + bu = -\tilde{a}y - \hat{b}u + bu = -\tilde{a}y - \tilde{b}u,$$

where $\tilde{a} = \hat{a} - a$, $\tilde{b} = \hat{b} - b$ are the parameter errors. Substituting in the SSPM, we obtain

$$\dot{y} = -a_m y - \tilde{a}y - \tilde{b}u, \tag{6.11}$$

which relates the regulation or estimation error y with the parameter errors $\tilde{a}, \tilde{b}$. The error equation (6.11) motivates the Lyapunov function

$$V = \frac{y^2}{2} + \frac{\tilde{a}^2}{2\gamma_1} + \frac{\tilde{b}^2}{2\gamma_2}$$

for some $\gamma_1, \gamma_2 > 0$. The time derivative $\dot{V}$ is given by

$$\dot{V} = -a_m y^2 - \tilde{a} y^2 - \tilde{b} u y + \frac{\tilde{a}\dot{\tilde{a}}}{\gamma_1} + \frac{\tilde{b}\dot{\tilde{b}}}{\gamma_2}.$$

Choosing

$$\dot{\tilde{a}} = \dot{\hat{a}} = \gamma_1 y^2, \qquad \dot{\tilde{b}} = \dot{\hat{b}} = \gamma_2 y u, \tag{6.12}$$

we have

$$\dot{V} = -a_m y^2 \leq 0,$$

which implies that $y, \hat{a}, \hat{b} \in \mathcal{L}_\infty$ and $y \in \mathcal{L}_2$. These properties, however, do not guarantee that $\hat{b}(t) \neq 0 \; \forall t \geq 0$, a condition that is required for k, given by (6.9), to be finite. In fact, for k to be uniformly bounded, we should have $|\hat{b}(t)| \geq b_0 > 0 \; \forall t \geq 0$ for some constant b_0. Since such a condition cannot be guaranteed by the adaptive law, we modify (6.12) assuming that $|b| \geq b_0 > 0$, where b_0 and $\text{sgn}(b)$ are known a priori, and use the projection techniques of Chapter 3 to obtain

$$\begin{aligned} \dot{\hat{a}} &= \gamma_1 y^2, \\ \dot{\hat{b}} &= \begin{cases} \gamma_2 y u & \text{if } |\hat{b}| > b_0 \text{ or if } |\hat{b}| = b_0 \text{ and } \text{sgn}(b) y u \geq 0, \\ 0 & \text{otherwise}, \end{cases} \end{aligned} \tag{6.13}$$

where $\hat{b}(0)$ is chosen so that $\hat{b}(0)\,\text{sgn}(b) \geq b_0$. The modified adaptive law guarantees that $|\hat{b}(t)| \geq b_0 \; \forall t \geq 0$. Furthermore, the time derivative $\dot{V}$ of V along the solution of (6.11) and (6.13) satisfies

$$\dot{V} = \begin{cases} -a_m y^2 & \text{if } |\hat{b}| > b_0 \text{ or if } |\hat{b}| = b_0 \text{ and } \text{sgn}(b) y u \geq 0, \\ -a_m y^2 - \tilde{b} y u & \text{if } |\hat{b}| = b_0 \text{ and } \text{sgn}(b) y u < 0. \end{cases}$$

The projection introduces the extra term $-\tilde{b} y u$ which is active when $|\hat{b}| = b_0$ and $\text{sgn}(b) y u < 0$. However, for $|\hat{b}| = b_0$ and $\text{sgn}(b) y u < 0$, we have

$$\tilde{b} y u = \hat{b} y u - b y u = (|\hat{b}| - |b|)\,\text{sgn}(b) y u = (b_0 - |b|)\,\text{sgn}(b) y u \geq 0,$$

where we use the fact that $|b| \geq b_0$. Hence the projection can only make $\dot{V}$ more negative and

$$\dot{V} \leq -a_m y^2 \quad \forall t \geq 0.$$

Hence $y, \tilde{a}, \tilde{b} \in \mathcal{L}_\infty$, $y \in \mathcal{L}_2$, and $|\hat{b}(t)| \geq b_0 \; \forall t \geq 0$, which implies that $\tilde{k} \in \mathcal{L}_\infty$. Using $y, \tilde{a}, \tilde{b}, u \in \mathcal{L}_\infty$ in (6.11), it follows that $\dot{y} \in \mathcal{L}_\infty$, which together with $y \in \mathcal{L}_2 \cap \mathcal{L}_\infty$ implies that $y(t) \to 0$ as $t \to \infty$.

The indirect adaptive pole placement scheme given by (6.8), (6.9), and (6.13) has, therefore, the same stability properties as the direct one. It has also several differences. The main difference is that the gain k is updated indirectly by solving an algebraic time-varying equation at each time t. According to (6.9), the control law (6.8) is designed to meet the pole placement objective for the estimated plant at each time t rather than for the actual plant. Therefore, for such a design to be possible, the estimated plant has to be controllable or at least stabilizable at each time t, which implies that $\hat{b}(t) \neq 0 \; \forall t \geq 0$. In addition, for uniform signal boundedness $\hat{b}$ should satisfy $|\hat{b}| \geq b_0 > 0$, where b_0 is a lower bound for $|b|$ that is known a priori. Another difference is that in the direct case we are estimating only one parameter, whereas in the indirect case we are estimating two.

6.2.2 Scalar Example: Adaptive Tracking

Let us consider the same plant (6.1) as in section 6.2.1, i.e.,

$$\dot{y} = ay + bu.$$

The control objective is modified to include tracking and is stated as follows: Choose the plant input u so that the closed-loop pole is at $-a_m$, $u, y \in \mathcal{L}_\infty$, and $y(t)$ tracks the reference signal $y_m(t) = c \; \forall t \geq 0$, where $c \neq 0$ is a known constant.

Let us first consider the case where a and b are known exactly. It follows from (6.1) that the tracking error $e = y - c$ satisfies

$$\dot{e} = ae + ac + bu. \tag{6.14}$$

Since a, b, c are known, we can choose

$$u = -k_1^* e - k_2^*, \tag{6.15}$$

where

$$k_1^* = \frac{a + a_m}{b}, \qquad k_2^* = \frac{ac}{b}$$

(provided $b \neq 0$), to obtain

$$\dot{e} = -a_m e. \tag{6.16}$$

It is clear from (6.16) that $e(t) = y(t) - y_m \to 0$ as $t \to \infty$ exponentially fast.

Let us now consider the design of an APPC scheme to meet the control objective when a and b are constant but unknown. The CE approach suggests the use of the same control law as in (6.15) but with k_1^* and k_2^* replaced with their online estimates $k_1(t)$ and $k_2(t)$, respectively, i.e.,

$$u(t) = -k_1(t)e(t) - k_2(t). \tag{6.17}$$

As in section 6.2.1, the updating of k_1 and k_2 may be done directly, or indirectly via calculation using the online estimates $\hat{a}$ and $\hat{b}$ of the plant parameters. We consider each case separately.

Direct adaptive tracking In this approach we develop an adaptive law that updates k_1 and k_2 in (6.17) directly without any intermediate calculation. In (6.14), we add and subtract $a_m e$ and substitute $a + a_m = bk_1^*$ and $ac = bk_2^*$ to obtain the B-SSPM

$$\dot{e} = -a_m e + b(u + k_1^* e + k_2^*).$$

As before, we can establish that the estimation error is the same as the tracking error and therefore there is no need to generate an estimation model. By substituting for the control law (6.17) in the above equation, we obtain the error equation

$$\dot{e} = -a_m e - b(\tilde{k}_1 e + \tilde{k}_2), \tag{6.18}$$

where $\tilde{k}_1 = k_1 - k_1^*$, $\tilde{k}_2 = k_2 - k_2^*$, which relates the tracking error e with the parameter errors $\tilde{k}_1, \tilde{k}_2$ and motivates the Lyapunov function

$$V = \frac{e^2}{2} + \frac{\tilde{k}_1^2 |b|}{2\gamma_1} + \frac{\tilde{k}_2^2 |b|}{2\gamma_2}, \tag{6.19}$$

whose time derivative $\dot{V}$ along the trajectory of (6.18) is forced to satisfy

$$\dot{V} = -a_m e^2 \tag{6.20}$$

by choosing

$$\dot{k}_1 = \gamma_1 e^2 \operatorname{sgn}(b), \qquad \dot{k}_2 = \gamma_2 e \operatorname{sgn}(b). \tag{6.21}$$

From (6.19) and (6.20) we have that $e, k_1, k_2 \in \mathcal{L}_\infty$ and $e \in \mathcal{L}_2$, which, in turn, imply that $y, u \in \mathcal{L}_\infty$ and $e(t) \to 0$ as $t \to \infty$ by following the usual arguments as in section 6.2.1.

The APPC scheme (6.17), (6.21) may be written as

$$\begin{aligned} u &= -k_1 e - \gamma_2 \operatorname{sgn}(b) \int_0^t e(\tau) d\tau, \\ \dot{k}_1 &= \gamma_1 e^2 \operatorname{sgn}(b). \end{aligned} \tag{6.22}$$

The control law (6.22) consists of the proportional control action for stabilization and the integral action for rejecting the constant term in the error equation (6.14). We refer to (6.22) as the *direct adaptive proportional plus integral (API) controller*.

The same approach may be repeated when y_m is a known bounded signal with known $\dot{y}_m \in \mathcal{L}_\infty$. In this case the reader may verify that the adaptive control scheme

$$\begin{aligned} &u = -k_1 e - k_2 y_m - k_3 \dot{y}_m, \\ &\dot{k}_1 = \gamma_1 e^2 \operatorname{sgn}(b), \qquad \dot{k}_2 = \gamma_2 e y_m \operatorname{sgn}(b), \qquad \dot{k}_3 = \gamma_3 e \dot{y}_m \operatorname{sgn}(b), \end{aligned} \tag{6.23}$$

where $e = y - y_m$, guarantees that all signals in the closed-loop plant in (6.1) and (6.23) are bounded, and $y(t) \to y_m(t)$ as $t \to \infty$.

Indirect adaptive tracking In this approach, we use the same control law as in the direct case, i.e.,

$$u = -k_1 e - k_2 \tag{6.24}$$

with k_1, k_2 calculated using the equations

$$k_1 = \frac{\hat{a} + a_m}{\hat{b}}, \qquad k_2 = \frac{\hat{a}c}{\hat{b}}, \tag{6.25}$$

provided $\hat{b} \neq 0$, where $\hat{a}$ and $\hat{b}$ are the online estimates of the plant parameters a and b, respectively.

We generate $\hat{a}$ and $\hat{b}$ by first obtaining a parametric model for the unknown parameters and then developing an error equation that relates the estimation error to the parameter errors $\tilde{a} = \hat{a} - a$, $\tilde{b} = \hat{b} - b$ as follows: In (6.14), we add and subtract $a_m e$ to obtain the SSPM

$$\dot{e} = -a_m e + (a + a_m)e + ac + bu.$$

The estimation model

$$\dot{\hat{e}} = -a_m \hat{e} + (\hat{a} + a_m)e + \hat{a}c + \hat{b}u, \quad \hat{e}(0) = 0,$$

together with (6.24), (6.25) implies that $\hat{e}(t) = 0 \; \forall t \geq 0$ and therefore the estimation error is simply the tracking error which satisfies the SSPM. Furthermore, there is no need to generate the estimation model. We now need to relate the tracking or estimation error with the parameter error. From (6.24), (6.25) we have

$$a_m e = (\hat{b}k_1 - \hat{a})e = -\hat{b}u - \hat{b}k_2 - \hat{a}e = -\hat{b}u - \hat{a}(e + c) = -\hat{b}u - \hat{a}y,$$

which we substitute into the SSPM equation to obtain

$$\dot{e} = -a_m e - \hat{b}u - \hat{a}y + ay + bu.$$

Substituting for the control law (6.24), (6.25), we obtain the estimation or tracking error equation

$$\dot{e} = -a_m e - \tilde{a}y - \tilde{b}u, \tag{6.26}$$

which relates the tracking or estimation error to the parameter errors $\tilde{a} = \hat{a} - a$, $\tilde{b} = \hat{b} - b$. Using (6.26), the following adaptive laws can now be derived by using the same approach as in section 6.2.1:

$$\begin{aligned} \dot{\hat{a}} &= \gamma_1 e y, \\ \dot{\hat{b}} &= \begin{cases} \gamma_2 e u & \text{if } |\hat{b}| > b_0 \text{ or if } |\hat{b}| = b_0 \text{ and } \operatorname{sgn}(b) e u \geq 0, \\ 0 & \text{otherwise,} \end{cases} \end{aligned} \tag{6.27}$$

where $\hat{b}(0)$ satisfies $\hat{b}(0) \operatorname{sgn}(b) \geq b_0$. The reader may verify that the time derivative of the Lyapunov function

$$V = \frac{e^2}{2} + \frac{\tilde{a}^2}{2\gamma_1} + \frac{\tilde{b}^2}{2\gamma_2}$$

along the trajectories of (6.26), (6.27) satisfies

$$\dot{V} \leq -a_m e^2,$$

and follow similar arguments as before to establish signal boundedness and convergence of the tracking error $e = y - c$ to zero as $t \to \infty$.

As in the case of MRAC, the above examples have the following characteristics:

(i) The adaptive laws are driven by the state tracking or regulation error that is equal to the estimation error without normalization. In this case the estimation model does not need to be generated, as its state is equal to zero.

(ii) A significant step for the design and analysis of the adaptive laws is the development of a scalar differential equation that relates the regulation or tracking error with the parameter error. This error equation motivates the selection of a quadratic Lyapunov function that is used to design the adaptive laws and establish signal boundedness simultaneously.

(iii) For the scalar examples considered, stable direct and indirect APPC schemes can be designed.

The extension of properties (i)–(iii) to higher-order plants where only the output rather than the full state is available for measurement is not possible in general for most pole placement structures. The difficulty arises because the nonlinear relationship between the plant and controller parameters does not allow us to relate the tracking or regulation error with the parameter error in a differential equation that has a particular form so that a single Lyapunov function can be used to establish stability. The nonlinear relationship between plant and controller parameters makes the design of direct APPC schemes impossible in general. The difficulty is due to the fact that a parametric model for the plant where the unknown controller parameter vector θ_c^* appears linearly or in the special bilinear form cannot be obtained in general. In the following sections we present the design and analysis of indirect APPC schemes that are applicable to SISO LTI plants of any order.

6.3 APPC Schemes: Polynomial Approach

Let us consider the SISO LTI plant

$$y_p = G_p(s)u_p, \quad G_p(s) = \frac{Z_p(s)}{R_p(s)}, \tag{6.28}$$

where $G_p(s)$ is proper and $R_p(s)$ is a monic polynomial. The control objective is to choose the plant input u_p so that the closed-loop poles are assigned to those of a given monic Hurwitz polynomial $A^*(s)$. The polynomial $A^*(s)$, referred to as the desired closed-loop characteristic polynomial, is chosen based on the closed-loop performance requirements. To meet the control objective, we make the following assumptions about the plant:

P1. $R_p(s)$ is a monic polynomial whose degree n is known.

P2. $Z_p(s)$, $R_p(s)$ are coprime and degree $(Z_p) < n$.

Assumptions P1 and P2 allow Z_p, R_p to be non-Hurwitz, in contrast to the MRC case where Z_p is required to be Hurwitz. If, however, Z_p is Hurwitz, the MRC problem is a special case of the general pole placement problem defined above with $A^*(s)$ restricted to have Z_p as a factor.

In general, by assigning the closed-loop poles to those of $A^*(s)$, we can guarantee closed-loop stability and convergence of the plant output y_p to zero, provided that there is no external input. We can also extend the pole placement control (PPC) objective to include tracking, where y_p is required to follow a certain class of reference signals y_m, by using the internal model principle as follows: The reference signal $y_m \in \mathcal{L}_\infty$ is assumed to satisfy

$$Q_m(s)y_m = 0, \tag{6.29}$$

where $Q_m(s)$, known as the internal model of y_m, is a known monic polynomial of degree q with all roots in $\Re[s] \leq 0$ and with no repeated roots on the $j\omega$-axis. The internal model $Q_m(s)$ is assumed to satisfy

P3. $Q_m(s)$, $Z_p(s)$ are coprime.

For example, if y_p is required to track the reference signal $y_m(t) = c + \sin(3t)$, where c is any constant, then $Q_m(s) = s(s^2+9)$ and, according to P3, $Z_p(s)$ should not have s or s^2+9 as a factor. Similarly, if y_p is required to track $y_m(t) = c + \sin(\omega_1 t) + \cos(\omega_2 t)$ for $\omega_1 \neq \omega_2$, then $Q_m(s) = s(s^2+\omega_1^2)(s^2+\omega_2^2)$.

The PPC objective with tracking can be stated as follows: Design the control input u_p so that the closed-loop poles are the same as the roots of $A^*(s) = 0$ and y_p tracks the reference signal y_m. Since the coefficients of $G_p(s)$ are unknown, an adaptive control scheme is used to meet the control objective. The design of the APPC scheme is based on the CE approach where the unknown parameters in the control law, designed to meet the control objective in the known parameter case, are replaced with their online estimates generated by an adaptive law.

The design of the APPC scheme is based on three steps presented in the following subsections. In the first step, a control law is developed that can meet the control objective in the known parameter case. In the second step, an adaptive law is designed to estimate the plant parameters online. The estimated plant parameters are then used to calculate the controller parameters at each time t, and the APPC is formed in Step 3 by replacing the controller parameters in Step 1 with their online estimates.

Step 1. PPC for known parameters We consider the control law

$$Q_m(s)L(s)u_p = -P(s)(y_p - y_m), \tag{6.30}$$

where $P(s)$, $L(s)$ are polynomials (with $L(s)$ monic) of degree $q+n-1, n-1$, respectively, chosen to satisfy the polynomial equation

$$L(s)Q_m(s)R_p(s) + P(s)Z_p(s) = A^*(s). \tag{6.31}$$

Assumptions P2 and P3 guarantee that $Q_m(s)R_p(s)$ and $Z_p(s)$ are coprime, which guarantees (see the Appendix) that $L(s)$, $P(s)$ satisfying (6.31) exist and are unique. The solution for the coefficients of $L(s)$, $P(s)$ of (6.31) may be obtained by solving the algebraic equation

$$S_l \beta_l = \alpha_l^*, \tag{6.32}$$

where S_l is the Sylvester matrix (see the Appendix) of $Q_m R_p$, Z_p of dimension $2(n+q) \times 2(n+q)$,

$$\beta_l = [l_q^T, p^T]^T, \qquad \alpha_l^* = [\underbrace{0, \dots, 0}_{q}, \alpha^{*T}]^T,$$

$$l_q = [\underbrace{0, \dots, 0}_{q}, 1, l^T]^T \in \mathcal{R}^{n+q},$$
$$l = [l_{n-2}, l_{n-3}, \dots, l_1, l_0]^T \in \mathcal{R}^{n-1},$$
$$p = [p_{n+q-1}, p_{n+q-2}, \dots, p_1, p_0]^T \in \mathcal{R}^{n+q},$$
$$\alpha^* = [a^*_{2n+q-2}, a^*_{2n+q-3}, \dots, a^*_1, a^*_0]^T \in \mathcal{R}^{2n+q-1},$$

and l_i, p_i, a_i^* are the coefficients of the polynomials

$$L(s) = s^{n-1} + l_{n-2}s^{n-2} + \cdots + l_1 s + l_0 = s^{n-1} + l^T \alpha_{n-2}(s),$$
$$P(s) = p_{n+q-1}s^{n+q-1} + p_{n+q-2}s^{n+q-2} + \cdots + p_1 s + p_0 = p^T \alpha_{n+q-1}(s),$$
$$A^*(s) = s^{2n+q-1} + a^*_{2n+q-2}s^{2n+q-2} + \cdots + a^*_1 s + a^*_0 = s^{2n+q-1} + \alpha^{*T}\alpha_{2n+q-2}(s),$$

respectively, where $\alpha_i(s) = [s^i, s^{i-1}, \dots, s, 1]^T$.

The coprimeness of $Q_m R_p$, Z_p guarantees that S_l is nonsingular; therefore, the coefficients of $L(s)$, $P(s)$ may be computed from the equation

$$\beta_l = S_l^{-1}\alpha_l^*. \tag{6.33}$$

Using (6.31) and (6.30), the closed-loop plant is described by

$$y_p = \frac{Z_p P}{A^*} y_m. \tag{6.34}$$

Similarly, from the plant equation in (6.28) and the control law in (6.30) and (6.31), we obtain

$$u_p = \frac{R_p P}{A^*} y_m. \tag{6.35}$$

Since $y_m \in \mathcal{L}_\infty$ and $\frac{Z_p P}{A^*}$, $\frac{R_p P}{A^*}$ are proper with stable poles, it follows that $y_p, u_p \in \mathcal{L}_\infty$. The tracking error $e_1 = y_p - y_m$ can be shown to satisfy

$$e_1 = -\frac{LR_p}{A^*} Q_m y_m = -\frac{LR_p}{A^*}[0]$$

by using (6.31), (6.34), and then (6.29). Since $\frac{LR_p}{A^*}$ is proper with poles in $\Re[s] < 0$, it follows that $e_1 \to 0$ as $t \to \infty$ exponentially fast. Therefore, the control law (6.30), (6.31) meets the control objective.

The control law (6.30) can be realized as

$$u_p = -C(s)(y_p - y_m), \quad C(s) = \frac{P(s)}{Q_m(s)L(s)},$$

using $n + q - 1$ integrators. Since $L(s)$ is not necessarily Hurwitz, $C(s)$ may have poles in $\Re[s] \geq 0$, which is not desirable in practice. An alternative realization of (6.30) is obtained by rewriting (6.30) as

$$\Lambda(s)u_p = \Lambda(s)u_p - Q_m(s)L(s)u_p - P(s)(y_p - y_m),$$

where $\Lambda(s)$ is any monic Hurwitz polynomial of degree $n+q-1$. Filtering each side with $\frac{1}{\Lambda(s)}$, we obtain

$$u_p = \frac{\Lambda - LQ_m}{\Lambda} u_p - \frac{P}{\Lambda}(y_p - y_m). \tag{6.36}$$

The control law is implemented using $2(n+q-1)$ integrators to realize the proper stable transfer functions $\frac{\Lambda - LQ_m}{\Lambda}$, $\frac{P}{\Lambda}$.

Step 2. Estimation of plant polynomials Using (6.28) and the results of Chapter 2, the following parametric model can be derived:

$$z = \theta_p^{*T}\phi,$$

where

$$z = \frac{s^n}{\Lambda_p(s)} y_p, \qquad \theta_p^* = [\theta_b^{*T}, \theta_a^{*T}]^T, \qquad \phi = \left[\frac{\alpha_{n-1}^T(s)}{\Lambda_p(s)} u_p, -\frac{\alpha_{n-1}^T(s)}{\Lambda_p(s)} y_p\right]^T,$$

$$\alpha_{n-1}(s) = [s^{n-1}, \ldots, s, 1]^T, \qquad \theta_a^* = [a_{n-1}, \ldots, a_0]^T, \qquad \theta_b^* = [b_{n-1}, \ldots, b_0]^T,$$

and $\Lambda_p(s)$ is a monic Hurwitz polynomial. If the degree of $Z_p(s)$ is less than $n-1$, then some of the first elements of $\theta_b^* = [b_{n-1}, \ldots, b_0]^T$ will be equal to zero, and the dimension of the parametric model can be reduced. Using the results of Chapter 3, a wide class of adaptive laws can be designed to estimate θ_p^*. As an example we consider the gradient algorithm

$$\dot{\theta}_p = \Gamma\varepsilon\phi, \qquad \varepsilon = \frac{z - \theta_p^T\phi}{m_s^2}, \qquad m_s^2 = 1 + \phi^T\phi, \tag{6.37}$$

where $\Gamma = \Gamma^T > 0$ is the adaptive gain and

$$\theta_p = [\hat{b}_{n-1}, \ldots, \hat{b}_0, \hat{a}_{n-1}, \ldots, \hat{a}_0]^T$$

are the estimated plant parameters which can be used to form the estimated plant polynomials

$$\hat{R}_p(s,t) = s^n + \hat{a}_{n-1}(t)s^{n-1} + \cdots + \hat{a}_1(t)s + \hat{a}_0(t),$$
$$\hat{Z}_p(s,t) = \hat{b}_{n-1}(t)s^{n-1} + \cdots + \hat{b}_1(t)s + \hat{b}_0(t),$$

of $R_p(s)$, $Z_p(s)$, respectively, at each time t.

Step 3. Adaptive control law The adaptive control law is obtained by replacing the unknown polynomials $L(s)$, $P(s)$ in (6.36) with their online estimates $\hat{L}(s,t)$, $\hat{P}(s,t)$ calculated at each frozen time[12] t using the polynomial equation

$$\hat{L}(s,t) \cdot Q_m(s) \cdot \hat{R}_p(s,t) + \hat{P}(s,t) \cdot \hat{Z}_p(s,t) = A^*(s); \tag{6.38}$$

[12]By frozen time we mean that the time-varying coefficients of the polynomials are treated as constants when two polynomials are multiplied.

i.e., the control law is formed as

$$u_p = (\Lambda(s) - \hat{L}(s,t)Q_m(s))\frac{1}{\Lambda(s)}u_p - \hat{P}(s,t)\frac{1}{\Lambda(s)}(y_p - y_m), \tag{6.39}$$

where $\hat{L}(s,t)$, $\hat{P}(s,t)$ are obtained using (6.38). In (6.38), the operation $X(s,t) \cdot Y(s,t)$ denotes a multiplication of polynomials, where s is simply treated as a variable. This is to distinguish it from the operation $X(s,t)Y(s,t)$, where s is a differential operator and the product $X(s,t)Y(s,t)$ involves time derivatives of the coefficients of the polynomials.

The APPC scheme is described by (6.37), (6.38), (6.39). Equation (6.38) can be solved for $\hat{L}(s,t)$, $\hat{P}(s,t)$ by equating the coefficients of powers of s on each side of the equation. This process leads to an algebraic equation of the form (6.32). The existence and uniqueness of $\hat{L}(s,t)$, $\hat{P}(s,t)$ is guaranteed, provided that $\hat{R}_p(s,t)Q_m(s)$, $\hat{Z}_p(s,t)$ are coprime at each frozen time t. The adaptive laws that generate the coefficients of $\hat{R}_p(s,t)$, $\hat{Z}_p(s,t)$ cannot guarantee this property, which means that at certain points in time the solution $\hat{L}(s,t)$, $\hat{P}(s,t)$ of (6.38) may not exist. This problem is known as the stabilizability problem in indirect APPC, and it will be discussed in section 6.6.

We first illustrate the design of the APPC scheme based on the polynomial approach using the following example. We then present the stability properties of the APPC scheme (6.37), (6.38), (6.39) for a general plant.

Example 6.3.1 Consider the plant

$$y_p = \frac{b}{s+a}u_p, \tag{6.40}$$

where a and b are constants. The control objective is to choose u_p such that the poles of the closed-loop plant are placed at the roots of $A^*(s) = (s+1)^2$ and y_p tracks the constant reference signal $y_m = 1$. Clearly the internal model of y_m is $Q_m(s) = s$, i.e., $q = 1$.

Step 1. PPC for known parameters Since $n = 1$, the polynomials L, P, Λ are of the form

$$L = 1, \qquad P = p_1 s + p_0, \qquad \Lambda(s) = s + \lambda_0,$$

where $\lambda_0 > 0$ is an arbitrary constant and p_0, p_1 are calculated by solving

$$s(s+a) + (p_1 s + p_0)b = (s+1)^2. \tag{6.41}$$

Equating the coefficients of the powers of s in (6.41), we obtain

$$p_1 = \frac{2-a}{b}, \qquad p_0 = \frac{1}{b}. \tag{6.42}$$

Therefore, the PPC law is given by

$$u_p = \frac{\lambda_0}{s+\lambda_0}u_p - \frac{p_1 s + p_0}{s+\lambda_0}e_1, \tag{6.43}$$

where $e_1 = y_p - y_m = y_p - 1$ and p_1, p_0 are given by (6.42).

Step 2. Estimation of plant parameters Since a, b are unknown, the control law (6.42), (6.43) cannot be implemented. Therefore, we use the gradient algorithm

$$\dot{\theta}_p = \Gamma \varepsilon \phi, \quad \varepsilon = \frac{z - \theta_p^T \phi}{m_s^2}, \quad m_s^2 = 1 + \phi^T \phi, \tag{6.44}$$

where

$$z = \frac{s}{s+\lambda} y_p, \qquad \phi = \frac{1}{s+\lambda}[u_p, -y_p]^T,$$

and $\lambda > 0$, $\Gamma = \Gamma^T > 0$ are design constants, to generate $\theta_p(t) = [\hat{b}(t), \hat{a}(t)]^T$, the estimate of $\theta_p^* = [b, a]^T$, at each time t.

Step 3. Adaptive control law We use the CE approach to form the adaptive control law by replacing the unknown parameters in the control law (6.42), (6.43) with their online estimates, i.e.,

$$\hat{p}_1 = \frac{2 - \hat{a}}{\hat{b}}, \qquad \hat{p}_0 = \frac{1}{\hat{b}}, \tag{6.45}$$

$$u_p = \frac{\lambda_0}{s+\lambda_0} u_p - (\hat{p}_1 s + \hat{p}_0) \frac{1}{s+\lambda_0} e_1. \tag{6.46}$$

The APPC scheme for the plant (6.40) is given by (6.44), (6.45), (6.46). The stabilizability problem arises in (6.45) in the form of $\hat{b}(t)$ becoming equal to zero at some time t. The adaptive law (6.44) cannot guarantee that $\hat{b}(t) \neq 0 \; \forall t \geq 0$. For computational reasons, we require $|\hat{b}(t)| \geq b_0 > 0$ for some constant b_0 (a lower bound for $|b|$) and $\forall t \geq 0$. For this scalar example, the stabilizability problem can be resolved by replacing the adaptive law (6.44) with

$$\begin{aligned}
\dot{\hat{b}} &= \begin{cases} \gamma_1 \varepsilon \phi_1 & \text{if } |\hat{b}| > b_0 \text{ or if } |\hat{b}| = b_0 \text{ and } \varepsilon \phi_1 \operatorname{sgn} \hat{b} \geq 0, \\ 0 & \text{otherwise,} \end{cases} \\
\dot{\hat{a}} &= \gamma_2 \varepsilon \phi_2, \\
\varepsilon &= \frac{z - \hat{b}\phi_1 - \hat{a}\phi_2}{m_s^2}, \quad m_s^2 = 1 + \phi^T \phi,
\end{aligned} \tag{6.47}$$

where $\phi = [\phi_1, \phi_2]^T$ and the initial condition for $\hat{b}$ is chosen to satisfy $\hat{b}(0)\operatorname{sgn}(b) \geq b_0$. Using the results of Chapter 3, we can establish that the adaptive law (6.47) guarantees that $|\hat{b}(t)| \geq b_0 \; \forall t \geq 0$ and the properties (i) $\hat{a}, \hat{b} \in \mathcal{L}_\infty$, (ii) $\varepsilon, \varepsilon m_s, \varepsilon\phi, \dot{\hat{a}}, \dot{\hat{b}} \in \mathcal{L}_2 \cap \mathcal{L}_\infty$ are satisfied independent of the boundedness of the plant input u_p and output y_p. ∎

For the scalar Example 6.3.1, the stabilizability problem is avoided by preventing $\hat{b}(t)$ from crossing zero using parameter projection. In the higher-order case, the stabilizability problem is far more complicated and is discussed in section 6.6. For the time being let us assume that the stabilizability condition holds at each time t and use the following theorem to describe the stability properties of the APPC scheme (6.37)–(6.39).

Theorem 6.3.2. *Assume that the estimated plant polynomials $\hat{R}_p Q_m$, $\hat{Z}_p$ are strongly coprime at each time t. Then all the signals in the closed loop (6.28), (6.37)–(6.39) are uniformly bounded, and the tracking error converges to zero asymptotically with time.*

Proof. The proof involves the use of a number of results presented in the Appendix and several complicated technical steps. Below we give an outline of the proof and present the full technical details in the web resource [94] for the ambitious reader.

We know from the results of Chapter 3 that the adaptive law (6.37) guarantees that

(i) $\theta_p \in \mathcal{L}_\infty$,

(ii) $\varepsilon, \varepsilon m_s, \dot{\theta}_p, \varepsilon\phi \in \mathcal{L}_2 \cap \mathcal{L}_\infty$,

independent of the boundedness of u_p, y_p, ϕ. The proof is carried out by using the above properties and the following four steps.

Step 1. Manipulate the estimation error and control law equations to express the plant input u_p and output y_p in terms of the estimation error This step leads to the equations

$$\begin{aligned} \dot{x} &= A(t)x + B_1(t)\varepsilon m_s^2 + B_2\bar{y}_m, \\ u_p &= C_1^T\dot{x} + D_1^T x, \\ y_p &= C_2^T\dot{x} + D_2^T x, \end{aligned} \tag{6.48}$$

where $\bar{y}_m \in \mathcal{L}_\infty$; $A(t)$, $B_1(t)$ are uniformly bounded because of the boundedness of the estimated plant and controller parameters (which is guaranteed by the adaptive law and the stabilizability assumption); and B_2, C_1, D_1, C_2, D_2 are constant vectors.

Step 2. Establish the exponential stability (e.s.) of the homogeneous part of (6.48) The matrix $A(t)$ has stable eigenvalues at each frozen time t that are equal to the roots of $A^*(s) = 0$. In addition, $\dot{\theta}_p, \dot{l}, \dot{p} \in \mathcal{L}_2$, where l, p are vectors with the coefficients of $\hat{L}$, $\hat{P}$, respectively (guaranteed by the adaptive law and the stabilizability assumption), imply that $\|\dot{A}(t)\| \in \mathcal{L}_2$. Therefore, using Theorem A.8.8, we conclude that the homogeneous part of (6.48) is uniformly asymptotically stable (u.a.s.), which is equivalent to e.s.

Step 3. Use the properties of the $\mathcal{L}_{2\delta}$ norm and the B–G lemma to establish boundedness Let $m_f^2 \triangleq 1 + \|u_p\|^2 + \|y_p\|^2$, where $\|\cdot\|$ denotes the $\mathcal{L}_{2\delta}$ norm. Using the results established in Steps 1 and 2 and the normalizing properties of m_f, we show that

$$m_f^2 \le c\int_0^t e^{-\delta(t-\tau)}\varepsilon^2 m_s^2 m_f^2 d\tau + c. \tag{6.49}$$

Since the adaptive law guarantees that $\varepsilon m_s \in \mathcal{L}_2$, the boundedness of m_f follows by applying the B–G lemma. Using the boundedness of m_f, we can establish the boundedness of all signals in the closed-loop plant.

Step 4. Establish that the tracking error e_1 converges to zero The convergence of e_1 to zero follows by using the control and estimation error equations to express e_1 as the output of proper stable LTI systems whose inputs are in $\mathcal{L}_2 \cap \mathcal{L}_\infty$ which implies, according to Corollary A.5.5 in the Appendix, that $e_1 \in \mathcal{L}_2$. By establishing $\dot{e}_1 \in \mathcal{L}_\infty$, it follows from $e_1 \in \mathcal{L}_2 \cap \mathcal{L}_\infty$ and Lemma A.4.7 that $e_1 \to 0$ as $t \to \infty$. $\square$

The assumption that $\hat{R}_p(s,t)Q_m(s)$, $\hat{Z}_p(s,t)$ are strongly coprime at each time t is crucial in establishing stability. By strongly coprime we mean that the polynomials do not have roots that are close to becoming common. For example, the polynomials $s + 2 + \varepsilon$, $s + 2$, where ε is arbitrarily small, are not strongly coprime. When the estimated plant polynomials $\hat{R}_P(s,t)Q(s)$, $\hat{Z}_p(s,t)$ lose coprimeness at a certain point in time t, the solution $\hat{L}(s,t)$, $\hat{P}(s,t)$ of the polynomial equation (6.38) may not exist. This coprimeness condition is equivalent to the condition that the estimated plant characterized by the polynomials $\hat{R}_p(s,t)Q_m(s)$, $\hat{Z}_p(s,t)$ be strongly controllable and observable at each time t. This is an expected requirement, since the control law is designed at each time t to meet the control objective for the estimated plant. In the scalar case this controllability condition is simply $|\hat{b}(t)| > b_0 > 0 \; \forall t$, which can be taken care of using projection and assuming that the sign of b and a lower bound b_0 for $|b| \geq b_0$ are known. In the general case it is difficult, if not impossible, to use projection to guarantee the strong coprimeness of the estimated plant polynomials. In section 6.6, we discuss certain fixes proposed in the literature at the expense of additional complexity in order to address the stabilizability problem.

6.4 APPC Schemes: State-Space Approach

We consider the SISO LTI plant (6.28), i.e.,

$$y_p = G_p(s)u_p, \quad G_p(s) = \frac{Z_p(s)}{R_p(s)},$$

where $G_p(s)$ is proper and $R_p(s)$ is monic. The control objective is the same as in section 6.3 and the same assumptions P1, P2, P3 apply. In this section, we consider a state-space approach to meet the control objective. As in section 6.3, we first solve the problem assuming known plant parameters. We then use the CE approach to replace the unknown parameters in the control law with their estimates calculated at each time t based on the plant parameter estimates generated by an adaptive law. The steps followed are presented below.

Step 1. PPC for known parameters We start by considering the expression

$$e_1 = \frac{Z_P(s)}{R_P(s)}u_p - y_m \tag{6.50}$$

for the tracking error. Filtering each side of (6.50) with $\frac{Q_m(s)}{Q_1(s)}$, where $Q_1(s)$ is an arbitrary monic Hurwitz polynomial of degree q, and using $Q_m(s)y_m = 0$, we obtain

$$e_1 = \frac{Z_p Q_1}{R_P Q_m}\bar{u}_p, \quad \bar{u}_p = \frac{Q_m}{Q_1}u_p. \tag{6.51}$$

With (6.51), we have converted the tracking problem into the regulation problem of choosing $\bar{u}_p$ to regulate e_1 to zero. Let (A, B, C) be a state-space realization of (6.51) in the observer canonical form, i.e.,

$$\dot{e} = Ae + B\bar{u}_p, \quad e_1 = C^T e, \tag{6.52}$$

where

$$A = \left[-\theta_1^* \mid \frac{I_{n+q-1}}{0} \right], \qquad B = \theta_2^*, \qquad C = [1, 0, \dots, 0]^T, \tag{6.53}$$

and $\theta_1^*, \theta_2^* \in \mathcal{R}^{n+q}$ are the coefficient vectors of the polynomials $R_p(s)Q_m(s) - s^{n+q}$ and $Z_p(s)Q_1(s)$, respectively. Because R_pQ_m, Z_p are coprime, any possible zero-pole cancellation in (6.51) between $Q_1(s)$ and $R_p(s)Q_m(s)$ will occur in $\Re[s] < 0$ due to $Q_1(s)$ being Hurwitz, which implies that (A, B) is always stabilizable. We consider the feedback control law

$$\bar{u}_p = -K_c\hat{e}, \quad u_p = \frac{Q_1}{Q_m}\bar{u}_p, \tag{6.54}$$

where $\hat{e}$ is the state of the full-order Luenberger observer [120]

$$\dot{\hat{e}} = A\hat{e} + B\bar{u}_p - K_o(C^T\hat{e} - e_1) \tag{6.55}$$

and K_c and K_o are solutions to the polynomial equations

$$\begin{aligned} \det(sI - A + BK_c) &= A_c^*(s), \\ \det(sI - A + K_oC^T) &= A_o^*(s), \end{aligned} \tag{6.56}$$

where A_c^* and A_o^* are given monic Hurwitz polynomials of degree $n + q$. The roots of $A_c^*(s) = 0$ represent the desired pole locations of the transfer function of the closed-loop plant, whereas the roots of $A_o^*(s)$ are equal to the poles of the observer dynamics. As in every observer design, the roots of $A_o^*(s) = 0$ are chosen to be faster than those of $A_c^*(s) = 0$ in order to reduce the effect of the observer dynamics on the transient response of the tracking error e_1. The existence of K_c in (6.56) follows from the controllability of (A, B). If (A, B) is stabilizable but not controllable because of common factors between $Q_1(s)$ and $R_p(s)$, the solution for K_c in (6.56) still exists, provided that $A_c^*(s)$ is chosen to contain the common factors of Q_1, R_p. Because $A_c^*(s)$ is chosen based on the desired closed-loop performance requirements and $Q_1(s)$ is an arbitrary monic Hurwitz polynomial of degree q, we can choose $Q_1(s)$ to be a factor of $A_c^*(s)$ and, therefore, guarantee the existence of K_c in (6.56) even when (A, B) is not controllable. The existence of θ_p in (6.56) follows from the observability of t. Because of the special form of (6.52) and (6.53), the solution of (6.56) is given by $K_o = \alpha_0^* - \theta_1^*$, where α_0^* is the coefficient vector of $A_o^*(s)$.

Theorem 6.4.1. *The PPC law* (6.54)–(6.56) *guarantees that all signals in the closed-loop plant are bounded and* e_1 *converges to zero exponentially fast.*

Proof. We define the observation error $e_o \stackrel{\Delta}{=} e - \hat{e}$. Subtracting (6.55) from (6.52), we have

$$\dot{e}_o = (A - K_oC^T)e_o. \tag{6.57}$$

Using (6.54) in (6.55), we obtain

$$\dot{\hat{e}} = (A - BK_c)\hat{e} + K_o C^T e_o. \tag{6.58}$$

Since the eigenvalues of the overall matrix in (6.57), (6.58) are equal to the eigenvalues of $A - K_o C^T$, $A - BK_c$, which are stable by design (see (6.56)), it follows that $\hat{e}, e_o \in \mathcal{L}_\infty$ and $\hat{e}(t), e_o(t) \to 0$ as $t \to \infty$ exponentially fast. From $e_0 = e - \hat{e}$ and $\bar{u}_p = -K_c\hat{e}$, it follows that $e, \bar{u}_p, e_1 \in \mathcal{L}_\infty$ and $e(t), \bar{u}_p(t), e_1(t) \to 0$ as $t \to \infty$, also exponentially fast. The boundedness of y_p follows from that of e_1 and y_m. We have

$$u_p = \frac{Q_1(s)}{Q_m(s)}\bar{u}_p,$$

where $\frac{Q_1}{Q_m}$ is proper with poles in $\Re[s] \le 0$ and nonrepeated poles on $\Re[s] = 0$. Furthermore, $\bar{u}_p$ is exponentially decaying to zero. Given these conditions it can be shown (see Problem 11) that $u_p \in \mathcal{L}_\infty$, and the proof is complete. □

Step 2. Estimation of plant parameters The adaptive law for estimating the plant parameters and generating $\hat{R}_p$, $\hat{Z}_p$ is the same as in section 6.3 and is given by

$$\begin{aligned} \dot{\theta}_p &= \Gamma\varepsilon\phi, \\ \varepsilon &= \frac{z - \theta^T\phi}{m_s^2}, \quad m_s^2 = 1 + \phi^T\phi; \end{aligned} \tag{6.59}$$

z, ϕ are as defined in (6.37) and $\theta_p = [\theta_a^T, \theta_b^T]^T$, where $\theta_b = [\hat{b}_{n-1}, \dots, \hat{b}_0]^T$, $\theta_a = [\hat{a}_{n-1}, \dots, \hat{a}_0]^T$ are the online estimates of the coefficient vectors $\theta_b^* = [b_{n-1}, \dots, b_0]^T$, $\theta_a^* = [a_{n-1}, \dots, a_0]^T$ of $Z_p(s)$, $R_P(s)$, respectively.

Step 3. Adaptive control law Using the CE approach, the adaptive control law is given by

$$\begin{aligned} \dot{\hat{e}} &= \hat{A}\hat{e} + \hat{B}\bar{u}_p - \hat{K}_0(c^T\hat{e} - e_1), \\ \bar{u}_p &= -\hat{K}_c\hat{e}, \quad u_p = \frac{Q_1(s)}{Q_m(s)}\bar{u}_p, \end{aligned} \tag{6.60}$$

where

$$\hat{A}(t) = \left[-\theta_1(t) \mid \frac{I_{n+q-1}}{0} \right], \qquad \hat{B}(t) = \theta_2(t); \tag{6.61}$$

$\theta_1(t)$ and $\theta_2(t)$ are the coefficient vectors of the polynomials

$$\begin{aligned} \hat{R}_p(s,t)Q_m(s) - s^{n+q} &= (s^n + \theta_a^T(t)\alpha_{n-1}(s))Q_m(s) - s^{n+q}, \\ \hat{Z}_p(s,t)Q_1(s) &= \theta_b^T(t)\alpha_{n-1}(s)Q_1(s), \end{aligned}$$

respectively;

$$\hat{K}_0(t) = \alpha_0^* - \theta_1(t); \tag{6.62}$$

α_0^* is the coefficient vector of $A_0^*(s)$; and $\hat{K}_c(t)$ is calculated at each time t by solving the polynomial equation

$$\det(sI - \hat{A} + \hat{B}\hat{K}_c) = A_c^*(s). \tag{6.63}$$

The APPC scheme is described by (6.59)–(6.63).

The stabilizability problem arises in (6.60), where for the calculation of $\hat{K}_c$ to be possible the pair $(\hat{A}(t), \hat{B}(t))$ has to be controllable at each time t and for implementation purposes strongly controllable. Let us first illustrate the design for a scalar plant and then present the stability properties of the APPC scheme (6.59)–(6.63) for the general plant.

Example 6.4.2 We consider the same plant as in Example 6.3.1, i.e.,

$$y_p = \frac{b}{s+a} u_p, \tag{6.64}$$

where a and b are unknown constants with $b \neq 0$. The input u_p is to be chosen so that the poles of the closed-loop plant are placed at the roots of $A^*(s) = (s+1)^2 = 0$, and y_p tracks the reference signal $y_m = 1$.

Step 1. PPC for known parameters If a, b are known, the following control law can be used to meet the control objective:

$$\begin{aligned} \dot{\hat{e}} &= \begin{bmatrix} -a & 1 \\ 0 & 0 \end{bmatrix} \hat{e} + \begin{bmatrix} 1 \\ 1 \end{bmatrix} b\bar{u}_p - K_o([1, 0]\hat{e} - e_1), \\ \bar{u}_p &= -K_c\hat{e}, \quad u_p = \frac{s+1}{s}\bar{u}_p, \end{aligned} \tag{6.65}$$

where K_o, K_c are calculated by solving the equations

$$\det(sI - A + BK_c) = (s+1)^2, \qquad \det(sI - A + K_oC^T) = (s+10)^2,$$

where the poles of the observer are chosen to be at $s = -10, -10$,

$$A = \begin{bmatrix} -a & 1 \\ 0 & 0 \end{bmatrix}, \qquad B = \begin{bmatrix} 1 \\ 1 \end{bmatrix} b, \qquad C^T = [1, 0].$$

The solution of the above polynomial equations is given by

$$K_c = \frac{1}{b}[1 - a, 1], \qquad K_o = [20 - a, 100]^T. \tag{6.66}$$

Step 2. Estimation of plant parameters The adaptive law uses the measurements of the plant input u_p and output y_p to generate the estimates $\hat{a}, \hat{b}$ of the unknown plant parameters a, b, respectively. It is therefore independent of the choice of the control law, and the same adaptive law (6.47) used in Example 6.3.1 is employed.

Step 3. Adaptive control law Using the CE approach, the adaptive control law is formed by replacing the unknown plant parameters a, b in (6.65), (6.66) with their online estimates $\hat{a}, \hat{b}$ generated by the adaptive law (6.47) from Step 2, i.e.,

$$\dot{\hat{e}} = \hat{A}(t)\hat{e} + \hat{B}(t)\bar{u}_p - \hat{K}_o(C^T\hat{e} - e_1), \tag{6.67}$$

$$\bar{u}_p = -\hat{K}_c\hat{e}, \quad u_p = \frac{s+1}{s}\bar{u}_p, \tag{6.68}$$

where

$$\begin{aligned} \hat{A} &= \begin{bmatrix} -\hat{a} & 1 \\ 0 & 0 \end{bmatrix}, \qquad \hat{B} = \begin{bmatrix} 1 \\ 1 \end{bmatrix}\hat{b}, \qquad C^T = [1, 0], \\ \hat{K}_c &= \frac{1}{\hat{b}}[1 - \hat{a}, 1], \qquad \hat{K}_o = [20 - \hat{a}, 100]^T. \end{aligned} \tag{6.69}$$

The APPC scheme for the plant (6.64) is described by (6.47), (6.67)–(6.69). The stabilizability problem is evident in (6.69), where for $\hat{K}_c$ to be finite at each time t, the estimated plant parameter $\hat{b}(t)$ should not cross zero. The adaptive law (6.47) has already been designed to guarantee that $|\hat{b}(t)| \geq b_0 > 0 \, \forall t \geq 0$ for some constant b_0 that satisfies $|b| \geq b_0$, and therefore $\hat{K}_c$ can be calculated at each time t using (6.69). ∎

The stability properties of the APPC scheme (6.59)–(6.63) are described by the following theorem.

Theorem 6.4.3. *Assume that the polynomials $\hat{Z}_p$, $\hat{R}_p Q_m$ are strongly coprime at each time t. Then all the signals in the closed-loop APPC scheme* (6.59)–(6.63) *are uniformly bounded, and the tracking error e_1 converges to zero asymptotically with time.*

Proof. The proof of Theorem 6.4.3 is a complicated one and involves the use of several results from the Appendix. Below we give an outline of the proof and present the technical details of the proof in the web resource [94].

Step 1. Develop the state error equations for the closed-loop APPC scheme

$$\begin{aligned} \dot{\hat{e}} &= A_c(t)\hat{e} + \hat{K}_c C^T e_o, \\ \dot{e}_o &= A_o e_o + \tilde{\theta}_1 e_1 - \tilde{\theta}_2 \bar{u}_p, \\ y_p &= C^T e_o + C^T\hat{e} + y_m, \\ u_p &= W_1(s)\hat{K}_c(t)\hat{e} + W_2(s)y_p, \\ \bar{u}_p &= -\hat{K}_c\hat{e}, \end{aligned} \tag{6.70}$$

where $e_o \stackrel{\Delta}{=} e - \hat{e}$ is the observation error, A_o is a constant stable matrix, $W_1(s)$ and $W_2(s)$ are strictly proper transfer functions with stable poles, and $A_c(t) = \hat{A} - \hat{B}\hat{K}_c$.

Step 2. Establish e.s. for the homogeneous part of the state equations in (6.70) The gain $\hat{K}_c$ is chosen so that the eigenvalues of $A_c(t)$ at each time t are equal to the roots of the Hurwitz polynomial $A_c^*(s)$. Since $\hat{A}, \hat{B} \in \mathcal{L}_\infty$ (guaranteed by the adaptive law) and $\hat{Z}_p, \hat{R}_p Q_m$ are strongly coprime (by assumption), we conclude that $(\hat{A}, \hat{B})$ is stabilizable in the strong sense and $\hat{K}_c, A_c \in \mathcal{L}_\infty$. Using $\dot{\theta}_a, \dot{\theta}_b \in \mathcal{L}_2$, guaranteed by the adaptive law, we can establish that $\dot{\hat{K}}_c, \dot{A}_c \in \mathcal{L}_2$. Therefore, applying Theorem A.8.8, we have that $A_c(t)$ is a u.a.s. matrix. Because A_o is a constant stable matrix, the e.s. of the homogeneous part of (6.70) follows.

Step 3. Use the properties of the $\mathcal{L}_{2\delta}$ norm and the B–G lemma to establish signal boundedness We use the properties of the $\mathcal{L}_{2\delta}$ norm and (6.70) to establish the inequality

$$m_f^2(t) \le c + c\int_0^t e^{-\delta(t-\tau)} g^2(\tau) m_f^2(\tau)d\tau,$$

where $g^2 \triangleq \varepsilon^2 m_s^2 + |\dot{\theta}_a|^2 + |\dot{\theta}_b|^2$, c is any finite constant, and $m_f^2 \triangleq 1 + \|u_p\|^2 + \|y_p\|^2$ is shown to bound all signals from above. Since $g \in \mathcal{L}_2$, it follows by applying the B–G lemma that $m_f \in \mathcal{L}_\infty$. Using $m_f \in \mathcal{L}_\infty$, we establish the boundedness of all signals in the closed-loop plant.

Step 4. Establish the convergence of the tracking error e_1 to zero We establish that $e_1 \in \mathcal{L}_2$ and $\dot{e}_1 \in \mathcal{L}_\infty$ and apply Lemma A.4.7 to conclude that $e_1 \to 0$ as $t \to \infty$. □

6.5 Adaptive Linear Quadratic Control (ALQC)

Another method for solving the PPC problem is using an optimization technique to design a control input that guarantees boundedness and regulation of the plant output or tracking error to zero by minimizing a certain cost function that reflects the performance of the closed-loop system. As we established in section 6.4, the tracking problem can be converted to the regulation problem of the system

$$\begin{aligned} \dot{e} &= Ae + B\bar{u}_p, \\ e_1 &= C^T e, \end{aligned} \tag{6.71}$$

where A, B, C are defined in (6.53), $u_p = \frac{Q_1(s)}{Q_m(s)}\bar{u}_p$, and $\bar{u}_p$ is to be chosen so that $e \in \mathcal{L}_\infty$ and $e_1 \to 0$ as $t \to \infty$. The desired $\bar{u}_p$ to meet this objective is chosen as the one that minimizes the quadratic cost

$$J = \int_0^\infty (e_1^2(t) + \lambda \bar{u}_p^2(t))dt,$$

where $\lambda > 0$, a weighting coefficient to be designed, penalizes the level of the control input signal. The optimum control input $\bar{u}_p$ that minimizes J is [120]

$$\bar{u}_p = -K_c e, \quad K_c = \lambda^{-1} B^T P, \tag{6.72}$$

where $P = P^T > 0$ satisfies the algebraic equation

$$A^T P + PA - PB\lambda^{-1}B^T P + CC^T = 0, \tag{6.73}$$

known as the *Riccati equation.*

Since D is stabilizable, due to Assumption P3 in section 6.3 and the fact that $Q_1(s)$ is Hurwitz, the existence and uniqueness of $P = P^T > 0$ satisfying (6.73) is guaranteed [120]. It is clear that as $\lambda \to 0$, a situation known as *low cost control*, $\|K_c\| \to \infty$, which implies that $\bar{u}_p$ may become unbounded. On the other hand, if $\lambda \to \infty$, a situation known as *high cost control*, $\bar{u}_p \to 0$ if the open-loop system is stable. With $\lambda > 0$ and finite, however, (6.72), (6.73) guarantee that $A - BK_c$ is a stable matrix, e, e_1 converge to zero exponentially fast, and $\bar{u}_p$ is bounded. The location of the eigenvalues of $A - BK_c$ depends on the particular choice of λ. In general, there is no guarantee that one can find a λ so that the closed-loop poles are equal to the roots of the desired polynomial $A^*(s)$. The importance of the linear quadratic (LQ) control design is that the resulting closed-loop system has good robustness properties.

As in section 6.4, the state e of (6.72) may not be available for measurement. Therefore, instead of (6.72), we use

$$\bar{u}_p = -K_c\hat{e}, \quad K_c = \lambda^{-1}B^T P, \tag{6.74}$$

where $\hat{e}$ is the state of the observer equation

$$\dot{\hat{e}} = A\hat{e} + B\bar{u}_p - K_o(C^T\hat{e} - e_1) \tag{6.75}$$

and K_o is, as in (6.55), calculated using (6.56). As in section 6.4, the control input is given by

$$u_p = \frac{Q_1(s)}{Q_m(s)}\bar{u}_p. \tag{6.76}$$

Theorem 6.5.1. *The LQ control law* (6.74)–(6.76) *guarantees that all the eigenvalues of $A - BK_c$ are in $\Re[s] < 0$, all signals in the closed-loop plant are bounded, and $e_1(t) \to 0$ as $t \to \infty$ exponentially fast.*

Proof. The proof is left as an exercise for the reader (see Problem 12). □

As in section 6.4, we can use the CE approach to form the adaptive control law

$$\begin{aligned}
\dot{\hat{e}} &= \hat{A}\hat{e} + \hat{B}\bar{u}_p - \hat{K}_0(c^T\hat{e} - e_1), \\
\bar{u}_p &= -\hat{K}_c\hat{e}, \quad \hat{K}_c = \lambda^{-1}\hat{B}^T P, \\
u_p &= \frac{Q_1(s)}{Q_m(s)}\bar{u}_p,
\end{aligned} \tag{6.77}$$

where $\hat{A}$, $\hat{B}$, $\hat{K}_0$ are as defined in section 6.4 and generated using the adaptive law (6.59), and $P(t)$ is calculated by solving the Riccati equation

$$\hat{A}^T(t)P(t) + P(t)\hat{A}(t) - P(t)\hat{B}(t)\lambda^{-1}\hat{B}^T(t)P(t) + CC^T = 0 \tag{6.78}$$

at each time t.

The ALQC scheme is described by (6.59), (6.61), (6.62), (6.77), (6.78).

Example 6.5.2 Let us consider the same plant as in Example 6.4.2, i.e.,

$$y_p = \frac{b}{s+a} u_p.$$

The control objective is to choose u_p so that the closed-loop poles are stable and y_p tracks the reference signal $y_m = 1$. The problem is converted to a regulation problem by considering the tracking error equation

$$e_1 = \frac{b(s+1)}{(s+a)s} \bar{u}_p, \quad \bar{u}_p = \frac{s}{s+1} u_p,$$

where $e_1 = y_p - y_m$, as shown in Example 6.4.2. The state-space representation of the tracking error equation is given by

$$\dot{e} = \begin{bmatrix} -a & 1 \\ 0 & 0 \end{bmatrix} e + \begin{bmatrix} 1 \\ 1 \end{bmatrix} b\bar{u}_p,$$
$$e_1 = [1, 0]e.$$

The observer equation is the same as in Example 6.4.2, i.e.,

$$\dot{\hat{e}} = \begin{bmatrix} -a & 1 \\ 0 & 0 \end{bmatrix} \hat{e} + \begin{bmatrix} 1 \\ 1 \end{bmatrix} b\bar{u}_p - K_o([1, 0]\hat{e} - e_1),$$

where $K_o = [20 - a, 100]^T$ is chosen so that the observer poles are equal to the roots of $A_o^*(s) = (s+10)^2 = 0$. The control law, according to (6.74)–(6.76), is given by

$$\bar{u}_p = -\lambda^{-1} b[1, 1]P\hat{e}, \quad u_p = \frac{s+1}{s} \bar{u}_p,$$

where P satisfies the Riccati equation

$$\begin{bmatrix} -a & 1 \\ 0 & 0 \end{bmatrix}^T P + P \begin{bmatrix} -a & 1 \\ 0 & 0 \end{bmatrix} - P \begin{bmatrix} b \\ b \end{bmatrix} \lambda^{-1} [b\ b] P + \begin{bmatrix} 1 & 0 \\ 0 & 0 \end{bmatrix} = 0$$

and $\lambda > 0$ is a design parameter to be chosen.

Let us now replace the unknown a, b with their estimates $\hat{a}, \hat{b}$ generated by the adaptive law (6.47) to form the adaptive control law

$$\begin{aligned} &\dot{\hat{e}} = \hat{A}\hat{e} + \hat{B}\bar{u}_p - \hat{K}_o([1, 0]\hat{e} - e_1), \\ &\hat{A} = \begin{bmatrix} -\hat{a} & 1 \\ 0 & 0 \end{bmatrix}, \quad \hat{B} = \hat{b} \begin{bmatrix} 1 \\ 1 \end{bmatrix}, \quad \hat{K}_o = \begin{bmatrix} 20 - \hat{a} \\ 100 \end{bmatrix}, \\ &\bar{u}_p = -\frac{\hat{b}}{\lambda}[1\ 1]P\hat{e}, \quad u_p = \frac{s+1}{s} \bar{u}_p, \end{aligned} \tag{6.79}$$

where $P(t) = P^T(t) > 0$ is calculated at each time t to satisfy

$$\hat{A}^T P + P\hat{A} - P\frac{\hat{B}\hat{B}^T}{\lambda} P + CC^T = 0, \quad C^T = [1, 0]. \tag{6.80}$$

Equations (6.47), (6.79), (6.80) describe the ALQC scheme. For the solution $P = P^T > 0$ of (6.80) to exist, the pair $(\hat{A}, \hat{B})$ has to be stabilizable. Since $(\hat{A}, \hat{B}, C)$ is the realization of $\frac{\hat{b}(s+1)}{(s+\hat{a})s}$, the stabilizability of $(\hat{A}, \hat{B})$ is guaranteed, provided $\hat{b} \neq 0$ (note that for $\hat{a} = 1$, the pair $(\hat{A}, \hat{B})$ is no longer controllable but is still stabilizable). In fact for $P(t)$ to be uniformly bounded, we require $|\hat{b}(t)| \geq b_0 > 0$, for some constant b_0, which is a lower bound for $|b|$. As in the previous examples, the adaptive law (6.47) guarantees that $|\hat{b}(t)| \geq b_0 \; \forall t \geq 0$ by assuming that b_0 and $\text{sgn}(b)$ are known a priori. ∎

As with the previous APPC schemes, the ALQC scheme depends on the solvability of the algebraic Riccati equation (6.78). The Riccati equation is solved for each time t by using the online estimates $\hat{A}$, $\hat{B}$ of the plant parameters. For the solution $P(t) = P^T(t) > 0$ to exist, the pair $(\hat{A}, \hat{B})$ has to be stabilizable at each time t. This implies that the polynomials $\hat{R}_p(s,t)Q_m(s)$ and $\hat{Z}_p(s,t)Q_1(s)$ should not have any common unstable zeros at each frozen time t. Because $Q_1(s)$ is Hurwitz, a sufficient condition for $(\hat{A}, \hat{B})$ to be stabilizable is that the polynomials $\hat{R}_p(s,t)Q_m(s)$ and $\hat{Z}_p(s,t)$ are coprime at each time t. For $P(t)$ to be uniformly bounded, however, we will require $\hat{R}_p(s,t)Q_m(s)$ and $\hat{Z}_p(s,t)$ to be strongly coprime at each time t.

In contrast to the simple example considered, the modification of the adaptive law to guarantee the strong coprimeness of $\hat{R}_p Q_m$ and $\hat{Z}_p$ without the use of additional a priori information about the unknown plant is not clear. The following theorem assumes that the stabilizability condition is satisfied at each time t and establishes the stability properties of the ALQC scheme.

Theorem 6.5.3. *Assume that the polynomials $\hat{R}_p(s,t)Q_m(s)$ and $\hat{Z}_p(s,t)$ are strongly coprime at each time t. Then the ALQC scheme described by* (6.59), (6.61), (6.62), (6.77), (6.78) *guarantees that all signals in the closed-loop plant are bounded and the tracking error e_1 converges to zero as $t \to \infty$.*

Proof. The proof is almost identical to that of Theorem 6.4.3, except for some minor details. The same error equations as in the proof of Theorem 6.4.3 that relate $\hat{e}$ and the observation error $e_o = e - \hat{e}$ with the plant input and output also hold here. The only difference is that in $A_c = \hat{A} - \hat{B}\hat{K}_c$ we have $\hat{K}_c = \frac{\hat{B}^T(t)P(t)}{\lambda}$. If we establish that $\hat{K}_c \in \mathcal{L}_\infty$ and A_c is u.a.s., then the rest of the proof is identical to that of Theorem 6.4.3. We do that as follows.

The strong coprimeness assumption about $\hat{R}_p Q_m$, $\hat{Z}_p$ guarantees that $(\hat{A}, \hat{B})$ is controllable at each time t, which implies that the solution $P(t) = P^T(t) > 0$ of the Riccati equation exists and $P \in \mathcal{L}_\infty$. This, together with the boundedness of the plant parameter estimates, guarantees that $\hat{B}$ and therefore $\hat{K}_c \in \mathcal{L}_\infty$. Furthermore, using the results of section 6.4, we can establish that $A_c(t)$ is a stable matrix at each frozen time t as follows. From $A_c = \hat{A} - \hat{B}\hat{K}_c$ we have

$$\|\dot{A}_c(t)\| \leq \|\dot{\hat{A}}(t)\| + \frac{2\|\dot{\hat{B}}(t)\|\|\hat{B}(t)\|\|P(t)\|}{\lambda} + \frac{\|\hat{B}(t)\|^2\|\dot{P}(t)\|}{\lambda}. \tag{6.81}$$

Differentiating each side of (6.78), we obtain

$$\dot{P}A_c + A_c^T\dot{P} = -Q, \tag{6.82}$$

where

$$Q = \dot{\hat{A}}^T P + P\dot{\hat{A}} - \frac{P\hat{B}\dot{\hat{B}}^T P}{\lambda} - \frac{P\dot{\hat{B}}\hat{B}^T P}{\lambda}. \tag{6.83}$$

Equation (6.82) is a Lyapunov equation, and its solution $\dot{P}$ exists and satisfies $\|\dot{P}(t)\| \leq c\|Q(t)\|$ for any given $Q(t)$, where $c > 0$ is a finite constant. Since $\|\dot{A}(t)\|, \|\dot{B}(t)\| \in \mathcal{L}_2$ due to the properties of the adaptive law and $\|P(t)\|, \|Q(t)\|, \|\dot{P}(t)\| \in L_\infty$, we can establish using (6.81) that $\|\dot{A}_c(t)\| \in \mathcal{L}_2$. The pointwise stability of A_c together with $\|\dot{A}_c(t)\| \in \mathcal{L}_2$ imply, using Theorem A.8.8, that A_c is a u.a.s. matrix. The rest of the proof is completed by following exactly the same steps as in the proof of Theorem 6.4.3. □

6.6 Stabilizability Issues and Modified APPC

The main common drawback of the APPC schemes of sections 6.3–6.5 is that the adaptive law cannot guarantee that the estimated plant parameters or polynomials satisfy the appropriate controllability or stabilizability condition at each time. Loss of stabilizability or controllability may lead to computational problems and instability.

In this section, we first demonstrate using a scalar example that the estimated plant could lose controllability at certain points in time leading to unbounded signals. We then describe briefly some of the approaches described in the literature to deal with the stabilizability problem.

6.6.1 Loss of Stabilizability: A Simple Example

Let us consider the first-order plant

$$\dot{y} = y + bu, \tag{6.84}$$

where $b \neq 0$ is an unknown constant. The control objective is to choose u such that $y, u \in \mathcal{L}_\infty$, and $y(t) \to 0$ as $t \to \infty$. If b were known, then the control law

$$u = -\frac{2}{b}y \tag{6.85}$$

would meet the control objective exactly. When b is unknown, we use the CE control law

$$u = -\frac{2}{\hat{b}}y, \tag{6.86}$$

where $\hat{b}(t)$ is the estimate of b at time t, generated online by an appropriate adaptive law. Let us consider the adaptive law

$$\begin{aligned} \dot{\hat{b}} &= \gamma\phi\varepsilon, & \hat{b}(0) &= \hat{b}_0 \neq 0, \\ \varepsilon &= \frac{z - \hat{b}\phi}{m_s^2}, & m_s^2 &= 1 + \phi^2, \\ \phi &= \frac{1}{s+1}u, & z &= \frac{s-1}{s+1}y, \end{aligned} \tag{6.87}$$

where $\gamma > 0$ is the constant adaptive gain.

It can be shown that the control law (6.86) with $\hat{b}$ generated by (6.87) meets the control objective, provided that $\hat{b}(t) \neq 0 \; \forall t \geq 0$. Let us now examine whether (6.87) can satisfy the condition $\hat{b}(t) \neq 0 \; \forall t \geq 0$. From (6.84) and (6.87), we obtain

$$\varepsilon = -\frac{\tilde{b}\phi}{m_s^2}, \tag{6.88}$$

where $\tilde{b} \triangleq \hat{b} - b$ is the parameter error. Using (6.88) in (6.87), we have

$$\dot{\hat{b}} = -\gamma \frac{\phi^2}{m_s^2}(\hat{b} - b), \quad \hat{b}(0) = \hat{b}_0. \tag{6.89}$$

It is clear from (6.89) that for $\hat{b}(0) = b$, $\dot{\hat{b}}(t) = 0$ and $\hat{b}(t) = b \; \forall t \geq 0$; therefore, the control objective can be met exactly with such initial condition for $\hat{b}$. For analysis purposes, let us assume that $b > 0$ (unknown to the designer). For $\phi \neq 0$, (6.89) implies that

$$\mathrm{sgn}(\dot{\hat{b}}) = -\,\mathrm{sgn}(\hat{b}(t) - b)$$

and, therefore, for $b > 0$ we have

$$\dot{\hat{b}}(0) > 0 \quad \text{if } \hat{b}(0) < b \quad \text{and} \quad b(t) > 0 \quad \text{if } \hat{b}(t) < b.$$

Hence, for $\hat{b}(0) < 0 < b$, $\hat{b}(t)$ is monotonically increasing and crosses zero, leading to an unbounded control u in (6.86).

The above example demonstrates that the CE control law (6.86) with (6.87) as the adaptive law for generating $\hat{b}$ is not guaranteed to meet the control objective. If the sign of b and a lower bound for $|b|$ are known, then the adaptive law (6.87) can be modified using projection to constrain $\hat{b}(t)$ from changing sign. This projection approach works for this simple example, but in general its extension to the higher-order case is awkward, if not impossible, due to the lack of any procedure for constructing the appropriate convex parameter sets with the following three properties:

- Stabilizability is guaranteed for every parameter vector in the set.
- The unknown parameter vector is located inside the set.
- The sets are convex and known.

In the following subsections we describe a number of methods proposed in the literature for dealing with the stabilizability problem.

6.6.2 Modified APPC Schemes

The stabilizability problem has attracted considerable interest in the adaptive control community, and several solutions have been proposed. We list the most important ones below with a brief explanation regarding their advantages and drawbacks.

(a) Stabilizability Stabilizability is assumed. In this case, no modifications are introduced, and stabilizability is assumed to hold for all $t \geq 0$. Even though there is no theoretical justification for such an assumption to hold, it has often been argued that in most simulation studies, the stabilizability problem rarely arises. The example presented above illustrates that no stabilizability problem would arise if the initial condition of $\hat{b}(0)$ happened to be in the region $\hat{b}(0) > b > 0$. In the higher-order case, loss of stabilizability occurs at certain isolated manifolds in the parameter space when visited by the estimated parameters. Therefore, one can easily argue that the loss of stabilizability is not a frequent phenomenon. If it does happen at a particular point in time, then heuristics may be used to deal with it, such as restarting the adaptive law with different initial conditions, ignoring the estimates that lead to loss of stabilizability, and changing the adaptive gain till the estimated parameters satisfy the stabilizability condition.

(b) Parameter Projection Methods [46, 121, 122] In this approach, the parameter estimates are constrained to lie inside a convex subset $\mathcal{C}_0$ of the parameter space that is assumed to have the following properties:

(i) The unknown plant parameter vector $\theta_p^* \in \mathcal{C}_0$.

(ii) Every member θ_p of $\mathcal{C}_0$ has a corresponding level of stabilizability greater than ε^* for some known constant $\varepsilon^* > 0$.

Given such a convex set $\mathcal{C}_0$, the stabilizability of the estimated parameters at each time t is ensured by incorporating a projection algorithm in the adaptive law to guarantee that the estimates are in $\mathcal{C}_0$ $\forall t \geq 0$. The projection is based on the gradient projection method and does not alter the usual properties of the adaptive law that are used in the stability analysis of the overall scheme. This approach is simple but relies on the rather strong assumption that the set $\mathcal{C}_0$ is known. No procedure has been proposed for constructing such a set $\mathcal{C}_0$ for a general class of plants.

An extension of this approach has been proposed in [123]. It is assumed that a finite number of convex subsets $\mathcal{C}_1, \ldots, \mathcal{C}_p$ are known such that

(i) $\theta_p^* \in \bigcup_{i=1}^p \mathcal{C}_i$ and the stabilizability degree of the corresponding plant is greater than some known $\varepsilon^* > 0$.

(ii) For every $\theta_p \in \bigcup_{i=1}^p \mathcal{C}_i$ the corresponding plant model is stabilizable with a stabilizability degree greater than ε^*.

In this case, p adaptive laws with a projection, one for each subset $\mathcal{C}_i$, are used in parallel. A suitable performance index is used to select the adaptive law at each time t whose parameter estimates are to be used to calculate the controller parameters. The price paid in this case is the use of p parallel adaptive laws with projection instead of one. As in the case of a single convex subset, there is no effective procedure for constructing $\mathcal{C}_i$, $i = 1, 2, \ldots, p$, with properties (i) and (ii) in general.

(c) Correction Approach [124] In this approach, a subset $\mathcal{D}$ in the parameter space is known with the following properties:

(i) $\theta_p^* \in \mathcal{D}$ and the stabilizability degree of the plant is greater than some known constant $\varepsilon^* > 0$.

(ii) For every $\theta_p \in \mathcal{D}$, the corresponding plant model is stabilizable with a degree greater than ε^*.

Two LS estimators with estimates $\hat{\theta}_p, \bar{\theta}_p$ of θ_p^* are run in parallel. The controller parameters are calculated from $\bar{\theta}_p$ as long as $\bar{\theta}_p \in \mathcal{D}$. When $\bar{\theta}_p \notin \mathcal{D}$, $\bar{\theta}_p$ is reinitialized as

$$\bar{\theta}_p = \hat{\theta}_p + P^{1/2}\gamma,$$

where P is the covariance matrix of the LS estimator of θ_p^*, and γ is a vector chosen so that $\bar{\theta}_p \in \mathcal{D}$. The search for the appropriate γ can be systematic or random.

The drawbacks of this approach are (1) added complexity due to the two parallel estimators, and (2) the search procedure for γ can be tedious and time-consuming. The advantage of this approach, when compared with the projection one, is that the subset D does not have to be convex. The importance of this advantage, however, is not clear, since no procedure is given for how to construct D to satisfy conditions (i)–(ii) above.

(d) Persistent Excitation Approach [119, 125] In this approach, the reference input signal or an external signal is chosen to be sufficiently rich in frequencies so that the signal information vector is PE over an interval. The PE property guarantees that the parameter estimate $\hat{\theta}_p$ of θ_p^* converges exponentially to θ_p^* (provided that the covariance matrix in the case of LS is prevented from becoming singular). Using this PE property, and assuming that a lower bound $\varepsilon^* > 0$ for the stabilizability degree of the plant is known, the following modification is used: When the stabilizability degree of the estimated plant is greater than ε^*, the controller parameters are computed using $\hat{\theta}_p$; otherwise the controller parameters are frozen to their previous value. Since $\hat{\theta}_p$ converges to θ_p^*, the stabilizability degree of the estimated plant is guaranteed to be greater than ε^* asymptotically with time.

The main drawback of this approach is that the reference signal or external signal has to be sufficiently rich and on all the time, which implies that accurate regulation or tracking of signals that are not rich is not possible. Thus the stabilizability problem is overcome at the expense of destroying the desired tracking or regulation properties of the adaptive scheme. Another less serious drawback is that a lower bound $\varepsilon^* > 0$ for the stabilizability degree of the unknown plant is assumed to be known a priori.

An interesting method related to PE is proposed in [126] for the stabilization of unknown plants. In this case the PE property of the signal information vector over an interval is generated by a "rich" nonlinear feedback term that disappears asymptotically with time. The scheme of [126] guarantees exact regulation of the plant output to zero. In contrast to other PE methods [119, 125], both the plant and the controller parameters are estimated online leading to a higher-order adaptive law.

(e) Cyclic Switching and Other Switching Methods In the cyclic switching approach [127] the control input is switched between the CE control law and a finite member of specially constructed controllers with fixed parameters. The fixed controllers have the property that at least one of them makes the resulting closed-loop plant observable through a certain error signal. The switching logic is based on a cyclic switching rule. The proposed scheme does not rely on persistent excitation and does not require the knowledge of a lower bound for the level of stabilizability. One can argue, however, that the concept of PE to help cross the points in the parameter space where stabilizability is weak or lost is implicitly used by the scheme, because the switching between different controllers which are not necessarily stabilizing may cause considerable excitation over intervals of time. Some of the drawbacks of the cyclic switching approach are the complexity and the possible bad transient of the plant or tracking error response during the initial stages of adaptation when switching is active. Another drawback is that there is no procedure for how to construct these fixed controllers to satisfy the required observability condition. This approach, however, led to subsequent approaches [18, 19, 128] based on switching without the use of any parameter estimator or adaptive law. These approaches, known as non–identifier-based, modify the adaptive control problem to the problem of finding the stabilizing controller from a known finite set of fixed controllers which are available to the designer. The assumption made is that at least one of the given fixed controllers is a stabilizing one.

(f) Switched Excitation Approach [129] This approach is based on the use of an open-loop rich excitation signal that is switched on whenever the calculation of the CE control law is not possible due to the loss of stabilizability of the estimated plant. It differs from the PE approach described in (d) in that the switching between the rich external input and the CE control law terminates in finite time, after which the CE control law is on and no stabilizability issues arise again. The method is very intuitive and relies on switching between the parameter identification and control objectives depending on the quality of parameter estimates. That is, when the parameter estimates are not suitable for control, the algorithm switches to the parameter identification objective by switching on an external sufficiently rich signal. When the quality of the parameter estimates improves, the algorithm switches back to the control objective.

Methods similar to those above have been proposed for APPC schemes for discrete-time plants [18, 130–137].

6.7 Robust APPC Schemes

As in the case of MRAC, the APPC schemes considered in the previous sections may lose stability in the presence of modeling errors and/or bounded disturbances. As we demonstrated using a simple example, even a small disturbance may lead to parameter drift in the adaptive law, which in turn may generate high gains in the feedback loop, leading to unbounded signals.

Robust APPC schemes can be designed by simply replacing the adaptive laws used in the previous sections with normalized robust adaptive laws. The normalizing signal is designed to bound from above the modeling error as in the case of robust MRAC. We demonstrate the design and analysis of robust APPC schemes for the plant

$$y_p = G_0(s)(1 + \Delta_m(s))[u_p + d_u], \tag{6.90}$$

where $G_0(s)$ satisfies P1–P3 given in section 6.3, $\Delta_m(s)$ is an unknown multiplicative uncertainty, d_u is a bounded input disturbance, and the overall plant transfer function $G(s) =$

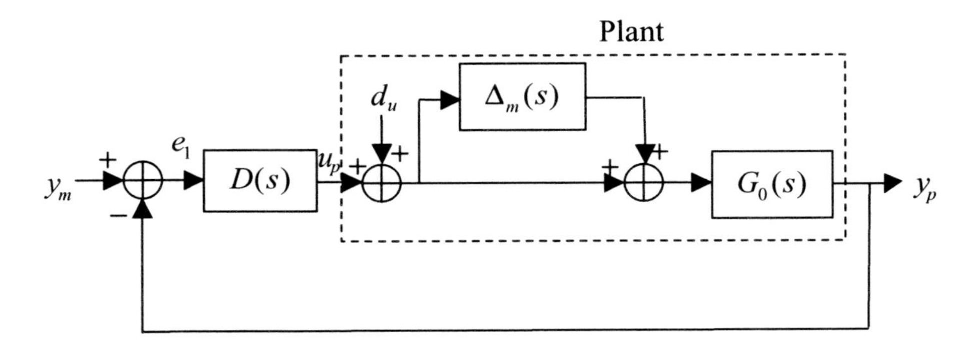

Figure 6.1. *Closed-loop PPC schemes with unmodeled dynamics and bounded disturbances.*

$G_0(s)(1 + \Delta_m(s))$ is strictly proper. We design APPC schemes for the lower-order plant model

$$y_p = G_0(s)u_p \tag{6.91}$$

but apply and analyze them for the higher-order plant model (6.90). The effect of the perturbation Δ_m and disturbance d_u on the stability and performance of the APPC schemes is investigated in the following sections. We first consider the nonadaptive case, where the parameters of $G_0(s)$ are known so that no adaptation is needed.

6.7.1 PPC: Known Parameters

Let us consider the control laws of sections 6.3–6.5 that are designed for the simplified plant model (6.91) with known plant parameters and apply them to the higher-order plant (6.90). The block diagram of the closed-loop plant is shown in Figure 6.1, where $D(s)$ is the transfer function of the controller. The expression for $D(s)$ for each of the PPC laws developed in section 6.3–6.5 is given as follows.

For the control law in (6.30), (6.31) of section 6.3, which is based on the polynomial approach,

$$D(s) = \frac{P(s)}{Q_m(s)L(s)}. \tag{6.92}$$

For the control law (6.54), (6.55) of section 6.4 and the control law (6.74)–(6.76) of section 6.5,

$$D(s) = \frac{K_c(sI - A + K_oC^T)^{-1}K_o}{(1 + K_c(sI - A + K_oC^T)^{-1}B)}\frac{Q_1(s)}{Q_m(s)}. \tag{6.93}$$

Theorem 6.7.1. *The closed-loop plant described in Figure* 6.1 *is stable in the sense that bounded y_m, d_u implies bounded u_p, y_p, provided*

$$\|T_0(s)\Delta_m(s)\|_\infty < 1,$$

where $T_0(s) = \frac{DG_0}{1+DG_0}$ is the closed-loop transfer function of the nominal plant. Furthermore, the tracking error e_1 converges exponentially to the residual set

$$\mathcal{D}_e = \{e_1 \mid |e_1| \le cd_0\}, \tag{6.94}$$

where d_0 is an upper bound for $|d_u|$ and $c > 0$ is a constant.

Proof. The characteristic equation of the feedback system in Figure 6.1 is

$$1 + DG_0(1 + \Delta_m) = 0,$$

which implies, due to the stable roots of $1 + DG_0 = 0$, that

$$1 + \frac{DG_0}{1 + DG_0}\Delta_m = 0.$$

Using the Nyquist criterion or the small gain theorem, the roots of the characteristic equation are in $\Re[s] < 0$ if

$$\left\| \frac{DG_0}{1 + DG_0}\Delta_m \right\|_\infty = \|T_0(s)\Delta_m\|_\infty < 1.$$

To establish (6.94), we use Figure 6.1 to show that

$$e_1 = -\frac{1}{1 + DG}y_m + \frac{G}{1 + DG}d_u.$$

It follows from (6.92), (6.93) that the controller $D(s)$ is of the form $D(s) = \frac{C_0(s)}{Q_m(s)}$ for some $C_0(s)$. Therefore,

$$e_1 = -\frac{Q_m}{Q_m + C_0G}y_m + \frac{GQ_m}{Q_m + C_0G}d_u,$$

where $G = G_0(1 + \Delta_m)$. Since $Q_m y_m = 0$ and the closed-loop plant is stable due to $\|T_0(s)\Delta_m(s)\|_\infty < 1$, we have

$$e_1 = \frac{(1 + \Delta_m)G_0Q_m}{Q_m + C_0G_0 + \Delta_m C_0G_0}d_u + \varepsilon_t, \tag{6.95}$$

where ε_t is a term exponentially decaying to zero. Therefore, (6.94) is implied by (6.95) and the stability of the closed-loop plant (see Problem 13). □

It should be pointed out that the tracking error at steady state is not affected by y_m despite the presence of the unmodeled dynamics. That is, if $d_u \equiv 0$ and $\Delta_m \neq 0$, we still have $e_1(t) \to 0$ as $t \to \infty$, provided that the closed-loop plant is stable. This is due to the incorporation of the internal model of y_m in the control law. If $Q_m(s)$ contains the internal model of d_u as a factor, i.e., $Q_m(s) = Q_d(s)\bar{Q}_m(s)$, where $Q_d(s)d_u = 0$ and $\bar{Q}_m(s)y_m = 0$, then it follows from (6.95) that $e_1 = \varepsilon_t$; i.e., the tracking error converges to zero exponentially fast despite the presence of the input disturbance. The internal model of d_u can be constructed if we know the frequencies of d_u. For example, if d_u is a slowly varying signal of unknown magnitude, we could choose $Q_d(s) = s$.

The robustness and performance properties of the PPC schemes given by Theorem 6.7.1 are based on the assumption that the parameters of the modeled part of the plant, i.e., the coefficients of $G_0(s)$, are known exactly. When the coefficients of $G_0(s)$ are unknown, the PPC laws (6.92) and (6.93) are combined with adaptive laws that provide online estimates for the unknown parameters leading to a wide class of APPC schemes. The design of these APPC schemes so that their robustness and performance properties are as close as possible to those described by Theorem 6.7.1 for the known parameter case is a challenging problem in robust adaptive control and is treated in the following sections.

6.7.2 Robust Adaptive Laws for APPC Schemes

We start by writing the plant equation (6.90) as

$$R_p y_p = Z_p(1 + \Delta_m)(u_p + d_u), \tag{6.96}$$

where $Z_p = \theta_b^{*T}\alpha_{n-1}(s)$, $R_p = s^n + \theta_a^{*T}\alpha_{n-1}(s)$; θ_b^*, θ_a^* are the coefficient vectors of Z_p, R_p, respectively; $\alpha_{n-1}(s) = [s^{n-1}, s^{n-2}, \dots, s, 1]^T$. Filtering each side of (6.96) with $\frac{1}{\Lambda_p(s)}$, where $\Lambda_p(s) = s^n + \lambda_1 s^{n-1} + \cdots + \lambda_n$ is Hurwitz, we obtain

$$z = \theta_p^{*T}\phi + \eta, \tag{6.97}$$

where

$$z = \frac{s^n}{\Lambda_p(s)} y_p, \qquad \theta_p^* = [\theta_b^{*T}, \theta_a^{*T}]^T,$$

$$\phi = \left[\frac{\alpha_{n-1}^T(s)}{\Lambda_p} u_p, -\frac{\alpha_{n-1}^T(s)}{\Lambda_p} y_p\right]^T, \qquad \eta = \frac{Z_p}{\Lambda_p}[\Delta_m u_p + (1 + \Delta_m) d_u].$$

As we have shown in Chapter 3, one of the main ingredients of a robust adaptive law is the normalizing signal m_s that is designed so that $\frac{|\phi|}{m_s}, \frac{\eta}{m_s} \in \mathcal{L}_\infty$. We apply Lemma A.5.9 and write

$$|\eta(t)| \le \left\|\frac{Z_p(s)}{\Lambda_p(s)}\Delta_m(s)\right\|_{2\delta} \|(u_p)_t\|_{2\delta} + \left\|\frac{Z_p(s)(1 + \Delta_m(s))}{\Lambda_p(s)}\right\|_{2\delta} \frac{d_0}{\sqrt{\delta}} + \varepsilon_t \tag{6.98}$$

for some $\delta > 0$, where ε_t is an exponentially decaying to zero term and $d_0 = \sup_t |d_u(t)|$. Similarly,

$$|\phi(t)| \le \sum_{i=1}^{n} \left\|\frac{s^{n-i}}{\Lambda_p(s)}\right\|_{2\delta} (\|(u_p)_t\|_{2\delta} + \|(y_p)_t\|_{2\delta}). \tag{6.99}$$

The above $H_{2\delta}$ norms exist, provided that $\frac{1}{\Lambda_p(s)}$ and $\Delta_m(s)$ are analytic in $\Re[s] \ge -\frac{\delta}{2}$ and $\frac{Z_p\Delta_m}{\Lambda_p}$ is strictly proper. Because the overall plant transfer function $G(s)$ and $G_0(s)$ is assumed to be strictly proper, it follows that $G_0\Delta_m$ and therefore $\frac{Z_p\Delta_m}{\Lambda_p}$ are strictly proper.

Let us now assume that $\Delta_m(s)$ is analytic in $\Re[s] \geq -\frac{\delta_0}{2}$ for some known $\delta_0 > 0$. If we design $\Lambda_p(s)$ to have roots in the region $\Re[s] < -\frac{\delta_0}{2}$, then it follows from (6.98), (6.99) by setting $\delta = \delta_0$ that the normalizing signal m_s given by

$$m_s^2 = 1 + \|(u_p)_t\|_{2\delta_0}^2 + \|(y_p)_t\|_{2\delta_0}^2$$

bounds η, ϕ from above. The signal m_s may be generated from the equations

$$\begin{aligned} m_s^2 &= 1 + n_d, \\ \dot{n}_d &= -\delta_0 n_d + u_p^2 + y_p^2, \quad n_d(0) = 0. \end{aligned} \tag{6.100}$$

Using (6.100) and the parametric model (6.97), a wide class of robust adaptive laws may be generated by employing the results of Chapter 3. As an example let us consider a robust adaptive law based on the gradient algorithm and switching σ-modification to generate online estimates of θ_p^* in (6.97). We have

$$\begin{aligned} \dot{\theta}_p &= \Gamma \varepsilon \phi - \sigma_s \Gamma \theta_p, \\ \varepsilon &= \frac{z - \theta_p^T \phi}{m_s^2}, \quad z = \frac{s^n}{\Lambda_p(s)} y_p, \quad m_s^2 = 1 + n_d, \\ \dot{n}_d &= -\delta_0 n_d + u_p^2 + y_p^2, \quad n_d(0) = 0, \\ \sigma_s &= \begin{cases} 0 & \text{if } |\theta_p| \leq M_0, \\ \sigma_0 \left(\frac{|\theta_p|}{M_0} - 1 \right) & \text{if } M_0 < |\theta_p| \leq 2M_0, \\ \sigma_0 & \text{if } |\theta_p| > M_0, \end{cases} \end{aligned} \tag{6.101}$$

where θ_p is the estimate of θ_p^*, $M_0 > |\theta_p^*|$, $\sigma_0 > 0$, and $\Gamma = \Gamma^T > 0$. As established in Chapter 3, the above adaptive law guarantees that (i) $\varepsilon, \varepsilon m_s, \theta_p, \dot{\theta}_p \in \mathcal{L}_\infty$, and (ii) $\varepsilon, \varepsilon m_s, \dot{\theta}_p \in \mathcal{S}(\frac{\eta^2}{m_s^2})$ independent of the boundedness of ϕ, z, m.

The adaptive law (6.101) or any other robust adaptive law based on the parametric model (6.97) can be combined with the PPC laws designed for the plant without uncertainties to generate a wide class of robust APPC schemes for the plant with uncertainties, as demonstrated in the following sections.

6.7.3 Robust APPC: Polynomial Approach

We consider the APPC scheme designed in section 6.3 for the ideal plant $y_p = G_0(s)u_p$. We replace the adaptive law with a robust one to obtain the following robust APPC scheme.

Robust adaptive law

$$
\begin{aligned}
\dot{\theta}_p &= \Gamma \varepsilon \phi - \sigma_s \Gamma \theta_p, \quad \Gamma = \Gamma^T > 0, \\
\varepsilon &= \frac{z - \theta_p^T \phi}{m_s^2}, \quad m_s^2 = 1 + n_d, \\
\dot{n}_d &= -\delta_0 n_d + u_p^2 + y_p^2, \quad n_d(0) = 0, \\
\sigma_s &= \begin{cases} 0 & \text{if } |\theta_p| \le M_0, \\ \sigma_0 \left(\frac{|\theta_p|}{M_0} - 1 \right) & \text{if } M_0 < |\theta_p| \le 2M_0, \\ \sigma_0 & \text{if } |\theta_p| > M_0, \end{cases}
\end{aligned} \tag{6.102}
$$

where $z = \frac{s^n}{\Lambda_p(s)} y_p$, $\phi = [\frac{\alpha_{n-1}^T(s)}{\Lambda_p(s)} u_p, -\frac{\alpha_{n-1}^T(s)}{\Lambda_p(s)} y_p]^T$, $\alpha_{n-1}(s) \triangleq [s^{n-1}, \dots, s, 1]^T$, $\Lambda_p(s)$ is a monic Hurwitz polynomial, and $\theta_p = [\hat{b}_{n-1}, \dots, \hat{b}_0, \hat{a}_{n-1}, \dots, \hat{a}_0]^T$ is used to form the estimated plant polynomials

$$
\hat{R}_p(s,t) = s^n + \hat{a}_{n-1} s^{n-1} + \cdots + \hat{a}_1 s + \hat{a}_0, \qquad \hat{Z}_p(s,t) = \hat{b}_{n-1} s^{n-1} + \cdots + \hat{b}_1 s + \hat{b}_0.
$$

Calculation of controller parameters Solve

$$
\hat{L}(s,t) \cdot Q_m(s) \cdot \hat{R}_p(s,t) + \hat{P}(s,t) \cdot \hat{Z}_p(s,t) = A^*(s) \tag{6.103}
$$

for $\hat{L}(s,t)$, $\hat{P}(s,t)$ at each time t.

Adaptive control law

$$
u_p = (\Lambda(s) - \hat{L}(s,t) Q_m(s)) \frac{1}{\Lambda(s)} u_p - \hat{P}(s,t) \frac{1}{\Lambda(s)} (y_p - y_m). \tag{6.104}
$$

The stability properties of (6.102)–(6.104) when applied to the actual plant (6.90) are given by the following theorem.

Theorem 6.7.2. *Assume that the estimated polynomials $\hat{R}_p(s,t)$, $\hat{Z}_p(s,t)$ of the plant model are such that $\hat{R}_p Q_m$, $\hat{Z}_p$ are strongly coprime at each time t. There exists a $\delta^* > 0$ such that if*

$$
\Delta_2^2 < \delta^*, \quad \textit{where } \Delta_2 \triangleq \left\| \frac{Z_p(s)}{\Lambda_p(s)} \Delta_m(s) \right\|_{2\delta_0},
$$

then the APPC scheme (6.102)–(6.104) *guarantees that all signals are bounded and the tracking error e_1 satisfies*[13]

$$
\int_0^t e_1^2 d\tau \le c(\Delta_2^2 + d_0^2)t + c \quad \forall t \ge 0.
$$

Proof. The proof is very similar to that in the ideal case. Below we present the main steps for the case of $d_u = 0$.

[13]The full details of the proof are presented in the web resource [94] and in [56].

Step 1. Express u_p, y_p in terms of the estimation error The same equations as in (6.48) can be established, i.e.,

$$\begin{aligned}\dot{x} &= A(t)x + B_1(t)\varepsilon m_s^2 + B_2\bar{y}_m,\\ u_p &= C_1^T\dot{x} + D_1^T x,\\ y_p &= C_2^T\dot{x} + D_2^T x.\end{aligned}$$

Step 2. Establish the exponential stability of $A(t)$ In this case, $A(t)$ has stable eigenvalues at each time t. However, due to the modeling error, $\|\dot{A}(t)\| \notin \mathcal{L}_2$. Instead, $\|\dot{A}(t)\| \in \mathcal{S}(\Delta_2^2)$. Applying Theorem A.8.8(b), it follows that for $\Delta_2 < \Delta^*$ and some $\Delta^* > 0$, $A(t)$ is e.s.

Step 3. Use the properties of the $\mathcal{L}_{2\delta}$ norm and the B–G lemma to establish boundedness of signals We show that

$$m_f^2(t) \triangleq 1 + \|u_{pt}\|_{2\delta}^2 + \|y_{pt}\|_{2\delta}^2 \leq c\int_0^t e^{-\delta(t-\tau)}\varepsilon^2 m_s^2 d\tau + c.$$

Since $\varepsilon m_s \in \mathcal{S}(\Delta_2^2)$, it follows by applying the B–G lemma that for $c\Delta_2^2 < \delta$ we have $m_f \in \mathcal{L}_\infty$. From $m_f \in \mathcal{L}_\infty$, we establish boundedness of all the signals in the closed-loop system, provided $\Delta_2^2 < \delta^*$, where $\delta^* = \min[\delta, \Delta^{*2}]$.

Step 4. Establish tracking error bound We express the tracking error e_1 as the sum of outputs of proper transfer functions with stable poles whose inputs belong to $\mathcal{S}(\Delta_2^2)$. Applying Corollary A.5.8, the result in Theorem 6.7.2 follows. □

In a similar way, the APPC scheme based on the LQ and state-space approaches can be made robust by replacing the adaptive law with robust adaptive laws such as (6.102), leading to a wide class of robust APPC schemes.

The following examples are used to illustrate the design of different robust APPC schemes.

Example 6.7.3 (polynomial approach) Let us consider the plant

$$y_p = \frac{b}{s+a}(1 + \Delta_m(s))u_p, \tag{6.105}$$

where a, b are unknown constants and $\Delta_m(s)$ is a multiplicative plant uncertainty. The input u_p has to be chosen so that the poles of the closed-loop modeled part of the plant (i.e., with $\Delta_m(s) \equiv 0$) are placed at the roots of $A^*(s) = (s+1)^2$ and y_p tracks the constant reference signal $y_m = 1$ as closely as possible. The same control problem has been considered and solved in Example 6.3.1 for the case where $\Delta_m(s) \equiv 0$. We design each block of the robust APPC for the above plant as follows.

Robust adaptive law The parametric model for the plant is

$$z = \theta_p^{*T}\phi + \eta,$$

where

$$z = \frac{s}{s+\lambda}y_p, \qquad \theta_p^* = [b, a]^T,$$
$$\phi = \frac{1}{s+\lambda}[u_p, -y_p]^T, \qquad \eta = \frac{b}{s+\lambda}\Delta_m(s)u_p.$$

We assume that $\Delta_m(s)$ is analytic in $\Re[s] \geq -\frac{\delta_0}{2}$ for some known $\delta_0 > 0$ and design $\lambda > \frac{\delta_0}{2}$. Using the results of Chapter 3, we develop the following robust adaptive law:

$$\begin{aligned}
\dot{\theta}_p &= \Pr(\Gamma(\varepsilon\phi - \sigma_s\theta_p)), \quad \Gamma = \Gamma^T > 0,\\
\varepsilon &= \frac{z - \theta_p^T\phi}{m_s^2}, \quad m_s^2 = 1 + n_d,\\
\dot{n}_d &= -\delta_0 n_d + u_p^2 + y_p^2, \quad n_d(0) = 0,\\
\sigma_s &= \begin{cases} 0 & \text{if } |\theta_p| \leq M_0,\\ \sigma_0\left(\frac{|\theta_p|}{M_0} - 1\right) & \text{if } M_0 < |\theta_p| \leq 2M_0,\\ \sigma_0 & \text{if } |\theta_p| > M_0, \end{cases}
\end{aligned} \tag{6.106}$$

where $\theta_p = [\hat{b}, \hat{a}]^T$, $\Pr(\cdot)$ is the projection operator which keeps the estimate $|\hat{b}(t)| \geq b_0 > 0$, b_0 is a known lower bound for $|b|$, and $M_0 > |\theta^*|$, $\sigma_0 > 0$ are design constants.

Calculation of controller parameters

$$\hat{p}_1 = \frac{2-\hat{a}}{\hat{b}}, \qquad \hat{p}_0 = \frac{1}{\hat{b}}.$$

Adaptive control law The control law is given by

$$u_p = \frac{\lambda}{s+\lambda}u_p - (\hat{p}_1 s + \hat{p}_0)\frac{1}{s+\lambda}[y_p - y_m].$$ ■

Example 6.7.4 (state feedback approach) Consider the same plant given by (6.105). The adaptive control law (6.67)–(6.69) designed for the plant (6.105) with $\Delta_m(s) = 0$ together with the robust adaptive law (6.106) is a robust APPC for the plant (6.105) with $\Delta_m(s) \neq 0$. ■

6.8 Case Study: ALQC Design for an F-16 Fighter Aircraft

The design of flight control systems for high-performance aircraft is a challenging problem, due to the sensitivity of the dynamical characteristics of the aircraft with respect to changes in flight conditions and the difficulty in measuring and estimating its aerodynamic characteristics. The traditional robust control techniques are often inadequate to deal with the wide range of flight conditions high-performance aircraft encounter due to the large parametric uncertainties and dynamical changes. Adaptive control techniques that can learn the dynamical changes online and adjust the controller parameters or structure appropriately in order to accommodate them offer a strong potential for meeting the stability and performance requirements of high-performance aircraft.

In this study the decoupled model of the longitudinal aircraft dynamics of an F-16 fighter aircraft is investigated. The full six-degrees-of-freedom nonlinear model [138, 139] of the F-16 aircraft is available in the web resource [94]. The complexity of the nonlinear model and the uncertainties involved make it difficult, if not impossible, to design a controller for the full-order nonlinear system. As in every practical control design the controller is designed based on a simplified model which is a good approximation of the full-order system in the frequency range of interest, but it is applied and evaluated using the full-order model. In this case we follow a similar approach.

In straight and level flight, the nonlinear equations of motion can be linearized at different operating points over the flight envelope defined by the aircraft's velocity and altitude. By trimming the aircraft at the specified velocity and altitude, and then numerically linearizing the equations of motion at the trim point, the longitudinal dynamics of the aircraft can be described by a family of LTI models:

$$\begin{aligned} \dot{x} &= Ax + Bu, \\ y &= Cx, \end{aligned} \tag{6.107}$$

where $\mathbf{x} = [V\ h\ \alpha\ \theta\ q\ P_{\text{pow}}]^T$ is the state vector, $u = [\delta_T\ \delta_E]^T$ is the control vector, and $\mathbf{y} = [V\ h]$ is the output vector, which is the vector of variables to be tracked. The physical meaning of the state and control variables is as follows: V: true velocity, ft/sec; h: altitude, ft; α: angle of attack, radian (range $-10° \sim 45°$); ζ: Euler (pitch) angle, rad; q: pitch rate, rad/sec; P_{pow}: power, 0–100%; δ_T: throttle setting, (0.0–1.0); δ_E: elevator setting, degree; V_m: desired true velocity, ft/sec; h_m: desired altitude, ft.

For each operating or trim point the nonlinear dynamics may be approximated by the linear model (6.107). As the aircraft moves from one operating point to another the parameters of the linear model will change. Consequently the values of the matrices A, B are functions of the operating point.

The control objective is to design the control signal u such that the output $y = [V\ h]$ tracks the vector $y_m = [V_m\ h_m]$. Without loss of generality we make the assumption that the pair (A, B) is stabilizable at each operating point. Let us first assume that we know the parameters.

We define the output tracking error as

$$e_v = V_m - V, \qquad e_h = h_m - h. \tag{6.108}$$

Since the trajectories to be tracked are constant we can use the internal model, which in this case is simply $Q_m(s) = s$, and the techniques of the chapter to convert the tracking problem to a regulation problem. Since all states are available for measurement we follow a slightly different approach [138]. We pass the tracking error state $e = [e_v, e_h]^T = y_m - Cx$ through an integrator whose state w satisfies

$$\dot{w} = e = y_m - Cx.$$

Define the augmented system as

$$\begin{aligned} \dot{x}_a &= A_a x_a + B_a u + \begin{bmatrix} 0 \\ I \end{bmatrix} y_m, \\ y &= \begin{bmatrix} C & 0 \end{bmatrix} x_a, \end{aligned} \tag{6.109}$$

where

$$x_a = \begin{bmatrix} x \\ w \end{bmatrix}, \qquad A_a = \begin{bmatrix} A & 0 \\ -C & 0 \end{bmatrix}, \qquad B_a = \begin{bmatrix} B \\ 0 \end{bmatrix}.$$

Applying the state feedback control law

$$u = -Kx_a, \tag{6.110}$$

the closed-loop system is given by

$$\begin{aligned} \dot{x}_a &= (A_a - B_a K)x_a + \begin{bmatrix} 0 \\ I \end{bmatrix} y_m \\ &= A^* x_a + B^* y_m, \end{aligned} \tag{6.111}$$

where $A^* = A_a - B_a K$, $B^* = [\begin{smallmatrix} 0 \\ I \end{smallmatrix}]$. Due to the controllability of A_a, B_a the control gain matrix K can be designed to assign the eigenvalues of A^* to be equal to any given desired set of eigenvalues. Let us first assume that the gain K is chosen so that the eigenvalues of A^* have all negative real parts. Let us define the vector

$$x_m = [V_m \ h_m \ \alpha_m \ \theta_m \ q_m \ P_{pow_m} \ w_m]^T.$$

The vector x_m corresponds to the steady-state value of x_a when $V = V_m$, $h = h_m$. Define the augmented tracking error $e_a = x_a - x_m$ which satisfies the state equation

$$\dot{e}_a = A^* e_a + B^* y_m + A^* x_m.$$

Since y_m, x_m are constant it follows that

$$\ddot{e}_a = A^* \dot{e}_a,$$

which implies due to the stability of A^* that $\dot{e}_a$ converges to zero exponentially fast. The last element of $\dot{e}_a$ is equal to $\dot{w} - \dot{w}_m = \dot{w} = e$, which implies that the output error $e = [e_v, e_h]^T = y_m - y$ converges to zero exponentially fast. Therefore, as long as the controller gain K is chosen to stabilize the system, the tracking error is guaranteed to converge to zero exponentially fast. The stabilizing gain can be chosen in different ways. In this study we follow the LQ control approach to obtain

$$K = R^{-1} B_a^T P, \tag{6.112}$$

where P is the positive definite solution of the algebraic Riccati equation

$$A_a^T P + P A_a - P B_a R^{-1} B_a^T P + Q = 0, \tag{6.113}$$

which is the result of minimizing the quadratic cost function

$$J = \int_0^\infty (x_a^T Q x_a^T + u^T R u) dt,$$

where $Q > 0$, $R > 0$ are design matrices. The problem we are faced with now is how to control the dynamics of the system as the aircraft moves from one operating point to another, which causes the parameters in (6.107) to change. In the following subsections we consider two approaches: a nonadaptive approach and an adaptive one.

6.8.1 LQ Control Design with Gain Scheduling

Since the parameters (A, B, C) in the linearized model (6.107) vary from one operating point (defined by the aircraft speed and altitude) to another, one way to handle these parameter changes is to estimate them offline and use them to design a fixed gain robust controller for each operating point. A table with the controller characteristics corresponding to each operating point is constructed. Between operating points an interpolation may be used to specify the controller characteristics. Once the operating point is identified the right controller is selected from the stored table. This method is known as *gain scheduling* and is used in many applications. The drawback of gain scheduling is that changes in parameters which have not been accounted for in the stored table may lead to deterioration of performance or even loss of stability. Such unpredictable parameter changes may be due to unexpected flight conditions, damage or failures of parts, etc. One of the advantages of the gain scheduling approach is that once the operating point is identified the switching to the right controller is done very fast [210, 211]. We present the main design steps of this approach below.

We develop a database of a number of possible operating points. For each operating point corresponding to the pair of speed and altitude (V_i, h_j), $j = 1, 2$, in a two-dimensional space we use linearization to obtain a linear model of the form (6.107). For each linearized model we apply the LQ procedure above to obtain the corresponding controller gain matrix K_{ij}. Between operating points we use the interpolation

$$\begin{aligned}
K_1 &= K_{11} + \frac{V - V_1}{V_2 - V_1}(K_{21} - K_{11}), \\
K_2 &= K_{12} + \frac{V - V_1}{V_2 - V_1}(K_{22} - K_{12}), \\
K &= K_1 + \frac{h - h_1}{h_2 - h_1}(K_2 - K_1)
\end{aligned} \tag{6.114}$$

to calculate the gain K as a function of current speed V and altitude h.

6.8.2 Adaptive LQ Control Design

The offline calculation of A, B in (6.107) often requires aerodynamic data that include stability derivatives which cannot be accurately determined in some flow regimes (e.g., transonic flight) by wind-tunnel experiments or computational fluid dynamic (CFD) studies. Experiments and measurements in wind-tunnel tests often contain modeling errors. These possible inaccuracies in the values of A, B make adaptive control a prime candidate for meeting the control objective despite large and unpredictable parametric uncertainties. In this subsection we use online parameter identification to estimate A, B and combine the estimator with the LQ control design we presented above to form an adaptive LQ controller [212].

Each row of (6.107) is given by

$$\dot{x}_i = a_{i1}x_1 + a_{i2}x_2 + \cdots + a_{i6}x_6 + b_{i1}\delta_T + b_{i2}\delta_E, \quad i = 1, 2, \ldots, 6, \tag{6.115}$$

where a_{ij}, b_{ij} are the elements of A and B matrices at the ith row and jth column. This leads to the static parametric models

$$z_i = \theta_i^{*T}\phi, \tag{6.116}$$

where $z_i = \dot{x}_I, \theta_i^* = [a_{i1}\ a_{i2}\ \cdots\ a_{i6}\ b_{i1}\ b_{i2}]^T, i = 1, 2, \ldots, 6$, and $\phi = [x_1\ x_2\ \cdots\ x_6\ \frac{\delta_T}{\delta_E}]^T$. In this application the first derivative of the state x can be measured or estimated indirectly. Therefore, in the parametric model we allow the use of the derivative of x. Using the parametric model (6.116), the estimate $\theta_i(t)$ of θ_i^* is given by

$$\dot{\theta}_i = \Pr[\Gamma\varepsilon_i\phi], \quad \varepsilon_i = \frac{z_i - \theta_i^T\phi}{50 + \phi^T\phi},$$

where $i = 1, 2, 3, \ldots, 6$ and $\Gamma = \Gamma^T$ is a positive definite matrix. The Pr operator is used to constrain the parameters to be within bounded sets defined based on a priori knowledge of the possible location of the parameters. The estimates $\theta_i(t)$ are used to form the estimated matrixes $\hat{A}$, $\hat{B}$. Using the CE approach the ALQC law is given as

$$\begin{aligned} u = -\hat{K}x_a, \quad \hat{K} = R^{-1}\hat{B}_{\text{aug}}^T\hat{P}, \\ \hat{A}_a^T\hat{P} + \hat{P}\hat{A}_a - \hat{P}\hat{B}_aR^{-1}\hat{B}_a^T\hat{P} + Q = 0, \end{aligned} \tag{6.117}$$

where the design parameters are chosen as

$$Q = \text{diag}\{4 \times 10^3\ 4 \times 10^3\ 10\ 10\ 10\ 10\ 10^4\ 10^4\}, \qquad R = \text{diag}\{5 \times 10^8\ 5 \times 10^7\}.$$

6.8.3 Simulations

The gain scheduling and adaptive LQ designs are implemented using the longitudinal dynamics of the F-16 fighter aircraft. The velocity and altitude range used for the gain scheduling design are V : 400 to 900 ft/sec; h : 0 to 40,000 ft.

Nominal case In this case the nonlinear model is free of uncertainties, and the linearized model is a good approximation of the nonlinear model at the operating point. Figure 6.2 shows the transient response of both the gain scheduling and adaptive LQ designs. It is clear that the two schemes demonstrate very similar performance.

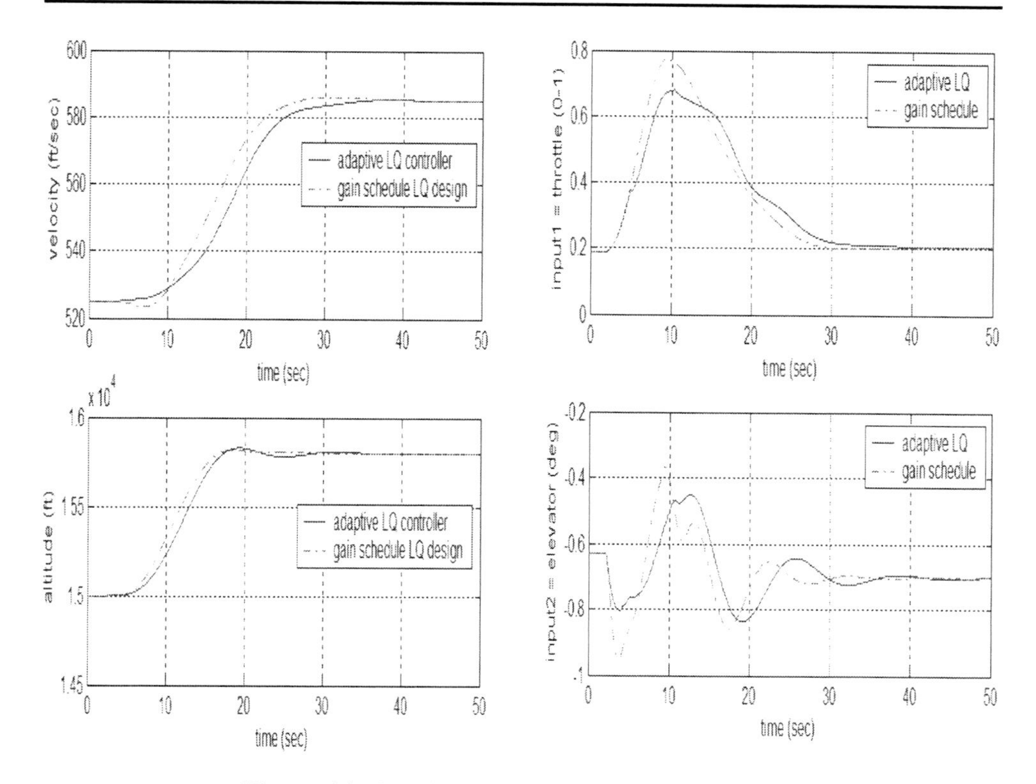

Figure 6.2. *Simulation results for the nominal case.*

Pitch Moment Coefficient Uncertainty Experimental data based on wind-tunnel testing are often used to generate lookup tables for all the aerodynamic coefficients. In these experiments, the accuracy of the measurements depends on the quality of the test and on the generated aerodynamic coefficient. Typically, the measurement of the pitching moment coefficient C_m is associated with some degree of uncertainty, as much as 70% and 80% on the lower and upper bounds, respectively. Let $C_{m,0}$ denote the lookup table value used to set the uncertainty bounds on the true value for C_m as

$$30\%C_{m,0} \leq C_m \leq 180\%C_{m,0}.$$

In this simulation the $C_m = C_{m,0}$ that is used to generate the linearized model (6.107) on which the control designs are based changes its value in the nonlinear model to $C_m = 30\%C_{m,0}$. Obviously this change, unknown to the designer, leads to a linearized model that is different from the one assumed during the control design. Figure 6.3 shows the simulation results for this scenario. It is clear that the gain scheduling scheme is affected by this uncertainty and its performance deteriorates, while the adaptive LQ design, which does not depend on the knowledge of the parameters, retains its performance.

Elevator Deflection Disturbance Since the system's dynamics do not behave exactly as predicted by the mathematical model, due to modeling errors and uncertainties, the steady-

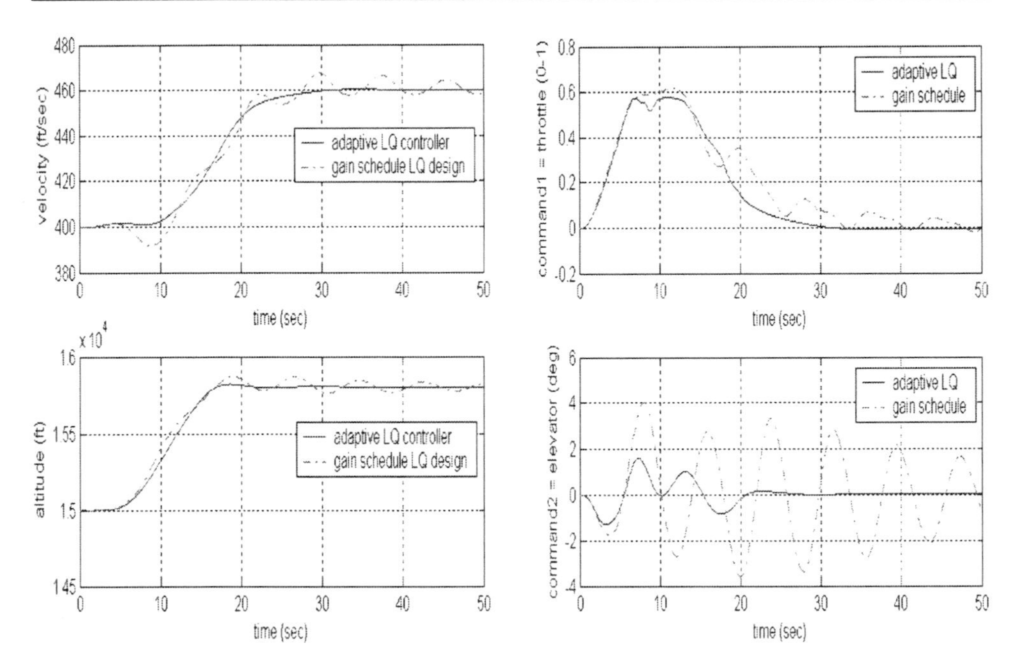

Figure 6.3. *Simulation results in the case of pitch moment coefficient uncertainty.*

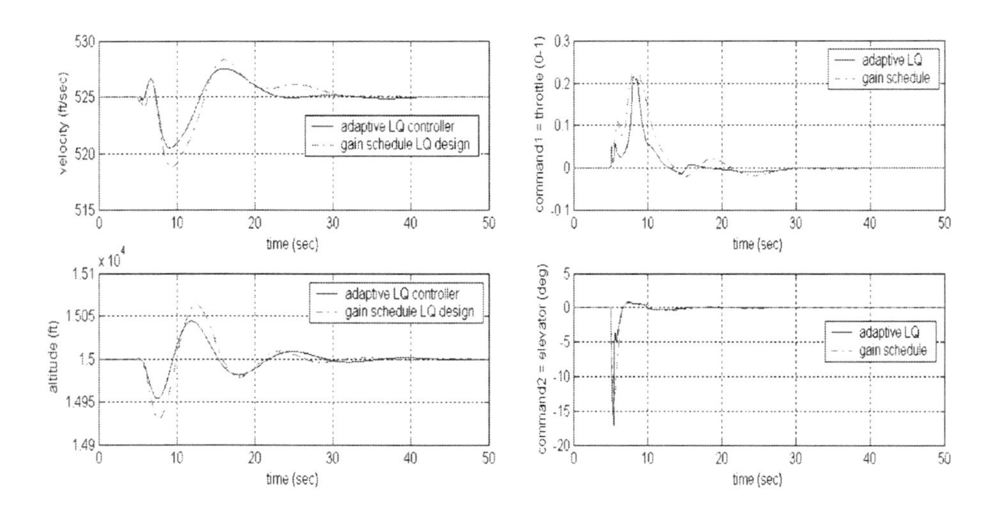

Figure 6.4. *Simulation for elevator deflection disturbance.*

state trim point is not exactly as calculated. In this simulation, the elevator trim value is disturbed by a large deflection angle of 15° (an impulse function at time = 5 seconds). The control objective is to return to the initial velocity and altitude. The simulation results are shown in Figure 6.4. It is clear that the adaptive LQ recovers faster and with smaller oscillations than the gain scheduling scheme.

The aircraft example demonstrates the effectiveness of the adaptive control scheme for accommodating large and unpredictable parametric uncertainties. It also demonstrates that any control scheme that meets the performance requirements in the case of known parameters can be combined with an online estimator to generate an adaptive control scheme for the case of unknown parameters. In the above example, the stabilizability issue which may arise in solving the Riccati equation is handled using an ad hoc technique. An ad hoc approach that will help reduce computations is to solve the Riccati equation at discrete points in time rather than continuously, or instead of the continuous-time algorithm develop a discrete-time adaptive LQ scheme of the type discussed in Chapter 7. Further information on recent applications of adaptive control to control high-performance aircraft can be found in [140–142].

See the web resource [94] for examples using the Adaptive Control Toolbox.

Problems

1. Consider the first-order plant

$$\dot{x} = ax + bu,$$
$$y = x.$$

The control objective is to choose u such that the closed-loop system is stable and $y(t)$ tracks the reference signal $y_m(t) = c$, where c is a finite constant.

   (a) Assume that a and b are known and $b < -1$. Design a controller using pole placement such that the closed-loop pole is located at -5.

   (b) Repeat (a) using LQ control. Determine the value of λ in the cost function so that the closed-loop system has a pole at -5.

   (c) Repeat (a) when a is known but b is unknown.

   (d) Repeat (a) when b is known but a is unknown.

2. Repeat Problem 1 for the reference signal $y_m(t) = c \sin \omega t$, where c and ω are positive finite constants. Simulate your controllers with $a = 5$, $b = -1$, $c = 1$, and $\omega = 3$.

3. Verify (6.23).

4. Consider the system

$$y = \frac{\omega_n^2}{s^2 + 2\varsigma\omega_n s + \omega_n^2} u,$$

where the parameter ω_n (the natural frequency) is known, but the damping ratio ς is unknown. The control objective is to choose u such that the closed-loop system is stable and $y(t)$ tracks the reference signal $y_m(t) = c$, where c is a finite constant. The performance specifications for the closed-loop system are given in terms of the unit step response as follows: (i) the peak overshoot is less than 5% and (ii) the settling time is less than 2 seconds.

(a) Design an estimation scheme to estimate ς when ω_n is known.

(b) Design an indirect APPC to meet the control objective.

5. Consider the plant

$$y = \frac{s+b}{s(s+a)}u.$$

(a) Design an adaptive law to generate $\hat{a}$ and $\hat{b}$, the estimates of a and b, respectively, online.

(b) Design an APPC scheme to stabilize the plant and regulate y to zero.

(c) Discuss the stabilizability condition that $\hat{a}$ and $\hat{b}$ have to satisfy at each time t.

(d) What additional assumptions do you need to impose on the parameters a and b so that the adaptive algorithm can be modified to guarantee the stabilizability condition? Use these assumptions to propose a modified APPC scheme.

6. Consider the speed control system given in Problem 3 in Chapter 5.

(a) Assume that a, b, and d are known. Design a controller to achieve the following performance specifications:

(i) No overshoot in the step response.

(ii) The time constant of the closed-loop system is less than 2 sec.

(iii) The steady-state error is zero for any constant d.

(b) Repeat (a) for unknown a, b, and d. Simulate the closed-loop system with $a = 0.02$, $b = 0.1$, and $d = 0.5$. Comment on your results.

7. Consider the plant

$$y = \frac{1}{s(s+p_1)(s+p_2)}u.$$

(a) Assuming that p_1 and p_2 are known, design a PPC law so that the closed-loop system is stable with a dominant pole at -1, and $y(t)$ tracks the reference signal $y_m(t) = c$, where c is a finite constant.

(b) Repeat (a) for unknown p_1 and p_2.

8. Consider the inequality (6.49), where $\varepsilon m_s \in \mathcal{L}_2$, $\delta > 0$, and c is a positive constant. Show that $m_f \in \mathcal{L}_\infty$.

9. Consider the plant

$$y = \frac{1}{(s+p_1)(s+p_2)}u$$

with unknown p_1 and p_2. The control objective is to choose u such that the closed-loop system is stable and $y(t)$ tracks the reference signal $y_m(t) = c$, where c is a finite constant.

(a) Obtain a state-space realization of the plant in the observer canonical form.

(b) Design a Luenberger observer to estimate the plant state in (a).

(c) Assuming that p_1 and p_2 are known, design a PPC law so that the closed-loop system is stable with a dominant pole at -5.

(d) Repeat (c) for unknown p_1 and p_2.

(e) Design an ALQC law.

10. Give the detailed stability proof for the APPC scheme of Example 6.4.2

11. Consider $u_p = \frac{Q_1(s)}{Q_m(s)}\bar{u}_p$, where $\bar{u}_p$ is exponentially decaying to zero, $\frac{Q_1(s)}{Q_m(s)}$ is biproper, and $Q_m(s)$ has no roots on the open right half-plane and nonrepeated roots on the $j\omega$-axis.

12. Complete the proof of Theorem 6.5.1.

13. Use (6.95) to establish the exponential convergence of the tracking error e_1 to the residual set (6.94).

14. Consider the plant

$$y_p = G_0(s)u_p + \Delta_a(s)u_p,$$

where $G_0(s)$ is the plant model and $\Delta_a(s)$ is an unknown additive perturbation. Consider the PPC law in (6.30), (6.31) designed based on the plant model with $\Delta_a(s) = 0$ but applied to the plant with $\Delta_a(s) \neq 0$. Obtain a bound for $\Delta_a(s)$ for robust stability. Repeat the analysis for the PPC law in (6.54), (6.55).

15. Consider the robust APPC scheme of Example 6.7.3 given by (6.106) designed for the plant model

$$y_p = \frac{b}{s+a}u_p$$

but applied to the following plants:

(i)

$$y_p = \frac{b}{s+a}(1+\Delta_m)u_p,$$

(ii)

$$y_p = \frac{b}{s+a}u_p + \Delta_a u_p,$$

where Δ_m and Δ_a are the multiplicative and additive perturbations, respectively.

(a) Obtain bounds and conditions for Δ_m and Δ_a for robust stability.

(b) Obtain bounds for the mean square values of the tracking errors.

(c) Simulate the APPC schemes for the plant (ii) when $a = -2\sin(0.02t)$, $b = 2(1+0.2\sin(0.01t))$, $\Delta_a = -\frac{\mu s}{(s+5)^2}$ for $\mu = 0, 0.1, 0.5, 1$. Comment on your results.

16. Use the Lyapunov-type function in (6.19) to verify (6.20) for the adaptive law choice given by (6.21).

17. Show that the adaptive law with projection given by (6.27) guarantees the properties of the corresponding adaptive law without projection.
18. Consider the time-varying system

$$\dot{x} = a(t)x + b(t)u,$$

where $a(t) = a_0 + a_1 e^{-\alpha t}$, $b(t) = b_0 + b_1 e^{-\alpha t}$, $b_0 > c_0 > 0$, $\alpha > 0$ are unknown, and c_0 is known. Design an APPC scheme based on the placement of the closed-loop pole at -5. Establish stability and convergence of the regulation error to zero. Demonstrate your result using simulations.
19. Repeat Problem 15 when
    (a) $a(t) = a_0 + a_1 \sin(\omega_1 t)$, $b(t) = b_0 + b_1 \sin(\omega_2 t)$, where the parameters are as defined in Problem 15 and ω_1, ω_2 are known.
    (b) Repeat (a) when ω_1, ω_2 are unknown but $|\dot{\omega}_1|, |\dot{\omega}_2| \leq \mu$, where $0 \leq \mu \ll 1$.
20. Consider the first-order plant

$$\dot{x} = ax + bu + d,$$
$$y = x,$$

where d is a bounded disturbance. The control objective is to choose u such that the closed-loop system is stable with a pole at -4 and $y(t)$ tracks the reference signal $y_m(t)$. Consider the following cases: (i) $y_m(t) = 2$, $d = 1.5$, (ii) $y_m(t) = 3\sin 2t$, $d = .1\sin 4t + d_0$, where d_0 is constant but unknown.
    (a) For each case design a PPC law when a, b are known.
    (b) Repeat (a) when a, b are unknown and the lower bound for b is 1.
    (c) Repeat (a) when b is known but a is unknown.
21. Consider the nonminimum-phase plant

$$y_p = \frac{b}{s+a}(1 + \Delta_m)(u_p + d),$$
$$\Delta_m(s) = -\frac{2\mu s}{1 + \mu s}, \quad 0 \leq \mu \ll 1,$$

where d is a bounded input disturbance.
    (a) Obtain a reduced-order model for the plant.
    (b) Design robust ALQC schemes based on projection, switching σ-modification, and dead zone.
    (c) Simulate your schemes in (b) for the following choices of parameters: $a = -2$, $b = 5$, $d = .1 + \sin .02t$ for the values of $\mu = 0, .01, 0.1, 0.5, 1$. Comment on your results.
22. Repeat Problem 18 for APPC schemes based on the polynomial approach.

23. Consider the following plant with delay,

$$y = \frac{e^{-\tau}b}{s-a}u, \quad 0 \le \tau \ll 1,$$

where a, b are unknown and b is positive and bounded from below by 1.5.

(a) Obtain a simplified model for the plant.

(b) Design an APPC scheme for the simple model in (a), and establish stability and convergence of the output to zero.

(c) Apply the APPC you obtained in (b) to the actual plant with delay.

(d) Demonstrate your results using simulations for $a = 2$, $b = 3.5$, and $\tau = 0, .01, 0.1, 0.5, 1$. Comment on your results.

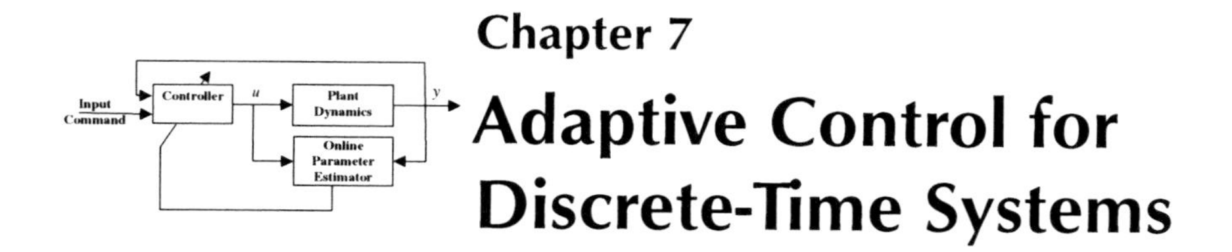

Chapter 7
Adaptive Control for Discrete-Time Systems

7.1 Introduction

The design and analysis of adaptive control schemes for discrete-time systems follows a procedure very similar to that used in continuous-time systems. The certainty equivalence (CE) approach is used to design an adaptive controller by combining a control law used in the known parameter case with an online parameter estimator or adaptive law. The way this combination is carried out, the type of controller structure used, and the kind of parameter estimator employed lead to a wide class of adaptive control schemes often with different stability and performance properties.

As in the continuous-time case, we can divide adaptive controllers into direct and indirect MRAC and APPC. While MRAC can be viewed as a special case of APPC, its restrictive structure and control objective allow us to design both stable direct and indirect MRAC. In the APPC case this is not possible in general due to the difficulty in obtaining a parametric model of the plant as a linear (or special bilinear) function of the desired controller parameters of the forms described in Chapter 2. Consequently we can design only indirect APPC schemes, which suffer from the same problem of stabilizability of the estimated plant as their counterparts in the continuous-time case. A special class of MRAC which received considerable attention in the literature is the adaptive one-step-ahead control [46, 88, 113] where the poles of the reference model and the filters are set equal to zero. This is possible in discrete time as poles at zero correspond to fastest stable response.

7.2 MRAC

We start with a simple example and then present the general case.

7.2.1 Scalar Example

Consider the scalar plant

$$x(k+1) = ax(k) + u(k), \quad x(0) = x_0, \tag{7.1}$$

where a is an unknown constant. The control objective is to determine a bounded sequence $u(k) = f(k, x)$ such that the state $x(k)$ is bounded and tracks the state $x_m(k)$ of the reference model

$$x_m(k+1) = a_m x_m(k) + r(k), \tag{7.2}$$

where $|a_m| < 1$ and $r(k)$ is any given bounded sequence referred to as the *reference signal*.

If the plant parameter a were known, we could use the control law

$$u(k) = K^* x(k) + r(k), \quad K^* = -a + a_m, \tag{7.3}$$

to obtain the closed-loop plant

$$x(k+1) = a_m x(k) + r(k). \tag{7.4}$$

Defining the tracking error $e(k) \triangleq x(k) - x_m(k)$, we have

$$e(k+1) = a_m e(k), \quad e(0) = x_0 - x_m(0),$$

which due to $|a_m| < 1$ implies that $e(k)$ converges to zero exponentially fast; i.e., $x(k)$ tracks $x_m(k)$ for any given reference input sequence $r(k)$.

Since a is unknown, instead of (7.3) we use the control law

$$u(k) = K(k)x(k) + r(k), \tag{7.5}$$

where $K(k)$ is the estimate of K^* at instant k to be generated by an adaptive law. The adaptive law for generating the sequence $K(k)$ is obtained by first getting a parametric model for K^* and then using the techniques of Chapter 4 to construct the adaptive law for generating $K(k)$. Substituting for $a = a_m - K^*$ in the plant equation (7.1), we obtain

$$x(k+1) = a_m x(k) - K^* x(k) + u(k), \tag{7.6}$$

which can be expressed in the form of the SPM by rewriting (7.6) as

$$x(k) = \frac{1}{z - a_m} u(k) - K^* \phi(k), \tag{7.7}$$

where

$$\phi(k) = \frac{1}{z - a_m}[x(k)] \tag{7.8}$$

or

$$\phi(k+1) = a_m \phi(k) + x(k). \tag{7.9}$$

Equation (7.8) is expressed in the form of the SPM

$$z(k) = K^* \phi(k) \tag{7.10}$$

by defining

$$z(k) = \frac{1}{z - a_m} u(k) - x(k). \tag{7.11}$$

Using (7.10), we can apply any adaptive law of the class developed in Chapter 4 to estimate K^* online. As an example let us choose the gradient algorithm

$$\varepsilon(k) = \frac{z(k) - K(k)\phi(k)}{m_s^2(k)}, \qquad m_s^2(k) = 1 + \phi^2(k), \tag{7.12}$$

$$K(k+1) = K(k) + \gamma\varepsilon(k)\phi(k), \tag{7.13}$$

where $0 < \gamma < 2$ is the adaptive gain. The adaptive law (7.12), (7.13) is in the form of (4.1), (4.2), and Lemma 4.1.1 guarantees the following properties:

(i) $K(k) \in \ell_\infty$.

(ii) $\varepsilon(k), \varepsilon(k)m_s(k), \varepsilon(k)\phi(k), |K(k+N) - K(k)| \in \ell_2 \cap \ell_\infty$ and converge to zero as $k \to \infty$, independent of the boundedness of $x(k)$, $u(k)$, $\phi(k)$, where N is any finite integer.

We now use the properties (i), (ii) of the adaptive law together with the control law (7.5) to first establish boundedness for all signals in the closed loop and show that $x(k) \to x_m(k)$ as $k \to \infty$. For simplicity let us take $a_m = 0$. This means that the pole of the reference model is placed at $z = 0$. Such a simplification is without loss of generality since (7.2) may be rewritten as

$$x_m(k+1) = \bar{r}(k), \tag{7.14}$$

where $\bar{r}(k) = \frac{1}{z-a_m}r(k)$ is considered as the new reference signal for the reference model (7.14) with pole at $z = 0$.

Stability Analysis The closed-loop plant (7.6) with $a_m = 0$ is given by

$$x(k+1) = (K(k) - K^*)x(k) + r(k), \tag{7.15}$$

and the tracking error equation (7.7) becomes

$$e(k+1) = (K(k) - K^*)x(k). \tag{7.16}$$

From (7.9) with $a_m = 0$ and (7.13), we have

$$m_s(k+1) = \sqrt{1 + \phi^2(k+1)} = \sqrt{1 + x^2(k)} \le 1 + |x(k)|. \tag{7.17}$$

From (7.13) we have

$$\begin{aligned} \varepsilon(k+1)m_s^2(k+1) &= z(k+1) - K(k+1)\phi(k+1) \\ &= -(K(k+1) - K^*)x(k) \\ &= -(K(k) - K^*)x(k) - (K(k+1) - K(k))x(k). \end{aligned} \tag{7.18}$$

From (7.18) we obtain

$$(K(k) - K^*)x(k) = -\varepsilon(k+1)m_s^2(k+1) - (K(k+1) - K(k))x(k),$$

which we substitute into (7.15) to obtain

$$x(k+1) - r(k) = -\varepsilon(k+1)m_s^2(k+1) - (K(k+1) - K(k))x(k). \tag{7.19}$$

From (7.17), (7.19) we can write

$$|x(k+1) - r(k)| \le |\varepsilon(k+1)m_s(k+1)|(1 + |x(k)|) + |K(k+1) - K(k)||x(k)|$$

or

$$|x(k+1) - r(k)| \le \gamma(k)|x(k)| + \Delta(k), \tag{7.20}$$

where $\gamma(k) \triangleq |\varepsilon(k+1)m_s(k+1)| + |K(k+1) - K(k)|$ and $\Delta(k) \triangleq |\varepsilon(k+1)m_s(k+1)|$. From the properties of the adaptive law we have $\gamma(k), \Delta(k) \in \ell_2 \cap \ell_\infty \to 0$ as $k \to \infty$. Applying Lemma A.12.32 (with $A(k) = 0, \mu = 0$) to (7.20) and using the boundedness of r_k, it follows that $x(k) \in \ell_\infty$. This in turn implies that $m_s(k), \phi(k), u(k) \in \ell_\infty$. Since $x(k) \in \ell_\infty$ and $|\varepsilon(k+1)m_s(k+1)|, |K(k+1) - K(k)| \to 0$ as $k \to \infty$, it follows from (7.20) that

$$\lim_{k\to\infty} (|x(k+1)| - r(k)) = 0. \tag{7.21}$$

Since $e(k+1) = x(k+1) - x_m(k+1) = x(k+1) - r(k)$, it follows from (7.21) that $e(k) \to 0$ as $k \to \infty$.

We have established that the adaptive control algorithm given by (7.5), (7.12), (7.13) guarantees signal boundedness and forces the plant output $x(k)$ to track the output $x_m(k)$ of the reference model. The question that remains open is whether $K(k)$ will converge to K^*. By Lemma 4.1.1, parameter convergence can be guaranteed if $\phi(k)$ is persistently exciting (PE). The PE property of $\phi(k)$ can be guaranteed in this case if $r(k)$ is a nonzero constant. Other choices for $r(k)$ are nonzero sinusoid signals, etc.

7.2.2 General Case: MRC

Consider the following plant model in the transfer function form

$$y_p = G_p(z)u_p, \quad G_p(z) = k_p \frac{Z_p(z)}{R_p(z)}, \tag{7.22}$$

where k_p is a constant; $Z_p(z)$, $R_p(z)$ are monic polynomials; and u_p and y_p are the scalar input and output of the plant, respectively. The control objective is to choose the input u_p so that the closed-loop system is stable in the sense that all signals are bounded and the plant output y_p tracks the output y_m of the reference model

$$y_m = W_m(z)r, \quad W_m(z) = k_m \frac{Z_m(z)}{R_m(z)}, \tag{7.23}$$

for any bounded reference input r, where $Z_m(z)$, $R_m(z)$ are monic polynomials and k_m is a constant. We refer to the control objective as the *model reference control* (*MRC*) *objective*. In order to meet the control objective we use the following assumptions:

- *Plant assumptions*:

  P1. $Z_p(z)$ is a monic Hurwitz polynomial.

  P2. An upper bound n of the degree n_p of $R_p(z)$ is known.

  P3. The relative degree $n^* = n_p - m_p$ of $G_p(z)$ is known, where m_p is the degree of $Z_p(z)$.

  P4. The sign of the high-frequency gain k_p is known.

- *Reference model assumptions*:

  M1. $Z_m(z)$, $R_m(z)$ are monic Hurwitz polynomials of degree q_m, p_m, respectively, where $p_m \leq n$.

  M2. The relative degree $n_m^* = p_m - q_m$ of $W_m(z)$ is the same as that of $G_p(z)$, i.e., $n_m^* = n^*$.

Remark 7.2.1 Assumption P1 requires that the plant transfer function $G_p(z)$ be minimum phase. We make no assumptions, however, about the location of poles of $G_p(z)$; i.e., the plant is allowed to have unstable poles. The minimum-phase assumption is a consequence of the control objective which requires the plant output to track the output of the reference model for *any* bounded reference input signal r. The only way to achieve such a strict objective is to make the closed-loop plant transfer function from r to y_p equal to that of the reference model. Such a plant-model transfer function matching requires changing not only the poles of the plant but also its zeros. The plant poles can be placed using feedback and pole placement techniques. The zeros, however, cannot be placed in a similar way as the poles. The only way to change the zeros of the plant is to cancel them by introducing poles at the same location, and replace them with those of the reference model. For stability, such cancellations are possible only when the plant zeros are stable. Unstable zero-pole cancellations may lead to unbounded signals driven by nonzero initial conditions. Furthermore, in practice (or even simulations) such cancellations cannot be exact, leading to unstable closed-loop poles.

The control law has the same form as that in the continuous-time case and is given by

$$u_p = \theta_1^{*T}\frac{\alpha(z)}{\Lambda(z)}u_p + \theta_2^{*T}\frac{\alpha(z)}{\Lambda(z)}y_p + \theta_3^* y_p + c_0^* r, \tag{7.24}$$

where

$$\begin{aligned} \alpha(z) &= \begin{cases} [z^{n-2}, z^{n-3}, \ldots, z, 1]^T & \text{if } n \geq 2, \\ 0 & \text{if } n = 1, \end{cases} \\ \Lambda(z) &= \Lambda_0(z) Z_m(z), \end{aligned} \tag{7.25}$$

$\Lambda_0(z)$ is a monic Hurwitz polynomial of degree $n_0 \triangleq n - 1 - q_m$, and $c_0^*, \theta_3^* \in \mathcal{R}$; $\theta_1^*, \theta_2^* \in \mathcal{R}^{n-1}$ are the controller parameters to be designed.

As in the continuous-time case, using the properties of polynomials, it can be shown that the control law (7.24) guarantees that the closed-loop plant transfer function from r

to y_p is equal to $W_m(z)$, the transfer function of the reference model, by choosing c_0^*, θ_i^*, $i = 1, 2, 3$, to satisfy the equalities

$$\begin{aligned} c_0^* &= \frac{k_m}{k_p}, \\ \theta_1^{*T}\alpha(z) &= \Lambda(z) - Z_p(z)Q(z), \\ \theta_2^{*T}\alpha(z) + \theta_3^*\Lambda(z) &= \frac{Q(z)R_p(z) - \Lambda_0(z)R_m(z)}{k_p}, \end{aligned} \tag{7.26}$$

where $Q(z)$ of degree $n - 1 - m_p$ is the quotient and $k_p\Delta^*(z)$ of degree at most $n_p - 1$ is the remainder in the division

$$\frac{\Lambda_0(z)R_m(z)}{R_p(z)} = Q(z) + k_p\frac{\Delta^*(z)}{R_p(z)}.$$

We can therefore establish, as in the continuous-time case, that the control law (7.24) with controller parameters satisfying the matching conditions (7.26) guarantees stability and convergence of the tracking error $e_1(k) \triangleq y_p(k) - y_m(k)$ to zero exponentially fast for any bounded reference input sequence $r(k)$ (see Problem 1).

If the plant parameters are unknown, then the controller parameters cannot be calculated using (7.26). In this case, instead of (7.24) we use

$$u_p(k) = \theta_1^T(k)\omega_1(k) + \theta_2^T(k)\omega_2(k) + \theta_3(k)y_p(k) + c_0(k)r(k), \tag{7.27}$$

where

$$\omega_1 = \frac{\alpha(z)}{\Lambda(z)}u_p, \qquad \omega_2 = \frac{\alpha(z)}{\Lambda(z)}y_p,$$

and

$$\theta(k) \triangleq [\theta_1^T(k), \theta_2^T(k), \theta_3(k), c_0(k)]^T$$

is the online estimate of $\theta^* \triangleq [\theta_1^{*^T}, \theta_2^{*^T}, \theta_3^*, c_0^*]^T$ to be generated online. The control law (7.27) can be implemented as

$$\begin{aligned} \omega_1(k+1) &= F\omega_1(k) + gu_p(k), \\ \omega_2(k+1) &= F\omega_2(k) + gy_p(k), \\ u_p(k) &= \theta^T(k)\omega(k), \end{aligned} \tag{7.28}$$

where

$$\omega(k) = [\omega_1^T(k), \omega_2^T(k), y_p(k), r(k)]^T, \qquad \theta(k) = [\theta_1^T(k), \theta_2^T(k), \theta_3(k), c_0(k)]^T,$$
$$(zI - F)^{-1}g = \frac{\alpha(z)}{\Lambda(z)}.$$

As in the continuous-time case, $\theta(k)$ can be generated directly using an adaptive law based on a parametric model for θ^*, leading to a direct MRAC scheme. Alternatively the plant parameters, i.e., the coefficients of $R_p(z)$, $k_pZ_p(z)$, can be estimated online and used together with (7.26) to calculate $\theta(k)$ at each instant k, leading to an indirect MRAC scheme. In the following sections we discuss each scheme separately.

7.2.3 Direct MRAC

The direct MRAC scheme consists of the control law (7.28) and an adaptive law to generate $\theta(k)$ at each instant k. The adaptive law for $\theta(k)$ is generated by first expressing the unknown controller parameter vector θ^* in the form of the parametric models discussed in Chapter 2 and then using Chapter 4 to choose the appropriate parameter estimator.

The parametric models for θ^* can be developed the same way as in continuous time and are given by the B-SPM

$$e_1(k) = \rho^*(\bar{u}_p(k) - \theta^{*^T}\phi(k)), \tag{7.29}$$

where

$$\begin{aligned} e_1(k) &= y_p(k) - y_m(k), \\ \rho^* &= \frac{1}{c_0^*}, \qquad \bar{u}_p = W_m(z)u_p, \qquad \phi = W_m(z)\omega, \\ \omega &= [\omega_1^T, \omega_2^T, y_p, r]^T = \left[\frac{\alpha^T(z)}{\Lambda(z)}u_p, \frac{\alpha^T(z)}{\Lambda(z)}y_p, y_p, r\right]^T, \end{aligned}$$

or the SPM

$$z(k) = \theta^{*^T}\phi_p(k), \tag{7.30}$$

where

$$\begin{aligned} z &= \bar{u}_p = W_m(z)u_p, \\ \phi_p &= \left[\frac{W_m(z)\alpha^T(z)}{\Lambda(z)}u_p, \frac{W_m(z)\alpha^T(z)}{\Lambda(z)}y_p, W_m(z)y_p, y_p\right]^T. \end{aligned}$$

In both cases $\theta^* \triangleq [\theta_1^{*^T}, \theta_2^{*^T}, \theta_3^*, c_0^*]^T$. Using (7.29) or (7.30), a wide class of adaptive laws may be selected to generate $\theta(k)$ to be used with (7.27) or (7.28) to form direct MRAC schemes. As an example, consider the direct MRAC scheme

$$u_p(k) = \theta^T(k)\omega(k), \tag{7.31}$$

$$\begin{aligned} \varepsilon(k) &= \frac{z(k) - \theta^T(k)\phi_p(k)}{m_s^2(k)}, \qquad m_s^2(k) = 1 + \phi_p^T(k)\phi_p(k), \\ \bar{\theta}(k+1) &= \bar{\theta}(k) + \gamma\varepsilon(k)\bar{\phi}_p(k), \\ \bar{c}_0(k+1) &= c_0(k) + \gamma\varepsilon(k)y_p(k), \\ c_0(k+1) &= \begin{cases} \bar{c}_0(k+1) & \text{if } \bar{c}_0(k+1)\,\mathrm{sgn}(c_0^*) \geq \beta_0, \\ \beta_0\,\mathrm{sgn}(c_0^*) & \text{otherwise,} \end{cases} \\ \text{where} \quad \bar{\omega}(k) &= [\omega_1^T(k), \omega_2^T(k), y_p(k)]^T, \qquad \bar{\phi}_p = W_m(z)\bar{\omega}, \\ \bar{\theta}(k) &= [\theta_1^T(k), \theta_2^T(k), \theta_3(k)]^T, \end{aligned} \tag{7.32}$$

where $\beta_0 > 0$ is a lower bound for $c_0^* = \frac{k_m}{k_p}$ which can be calculated using the knowledge of an upper bound for k_p. The adaptive gain satisfies $0 < \gamma < 2$. The modification in (7.32) prevents the estimate $c_0(k)$ of c_0^* from taking values close to zero. This condition is used in the analysis of the adaptive controller in order to establish boundedness of signals in the closed-loop system.

Theorem 7.2.2 (ideal case). *The direct MRAC scheme* (7.31), (7.32) *has the following properties:*

(i) *All signals are bounded.*

(ii) *The tracking error $e_1(k) = y_p(k) - y_m(k)$ converges to zero as $k \to \infty$.*[14]

As in the continuous-time case, the adaptive control law (7.31), (7.32) cannot guarantee stability if it is applied to the plant (7.22) in the presence of a small perturbation $\eta(k)$, i.e., to the plant

$$y_p(k) = G_p(z)[u_p(k)] + \eta(k). \tag{7.33}$$

In general, $\eta(k)$ has the form

$$\eta(k) = \mu\Delta_1(z)[u_p(k)] + \mu\Delta_2(z)[y_p(k)] + d(k), \tag{7.34}$$

where $d(k)$ is a bounded disturbance, μ is a small scalar, $\Delta_1(z)$, $\Delta_2(z)$ are unknown proper transfer functions with poles q_j, say, that satisfy

$$|q_j| < \sqrt{p_0}, \tag{7.35}$$

where p_0 is a constant that satisfies

$$0 < p_0 < \delta_0 < 1 \tag{7.36}$$

with δ_0 known. The orders of $\Delta_1(z)$, $\Delta_2(z)$ are also unknown. Note that (7.35) is equivalent to $\Delta_1(z\sqrt{p_0})$, $\Delta_2(z\sqrt{p_0})$ having poles inside the unit circle. For the class of plant perturbations described by (7.33), (7.34), the adaptive control law (7.31), (7.32) can be made robust by replacing the parameter estimator with a robust one from Chapter 4. As an example let us choose the gradient-based robust adaptive law with switching σ-modification. We obtain the following direct MRAC scheme:

$$u_p(k) = \theta^T(k)\omega(k), \tag{7.37}$$

$$\begin{aligned}
\varepsilon(k) &= \frac{z(k) - \theta^T(k)\phi_p(k)}{m_s^2(k)}, \\
m_s^2(k) &= 1 + \phi_p^T(k)\phi_p(k) + n_d(k), \\
\bar{\theta}(k+1) &= (1 - \sigma_s(k))\bar{\theta}(k) + \gamma\varepsilon(k)\bar{\phi}_p(k), \\
\bar{c}_0(k+1) &= (1 - \sigma_s(k))c_0(k) + \gamma\varepsilon(k)y_p(k), \\
c_0(k+1) &= \begin{cases} \bar{c}_0(k+1) & \text{if } \bar{c}_0(k+1)\,\mathrm{sgn}(c_0^*) \geq \beta_0, \\ \beta_0\,\mathrm{sgn}(c_0^*) & \text{otherwise,} \end{cases}
\end{aligned} \tag{7.38}$$

[14]The proof of Theorem 7.2.2 is presented in the web resource [94]. It can also be found in [91, 113, 143].

$$n_d(k+1) = \delta_0 n_d(k) + |u_p(k)|^2 + |y_p(k)|^2, \quad n_d(0) = 0,$$
$$\sigma_s(k) = \begin{cases} 0 & \text{if } |\theta(k)| \le M_0, \\ \sigma_0 & \text{if } |\theta(k)| > M_0, \end{cases} \tag{7.39}$$

where M_0, δ_0, γ, σ_0 are positive design constants which, according to Theorem 4.11.3, should satisfy the following conditions:

$$0 < \sigma_0 < 2, \qquad 0 < \gamma < \frac{2-\sigma_0}{2+\sigma_0}, \qquad M_0 \ge \frac{4(1+\gamma)|\theta^*|}{2-\sigma_0-\gamma(2+\sigma_0)}.$$

Given the properties of the dynamic normalization (7.39), we can establish as in Chapter 4 that

$$\frac{\eta(k)}{m_s(k)} \le c\mu + \frac{d_0}{m_s(k)}, \tag{7.40}$$

where $c > 0$ denotes a generic constant and d_0 is an upper bound for $|d(k)|$. We use this inequality in the analysis of the robust adaptive controller and establish the following theorem.

Theorem 7.2.3 (robust stability). *Consider the direct MRAC scheme (7.37)–(7.39) designed for the simplified plant (7.33) with $\eta(k) = 0$ but applied to the actual plant (7.33), (7.34) with $\eta(k) \neq 0$: For any $\mu \in [0, \mu^*]$ and some $\mu^* > 0$ all the signals in the closed-loop plant are bounded. Furthermore, the tracking error converges to the residual set*

$$\mathcal{D} = \left\{ e_1(k) \mid \lim_{\bar{N}\to\infty} \sup_{0<N\le\bar{N}} \frac{1}{N}\sum_{k=0}^{N-1} |e_1(k)| \le c\left(\mu + \sqrt{\varepsilon_0} + d_0\right) \right\},$$

where $\varepsilon_0 > 0$ is an arbitrarily small number, d_0 is an upper bound for $|d(k)|$, and $c > 0$ denotes a generic constant.[15]

The robustness results are very similar to those in continuous time. If the plant perturbations satisfy certain bounds, i.e., they are small relative to the dominant part of the plant, and if the external disturbance is bounded, then the robust adaptive controller guarantees boundedness and convergence of the tracking error in the mean square sense (m.s.s.) to a bound that is of the order of the modeling error. As indicated in the continuous-time case this m.s.s. bound does not guarantee that the tracking error is of the order of the modeling error at each time t. A phenomenon known as "bursting" [105], where at steady state the tracking error suddenly begins to oscillate with large amplitude over short intervals of time, is possible and is not excluded by Theorem 7.2.3. This bursting phenomenon can be eliminated if we include a dead zone in the adaptive law or if we choose the reference input to be dominantly rich.

[15]The proof of Theorem 7.2.3 is presented in the web resource [94]. Variations of the proof may be found in [41, 57, 91, 112, 113, 144].

7.2.4 Indirect MRAC

The indirect MRAC is described as

$$u_p(k) = \theta^T(k)\omega(k), \tag{7.41}$$

where $\theta(k) = [\theta_1^T(k), \theta_2^T(k), \theta_3(k), c_0(k)]^T$ is generated by solving the equations

$$\begin{aligned} c_0(k) &= \frac{k_m}{\hat{k}_p(k)}, \\ \theta_1^T(k)\alpha(z) &= \Lambda(z) - \hat{Z}_p(z,k)\cdot Q(z), \\ \theta_2^T(k)\alpha(z) + \theta_3(k)\Lambda(z) &= \frac{1}{\hat{k}_p(k)}(Q(z)\cdot \hat{R}_p(z,k) - \Lambda_0(z)R_m(z)), \end{aligned} \tag{7.42}$$

where $Q(z)$ is the quotient of the division

$$\frac{\Lambda_0(az)R_m(z)}{\hat{R}_p(z,k)}$$

and $A(z,k)\cdot B(z,k)$ denotes the frozen time product of the two operators $A(z,k)$, $B(z,k)$; i.e., in the multiplication $A(z,k)$, $B(z,k)$ are treated as polynomials with constant coefficients.

The polynomials $\hat{R}_p(z,k)$, $\hat{Z}_p(z,k)$ are generated online as follows.

The plant

$$y_p = \frac{\bar{Z}_p(z)}{R_p(z)}u_p = \frac{b_m z^m + b_{m-1}z^{m-1} + \cdots + b_1 z + b_0}{z^n + a_{n-1}z^{n-1} + \cdots + a_1 z + a_0}u_p,$$

where $\bar{Z}_p(z) = k_p Z_p(z)$ and $b_m = k_p$, is expressed in the form of the SPM

$$z(k) = \theta_p^{*^T}\phi(k), \tag{7.43}$$

where

$$\begin{aligned} \theta_p^* &= [b_m, \ldots, b_0, a_{n-1}, \ldots, a_0]^T, \\ z &= \frac{z^n}{\Lambda_p(z)}y_p, \\ \phi &= \left[\frac{\alpha_m^T(z)}{\Lambda_p(z)}u_p, -\frac{\alpha_{n-1}^T(z)}{\Lambda_p(z)}y_p\right]^T. \end{aligned}$$

$\Lambda_p(z)$ is a monic Hurwitz polynomial of degree n, and $\alpha_{i-1}(z) = [z^{i-1}, z^{i-2}, \ldots, z, 1]^T$.

We can use (7.43) to generate

$$\theta_p(k) = [\hat{b}_m(k), \ldots, \hat{b}_0(k), \hat{a}_{n-1}(k), \ldots, \hat{a}_0(k)]^T$$

and form

$$
\begin{aligned}
\hat{\bar{Z}}_p(z,k) &= \hat{b}_m(k)z^m + \hat{b}_{m-1}(k)z^{m-1} + \cdots + \hat{b}_1(k)z + \hat{b}_0(k),\\
\hat{R}_p(z,k) &= z^n + \hat{a}_{n-1}(k)z^{n-1} + \cdots + \hat{a}_1(k)z + \hat{a}_0(k),\\
\hat{k}_p(k) &= \hat{b}_m(k),\\
\hat{Z}_p(z,k) &= \frac{1}{\hat{k}_p(k)}\hat{\bar{Z}}_p(z,k)
\end{aligned}
\tag{7.44}
$$

by using any of the parameter estimators of Chapter 4. As an example we use the gradient algorithm

$$
\begin{aligned}
\varepsilon(k) &= \frac{z(k) - \theta_p^T(k)\phi(k)}{m_s^2(k)},\\
m_s^2(k) &= 1 + \phi^T(k)\phi(k),\\
\theta_p(k+1) &= \theta_p(k) + \gamma\varepsilon(k)\phi(k),
\end{aligned}
\tag{7.45}
$$

where $0 < \gamma < 2$. Since in (7.42) we have division by $\hat{k}_p(k)$, the adaptive law for $\hat{k}_p(k)$ is modified using the knowledge of the sign of k_p and a lower bound $|k_p| \geq k_0 > 0$ in order to guarantee that $\hat{k}_p(k)$ does not pass through zero. The modification involves only the first element $\theta_{p1}(k) = \hat{b}_m(k) = \hat{k}_p(k)$ of $\theta_p(k)$ as follows:

$$
\begin{aligned}
\bar{k}_p(k+1) &= \hat{k}_p(k) + \gamma\varepsilon(k)\phi_1(k),\\
\hat{k}_p(k+1) &= \begin{cases} \bar{k}_p(k+1) & \text{if } \bar{k}_p(k+1)\,\mathrm{sgn}(k_p) \geq k_0,\\ \hat{k}_p(k) & \text{otherwise,} \end{cases}\\
\hat{k}_p(0)\,\mathrm{sgn}(k_p) &> k_0.
\end{aligned}
\tag{7.46}
$$

The other elements of $\theta_p(k)$ are updated as in (7.45).

Theorem 7.2.4 (ideal case). *The indirect MRAC scheme described by* (7.41), (7.42), (7.44), (7.45), (7.46) *guarantees that*

(i) *All signals are bounded.*

(ii) *The tracking error $e_1(k) = y_p(k) - y_m(k)$ converges to zero as $k \to \infty$.*[16]

As in the case of the direct MRAC, the indirect MRAC can be modified for robustness using the robust adaptive law

$$
\begin{aligned}
\theta_p(k+1) &= (1 - \sigma_s(k))\theta_p(k) + \gamma\varepsilon(k)\phi(k),\\
\sigma_s(k) &= \begin{cases} 0 & \text{if } |\theta_p(k)| \leq M_0,\\ \sigma_0 & \text{if } |\theta_p(k)| > M_0 \end{cases}
\end{aligned}
\tag{7.47}
$$

[16]The proof of Theorem 7.2.4 is presented in the web resource [94].

and the normalization signal

$$\begin{aligned} m_s^2(k) &= 1 + \phi^T(k)\phi(k) + n_d(k), \\ n_d(k+1) &= \delta_0 n_d(k) + |u_p(k)|^2 + |y_p(k)|^2, \quad n_d(0) = 0, \end{aligned} \tag{7.48}$$

where M_0, δ_0, γ, σ_0 are positive design constants and satisfy the same conditions as in the case of the direct MRAC in (7.39).

Theorem 7.2.5 (robust stability). *The indirect MRAC scheme described by* (7.41), (7.47), (7.46), (7.48) *has the following properties when applied to the plant* (7.33), (7.34) *with modeling error: For any $\mu \in [0, \mu^*]$ and some $\mu^* > 0$ all the signals in the closed-loop plant are bounded. Furthermore, the tracking error $e_1(k) = y_p(k) - y_m(k)$ converges to the residual set*

$$\mathcal{D} = \left\{ e_1(k) \mid \lim_{\bar{N}\to\infty} \sup_{0<N\le\bar{N}} \frac{1}{N} \sum_{k=0}^{N-1} |e_1(k)| \le c(\mu + \sqrt{\varepsilon_0} + d_0) \right\},$$

where $\varepsilon_0 > 0$ is an arbitrarily small number, d_0 is an upper bound for the disturbance $|d(k)|$, and $c > 0$ denotes a generic constant.[17]

The same comments as in the case of the direct MRAC apply here, too.

7.3 Adaptive Prediction and Control

Let us consider the third-order plant

$$y_p = \frac{b_1 z + b_0}{z^3 + a_2 z^2 + a_1 z + a_0} u_p \tag{7.49}$$

whose relative degree is $n^* = 2$. It can also be represented as

$$y_p(k+3) = -a_2 y_p(k+2) - a_1 y_p(k+1) - a_0 y_p(k) + b_1 u_p(k+1) + b_0 u_p(k). \tag{7.50}$$

We can express (7.50) in an alternative predictor form that will allow us to predict the value of $y_p(k+n^*)$, $n^* = 2$, at instant k using only the values of y_p, u_p at instant k and before.

We write

$$y_p(k+2) = -a_2 y_p(k+1) - a_1 y_p(k) - a_0 y_p(k-1) + b_1 u_p(k) + b_0 u_p(k-1). \tag{7.51}$$

Similarly

$$y_p(k+1) = -a_2 y_p(k) - a_1 y_p(k-1) - a_0 y_p(k-2) + b_1 u_p(k-1) + b_0 u_p(k-2). \tag{7.52}$$

If we now substitute for $y_p(k+1)$ from (7.52) into (7.51), we obtain

$$\begin{aligned} y_p(k+2) = {} & (a_2^2 - a_1) y_p(k) + (a_1 a_2 - a_0) y_p(k-1) + a_0 a_2 y_p(k-2) \\ & + b_1 u_p(k) + (b_0 - a_2 b_1) u_p(k-1) - a_2 b_0 u_p(k-2), \end{aligned} \tag{7.53}$$

[17]The proof of Theorem 7.2.5 is given in the web resource [94].

which is in the predictor form, since we can calculate $y_p(k+2)$ at instant k from the measurements of y_p, u_p at instants $k, k-1, k-2$.

The predictor form (7.53) can be generated for any LTI plant by using Lemma 7.3.1 (below) as follows.

We consider the plant (7.22), i.e.,

$$y_p = \frac{\bar{Z}_p(z)}{R_p(z)} u_p, \tag{7.54}$$

where $\bar{Z}_p(z) = b_m z^m + b_{m-1} z^{m-1} + \cdots + b_1 z + b_0$, $R_p(z) = z^n + a_{n-1} z^{n-1} + \cdots + a_1 z + a_0$, $n > m$. Let $q^{-1} = z^{-1}$ be the shift operator and define

$$\begin{aligned} A(q^{-1}) &\triangleq z^{-n} R_p(z)|_{z^{-1}=q^{-1}} = 1 + a_{n-1} q^{-1} + \cdots + a_1 q^{-n+1} + a_0 q^{-n}, \\ B(q^{-1}) &\triangleq z^{-n} \bar{Z}_p(z)|_{z^{-1}=q^{-1}} = q^{-n} q^m (b_m + b_{m-1} q^{-1} + \cdots + b_0 q^{-m}) = q^{-n^*} \bar{B}(q^{-1}), \end{aligned} \tag{7.55}$$

where $\bar{B}(q^{-1}) = b_m + b_{m-1} q^{-1} + \cdots + b_0 q^{-m}$ and $n^* = n - m$ is the relative degree of the plant. Using (7.55), the plant can be represented in the *autoregressive moving average (ARMA) form*

$$A(q^{-1}) y_p(k) = q^{-n^*} \bar{B}(q^{-1}) u_p(k). \tag{7.56}$$

Lemma 7.3.1. *The plant* (7.56) *can be expressed in the predictor form*

$$y_p(k+n^*) = \alpha(q^{-1}) y_p(k) + \beta(q^{-1}) u_p(k), \tag{7.57}$$

where

$$\begin{aligned} \beta(q^{-1}) &= b_m + \beta_1 q^{-1} + \cdots + \beta_{n-1} q^{-n+1} = f(q^{-1}) \bar{B}(q^{-1}), \\ \alpha(q^{-1}) &= \alpha_0 + \alpha_1 q^{-1} + \cdots + \alpha_{n-1} q^{-n+1}, \\ f(q^{-1}) &= 1 + f_1 q^{-1} + \cdots + f_{n^*-1} q^{-n^*+1} \end{aligned}$$

are the unique polynomials that satisfy the equation

$$f(q^{-1}) A(q^{-1}) + q^{-n^*} \alpha(q^{-1}) = 1, \tag{7.58}$$

whose coefficients are computed as follows:

$$\begin{aligned} f_0 &= 1, \\ f_i &= -\sum_{j=0}^{i-1} f_j a_{n-i+j}, \qquad i = 1, \ldots, n^* - 1, \\ \alpha_i &= -\sum_{j=0}^{n^*-1} f_j a_{n-i-n^*+j}, \quad i = 1, \ldots, n-1, \end{aligned} \tag{7.59}$$

where $a_j = 0$ *for* $j < 0$.

Proof. Multiplying (7.56) by $f(q^{-1})$, we have

$$f(q^{-1})A(q^{-1})y_p(k) = q^{-n^*} f(q^{-1})\bar{B}(q^{-1})u_p(k). \tag{7.60}$$

Substituting $f(q^{-1})A(q^{-1}) = 1 - q^{-n^*}\alpha(q^{-1})$ from (7.58) into (7.60), we obtain

$$[1 - q^{-n^*}\alpha(q^{-1})]y_p(k) = q^{-n^*} f(q^{-1})\bar{B}(q^{-1})u_p(k).$$

Multiplying each side by q^{n^*}, we have

$$y_p(k+n^*) = \alpha(q^{-1})y_p(k) + \beta(q^{-1})u_p(k),$$

where $\beta(q^{-1}) = f(q^{-1})\bar{B}(q^{-1})$. The coefficients α_i in (7.59) are obtained by equating the coefficients of q^{-i}, $i = 0, \dots, n-1$, in (7.58). □

It is clear from (7.57) that a parametric model can be constructed to estimate the coefficients of $\alpha(q^{-1})$, $\beta(q^{-1})$, which are related to the coefficients of $\bar{Z}_p(z)$, $R_p(z)$ in (7.54) via (7.59). From (7.57) we have

$$y_p(k+n^*) = \theta^{*^T}\phi_p(k), \tag{7.61}$$

where

$$\phi_p(k) = [y_p(k), y_p(k-1), \dots, y_p(k-n+1), u_p(k), u_p(k-1), \dots, u_p(k-n+1)]^T,$$
$$\theta^* = [\alpha_0, \alpha_1, \dots, \alpha_{n-1}, b_m, \beta_1, \dots, \beta_{n-1}]^T,$$

and α_i, $i = 0, \dots, n-1$, and β_j, $j = 1, \dots, n-1$, are defined in Lemma 7.3.1.

Since at instant k, $y_p(k+n^*)$ is not available for measurement, we express (7.61) as

$$y_p(k) = \theta^{*^T}\phi_p(k-n^*), \tag{7.62}$$

which is in the form of the SPM and can be used to estimate θ^* online using the parameter estimators presented in Chapter 4.

Let us now consider the plant (7.54) in the predictor form (7.57) with the control objective of designing $u_p(k)$ so that all the signals are bounded and $y_p(k)$ tracks the reference sequence $y_m(k)$. It follows from (7.57) that the control law

$$\beta(q^{-1})u_p(k) = y_m(k+n^*) - \alpha(q^{-1})y_p(k), \tag{7.63}$$

which can be implemented as

$$u_p(k) = \frac{1}{b_m}(y_m(k+n^*) - \alpha(q^{-1})y_p(k) - \bar{\beta}(q^{-1})u_p(k-1)), \tag{7.64}$$

where

$$\bar{\beta}(q^{-1}) \stackrel{\Delta}{=} q(\beta(q^{-1}) - b_m)$$

leads to the closed-loop plant

$$y_p(k+n^*) = y_m(k+n^*),$$

which implies that in n^* steps y_p is equal to y_m, i.e., $y_p(k) = y_m(k)\ \forall k \geq n^*$. The stability properties of the control law (7.63), (7.64) are presented in the following theorem.

Theorem 7.3.2. *Consider the plant* (7.54) *where* $\bar{Z}_p(z)$ *is a Hurwitz polynomial, i.e., the plant is minimum phase. The control law* (7.63) *guarantees that all signals in the closed loop are bounded and* $y_p(k) = y_m(k)\ \forall k \geq n^*$.

Proof. Substituting (7.63) into (7.57), we obtain

$$y_p(k+n^*) = y_m(k+n^*) \quad \forall k \geq 0,$$

which implies that $y_p \in \ell_\infty$ and $y_p(k) = y_m(k)\ \forall k \geq n^*$. We now need to establish that $u_p \in \ell_\infty$. From (7.54) we have

$$u_p = \frac{R_p(z)}{\bar{Z}_p(z)} y_p.$$

Since $y_p \in \ell_\infty$ and $\bar{Z}_p(z)$ is Hurwitz, it follows that $u_p(k)$ is a bounded sequence. Therefore, all signals in the closed-loop system are bounded. $\square$

The control law (7.63) is referred to as the *one-step-ahead controller* [46]. It can be shown (see Problem 1) that the same control law can be developed by choosing u_p to minimize the square of the prediction tracking error

$$J(k) = \frac{(y_p(k+n^*) - y_m(k+n^*))^2}{2}. \tag{7.65}$$

The one-step-ahead controller may result in large values of u_p in an effort to force y_p to be equal to y_m in n^* steps. Instead of the cost (7.65) we can use

$$J_\lambda(k) = \frac{(y_p(k+n^*) - y_m(k+n^*))^2}{2} + \frac{\lambda}{2} u_p^2(k), \tag{7.66}$$

where $\lambda > 0$ is a design constant. This modified cost also penalizes u_p. From (7.57) we have that

$$y_p(k+n^*) = \alpha(q^{-1}) y_p(k) + b_m u_p(k) + \bar{\beta}(q^{-1}) u_p(k-1), \tag{7.67}$$

where $\bar{\beta}(q^{-1}) = q(\beta(q^{-1}) - b_m)$ and b_m is the leading coefficient of $B(q^{-1})$, which is also the leading coefficient of $\bar{Z}_p(z)$ in (7.54). Substituting (7.67) into (7.66), we can show (see Problem 3) that the value of $u_p(k)$ that makes $\frac{dJ_p(k)}{du_p(k)} = 0$ is given by

$$u_p(k) = \frac{b_m}{b_m^2 + \lambda}(y_m(k+n^*) - \alpha(q^{-1}) y_p(k) - \bar{\beta}(q^{-1}) u_p(k-1)). \tag{7.68}$$

The control law (7.68) is known as the *weighted one-step-ahead controller*, and its stability properties are given by the following theorem.

Theorem 7.3.3. *If the roots of* $\bar{B}(z^{-1}) + \frac{\lambda}{b_m} A(z^{-1})$ *lie inside the unit circle* ($|z| < 1$), *then all signals in the closed-loop system* (7.54), (7.68) *are bounded, and the tracking error*

$e_1(k) = y_p(k) - y_m(k)$ *satisfies*

$$e_1(k+n^*) = \frac{-\frac{\lambda}{b_m}A(z^{-1})}{\bar{B}(z^{-1}) + \frac{\lambda}{b_m}A(z^{-1})} y_m(k+n^*).$$

Proof. The control law (7.68) may be expressed as

$$b_m(\alpha(q^{-1})y_p(k) + \beta(q^{-1})u_p(k) - y_m(k+n^*)) + \lambda u_p(k) = 0.$$

Using the plant representation (7.57), we have

$$b_m(y_p(k+n^*) - y_m(k+n^*)) + \lambda u_p(k) = 0. \tag{7.69}$$

Multiplying (7.69) by $A(q^{-1})$, we obtain

$$b_m\left(A(q^{-1})y_p(k+n^*) - A(q^{-1})y_m(k+n^*) + \frac{\lambda}{b_m}A(q^{-1})u_p(k)\right) = 0.$$

From (7.56) we have that

$$A(q^{-1})y_p(k+n^*) = \bar{B}(q^{-1})u_p(k). \tag{7.70}$$

Therefore,

$$b_m\left(\bar{B}(q^{-1})u_p(k) - A(q^{-1})y_m(k+n^*) + \frac{\lambda}{b_m}A(q^{-1})u_p(k)\right) = 0$$

or

$$\left[\bar{B}(q^{-1}) + \frac{\lambda}{b_m}A(q^{-1})\right]u_p(k) = A(q^{-1})y_m(k+n^*). \tag{7.71}$$

Since $y_m(k)$ is a bounded sequence, the boundedness of $u_p(k)$ follows from the assumption that the roots of $\bar{B}(z^{-1}) + \frac{\lambda}{b_m}A(z^{-1})$ are within the unit circle. Similarly, multiplying (7.69) by $\bar{B}(q^{-1})$, we get

$$b_m(\bar{B}(q^{-1})y_p(k+n^*) - \bar{B}(q^{-1})y_m(k+n^*)) + \lambda\bar{B}(q^{-1})u_p(k) = 0.$$

Using (7.70), we obtain

$$\left[\bar{B}(q^{-1}) + \frac{\lambda}{b_m}A(q^{-1})\right]y_p(k+n^*) = \bar{B}(q^{-1})y_m(k+n^*), \tag{7.72}$$

which can be used to conclude that $y_p(k)$ is bounded as in the case of $u_p(k)$.

From (7.72) we have

$$y_p(k+n^*) = \frac{\bar{B}(z^{-1})}{\bar{B}(z^{-1}) + \frac{\lambda}{b_m}A(z^{-1})} y_m(k+n^*),$$

which implies that the tracking error $e_1(k) = y_p(k) - y_m(k)$ satisfies (show it!)

$$e_1(k+n^*) = \frac{-\frac{\lambda}{b_m}A(z^{-1})}{\bar{B}(z^{-1}) + \frac{\lambda}{b_m}A(z^{-1})} y_m(k+n^*).$$

In this case the tracking error may not converge to zero in general, which is one of the drawbacks of the weighted one-step-ahead controller. On the other hand, the condition $\bar{B}(z^{-1})$ being Hurwitz in the case of the one-step-ahead controller is relaxed to $\bar{B}(z^{-1}) + \frac{\lambda}{b_m}A(z^{-1})$ being Hurwitz. Since λ is a design constant in some cases (see Problem 4), λ could be chosen to guarantee that $\bar{B}(z^{-1}) + \frac{\lambda}{b_m}A(z^{-1})$ is Hurwitz even when $\bar{B}(z^{-1})$ is not. □

The one-step-ahead controller (7.64) is a special case of the model reference control law (7.24) with $W_m(z) = z^{-n^*}$ and $\Lambda(z) = z^{n-1}$ (see Problem 5).

7.3.1 Adaptive One-Step-Ahead Control

In the case of unknown plant parameters the control law (7.64) is replaced with

$$u_p(k) = \frac{1}{\hat{b}_m(k)}(y_m(k+n^*) - \hat{\alpha}(q^{-1},k)y_p(k) - \hat{\bar{\beta}}(q^{-1},k)u_p(k-1)), \tag{7.73}$$

where

$$\begin{aligned} \hat{\alpha}(q^{-1},k) &= \hat{\alpha}_0(k) + \hat{\alpha}_1(k)q^{-1} + \cdots + \hat{\alpha}_{n-1}(k)q^{-n+1}, \\ \hat{\bar{\beta}}(q^{-1},k) &= \hat{\beta}_1(k) + \hat{\beta}_2(k)q^{-1} + \cdots + \hat{\beta}_{n-1}(k)q^{-n+2}, \end{aligned} \tag{7.74}$$

and

$$\theta(k) = [\hat{\alpha}_0(k), \hat{\alpha}_1(k), \ldots, \hat{\alpha}_{n-1}(k), \hat{b}_m(k), \hat{\beta}_1(k), \ldots, \hat{\beta}_{n-1}(k)]^T$$

is the estimate of θ^* in the parametric model (7.62), i.e.,

$$y_p(k) = \theta^{*^T}\phi_p(k-n^*), \tag{7.75}$$

where

$$\phi_p(k) = [y_p(k), y_p(k-1), \ldots, y_p(k-n+1), u_p(k), u_p(k-1), \ldots, u_p(k-n+1)]^T.$$

Since we have division by $\hat{b}_m(k)$, the adaptive law for $\hat{b}_m(k)$ has to be modified using the knowledge of the $\text{sgn}(b_m)$ and the lower bound b_0 of $|b_m| \geq b_0 > 0$.

A wide class of parameter estimators presented in Chapter 4 can be used to generate $\theta(k)$ based on the parametric model (7.75). As an example, consider the gradient algorithm

$$\begin{aligned}
\varepsilon(k) &= \frac{y_p(k) - \theta^T(k)\phi_p(k-n^*)}{m_s^2(k-n^*)},\\
\bar{\theta}(k+1) &= \bar{\theta}(k) + \gamma\varepsilon(k)\bar{\phi}_p(k-n^*),\\
\bar{b}_m(k+1) &= \hat{b}_m(k) + \gamma\varepsilon(k)\phi_{(n-m)}(k),\\
\hat{b}_m(k+1) &= \begin{cases} \bar{b}_m(k+1) & \text{if } \bar{b}_m(k+1)\,\mathrm{sgn}(b_m) \geq b_0,\\ \hat{b}_m(k) & \text{otherwise,}\end{cases}\\
m_s^2(k) &= 1 + \phi^T(k)\phi(k),\\
\bar{\theta}(k) &= [\hat{\alpha}_0, \hat{\alpha}_1(k), \dots, \hat{\alpha}_{n-1}(k), \hat{\beta}_1(k), \dots, \hat{\beta}_{n-1}(k)]^T,\\
\bar{\phi}_p(k) &= [y_p(k), y_p(k-1), \dots, y_p(k-n+1), u_p(k-1), \dots, u_p(k-n+1)]^T.
\end{aligned} \tag{7.76}$$

The modification for $\hat{b}_m(k)$ guarantees that $|\hat{b}_m(k)| \geq b_0\ \forall k$, which in turn guarantees that no division by zero takes place in (7.73).

The adaptive control law (7.73), (7.76) is referred to as the *direct one-step-ahead adaptive controller* because the controller parameters are estimated directly without any intermediate calculations involving the plant parameters in the transfer function (7.54).

An *indirect one-step-ahead adaptive controller* can be designed as follows: The plant parameters, i.e., the coefficients of $\bar{Z}_p(z)$, $R_p(z)$ in (7.54), are estimated online using the techniques of Chapter 4 and used to form the estimated polynomials $\hat{A}(q^{-1}, k)$, $\hat{\bar{B}}(q^{-1}, k)$ in (7.55). The controller parameters can then be calculated using Lemma 7.3.1, which relates the coefficients of $A(q^{-1})$ and $\bar{B}(q^{-1})$ with the coefficients of $\alpha(q^{-1})$ and $\beta(q^{-1})$.

Since the one-step-ahead controller is a special case of the MRC, the rest of the details are left as an exercise for the reader.

7.4 APPC

Consider the discrete-time LTI plant

$$y_p = \frac{Z_p(z)}{R_p(z)} u_p, \tag{7.77}$$

where

$$\begin{aligned}
Z_p(z) &= b_m z^m + b_{m-1} z^{m-1} + \cdots + b_1 z + b_0,\\
R_p(z) &= z^n + a_{n-1} z^{n-1} + \cdots + a_1 z + a_0,
\end{aligned}$$

and $m \leq n-1$. The control objective is to choose the input $u_p(k)$ so that all signals are bounded and $y_p(k)$ tracks the reference sequence $y_m(k)$ which satisfies

$$Q_m(z) y_m = 0, \tag{7.78}$$

where $Q_m(z)$ is a monic polynomial of degree q with all roots in $|z| \leq 1$ and with nonrepeated roots on $|z| = 1$.

$Q_m(z)$ is often referred to as the *internal model* of $y_m(k)$. Examples of $y_m(k)$, $Q_m(z)$ that satisfy (7.78) are

- $y_m(k) = 0 \Rightarrow Q_m(z) = 1$,
- $y_m(k) = c \neq 0 \Rightarrow Q_m(z) = z - 1$,
- $y_m(k) = c_1 \sin \omega k \Rightarrow Q_m(z) = z^2 - (2\cos\omega)z + 1$,
- $y_m(k) = c + c_1 \sin \omega k \Rightarrow Q_m(z) = (z-1)(z^2 - (2\cos\omega)z + 1)$.

As in the continuous-time case we make the following assumptions:

A1. The order n of $R_p(z)$ is known.

A2. $Q_m(z)R_p(z)$ and $Z_p(z)$ are coprime.

We present below an example of a control law which can be combined with an adaptive law that generates estimates for the plant parameters, i.e., the coefficients of $Z_p(z)$, $R_p(z)$, to form an indirect APPC.

Consider the pole placement objective of placing the closed-loop poles at the roots of the desired polynomial $A^*(z)$. As in the continuous-time case, $A^*(z)$ could be chosen based on the performance requirements of the closed-loop plant. The control law is given as

$$Q_m(z)L(z)[u_p(k)] = -P(z)[y_p(k) - y_m(k)], \tag{7.79}$$

where $L(z)$, $P(z)$ satisfy

$$L(z)Q_m(z)R_p(z) + P(z)Z_p(z) = A^*(z). \tag{7.80}$$

The polynomials $L(z)$, $P(z)$ are of degree $n-1$, $n+q-1$, respectively, and $L(z)$ is monic. The control law (7.79) can be implemented as

$$u_p(k) = \frac{\Lambda(z) - Q_m(z)L(z)}{\Lambda(z)}[u_p(k)] - \frac{P(z)}{\Lambda(z)}[y_p(k) - y_m(k)], \tag{7.81}$$

where $\Lambda(z)$ is a monic Hurwitzpolynomial of degree $n + q - 1$.

The control law (7.80), (7.81) relies on the knowledge of $Z_p(z)$ and $R_p(z)$. In the case of unknown coefficients of $Z_p(z)$ and $R_p(z)$ an adaptive law is used to generate estimates of the coefficients of $Z_p(z)$, $R_p(z)$ and form the estimated polynomials $\hat{Z}_p(z,k)$, $\hat{R}_p(z,k)$, which replace $Z_p(z)$, $R_p(z)$ in (7.80). The design of the APPC law to be used in the unknown parameter case is developed as follows.

The plant parametric model (7.43) may be used to generate online estimates for the polynomial coefficient vector $\theta_p^* = [b_m, \ldots, b_0, a_{n-1}, \ldots, a_0]^T$. Using (7.43) and the gradient algorithm from Chapter 4, we have

$$\begin{aligned} \varepsilon(k) &= \frac{z(k) - \theta_p^T(k)\phi(k)}{m_s^2(k)}, \\ m_s^2(k) &= 1 + \phi^T(k)\phi(k), \\ \theta_p(k+1) &= \theta_p(k) + \gamma\varepsilon(k)\phi(k), \end{aligned} \tag{7.82}$$

where $0 < \gamma < 2$,

$$\theta_p(k) = [\hat{b}_m(k), \ldots, \hat{b}_0(k), \hat{a}_{n-1}(k), \ldots, \hat{a}_0(k)]^T,$$
$$z = \frac{z^n}{\Lambda_p(z)} y_p,$$
$$\phi = \left[\frac{\alpha_m^T(z)}{\Lambda_p(z)} u_p, -\frac{\alpha_{n-1}^T(z)}{\Lambda_p(z)} y_p \right]^T,$$

and $\Lambda_p(z)$ is a monic Hurwitzpolynomial of degree n. For simplicity, one could take $\Lambda_p(z) = z^n$, in which case $\phi(k)$ becomes the vector of previous values of $u_p(k)$, $y_p(k)$.

Using $\theta_p(k)$, we form

$$\begin{aligned}\hat{Z}_p(z,k) &= \hat{b}_m(k)z^m + \hat{b}_{m-1}(k)z^{m-1} + \cdots + \hat{b}_1(k)z + \hat{b}_0(k),\\ \hat{R}_p(z,k) &= z^n + \hat{a}_{n-1}(k)z^{n-1} + \cdots + \hat{a}_1(k)z + \hat{a}_0(k)\end{aligned} \tag{7.83}$$

and solve for $\hat{L}(z,k)$, $\hat{P}(z,k)$ the polynomial equation

$$\hat{L}(z,k) \cdot Q_m(z) \cdot \hat{R}_p(z,k) + \hat{P}(z,k) \cdot \hat{Z}_p(z,k) = A^*(z), \tag{7.84}$$

where $A(z,k) \cdot B(z,k)$ denotes the product of two polynomials with z treated as a dummy variable rather than the shift operator, i.e., $z \cdot \alpha(k) = \alpha(k)z$.

The adaptive control law is now given by

$$u_p(k) = \{\Lambda(z) - \hat{L}(z,k) \cdot Q_m(z)\} \frac{1}{\Lambda(z)} [u_p(k)] - \hat{P}(z,k) \frac{1}{\Lambda(z)} [y_p(k) - y_m(k)], \tag{7.85}$$

which together with (7.82)–(7.84) forms the APPC scheme. One problem that arises in this case is the solvability of the pole placement equation (7.84). The solution for $\hat{L}(z,k)$, $\hat{P}(z,k)$ exists and it is unique, provided that $Q_m(z) \cdot \hat{R}_p(z,k)$ and $\hat{Z}_p(z,k)$ are coprime. Such an assumption cannot be guaranteed by the adaptive law (7.82) leading to the so-called stabilizability problem, which we also encountered in APPC for continuous-time plants.

Another problem is the robustness of the APPC scheme in the presence of modeling errors. As in the continuous-time case the robustness of the APPC scheme can be established by replacing the adaptive law with the robust adaptive law

$$\begin{aligned}\varepsilon(k) &= \frac{z(k) - \theta_p^T(k)\phi(k)}{m_s^2(k)},\\ m_s^2(k) &= 1 + \phi^T(k)\phi(k) + n_d(k),\\ \theta_p(k+1) &= (1 - \sigma_s(k))\theta_p(k) + \gamma\varepsilon(k)\phi(k),\\ n_d(k+1) &= \delta_0 n_d(k) + |u_p(k)|^2 + |y_p(k)|^2, \quad n_d(0) = 0,\\ \sigma_s(k) &= \begin{cases} 0 & \text{if } |\theta_p(k)| \le M_0, \\ \sigma_0 & \text{if } |\theta_p(k)| > M_0, \end{cases}\end{aligned} \tag{7.86}$$

where $1 > \delta_0 > 0$ is chosen according to the stability margins of the unmodeled dynamics, $M_0 \ge 2|\theta^*|$, and $0 < \gamma < 2 - 2\delta_0$.

Theorem 7.4.1. *Consider the robust APPC scheme (7.83)–(7.86) applied to the plant*

$$\begin{aligned} y_p(k) &= G_p(z)[u_p(k)] + \eta(k), \\ \eta(k) &= \mu\Delta_1(z)[u_p(k)] + \mu\Delta_2(z)[y_p(k)] + d(k), \end{aligned} \tag{7.87}$$

where $\Delta_1(z)$, $\Delta_2(z)$ are proper with the property that $\Delta_1(\rho_0 z)$, $\Delta_2(\rho_0 z)$ have stable poles for some $0 < \rho_0 < \delta_0 < 1$, and $d(k)$ is a bounded disturbance. Assume that the polynomials $Q_m(z) \cdot \hat{R}_p(z,k)$ and $\hat{Z}_p(z,k)$ are strongly coprime at each time k. Then all the closed-loop signals are bounded. Furthermore, the following is guaranteed:

(i) *The tracking error $e_1(k) = y_p(k) - y_m(k)$ converges to the residual set*

$$\mathcal{D} = \left\{ e_1(k) | \lim_{\bar{N}\to\infty} \sup_{0<N\le\bar{N}} \frac{1}{N} \sum_{k=0}^{N-1} |e_1(k)| \le c(\mu + \sqrt{\varepsilon_0} + d_0) \right\},$$

where $\varepsilon_0 > 0$ is an arbitrarily small number, d_0 is an upper bound for the disturbance $|d(k)|$, and $c > 0$ denotes a generic constant.

(ii) *In the absence of modeling error effects, i.e., $\eta(k) = 0\ \forall k$, the robust APPC (7.83)–(7.86) or the APPC (7.82)–(7.85) guarantees signal boundedness and convergence of the tracking error $e_1(k)$ to zero.*[18]

See the web resource [94] for examples using the Adaptive Control Toolbox.

Problems

1. Show that the MRC law (7.24) with matching equations (7.26) guarantees closed-loop stability and exponential convergence of the tracking error to zero.
2. Show that the one-step-ahead control law (7.63) minimizes the cost function J defined in (7.65).
3. Show that the weighted one-step-ahead control law (7.68) minimizes the cost function J_p defined in (7.66).
4. Consider the polynomials

$$\begin{aligned} A(z^{-1}) &= 1 + a_{n-1}z^{-1} + \cdots + a_1 z^{-n+1} + a_0 z^{-n}, \\ \bar{B}(z^{-1}) &= b_m + b_{m-1}z^{-1} + \cdots + b_0 z^{-m}, \end{aligned}$$

   where $n > m \ge 1$ and $\bar{B}(z^{-1})$ is not Hurwitz. Is it always possible/impossible to find a positive constant λ so that $\bar{B}(z^{-1}) + \frac{\lambda}{b_m}A(z^{-1})$ is Hurwitzif (i) $A(z^{-1})$ is Hurwitz, (ii) $A(z^{-1})$ is not Hurwitz? (*Hint*: First, assume that $n = 2$, $m = 1$.)
5. Show that the one-step-ahead controller (7.64) is a special case of the model reference control law (7.24) with $W_m(z) = z^{-n^*}$ and $\Lambda(z) = z^{n-1}$.

[18]The proof of Theorem 7.4.1 is presented in the web resource [94].

6. Design an indirect one-step-ahead adaptive controller for the plant given by (7.54).

7. Consider the scalar plant

$$x(k+1) = ax(k) + bu(k), \quad x(0) = x_0,$$

where a and b are unknown constants with the knowledge that $b \geq b_0 > 0$. Design and analyze an MRAC scheme that can stabilize the plant and force x to follow the output x_m of the reference model

$$x_m(k+1) = a_m x_m(k) + r(k),$$

where $|a_m| < 1$ and $r(k)$ is any given bounded reference signal.

8. Consider the plant and the reference model given in Problem 7 with $b_0 = 0.2$ and $a_m = 0.5$. Design two different MRAC schemes and simulate them with $r(k) = 1 + \sin(3k)$. Use $a = 2$ and $b = 1$ in your simulations.

9. Consider the second-order plant

$$y_p = \frac{b_1 z + b_0}{z^2 + a_1 z + a_0} u_p,$$

where a_0, a_1, b_0, and b_1 are unknown constants. The reference model is given by

$$y_m = \frac{1}{z + 0.1} r.$$

The control objective is to guarantee closed-loop stability and force y_p to track y_m.

   (a) What assumptions will you make in order to achieve the control objective?

   (b) Assume that a_0, a_1, b_0, and b_1 are known. Design an MRC law to achieve the control objective.

   (c) Design a direct MRAC law when a_0, a_1, b_0, and b_1 are unknown.

10. Consider the plant and control objective of Problem 7. Design a control law to meet the control objective when the parameters are known. Use analysis to establish signal boundedness and convergence of the tracking error to zero. Present all the details of the analysis.

11. Consider the third-order plant

$$y_p = \frac{b_1 z + b_0}{z^3 + a_2 z^2 + a_1 z + a_0} u_p,$$

where a_0, a_1, a_2, b_0, and b_1 are unknown constants. The reference model is given by

$$y_m = \frac{1}{(z + 0.1)^2} r.$$

The control objective is to guarantee closed-loop stability and force y_p to track y_m.

   (a) Design an indirect MRAC law using the a priori knowledge that $b_1 \geq 2$ and $-1 \leq b_0 \leq 1$.

(b) Repeat (a) with the knowledge that $b_1 = 5$ and $b_0 = 1$.

(c) Simulate and compare your MRAC laws designed in (a) and (b) with (i) $r(k) = 1$ and (ii) $r(k) = 3\sin(2k)$. In your simulations, use $a_0 = 0$, $a_1 = 2$, $a_2 = 3$, $b_0 = 1$, and $b_1 = 5$.

12. Consider the second-order plant

$$y_p = \frac{b_1 s + b_0}{s^2 + a_1 s + a_0} u_p,$$

where a_0, a_1, b_0, and b_1 are unknown constants. The reference model is given by

$$y_m = \frac{1}{s+1} r.$$

The control objective is to choose up to guarantee closed-loop stability and force y_p to track y_m.

(a) Discretize the plant and the reference model with the sampling time T_s.

(b) Design an indirect MRAC law using the discretized plant and reference model. What assumptions are required so that the control objective can be achieved?

(c) Simulate your MRAC law for $T_s = 0.01$ sec, 0.1 sec, and 1 sec with (i) $r(t) = 1$ and (ii) $r(t) = 2 + 5\sin(3t)$. In your simulations, use $a_0 = -5$, $a_1 = -2$, $b_0 = 1$, and $b_1 = 2$.

13. Consider the plant and reference model in Problem 12.

(a) Design an indirect MRAC law in the continuous-time domain. What assumptions will you make in order to meet the control objective?

(b) Discretize the indirect MRAC law designed in (a) with a sampling time T_s, and compare it with the MRAC law designed in Problem 12(b).

(c) Simulate your MRAC law for $T_s = 0.01$ sec, 0.1 sec, and 1 sec with (i) $r(t) = 1$ and (ii) $r(t) = 2 + 5\sin(3t)$. In your simulations, use $a_0 = -5$, $a_1 = -2$, $b_0 = 1$, and $b_1 = 2$. Compare the results with those obtained in Problem 12(c).

14. Consider the plant

$$y_p = G_0(z)(1 + \Delta_m(z))u_p,$$

where $G_0(z) = \frac{b}{z+a}$ is the plant model with unknown constants a and $b \geq 0.1$, and $\Delta_m(z)$ is the multiplicative perturbation with known stability margin $\delta_0 = 0.5$. The reference model is given as

$$y_m = \frac{1}{z + 0.1} r.$$

(a) Design a robust direct MRAC law.

(b) Design a robust indirect MRAC law.

(c) Simulate your designs for the plant with $a = -2$, $b = 1.5$, and $\Delta_m(z) = \frac{\mu z}{z+\mu}$ for $\mu = 0, 0.1, 0.4$. Comment on your results.

15. Consider the plant

$$y_p = \frac{b_1 z + b_0}{z^2 + a_1 z + a_0} u_p,$$

where a_0, a_1, b_0, and $b_1 \geq b_0 > 0$ are unknown constants.

   (a) Suppose all the parameters are known. Design a one-step-ahead controller. What assumptions will you use to establish that y_p can perfectly track the reference signal in one step?

   (b) Repeat (a) when the parameters are unknown.

   (c) Simulate your design in (b) for the plant using (i) $r(k) = 1$ and (ii) $r(k) = 1 + 3\sin(2k)$. You can assume that $a_0 = 3$, $a_1 = 1$, $b_0 = 1$, and $b_1 = 2$ in your simulations.

16. Consider the plant

$$y_p = \frac{b_2 z^2 + b_1 z + b_0}{z^3 + a_2 z^2 + a_1 z + a_0} u_p,$$

where a_0, a_1, a_2, b_0, b_1, and b_2 are unknown constants.

   (a) Suppose all the parameters are known. Design a weighted one-step-ahead controller with the weighting factor λ. What assumptions will you make so that the designed controller guarantees closed-loop stability?

   (b) Repeat (a) when the parameters are unknown.

   (c) Simulate your design in (b) for the plant using (i) $r(k) = 3$ and (ii) $r(k) = 3\cos(5k)$. You can assume that $a_0 = 6$, $a_1 = 11$, $a_2 = 5$, $b_0 = 0.02$, $b_1 = 0.3$, and $b_2 = 1$ in your simulations. Choose different values for λ and comment on your results.

17. Consider the plant

$$y_p = k_m \frac{(z - \rho)}{z^2 + a_1 z + a_0} u_p,$$

where a_0, a_1, $k_m > 1$, and ρ are unknown constants.

   (a) Suppose all the parameters are known and $\rho > 1$. Design a PPC law so that the closed-loop system is stable with one pole at 0.1 (dominant pole) and the rest at 0, and y_p can track any constant reference signal.

   (b) Repeat (a) for the case where $\rho = 1$.

   (c) Suppose $\rho \neq 1$ is known. Design an APPC law so that the closed-loop system is stable with one pole at 0.1 (dominant pole) and the rest at 0, and y_p can track any constant reference signal.

   (d) Simulate your design in (c) for the plant when $a_0 = 3$, $a_1 = 2$, $k_m = 2$, $\rho = 1.2$.

18. Consider the LTI plant

$$y_p = \frac{b_3 z^3 + b_2 z^2 + b_1 z + b_0}{z^4 + a_3 z^3 + a_2 z^2 + a_1 z + a_0} u_p,$$

where the coefficients are unknown. The control objective is to design a stabilizing control law so that the closed-loop system has all poles at 0, and y_p can track any constant reference signal. Assumptions A1 and A2 for PPC (section 7.4) are assumed to be satisfied.

   (a) Design a robust APPC law.

   (b) Simulate your design with the plant

$$y_p = \frac{1}{z^4 + z^3 + 3z^2 + 2z + 1} u_p.$$

19. Consider the plant

$$y_p = \frac{s + b_0}{s^2 + a_1 s + a_0} u_p,$$

where a_0, a_1, and b_0 are unknown constants. The control objective is to design a stabilizing control law so that the closed-loop system has a desired dominant pole which corresponds to a damping ratio of 0.707 and natural frequency of 30 rad/sec, the rest of the poles are fast (in the continuous-time domain), and y_p can track any constant reference signal.

   (a) Discretize the plant with the sampling time T_s.

   (b) If we design the controller in the discrete-time domain with the model obtained in (a), where should we place the dominant and the rest of the poles?

   (c) Suppose $b_0 \neq 0$ is known. What additional assumptions will you make in order to implement an APPC scheme in the discrete-time domain?

   (d) With the knowledge $b_0 \neq 0$ and the additional assumptions you made in (c), design a robust APPC law in the discrete-time domain. Simulate your design for the plant when $a_0 = 3$, $a_1 = -2$, and $b_0 = 1$. Try different values for T_s and comment on your results.

20. Consider the plant given in Problem 14 with $\Delta_m(z) = \frac{\mu z}{z+\mu}$. Suppose it is known that $\delta_0 = 0.5$.

   (a) Design a robust APPC for the plant with $\mu = 0$.

   (b) Simulate your design for the plant when $a = -2$, $b = 1.5$, and $\Delta_m(z) = \frac{\mu z}{z+\mu}$ for $\mu = 0, 0.1, 0.4$. Comment on your results.

21. Consider the plant

$$y_p = e^{-\tau s} \frac{1}{s + a_0} u_p,$$

where a_0 is an unknown constant and τ is the time delay that is much smaller than the sampling time T_s. In the control design we consider the nominal model

$$y_p = \frac{1}{s + a_0} u_p.$$

The control objective is to design a stabilizing control law so that y_p can track any constant reference signals.

(a) Discretize the nominal plant with the sampling time T_s. Design a robust APPC law based on the discretized nominal plant, with the desired pole placed at 0.1.

(b) Simulate your designs for the plant when $a_0 = -1$ for $T_s = 0.01$, 0.1, and 1.0. Try different values for τ and comment on your results.

Chapter 8

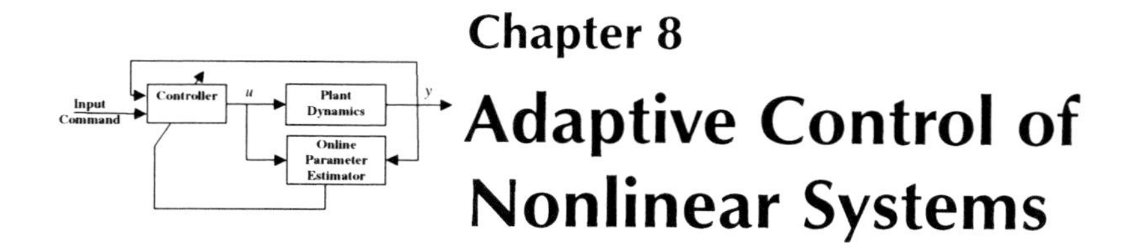

Adaptive Control of Nonlinear Systems

8.1 Introduction

In the previous chapters we considered the design, analysis, and robustness of adaptive systems for LTI plants. We have shown that adaptive systems can deal with any size of parametric uncertainty as well as with dynamic uncertainties due to neglected dynamics, provided the proper robust algorithms are used. The bounds for the allowable dynamic uncertainties cannot be calculated as easily as in the nonadaptive case due to the nonlinear nature of the adaptive system as well as the fact that the plant parameters are considered to be unknown. As we have shown in previous chapters, adaptive control designed for LTI plants leads to a closed-loop system that is nonlinear. Therefore, poles and zeros, frequency domain phase, and gain margins no longer make sense, since they are not defined for nonlinear systems. Nonlinear techniques based on Lyapunov and passivity arguments in addition to linear system theory are used to establish stability and robustness and develop stability/robustness margins that are not as easy to compute a priori as in the LTI case. Qualitatively, however, LTI intuition can be used to understand the robust modifications of adaptive systems. For example, the dynamic normalizing signal limits the rate of adaptation to be finite and small relative to the level of dynamic uncertainty. The limiting of the estimated controller parameters to assume large values eliminates the possibility of high gain control. Both high gain and high speed of adaptation may increase the controller bandwidth and can lead to instability in the presence of unmodeled dynamics.

As we indicated in Chapter 1, adaptive control was motivated to deal with plants whose parameters vary with time. The adaptive systems of the previous chapters are designed for LTI systems. Their extension to linear plants with time-varying (TV) parameters was a major obstacle until the basic robustness questions were answered in the early 80s. In a series of papers and a monograph [5, 61, 62, 72, 73, 145] it was established that the same robustness modifications which include the dynamic normalizing signal together with leakage, dead zone, or parameter projection could be used to deal with a wide class of parameter variations. This class includes slowly varying parameters as well as infrequent jumps in the parameter values. In several cases the error due to time variations can be reduced by proper parameterization of the TV plant model used for control design [5, 61, 62, 72, 73,

145]. In the linear TV case stability margins, bandwidth, frequency domain characteristics, poles, and zeros do not make much sense even in the case of known TV parameters unless approximations are made using the assumption of slowly varying parameters, etc.

A natural extension of the adaptive control results for LTI and linear TV (LTV) plants is their use to control nonlinear systems. Nonlinear systems belong to a wider class of systems which also includes the LTI and LTV plants. Adaptive systems, however, are designed to deal with unknown parameters, and in most nonlinear systems the uncertainties are the unknown nonlinear functions in addition to parameters. Furthermore, the design of stabilizing controllers for nonlinear plants where all the parameters and nonlinear functions are known exactly is in its infancy despite the remarkable results generated in the area during the last two decades [146]. Consequently, the extension of the adaptive control techniques for LTI and LTV systems to the nonlinear ones is still a challenging problem for general classes of nonlinear systems.

In this chapter we present some of the efforts made to extend the adaptive control designs for LTI systems to classes of nonlinear plants. These approaches can be divided into two major classes. The first class includes nonlinear systems whose nonlinear functions are known and whose unknown parameters appear linearly. The approach followed in this case has similarities to the case of LTI plants but also differences. In the second class the unknown nonlinear functions are assumed to be approximated with unknown parameters multiplying known basis functions. These basis functions include those used in neural networks, and the approach followed is similar to that in the neural network area. The only difference is that the unknown parameters or weights, as they are referred to in the neural network literature, appear linearly or in a single layer. This property is fundamental in developing analytical stability results with large regions of attraction.

The purpose of this chapter is to give a flavor of the complexity of controlling nonlinear systems with unknown parameters and/or unknown nonlinearities. It is not meant to cover all approaches and theories related to this area, as that would require much more space than available and is outside the scope of this book.

The chapter is organized as follows: We present some basic results in nonlinear control based on the assumption that all functions and parameters are known exactly. Then we extend these results to the case of unknown parameters, using simple examples to demonstrate new concepts as well as difficulties in extending the results for LTI systems to classes of nonlinear systems. Finally, we discuss techniques used to deal with unknown nonlinearities in addition to unknown parameters.

8.2 Feedback Linearization

Feedback linearization is based on the idea of using change of coordinates and feedback control to cancel all or most of the nonlinear terms so that the system behaves as a linear or partially linear system. Considerable research has been performed in this area, described in a wide range of papers and books [66, 146–150]. We review some of the basic concepts in feedback linearization and demonstrate this method using simple examples.

Let us consider the nonlinear system

$$\begin{aligned} \dot{x} &= f(x) + g(x)u, \\ y &= h(x), \end{aligned} \tag{8.1}$$

where $x \in \mathcal{R}^n$, $u, y \in \mathcal{R}$, and f, g, h are smooth (differentiable infinitely many times) nonlinear functions. The description in (8.1) is considered without loss of generality. For example, the general nonlinear system

$$\dot{x} = f(x, u), \qquad y = h(x)$$

may be expressed in the form of (8.1) by augmenting it with an input integrator and defining a new input v, i.e.,

$$\begin{aligned}\dot{x} &= f(x, u),\\ \dot{u} &= v,\\ y &= h(x).\end{aligned}$$

By defining the augmented state $x_a = [x^T, u]^T$ and considering v as the new input, the above system is in the form of (8.1).

Let us now differentiate y in (8.1) with respect to time. We obtain

$$\dot{y} = \frac{\partial h}{\partial x}(x) f(x) + \frac{\partial h}{\partial x}(x) g(x) u,$$

where

$$\frac{\partial h}{\partial x} f = \frac{\partial h}{\partial x_1} f_1 + \cdots + \frac{\partial h}{\partial x_n} f_n \triangleq L_f f$$

and $L_f f$ is known as the *Lie derivative*. If $\frac{\partial h}{\partial x}(x_0) g(x_0) \neq 0$ at some point x_0, then we say that the system (8.1) has *relative degree* 1 at x_0. In the LTI case this means that the output is separated from the input by one integrator only; i.e., the system has one pole more than zeros or has relative degree 1. Another interpretation is that the relative degree is equal to the number of times the output has to be differentiated until the input appears in the expression of the differentiated output.

If $\frac{\partial h}{\partial x}(x) g(x) = 0 \ \forall x \in B_{x_0}$ in some neighborhood B_{x_0} of x_0, then we take the second derivative of y to obtain

$$\ddot{y} = \frac{\partial}{\partial x}\left(\frac{\partial h}{\partial x} f\right) f + \frac{\partial}{\partial x}\left(\frac{\partial h}{\partial x} f\right) g u.$$

If $\frac{\partial}{\partial x}(\frac{\partial h}{\partial x}(x) f(x)) g(x) \mid_{x=x_0} \neq 0$, then (8.1) is said to have relative degree 2 at x_0. If $\frac{\partial}{\partial x}(\frac{\partial h}{\partial x}(x) f(x)) g(x) = 0$ in a neighborhood of x_0, then the differentiation procedure is continued and the relative degree is found (if it is well defined) in a similar fashion. In order to give a general definition of the relative degree of (8.1), we use the following notation and identities:

$$\begin{aligned}L_f^0 h &\triangleq h,\\ L_f^1 h &\triangleq L_f h \triangleq \frac{\partial h}{\partial x} f,\\ L_f^2 h &\triangleq L_f(L_f h) = \frac{\partial}{\partial x}\left(\frac{\partial h}{\partial x} f\right) \cdot f,\end{aligned}$$

$$L_f^3 h \triangleq L_f(L_f^2 h) = \frac{\partial}{\partial x}\left(\frac{\partial}{\partial x}\left(\frac{\partial h}{\partial x} f\right)\cdot f\right)\cdot f,$$

$$\vdots$$

i.e.,

$$L_f^{i+1} h \triangleq L_f(L_f^i h) = \frac{\partial (L_f^i h)}{\partial x}\cdot f, \quad i = 0, 1, 2, 3, \dots, \tag{8.2}$$

where

$$L_f^i h = \underbrace{\frac{\partial}{\partial x}\left(\frac{\partial}{\partial x}\left(\cdots\frac{\partial}{\partial x}\left(\frac{\partial h}{\partial x}\right.\right.\right.}_{i\text{-derivatives}} f\Big)\cdot f\cdots\Big)\cdot f\Big)\cdot f.$$

In addition we define

$$L_g L_f h \triangleq \frac{\partial (L_f h)}{\partial x}\cdot g. \tag{8.3}$$

The SISO nonlinear system (8.1) has *relative degree* ρ at a point x_0 if

(i) $L_g L_f^i h(x) = 0 \; \forall x \in B_{x_0}$, where B_{x_0} is some neighborhood of x_0 $\forall i = 1, 2, 3, \dots, \rho - 2$.

(ii) $L_g L_f^{\rho-1} h(x_0) \neq 0$.

Assume that the system (8.1) has *relative degree* $\rho = n$ at x, where n is the order of (8.1), and consider the transformation

$$\begin{aligned} &z_1 = y = h(x), \qquad z_2 = \dot{y} = L_f h(x), \qquad z_3 = \ddot{y} = L_f^2 h(x), \qquad \dots, \\ &z_i = y^{(i-1)} = L_f^{i-1} h(x), \dots, z_n = y^{(n-1)} = L_f^{n-1} h(x). \end{aligned} \tag{8.4}$$

It follows that

$$\begin{aligned} \dot{z}_1 &= z_2, \\ \dot{z}_2 &= z_3, \\ &\vdots \\ \dot{z}_{n-1} &= z_n, \\ \dot{z}_n &= L_f^n h(x) + (L_g L_f^{n-1} h(x))u, \\ y &= z_1, \end{aligned} \tag{8.5}$$

which is known as the *canonical* or *normal form* of the system with *no zero dynamics*.

The feedback control law

$$u = \frac{1}{L_g L_f^{n-1} h(x)}[v - L_f^n h(x)], \tag{8.6}$$

where $v \in \mathcal{R}$ is a new input, leads to the LTI system

$$\begin{aligned}\dot{z}_1 &= z_2,\\ \dot{z}_2 &= z_3,\\ &\vdots\\ \dot{z}_{n-1} &= z_n,\\ \dot{z}_n &= v,\\ y &= z_1,\end{aligned}$$

which can be put in the matrix form

$$\dot{z} = Az + Bv, \qquad y = C^T z, \tag{8.7}$$

where

$$A = \begin{bmatrix} 0 & 1 & 0 & \cdots & 0 \\ \vdots & \ddots & \ddots & \ddots & \vdots \\ \vdots & & \ddots & \ddots & 0 \\ \vdots & & & \ddots & 1 \\ 0 & \cdots & \cdots & \cdots & 0 \end{bmatrix}, \qquad B = \begin{bmatrix} 0 \\ \vdots \\ \vdots \\ 0 \\ 1 \end{bmatrix}, \qquad C = \begin{bmatrix} 1 \\ 0 \\ \vdots \\ \vdots \\ 0 \end{bmatrix}.$$

Since (8.7) is an observable and controllable LTI system, the input v can be easily selected to meet regulation or tracking objectives for the plant output y. In this case the control law (8.6) cancels all the nonlinearities via feedback and forces the closed-loop system to behave as an LTI system.

If the system (8.1) has relative degree $\rho < n$, then the change of coordinates

$$z_1 = y, \qquad z_2 = \dot{y}, \ldots, \qquad z_\rho = y^{(\rho-1)} = L_f^{(\rho-1)}h(x)$$

will lead to

$$\begin{aligned}\dot{z}_1 &= z_2,\\ \dot{z}_2 &= z_3,\\ &\vdots\\ \dot{z}_{\rho-1} &= z_\rho,\\ \dot{z}_\rho &= L_f^\rho h(x) + (L_g L_f^{\rho-1} h(x))u.\end{aligned} \tag{8.8}$$

Since the order of the system is n, additional $n - \rho$ states are needed. In this case it is possible to find functions $h_{\rho+1}(x), \ldots, h_n(x)$ with $\frac{\partial h_i(x)}{\partial x} g(x) = 0$, $i = \rho + 1, \ldots, n$, and define the $n - \rho$ states as

$$z_{\rho+1} = h_{\rho+1}(x), \ldots, z_n = h_n(x)$$

to obtain the additional states

$$\begin{aligned}\dot{z}_{\rho+1} &= \frac{\partial h_{\rho+1}(x)}{\partial x} \cdot f(x) \triangleq \varphi_{\rho+1}(z),\\ &\vdots\\ \dot{z}_n &= \frac{\partial h_n(x)}{\partial x} \cdot f(x) \triangleq \varphi_n(z),\\ y &= z_1,\end{aligned} \tag{8.9}$$

where $z = [z_1, z_2, \dots, z_n]^T$ is the overall new state. If we use the feedback control law

$$u = \frac{1}{L_g L_f^{\rho-1} h(x)}[v - L_f^{\rho} h(x)], \tag{8.10}$$

we obtain the system

$$\begin{aligned}\dot{z}_1 &= z_2,\\ &\vdots\\ \dot{z}_{\rho-1} &= z_\rho,\\ \dot{z}_\rho &= v,\\ \dot{z}_{\rho+1} &= \varphi_{\rho+1}(z),\\ &\vdots\\ \dot{z}_n &= \varphi_n(z),\\ y &= z_1.\end{aligned} \tag{8.11}$$

In this case the input v may be chosen to drive the output y and states $z_1, \dots, z_\rho$ to zero or meet some tracking objective for y. The choice of the input v, however, may not guarantee that the states $z_{\rho+1}, \dots, z_n$ are bounded even when $z_1, \dots, z_\rho$ are driven to zero. When $z_1 = z_2 = \cdots = z_\rho = 0$, the dynamics

$$\begin{aligned}\dot{z}_{\rho+1} &= \varphi_{\rho+1}(0, \dots, 0, z_{\rho+1}, \dots, z_n),\\ &\vdots\\ \dot{z}_n &= \varphi_n(0, \dots, 0, z_{\rho+1}, \dots, z_n)\end{aligned} \tag{8.12}$$

are called the *zero dynamics* of (8.1). These are the dynamics of the system (8.1) when y and its first ρ derivatives are equal to zero. If the equilibrium $z_{\rho+1} = 0, \dots, z_n = 0$ of (8.12) is asymptotically stable, the system (8.1) is said to be *minimum-phase*.

The process of using feedback to transform the nonlinear system to a linear system from the input v to the output y is called *I/O feedback linearization* [147–150]. In the case of $\rho = n$, the system (8.1) is linearized to obtain (8.7) without zero dynamics, and this process is called *full-state feedback linearization*.

Example 8.2.1 Let us apply the above theory to an LTI example which is a special case of a nonlinear system. Consider the third-order system

$$\begin{aligned}\dot{x}_1 &= x_2,\\ \dot{x}_2 &= x_3 + u,\\ \dot{x}_3 &= -u,\\ y &= x_1,\end{aligned}$$

which has the transfer function

$$\frac{Y(s)}{U(s)} = \frac{s-1}{s^3},$$

i.e., it is nonminimum-phase with relative degree 2. In this case, $f(x) = [x_2, x_3, 0]^T$, $g(x) = [0, 1, -1]^T$, and $h(x) = x_1$. Using the definition of relative degree for nonlinear systems (LTI systems also belong to this class), we have

$$\begin{aligned}\dot{y} &= \dot{x}_1 = x_2,\\ \ddot{y} &= \dot{x}_2 = x_3 + u.\end{aligned}$$

Since u appears in the second derivative of y, we conclude that the relative degree is equal to 2 $\forall x$. Let us now put the LTI system into the form of (8.11). We define the new coordinates as

$$z_1 = y, \qquad z_2 = \dot{y}, \qquad z_3 = h_3(x),$$

where $h_3(x)$ is a function that satisfies

$$\frac{\partial h_3(x)}{\partial x} \cdot \begin{bmatrix} 0 \\ 1 \\ -1 \end{bmatrix} = \frac{\partial h_3}{\partial x_1} \cdot 0 + \frac{\partial h_3}{\partial x_2} - \frac{\partial h_3}{\partial x_3} = 0.$$

It follows that $h_3(x) = x_1 + x_2 + x_3$ satisfies the above equality, and therefore the new coordinates are

$$z_1 = y, \qquad z_2 = \dot{y}, \qquad z_3 = x_1 + x_2 + x_3,$$

which lead to the system

$$\dot{z}_1 = x_2, \qquad \dot{z}_2 = x_3 + u, \qquad \dot{z}_3 = x_2 + x_3.$$

Since $z_1 = x_1, z_2 = x_2, z_3 = x_1 + x_2 + x_3$, the system in the z-coordinates is given by

$$\begin{aligned}\dot{z}_1 &= z_2,\\ \dot{z}_2 &= z_3 - z_1 - z_2 + u,\\ \dot{z}_3 &= z_3 - z_1.\end{aligned}$$

The zero dynamics are described by $\dot{z}_3 = z_3$ which has an eigenvalue at 1, and therefore the equilibrium $z_{3e} = 0$ is unstable; i.e., the LTI system is nonminimum-phase. We should note that the eigenvalue of the zero dynamics coincide with the unstable zero of the transfer function of the system, as one would expect. Since the system is LTI the choice of the control law to drive all the states to zero is straightforward. ∎

Example 8.2.2 Consider the controlled van der Pol equation

$$\begin{aligned}\dot{x}_1 &= x_2,\\ \dot{x}_2 &= -x_1 + \varepsilon(1 - x_1^2)x_2 + u,\\ y &= x_2,\end{aligned}$$

where $\varepsilon > 0$ is a constant. The first derivative of y is

$$\dot{y} = -x_1 + \varepsilon(1 - x_1^2)x_2 + u.$$

Hence the second-order system has relative degree 1. Using the change of coordinates $z_1 = x_2$, $z_2 = x_1$, we obtain

$$\begin{aligned}\dot{z}_1 &= -z_2 + \varepsilon(1 - z_2^2)z_1 + u,\\ \dot{z}_2 &= z_1,\\ y &= z_1.\end{aligned}$$

The zero dynamics of the system are given by $\dot{z}_2 = 0$, which does not have an asymptotically stable equilibrium point. Therefore, the nonlinear system is not minimum-phase. In this case a feedback linearizing control law could be

$$u = z_2 - \varepsilon(1 - z_2^2)z_1 - \lambda z_1,$$

where $\lambda > 0$ is a design constant leading to the closed-loop system

$$\begin{aligned}\dot{z}_1 &= -\lambda z_1,\\ \dot{z}_2 &= z_1,\end{aligned}$$

which has a nonisolated equilibrium $z_{1e} = 0$, $z_{2e} = c$, where c is any constant. We can establish that from any initial condition $z_1 \to 0$, $z_2 \to c$ exponentially with time. We should note that if instead of $y = x_2$ the output is $y = x_1$, then we can show that the system has relative degree 2 with no zero dynamics. ∎

8.3 Control Lyapunov Functions

Lyapunov functions and associated theory discussed in the Appendix are used to examine stability of the equilibrium points of differential equations of the form

$$\dot{x} = f(x, t), \quad x(t_0) = x_0, \tag{8.13}$$

where $x \in \mathcal{R}^n$ and f is such that the differential equation has a unique solution for each initial condition x_0. As we have already shown in Chapter 3, in the case of adaptive laws based on the SPR-Lyapunov synthesis approach, Lyapunov functions are also used to design adaptive laws in addition to analysis. Along this spirit, Lyapunov functions can be used to design stabilizing feedback control laws. This extended concept involves the design of the control law so that the selected Lyapunov function V and its time derivative satisfy certain properties that imply stability of the equilibrium of the resulting closed-loop system.

The Lyapunov function which can serve this purpose is referred to as the *control Lyapunov function* (*CLF*).

Let us consider the nonlinear system

$$\dot{x} = f(x, u), \tag{8.14}$$

where $x \in \mathcal{R}^n$, $u \in \mathcal{R}$, and $f(0,0) = 0$. We want to choose the control input $u = q(x)$ for some function $q(x)$ with $q(0) = 0$ so that the equilibrium $x_e = 0$ of the closed-loop system

$$\dot{x} = f(x, q(x)) \tag{8.15}$$

is asymptotically stable in the large. The CLF method involves the selection of a function $V(x)$ as a Lyapunov candidate and the design of $q(x)$ so that $\dot{V}(x) \leq -Q(x)$, where $Q(x)$ is a positive definite function.[19] We therefore need to find $q(x)$ so that

$$\dot{V} = \frac{\partial V}{\partial x}\dot{x} = \frac{\partial V}{\partial x} f(x, q(x)) \leq -Q(x). \tag{8.16}$$

It is obvious that the success of the approach depends very much on the choice of $V(x)$. A poor choice of $V(x)$ may make it impossible to satisfy (8.16) even when a stabilizing controller for (8.14) exists. A formal definition of CLF is given below.

Definition 8.3.1 ([66]). *A smooth positive definite and radially unbounded function* $V(x)$ *is called a CLF for* (8.14) *if*

$$\inf_{u} \left\{ \frac{\partial V}{\partial x}(x) f(x, u) \right\} \leq -Q(x) \quad \forall x \neq 0 \tag{8.17}$$

for some positive definite function $Q(x)$.

If the system is in the form of

$$\dot{x} = f(x) + g(x)u, \tag{8.18}$$

then the inequality (8.16) becomes

$$\frac{\partial V}{\partial x} f(x) + \frac{\partial V}{\partial x} g(x) q(x) \leq -Q(x). \tag{8.19}$$

In this case, it is clear that for (8.19) to be satisfied

$$\frac{\partial V}{\partial x} f(x) < 0 \quad \text{whenever } \frac{\partial V}{\partial x} g(x) = 0$$

$\forall x \neq 0$.

[19] Note that the condition of positive definiteness for $Q(x)$ can be relaxed to positive semidefinite by applying LaSalle's invariance theorem for stability [66, 72, 73, 146, 147].

Example 8.3.2 Consider the system

$$\begin{aligned}\dot{x}_1 &= x_2 - x_1,\\ \dot{x}_2 &= \sin x_1 \cos x_2 - x_2^3 + u,\\ y &= x_1.\end{aligned} \tag{8.20}$$

The feedback linearization control law

$$u = -\sin x_1 \cos x_2 + x_2^3 - x_1 - x_2 \tag{8.21}$$

leads to the LTI system

$$\begin{aligned}\dot{x}_1 &= x_2 - x_1,\\ \dot{x}_2 &= -x_1 - x_2,\end{aligned}$$

whose equilibrium $x_{1e} = 0$, $x_{2e} = 0$ is e.s. Let us apply the CLF approach and consider

$$V(x) = \frac{x_1^2}{2} + \frac{x_2^2}{2}. \tag{8.22}$$

Then

$$\begin{aligned}\dot{V} &= x_1(x_2 - x_1) + x_2(\sin x_1 \cos x_2 - x_2^3 + u)\\ &= -x_1^2 + x_1 x_2 - x_2^4 + x_2 \sin x_1 \cos x_2 + x_2 u.\end{aligned}$$

It is clear that by choosing

$$u = -\sin x_1 \cos x_2 - x_1, \tag{8.23}$$

we obtain

$$\dot{V} = -x_1^2 - x_2^4 \leq -(x_1^2 + x_2^4) = -Q(x),$$

and therefore (8.22) is a CLF for the system (8.20). Comparing the control law (8.23) obtained using the CLF (8.22) with (8.21) obtained using feedback linearization, it is clear that (8.23) requires less control energy than (8.21) and is therefore superior to (8.21). In this case the nonlinear term $-x_2^3$ is beneficial to stability and is kept in the CLF-based design approach, whereas in the feedback linearization case it is canceled. Additional examples and explanations with regard to the advantages of the use of CLF may be found in [66]. ∎

8.4 Backstepping

Let us consider the first-order nonlinear system

$$\dot{x} = x^3 u, \tag{8.24}$$

which is uncontrollable at $x = 0$. We consider

$$V(x) = \frac{x^2}{2}$$

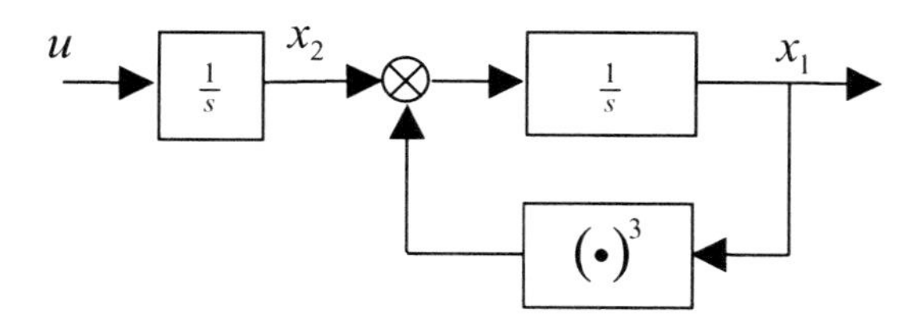

Figure 8.1. *Closed-loop PPC schemes with unmodeled dynamics and bounded disturbances.*

as a potential CLF. Then $\dot{V} = x^4 u$, and for

$$u = -x^2 \tag{8.25}$$

we obtain

$$\dot{V} = -x^6 \leq -Q(x) = -x^6.$$

Therefore, $V(x)$ is a CLF, and the equilibrium $x_e = 0$ of the closed-loop system (8.24), (8.25) described by $\dot{x} = -x^5$ is a.s. The choice of the CLF in this case was easy. Let us now consider the second-order nonlinear system

$$\begin{aligned} \dot{x}_1 &= x_1^3 x_2, \\ \dot{x}_2 &= u. \end{aligned} \tag{8.26}$$

The block diagram of (8.26) is shown in Figure 8.1.

Comparing the first equation of (8.26) with (8.24), we see that x_2 took the place of u. If x_2 was the control variable, then $x_2 = -x_1^2$ would guarantee that x_1 converges to zero. Let us denote this desired value of x_2 as $x_{2d} \overset{\Delta}{=} q(x_1) = -x_1^2$ and define the error between x_2 and x_{2d} as

$$z = x_2 - x_{2d} = x_2 - q(x_1).$$

Then the system (8.26) may be represented as

$$\begin{aligned} \dot{x}_1 &= x_1^3(z + q(x_1)), \quad z = x_2 - q(x_1), \\ \dot{x}_2 &= u, \end{aligned} \tag{8.27}$$

shown in Figure 8.2.

If we treat z as the new state, we obtain the representation

$$\begin{aligned} \dot{x}_1 &= x_1^3(z + q(x_1)), \\ \dot{z} &= u - \dot{q}, \end{aligned} \tag{8.28}$$

shown in Figure 8.3.

The new representation (8.28) is obtained by using $q(x_1)$ as a feedback control loop while we "back step" $-q(x_1)$ through the integrator as shown in Figure 8.3. This process motivated the naming of the subsequent control design procedure as *backstepping* [66] and became a popular method in nonlinear control design for certain classes of nonlinear

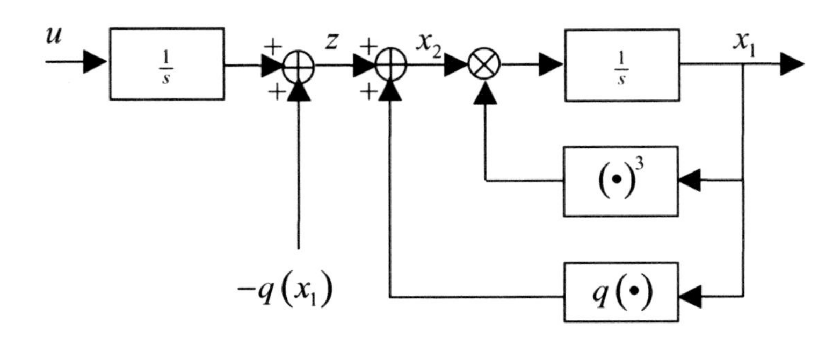

Figure 8.2. *Nonlinear system* (8.26) *with error variable z.*

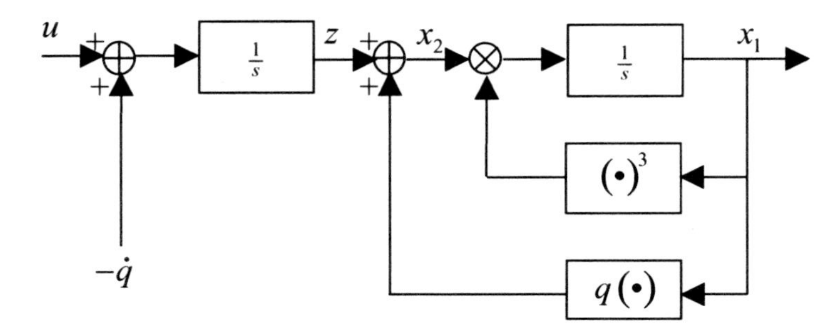

Figure 8.3. *Nonlinear system* (8.26) *with "backstepping" $-q$ through the integrator.*

systems. One of the significant features of the backstepping approach is that a CLF for (8.28) can be obtained by augmenting the CLF for the system $\dot{x}_1 = x_1^3 u$ with a quadratic term in the error variable z, i.e.,

$$V_a(x, z) = V(x_1) + \frac{z^2}{2},$$

where $V(x_1) = \frac{x_1^2}{2}$. Then

$$\dot{V}_a = x_1^4 z + x_1^4 q(x_1) + zu - z\dot{q}(x_1),$$

where $q(x_1) = -x_1^2$. Hence

$$\dot{V}_a = -x_1^6 + z[x_1^4 - \dot{q}(x_1) + u].$$

Choosing

$$u = \dot{q}(x_1) - x_1^4 - z, \tag{8.29}$$

we have

$$\dot{V}_a = -x_1^6 - z^2 \leq -(x_1^6 + z^2) = -Q(x, z),$$

which implies that the equilibrium $x_{1e} = 0$, $z_e = 0$ of the closed-loop system

$$\dot{x}_1 = x_1^3(z - x_1^2),$$

$$\dot{z} = -x_1^4 - z$$

is asymptotically stable. Since

$$\dot{q}(x_1) = -\frac{dx_1^2}{dt} = -2x_1\dot{x}_1 = -2x_1^4(z - x_1^2),$$

the control law (8.29) can be implemented as

$$\begin{aligned} u &= -2x_1^4(z - x_1^2) - x_1^4 - z \\ &= -x_2 - x_1^2 - x_1^4 - 2x_1^4 x_2. \end{aligned}$$

The above example demonstrates that if we know the CLF of the system, we can generate the CLF of the same system augmented with a pure integrator.

Let us extend the backstepping procedure to the more general nonlinear system

$$\begin{aligned} \dot{x}_1 &= f(x_1) + g(x_1)x_2, \\ \dot{x}_2 &= u, \end{aligned} \tag{8.30}$$

where $f(0) = 0$; $x_1 \in \mathcal{R}^n$; $x_2, u \in \mathcal{R}$. We consider the following assumption.

Assumption 8.4.1. *Consider the system*

$$\dot{x}_1 = f(x_1) + g(x_1)u, \quad f(0) = 0, \tag{8.31}$$

where $x_1 \in \mathcal{R}^n$, $u \in \mathcal{R}$, and the functions f, g are the same as in (8.30). *There exists a feedback control input $u = q_1(x_1)$, $q_1(0) = 0$ which satisfies*

$$\frac{\partial V}{\partial x_1}(x_1)(f(x_1) + g(x_1)q_1(x_1)) \le -Q(x_1) < 0$$

for some smooth, positive definite, radially unbounded function $V : \mathcal{R}^n \to \mathcal{R}$, where $Q : \mathcal{R}^n \to \mathcal{R}$ is a positive definite function; i.e., $V(x)$ is a CLF for the feedback system (8.31).

Lemma 8.4.2. *Consider the nonlinear $(n+1)$-order system* (8.30) *which satisfies Assumption* 8.4.1 *with x_2 as its control input. Then the function*

$$V_a(x_1, x_2) = V(x_1) + \frac{z^2}{2}, \tag{8.32}$$

where $z = x_2 - q_1(x_1)$ and $q_1(x_1)$ is defined in Assumption 8.4.1, *is a CLF for the system* (8.30) *in the sense that there exists a feedback control law*

$$u = q_2(x_1, x_2)$$

which guarantees that the equilibrium $x_{1e} = 0$, $x_{2e} = 0$ is a.s. in the large. Furthermore, one choice of such a feedback control law is

$$u = -c(x_2 - q_1(x_1)) + \frac{\partial q_1}{\partial x_1}(x_1)(f(x_1) + g(x_1)x_2) - \frac{\partial V}{\partial x_1}(x_1)g(x_1) \stackrel{\Delta}{=} q_2(x_1, x_2), \tag{8.33}$$

where $c > 0$ is an arbitrary constant.

Proof. We treat x_2 as the virtual control with the desired value of $q_1(x_1)$. The error variable $z = x_2 - q_1(x_1)$ is used to transform the system to

$$\begin{aligned}\dot{x}_1 &= f(x_1) + g(x_1)[q_1(x_1) + z],\\ \dot{z} &= u - \frac{\partial q_1}{\partial x_1}(x_1)[f(x_1) + g(x_1)(q_1(x_1) + z)].\end{aligned} \tag{8.34}$$

We consider the function (8.32) whose time derivative along (8.34) is given by

$$\begin{aligned}\dot{V}_a &= \frac{\partial V}{\partial x_1}(x_1)(f(x_1) + g(x_1)(q_1(x_1) + z))\\ &\quad + z\left(u - \frac{\partial q_1}{\partial x_1}(x_1)(f(x_1) + g(x_1)(q_1(x_1) + z))\right)\\ &\leq -Q(x_1) + z\left(u - \frac{\partial q_1}{\partial x_1}(f + g(q_1 + z)) + \frac{\partial V}{\partial x_1}g\right),\end{aligned}$$

where the last inequality is obtained by using Assumption 8.4.1. Choosing u as in (8.33), it follows that

$$\dot{V}_a \leq -Q(x_1) - cz^2 \leq -Q_a(x_1, z),$$

where $Q_a(x_1, z) = Q(x_1) + cz^2$ is positive definite in $\mathcal{R}^{n+1}$, which implies that V_a is CLF for the system (8.30) and that the equilibrium $x_{1e} = 0$, $z_e = 0$ is a.s. in the large. This in turn implies that the equilibrium point $x_{2e} = 0$ is also a.s. in the large due to $x_{2e} = z_e - q_1(x_{1e})$ and $q_1(0) = 0$. $\square$

We should note that the choice of the control law (8.33) is not unique, and in some cases it may not be the desirable one since it cancels all the nonlinear terms in the expression of $\dot{V}_a$ inside the brackets. Some of these nonlinear terms may make $\dot{V}_a$ more negative, leading to faster convergence. Furthermore, by not canceling them, less control energy may be needed.

The backstepping methodology may be extended to the more general system

$$\begin{aligned}\dot{x}_1 &= f(x_1) + g(x_1)x_2,\\ \dot{x}_2 &= x_3,\\ &\vdots\\ \dot{x}_{k-1} &= x_k,\\ \dot{x}_k &= u,\end{aligned}$$

where f, g satisfy Assumption 8.4.1. In this case the repeated application of Lemma 8.4.2 results in the CLF

$$V_a(x_1, \ldots, x_k) = V(x_1) + \frac{1}{2}\sum_{i=2}^{k} z_i^2,$$

where

$$z_i = x_i - q_{i-1}(x_1, x_2, \ldots, x_{i-1}), \quad i = 2, 3, \ldots, k.$$

In this case, $q_1(x_1)$ is the stabilizing control input for the system

$$\dot{x}_1 = f(x_1) + g(x_1)u,$$

$q_2(x_1, x_2)$ is the stabilizing control input for the system

$$\begin{aligned}\dot{x}_1 &= f(x_1) + g(x_1)x_2,\\ \dot{x}_2 &= u,\end{aligned}$$

$q_r(x_1, x_2, \dots, x_r)$ is the stabilizing control input for the system

$$\begin{aligned}\dot{x}_1 &= f(x_1) + g(x_1)x_2,\\ \dot{x}_2 &= x_3,\\ &\vdots\\ \dot{x}_r &= u,\end{aligned}$$

and so on. The backstepping methodology is a recursive one which allows the generation of the stabilizing functions $q_r(x_1, x_2, \dots, x_r)$, $r = 1, 2, \dots, k-1$, at each step, which are then used to construct the overall CLF.

8.5 Adaptive Backstepping with Tuning Functions

In the backstepping methodology presented in section 8.4, all the nonlinear functions and parameters are assumed to be known. In this section we consider nonlinear plants where the nonlinear functions are made up of products of unknown parameters with known nonlinearities. The unknown parameters are assumed to appear linearly, a property that is crucial in developing adaptive controllers with global stability properties, as we have already discussed in the LTI case. We start with simple examples to demonstrate the adaptive backstepping control design methodology.

Example 8.5.1 Consider the nonlinear system

$$\dot{x} = \theta^{*^T} f(x) + u,$$

where $u, x \in \mathcal{R}$; $\theta^* \in \mathcal{R}^n$ is a vector of unknown constants and $f(x) \in \mathcal{R}^n$ is a vector of known smooth functions with the property $f(0) = 0$; and x is available for measurement. If θ^* is known, then the control law

$$u = -\theta^{*^T} f(x) - cx$$

for some constant $c > 0$ leads to the closed-loop system

$$\dot{x} = -cx,$$

whose equilibrium $x_e = 0$ is e.s. Since θ^* is unknown, we use the certainty equivalence (CE) control law

$$u = -\theta^T f(x) - cx, \tag{8.35}$$

where $\theta(t)$ is the estimate of θ^* at time t to be generated by an adaptive law. The closed-loop system becomes

$$\dot{x} = -cx - \tilde{\theta}^T f(x),$$

where $\tilde{\theta} \triangleq \theta - \theta^*$ is the parameter error vector. The above equation relates the parameter error with the regulation error (which in this case can also be interpreted as the estimation error) in a way that motivates the Lyapunov-like function

$$V(x, \tilde{\theta}) = \frac{x^2}{2} + \frac{\tilde{\theta}^T \Gamma^{-1} \tilde{\theta}}{2},$$

whose time derivative satisfies

$$\dot{V} = -cx^2 - \tilde{\theta}^T f(x)x + \tilde{\theta}^T \Gamma^{-1} \dot{\tilde{\theta}}.$$

Choosing the adaptive law

$$\dot{\theta} = \dot{\tilde{\theta}} = \Gamma f(x)x, \tag{8.36}$$

we obtain

$$\dot{V} = -cx^2 \leq 0,$$

which implies that $\theta, x \in \mathcal{L}_\infty$ and $x \in \mathcal{L}_2$. By following the same arguments as in the linear case, we can establish that $u \in \mathcal{L}_\infty$ and $x \to 0$ as $t \to \infty$. ∎

Example 8.5.2 Let us consider the second-order system

$$\dot{x}_1 = \theta^{*^T} f(x_1) + x_2,$$
$$\dot{x}_2 = u.$$

If θ^* is known, then the following backstepping procedure can be used to design a stabilizing controller: The first stabilizing function $q_1(x_1, \theta^*) = -c_1 x_1 - \theta^{*^T} f(x_1)$, where $c_1 > 0$ is a constant, is used to define the new state variables

$$\begin{aligned} z_1 &= x_1, \\ z_2 &= x_2 - q_1(x_1, \theta^*) = x_2 - c_1 x_1 - \theta^{*^T} f(x_1) \end{aligned} \tag{8.37}$$

and obtain the system

$$\dot{z}_1 = -c_1 z_1 + z_2,$$
$$\dot{z}_2 = -\frac{\partial q_1}{\partial x_1}(x_2 + \theta^{*^T} f) + u.$$

We consider the Lyapunov function

$$V(z_1, z_2) = \frac{z_1^2}{2} + \frac{z_2^2}{2}. \tag{8.38}$$

Its time derivative is given by

$$\dot{V} = -c_1 z_1^2 + z_2 \left(z_1 - \frac{\partial q}{\partial x_1}(x_2 + \theta^{*^T} f) + u \right).$$

The control law

$$u = -z_1 + \frac{\partial q}{\partial x_1}(x_2 + \theta^{*^T} f) - c_2 z_2,$$

where $c_2 > 0$ is a design constant, leads to

$$\dot{V} = -c_1 z_1^2 - c_2 z_2^2 \leq -Q(z_1, z_2) < 0,$$

which implies that the equilibrium $z_{1e} = 0$, $z_{2e} = 0$ is a.s. in the large. This in turn implies that $x_1, x_2 \to 0$ as $t \to \infty$ by using (8.37) and the property $f(0) = 0$.

Let us now consider the case where θ^* is unknown. We use the CE approach and replace the unknown θ^* in the definition of the state variables (8.37) with its estimate $\theta(t)$ to be generated by an adaptive law, i.e.,

$$\begin{aligned} z_1 &= x_1, \\ z_2 &= x_2 - q_1(x_1, \theta), \qquad q_1(x_1, \theta) = -c_1 x_1 - \theta^T f(x_1). \end{aligned}$$

The resulting system in the new coordinate system is given by

$$\begin{aligned} \dot{z}_1 &= -c_1 z_1 + z_2 - \tilde{\theta}^T f(x_1), \\ \dot{z}_2 &= u + \tilde{\theta}^T f \frac{\partial q_1}{\partial x_1} - \frac{\partial q_1}{\partial x_1}(z_2 - c_1 z_1) + \dot{\theta}^T f. \end{aligned} \tag{8.39}$$

Let us consider the Lyapunov-like function

$$V(z_1, z_2, \tilde{\theta}) = \frac{z_1^2}{2} + \frac{z_2^2}{2} + \frac{\tilde{\theta}^T \Gamma^{-1} \tilde{\theta}}{2},$$

which is equal to the function used in the known parameter case plus the additional term for the parameter error. The time derivative along the solution of (8.39) is given by

$$\dot{V} = -c_1 z_1^2 + z_1 z_2 - \tilde{\theta}^T f z_1 + z_2 \left(u + \tilde{\theta}^T f \frac{\partial q_1}{\partial x_1} - \frac{\partial q_1}{\partial x_1}(z_2 - c_1 z_1) + \dot{\theta}^T f \right) + \tilde{\theta}^T \Gamma^{-1} \dot{\tilde{\theta}}.$$

After rearranging the various terms, we obtain

$$\dot{V} = -c_1 z_1^2 - \tilde{\theta}^T \left(f z_1 - f \frac{\partial q_1}{\partial x_1} z_2 \right) + z_2 \left(u + z_1 - \frac{\partial q_1}{\partial x_1}(z_2 - c_1 z_1) + \dot{\theta}^T f \right) + \tilde{\theta}^T \Gamma^{-1} \dot{\tilde{\theta}}.$$

Choosing the control law

$$u = -z_1 + \frac{\partial q_1}{\partial x_1}(z_2 - c_1 z_1) - \dot{\theta}^T f - c_2 z_2 \tag{8.40}$$

and the adaptive law

$$\dot{\theta} = \dot{\tilde{\theta}} = \Gamma \left(f z_1 - f \frac{\partial q_1}{\partial x_1} z_2 \right), \tag{8.41}$$

we obtain

$$\dot{V} = -c_1 z_1^2 - c_2 z_2^2 \leq 0,$$

which implies that $z_1, z_2, \theta \in \mathcal{L}_\infty$ and $z_1, z_2 \in \mathcal{L}_2$. Using these properties, we can show that $\dot{z}_1, \dot{z}_2 \in \mathcal{L}_\infty$, which together with $z_1, z_2 \in \mathcal{L}_2$ imply that $z_1, z_2 \to 0$ as $t \to \infty$. The implementation of the control law (8.39) does not require the use of differentiators since $\dot{\theta}$ is available from (8.41). Furthermore,

$$\frac{\partial q_1}{\partial x_1} = -c_1 - \theta^T \frac{\partial f}{\partial x_1}$$

is also available for measurement. Comparing (8.41) with (8.36) we can see that the term fz_1 in (8.41) is of the same form as that in (8.36) and can be considered as the adaptive law term for the z_1 system. The functions

$$\tau_1(x_1) = fz_1,$$
$$\tau_2(x_1, x_2, \theta) = \tau_1(x_1) - f\frac{\partial q_1}{\partial x_1}z_2$$

are referred to as the *tuning functions* and play a role as terms in the adaptive laws for intermediate systems in the backstepping procedure. ∎

We should also note that the adaptive control law (8.40), (8.41) involves no normalization and has similarities to the MRAC schemes developed in Chapter 5 for SISO LTI plants with relative degree 1 and 2. As the relative degree of the plant increases, the complexity of the nonlinear adaptive control law also increases but the procedure remains the same. For more details on adaptive backstepping for higher-order plants and other classes of nonlinear systems the reader is referred to the book [66].

8.6 Adaptive Backstepping with Nonlinear Damping: Modular Design

The adaptive control schemes developed in section 8.5 are based on the Lyapunov synthesis approach, and the design of the control and adaptive laws is done simultaneously in order to make the derivative of the Lyapunov function negative and establish stability. In this case the adaptive law is restricted to be of the gradient type. The question is whether the design of the control and adaptive laws can be established independently and then combined using the CE approach to form the adaptive control law as done in the case of LTI systems. In such cases different adaptive laws such as LS, gradient, and their variations may be employed. The following example shows that in the nonlinear case the direct use of the CE approach as in the LTI case may not work, and additional modifications are needed in order to establish stability.

Example 8.6.1 Consider the same first-order system as in Example 8.5.1, i.e.,

$$\dot{x} = \theta^{*^T} f(x) + u, \quad x(0) = x_0.$$

We have established using the Lyapunov synthesis approach that the adaptive control law

$$u = -\theta^T f(x) - cx, \quad \dot{\theta} = \dot{\tilde{\theta}} = \Gamma f(x)x \tag{8.42}$$

with $c > 0$ guarantees that all signals are bounded and x goes to zero asymptotically with time. Let us now use the same control law, which is the same as the CE control law, but design the adaptive law independently using the methods of Chapter 3. We express the system in the form of the SPM

$$z = \theta^{*^T}\phi,$$

$$z = \frac{s}{s+1}x - \frac{1}{s+1}u, \qquad \phi = \frac{1}{s+1}f(x)$$

and use the results of Chapter 3 to write the adaptive law which is of the form

$$\dot{\theta} = \dot{\tilde{\theta}} = \Gamma\phi\varepsilon, \qquad \varepsilon = \frac{z - \theta^T\phi}{m_s^2}, \tag{8.43}$$

where Γ is fixed in the case of gradient algorithms or is the covariance matrix in the case of LS algorithms and m_s is the normalizing signal designed to bound φ from above. Let us assume that $f(x) = x^2$. Then the closed-loop system may be written as

$$\dot{x} = -cx + \tilde{\theta}x^2, \quad \dot{\tilde{\theta}} = \Gamma\phi\varepsilon. \tag{8.44}$$

The adaptive law guarantees independently the boundedness of x, u, f, that $\theta \in \mathcal{L}_\infty$, and $\dot{\theta}, \varepsilon, \varepsilon\phi, \varepsilon m_s \in \mathcal{L}_\infty \cap \mathcal{L}_2$. The question is whether these properties of the adaptive law guarantee that x is bounded and converges to zero with time. It can be shown that $\tilde{\theta}$ cannot converge to zero faster than an exponential. So let us take the most optimistic scenario and consider the case where the parameter error $\tilde{\theta}$ converges to zero exponentially fast, i.e., $\tilde{\theta}(t) = c_0 e^{-ct}$ for some constant c_0 which depends on the initial condition of $\tilde{\theta}$. Then the closed-loop error equation becomes

$$\dot{x} = -cx + x^2 c_0 e^{-ct}.$$

The solution of the above equation can be found by first using the transformation $y = xe^{ct}$ and then direct integration. It is given by

$$x(t) = \frac{2cx_0 e^{-ct}}{2c - c_0 x_0(1 - e^{-2ct})}.$$

It is clear that for $c_0 x_0 < 2c$, x is bounded and converges to zero with time. For $c_0 x_0 > 2c$, however, $2c - c_0 x_0(1 - e^{-2ct}) \to 0$ and $x(t) \to \infty$ as $t \to t_f$, where

$$t_f = \frac{1}{2c}\ln\frac{c_0 x_0}{c_0 x_0 - 2c}.$$

In other words, for some initial conditions, the closed-loop system has a finite escape time. Examples similar to the above, presented in [66], demonstrate that the CE approach used for LTI systems may not work in the case of nonlinear systems, and additional modifications may be required if the design of the adaptive laws is to be separated from that of the control law.

Let us examine the adaptive law in the Lyapunov synthesis approach given by (8.42). We can see that the update of the parameters is driven by the level of the nonlinearity, i.e.,

$f(x)x$, which means that the parameters are adjusted rapidly in order to catch up with the nonlinear dynamics. In the case of the CE adaptive law, the speed of adaptation is controlled by normalization to be "small"; as a result, the adaptive controller cannot catch up with the effect of the nonlinearities. One way to compensate for this deficiency of the adaptive law with normalization is to modify the CE control law by using additional terms in order to accommodate the lower speed of adaptation. This modification involves an additive term in the control law referred to as "*nonlinear damping*." We demonstrate the use of nonlinear damping by modifying the CE control law in (8.42) to

$$u = -\theta^T f(x) - cx - f^T f x. \tag{8.45}$$

With the above control law the closed-loop system becomes

$$\dot{x} = -x - f^T f x - \tilde{\theta}^T f. \tag{8.46}$$

We consider the Lyapunov-like function

$$V(x) = \frac{x^2}{2},$$

whose time derivative along the solution of (8.46) is given by

$$\dot{V} = -x^2 - x^2 f^T f - \tilde{\theta}^T f x \leq -x^2 - \left(xf + \frac{\tilde{\theta}}{2}\right)^T \left(xf + \frac{\tilde{\theta}}{2}\right) + \frac{\tilde{\theta}^T\tilde{\theta}}{4}. \tag{8.47}$$

Since the normalized adaptive law (8.43) guarantees that $\theta, \tilde{\theta}$ are bounded, it follows that whenever x^2 exceeds the bound for $\frac{\tilde{\theta}^T\tilde{\theta}}{4}$ we have $\dot{V} < 0$, which implies that x is bounded. This together with the boundedness of the parameter estimates guarantees that all signals in the closed loop are bounded. To establish that x converges to zero, we use the identity

$$(s+1)\tilde{\theta}^T\phi = \dot{\tilde{\theta}}^T\phi + \tilde{\theta}^T f,$$

where

$$\phi = \frac{1}{s+1} f(x),$$

which implies the result of the more general swapping lemma, i.e.,

$$\chi \triangleq \frac{1}{s+1}\tilde{\theta}^T f = \tilde{\theta}^T\phi - \frac{1}{s+1}\dot{\tilde{\theta}}^T\phi.$$

The nonlinear damping guarantees that x is bounded, which implies that ϕ, f are bounded. The adaptive law guarantees that $\varepsilon m_s, \dot{\tilde{\theta}} \in \mathcal{L}_2 \cap \mathcal{L}_\infty$. Since x is bounded we have m_s to be bounded, and therefore $\varepsilon m_s^2 = -\tilde{\theta}^T\phi \in \mathcal{L}_2 \cap \mathcal{L}_\infty$. Therefore, $\chi \in \mathcal{L}_2 \cap \mathcal{L}_\infty$ and since $\dot{\chi}$ can be shown to be bounded we have that χ converges to zero with time. The solution of (8.46) is given by

$$x(t) = e^{-t}e^{-\int_0^t f^T f d\tau}x(0) - \int_0^t e^{-(t-\tau)}e^{-\int_\tau^t f^T f d\sigma}\tilde{\theta}^T f d\tau,$$

which implies that

$$|x(t)| \le e^{-t}|x(0)| + \left|\int_0^t e^{-(t-\tau)}\tilde{\theta}^T f d\tau\right| = e^{-t}|x(0)| + |\chi(t)|.$$

Since $\chi \in \mathcal{L}_2 \cap \mathcal{L}_\infty$ and goes to zero with time, it follows that x converges to zero, too. ∎

In [66], the nonlinear adaptive control schemes where the adaptive laws are designed independently of the control law are referred to as *modular designs*. The generalization of the above example to higher-order plants and wider classes of nonlinear systems is treated extensively in [66]. One of the significant features of the modular designs is that in addition to having the freedom to choose from different classes of adaptive laws they also allow the design of robust adaptive laws as done in the LTI case. Consequently the use of robust adaptive laws in modular designs is expected to enhance robustness in the presence of dynamic uncertainties.

In the modular design approach the control law is designed to guarantee signal boundedness for bounded parameter estimates by using nonlinear terms referred to as nonlinear damping. The adaptive law is designed independently of the control law by simply following the same techniques as in the LTI case presented in previous chapters. The only requirement for closed-loop stability and convergence of the regulation or tracking error is that these adaptive laws guarantee that the estimated parameters, estimation error, and speed of adaptation satisfy the usual $\mathcal{L}_2$ and $\mathcal{L}_\infty$ properties described in Chapter 3.

8.7 Neuroadaptive Control

In the previous sections, we used simple examples to introduce some of the basic concepts of feedback linearization, CLF, backstepping, and adaptive backstepping used in the design of stabilizing controllers for classes of nonlinear systems. In all these approaches we made the fundamental assumption that all the nonlinearities are known exactly. In the adaptive case we allow the presence of unknown parameters multiplying known nonlinearities in a way that leads to parametric models of the type introduced in Chapter 2. In most practical situations exact knowledge of the nonlinearities present in the system is not easy, if at all possible, and the identification of nonlinearities in general is by itself an important research topic. One approach that has attracted considerate interest is the use of artificial neural networks to match or fit unknown nonlinearities either offline or online. The matched or trained neural network can then be used in the place of the unknown nonlinearities for control design purposes. This approach has strong similarities to the adaptive control and CE methodology used in the previous chapters. Due to the nonlinear nature, not only of the plant but also the neural network itself, rigorous stability results are few and often based on restrictive assumptions despite the publication of numerous books [151–154] and papers [155–179] in the area of identification and control of nonlinear systems using neural networks.

In the following subsections, we present some of the concepts of neural networks used in the identification and control of nonlinear dynamical systems.

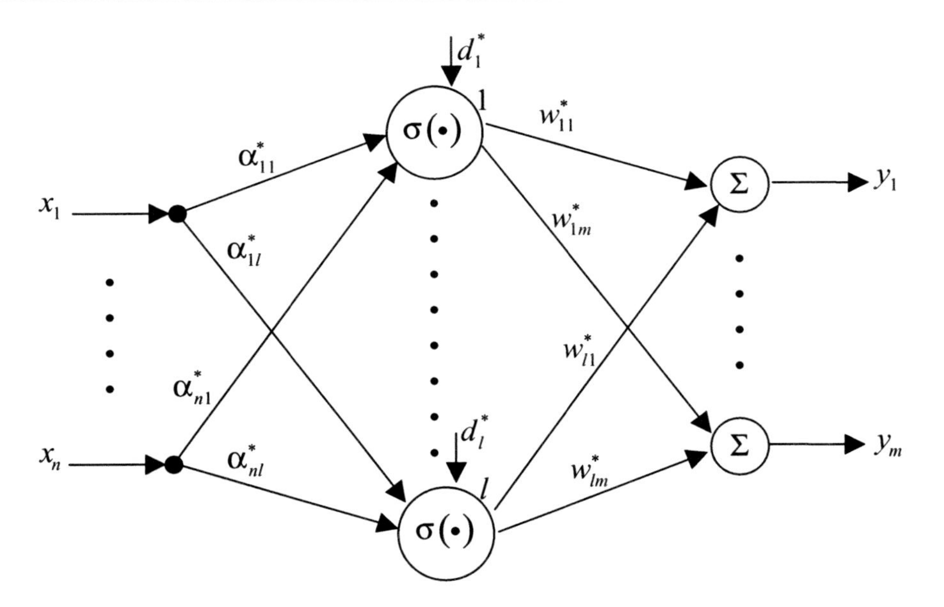

Figure 8.4. *A one-hidden-layer neural network.*

8.7.1 Neural Networks for Identification

Neural networks are motivated from biological systems, where networks of neurons in living organisms respond to external excitations by learning and producing the desired outputs. For example, after several trials a human learns to ride a bicycle. A mosquito has a relatively small number of neurons, yet it learns to perform many operations. These concepts led to the development of artificial neural networks which can be trained in a conceptually similar manner to match behaviors, I/O maps, etc.

Figure 8.4 shows a *one-hidden-layer neural network* with n-inputs, l *hidden units* or *neurons*, and m-outputs. In this figure, a_{ij}^*, w_{jk}^* $(i = 1, 2, \dots, n,\ j = 1, 2, \dots, l,$ $k = 1, 2, \dots, m)$ are constants called the *weights*, and $\sigma(\cdot)$ is a continuous nonlinear function. Examples of such functions are the sigmoidal functions which include the hyperbolic tangent, the logistic function

$$\sigma(\beta) = \frac{1}{1 - e^{-\beta}},$$

the Gaussian function

$$\sigma(\beta) = e^{-\beta^2},$$

and others.

The network shown in Figure 8.4 is a one-hidden-layer network. One could add more layers, leading to more complex *multilayer neural networks* Mathematically, the I/O properties of the neural network in Figure 8.4 may be expressed as

$$y_i = \sum_{j=1}^{l} w_{ji}^* \sigma \left(\sum_{k=1}^{n} a_{kj}^* x_k + d_j^* \right), \quad i = 1, 2, \dots, m. \tag{8.48}$$

In multilayer networks this function becomes more complex. For simplicity and clarity of presentation let us assume that $m = 1$ so that

$$y = \sum_{i=1}^{l} w_i^* \sigma \left(\sum_{j=1}^{n} a_{ji}^* x_j + d_i^* \right). \tag{8.49}$$

It has been established [180–183] that if l is sufficiently large, i.e., enough neurons are used, then there exist weight values w_i^*, a_{ji}^*, d_i^* such that the neural network shown in Figure 8.4 with $m = 1$ can approximate any continuous function $y = f(x)$ to any degree of accuracy for all x in some compact set $\Omega \subset \mathcal{R}^n$. This existence result is important but provides no clue regarding how to choose w_i^*, a_{ji}^*, d_i^* to approximate any function $y = f(x)$ given the I/O data $x(t)$, $y(t)$. In an effort to find these weights researchers formulated the problem as an online parameter identification problem as follows.

The same network as in Figure 8.4 is built but with the unknown weights w_i^*, a_{ji}^*, d_i^* replaced with their online estimates w_i, a_{ji}, d_i, producing the output

$$\hat{y} = \sum_{i=1}^{l} w_i(t) \sigma \left(\sum_{j=1}^{n} a_{ji}(t) x_j + d_j(t) \right). \tag{8.50}$$

The adjustments of the weights can then be done by following the same approach as in PI of LTI systems.

We define the estimation error $\varepsilon = y - \hat{y}$ and use the gradient method to minimize the cost,

$$J = \frac{\varepsilon^2}{2} = \frac{1}{2} \left[y - \sum_{i=1}^{l} w_i(t) \sigma \left(\sum_{j=1}^{n} a_{ji}(t) x_j + d_j(t) \right) \right]^2,$$

leading to

$$\dot{w}_i = -\gamma_i \frac{\partial J}{\partial w_i}, \qquad \dot{a}_{ji} = -\gamma_{ji} \frac{\partial J}{\partial a_{ji}}, \qquad \dot{d}_i = -\gamma_{di} \frac{\partial J}{\partial d_i}, \tag{8.51}$$

where $\gamma_i, \gamma_{ji}, \gamma_{di}$ are positive constants. In the neural network literature the above update laws and their variations are used to "train" the neural network. By *training*, it is implied that after presenting the network with the input data x_i the weights will settle or converge to some values indicating that the network learned from the data and can now reproduce the unknown map or function for new sets of inputs x. While this wishful thinking, supported by numerous simulation studies, prevailed for some time in the neural network literature, it has not been supported by analysis for the reasons we have already covered in Chapters 3 and 4. For example, the unknown parameters in (8.48) appear nonlinearly, and the gradient algorithm (8.51) cannot be shown to have the desired properties that we have in the case of linear in the parameters parameterizations. Furthermore, as in the LTI case, for parameter convergence, the regressor vector has to satisfy the persistence of excitation (PE) condition. The regressor vector in this case consists of nonlinear elements, and there is no theory that guarantees the PE property of the regressor even when the input x is rich in frequencies. The problem of linear in the parameters parameterization can be resolved by assigning values to a_{ji}^*, d_i^* so that in (8.48) only w_i^* is unknown. One class of neural networks where this approach has been followed is that of *radial basis functions* (*RBFs*), where the activation function is of the form

$$\sigma_i\left(\frac{|x-c_i|}{\lambda_i}\right), \quad i=1,2,\dots,l,$$

where $c_i \in \mathcal{R}^n$ is the center, $\lambda_i > 0$ is the width of the kernel unit, and $|\cdot|$ is the Euclidean vector norm. By fixing the centers c_i to be evenly distributed within the compact set of interest and the $\lambda_i > 0$ to provide sufficient overlap with the neighboring functions, the approximation properties of the linear in the parameters or weights neural network may be less affected by fixing c_i and λ_i a priori. In such a case, the parametric model

$$y=\sum_{i=1}^{l} w_i^* \sigma_i\left(\frac{|x-c_i|}{\lambda_i}\right) \tag{8.52}$$

is linear in the unknown parameters w_i^*, and the adaptive law for w_i is given by

$$\dot{w}_i=-\gamma_i\frac{\partial J}{\partial w_i}=\gamma_i\varepsilon\sigma_i\left(\frac{|x-c_i|}{\lambda_i}\right), \quad i=1,2,\dots,l, \tag{8.53}$$

and

$$\varepsilon=y-\hat{y}=-\sum_{i=1}^{l}\tilde{w}_i\sigma_i\left(\frac{|x-c_i|}{\lambda_i}\right)=-\tilde{\theta}^T\phi,$$

where

$$\tilde{\theta}=[\tilde{w}_1,\dots,\tilde{w}_l]^T, \qquad \phi=[z_1,\dots,z_l]^T, \qquad z_i=\sigma_i\left(\frac{|x-c_i|}{\lambda_i}\right).$$

Since ϕ is bounded, no normalization is needed, and the adaptive law (8.53) guarantees that $w_i \in \mathcal{L}_\infty$, $\dot{w}_i, \varepsilon \in \mathcal{L}_2 \cap \mathcal{L}_\infty$. For convergence of w_i to w_i^*, however, ϕ is required to be PE, something that cannot be guaranteed by choosing the inputs x_i due to the nonlinearities in ϕ. As in the LTI case, one can ignore parameter convergence when the adaptive law is used as part of the feedback control law and focus on the performance of the closed-loop system.

Let us now consider the nonlinear system

$$\dot{x}=f(x)+g(x)u, \tag{8.54}$$

where $x \in \mathcal{R}^n$, $u \in \mathcal{R}$, and $f(x)$, $g(x)$ are vectors of smooth functions. We approximate $f(x)$, $g(x)$ with neural networks, i.e.,

$$f(x)=f_N(x,\theta_f^*), \qquad g(x)=g_N(x,\theta_g^*),$$

where θ_f^*, θ_g^* are the desired values of the weights of the respective neural networks defined as

$$\theta_f^*=\arg\min_{\theta_f}\left\{\sup_{x\in\Omega}|f(x)-f_N(x,\theta_f)|\right\},$$

$$\theta_g^*=\arg\min_{\theta_g}\left\{\sup_{x\in\Omega}|g(x)-g_N(x,\theta_g)|\right\},$$

i.e., θ_f^*, θ_g^* are the desired values of the weights which produce the smallest error in approximating the unknown functions f, g for a given neural network architecture.

The system (8.54) may be written as

$$\dot{x} = f_N(x, \theta_f^*) + g_N(x, \theta_g^*)u + v,$$

where $v \in \mathcal{R}^n$ is the approximation error vector given by

$$v = f(x) - f_N(x, \theta_f^*) + [g(x) - g_N(x, \theta_g^*)]u.$$

If the neural network is designed to be linear in the unknown parameters θ_f^*, θ_g^*, then we have

$$\dot{x}_i = \theta_{fi}^{*^T} \phi_f + \theta_{gi}^{*^T} \phi_g u + v_i, \quad i = 1, 2, \ldots, n,$$

where $\phi_f \in \mathcal{R}^{n_f}$, $\phi_g \in \mathcal{R}^{n_g}$ (n_f, n_g are some positive integer design constants) contain the nonlinear functions excited by x; $\theta_f^* = [\theta_{f1}^{*T}, \ldots, \theta_{fn}^{*T}]^T$, $\theta_g^* = [\theta_{g1}^{*T}, \ldots, \theta_{gn}^{*T}]^T$, and $\theta_{fi} \in \mathcal{R}^{n_f}, \theta_{gi} \in \mathcal{R}^{n_g}$ $\forall i = 1, 2, \ldots, n$. In the above form, the parameters appear linearly, and a wide range of robust adaptive laws can be generated using the results of Chapters 3 and 4. These adaptive laws will have guaranteed stability properties despite the presence of the uncertainty v, provided the normalizing signal bounds ϕ_f, $\phi_g u$, and v from above. We leave the details of these designs to the reader since they follow directly from the results of the previous chapters.

8.7.2 Neuroadaptive Control

Since neural networks can be used to approximate nonlinear functions online, the control laws designed for nonlinear plants with known nonlinearities can be used as the basis for developing adaptive control laws to deal with the same class of plants with unknown nonlinearities. We demonstrate this neuroadaptive control approach for the following nonlinear plant in the normal form with zero dynamics:

$$\begin{aligned} \dot{x}_1 &= x_2, \\ \dot{x}_2 &= x_3, \\ &\vdots \\ \dot{x}_n &= f(x) + g(x)u, \\ y &= x_1, \end{aligned} \tag{8.55}$$

where $x = [x_1, \ldots, x_n]^T$ is available for measurement, $x_2 = \dot{y}, \ldots, x_n = y^{(n-1)}$, f, g are smooth functions, and u is the scalar input. The problem is to design a control law u such that the output $y(t)$ tracks a given desired trajectory $y_d(t)$, a known smooth function of time, with known derivatives.

Assumption 8.7.1. *$g(x)$ is bounded from below by a constant $\bar{b}$; i.e., $g(x) \geq \bar{b}$ $\forall x \in \mathcal{R}^n$.*

This assumption guarantees that the system (8.55) is controllable $\forall x \in \mathcal{R}^n$. We define the scalar function $S(t)$ as the metric that describes the tracking error dynamics,

$$S(t) = \left(\frac{d}{dt} + \lambda\right)^{n-1} e(t), \tag{8.56}$$

where $\lambda > 0$ is a design constant. $S(t) = 0$ represents a linear differential equation whose solution implies that $e(t)$ and its $(n-1)$ derivatives converge to zero exponentially fast. Differentiating $S(t)$ with respect to time, we obtain

$$\begin{aligned}\dot{S}(t) &= e^{(n)} + \alpha_{n-1}e^{(n-1)} + \cdots + \alpha_1\dot{e} \\ &= f(x) + g(x)u - y_d^{(n)} + (\alpha_{n-1}e^{(n-1)} + \cdots + \alpha_1\dot{e}),\end{aligned}$$

where $\alpha_{n-1}, \ldots, \alpha_1$ are the coefficients of the polynomial $(s+\lambda)^{n-1}$.

Let

$$v = -y_d^{(n)} + (\alpha_{n-1}e^{(n-1)} + \cdots + \alpha_1\dot{e}).$$

Then $\dot{S}(t)$ can be written in the compact form

$$\dot{S}(t) = f(x) + g(x)u + v. \tag{8.57}$$

If $f(x)$ and $g(x)$ are known functions, then the control law

$$u = \frac{1}{g(x)}(-f(x) - v - k_S S(t)), \tag{8.58}$$

where $k_S > 0$ is a design constant, can be used to meet the control objective. Indeed, by applying (8.58) to (8.57), we obtain

$$\dot{S} = -k_S S(t),$$

which implies that $S(t)$ converges to zero exponentially fast. Since $S = 0$ implies that the tracking error $e(t)$ and its first $(n-1)$ derivatives converge to zero, the control objective is achieved.

Let us consider the case where $f(x)$ and $g(x)$ are unknown and approximated as

$$\begin{aligned}f(x) &\approx f^a(x) = \sum_{i=1}^{l_f} \theta_{fi}^* \phi_{fi}(x) = \theta_f^{*^T} \phi_f, \\ g(x) &\approx g^a(x) = \sum_{i=1}^{l_g} \theta_{gi}^* \phi_{gi}(x) = \theta_g^{*^T} \phi_g,\end{aligned}$$

where θ_f^*, θ_g^* are the unknown "optimal" weights; $\phi_f(x), \phi_g(x)$ are selected basis functions of a neural network, which approximates the unknown functions in a compact set Ω of the vector x; and l_f, l_g, are the number of the nodes. The neural network approximation errors $d_f(x), d_g(x)$ are given by

$$d_f(x) := f(x) - f^a(x), \qquad d_g(x) := g(x) - g^a(x).$$

The unknown "optimal" weights θ_f^*, θ_g^* minimize $d_f(x)$, $d_g(x)$ $\forall x \in \Omega$, respectively, i.e.,

$$\theta_f^* := \arg\min_{\theta_f \in \Re^{l_f}} \left\{ \sup_{x\in\Omega} |f(x) - f^a(x)| \right\}, \qquad \theta_g^* := \arg\min_{\theta_g \in \Re^{l_g}} \left\{ \sup_{x\in\Omega} |g(x) - g^a(x)| \right\}.$$

Assumption 8.7.2. *The approximation errors are upper bounded by some known constants $\psi_f > 0$ and $\psi_g > 0$ over the compact set $\Omega \subset \Re^n$, i.e.,*

$$\sup_{x\in\Omega} |d_f(x)| \le \psi_f, \qquad \sup_{x\in\Omega} |d_g(x)| \le \psi_g.$$

The online estimate of the unknown functions generated by the neural network is then given as

$$\begin{aligned} \hat{f}(x,t) &= \sum_{i=1}^{l_f} \theta_{fi}(t)\phi_{fi}(x) = \theta_f^T(t)\phi_f, \\ \hat{g}(x,t) &= \sum_{i=1}^{l_f} \theta_{gi}(t)\phi_{gi}(x) = \theta_g^T(t)\phi_g, \end{aligned} \tag{8.59}$$

where $\theta_f(t)$, $\theta_g(t)$ are to be generated by an adaptive law online. Following the CE approach, we can propose the control law

$$u = \frac{1}{\hat{g}(x,t)}(-\hat{f}(x,t) - \upsilon - k_S S(t)). \tag{8.60}$$

The problem that arises with (8.60) is the possibility of $\hat{g}$ approaching or crossing zero, giving rise to the so-called stabilizability problem, which is also present in indirect APPC. Another problem is the effect of the deviations of $\hat{f}$ and $\hat{g}$ from the actual nonlinearities f and g on the stability of the closed-loop system. A considerable number of research efforts have concentrated on resolving the above two problems [156–179]. In most of these approaches fixes such as switching, projection, and additional nonlinear terms are used to modify (8.60) for stability. Since the approximation of the nonlinear functions f, g with neural networks is valid as long as x is inside a compact set, the updating of the weights in $\hat{f}$, $\hat{g}$ as well as the control law should not push x outside the set. This consideration has often been ignored in most papers. An approach that resolves the above issues and guarantees that the tracking error at steady state remains within prespecified bounds has been presented for single input systems in [177] and for multiple input systems in [178] for the case of a nonlinear system in the normal form with no zero dynamics. The approach involves continuous switching dead zones and additional nonlinear terms in a modified CE control law with robust adaptive laws. We present the algorithm of [177] and its properties below.

Instead of the CE control law (8.60), we choose the control law

$$u = \frac{\hat{g}(x,t)}{(\hat{g}(x,t))^2 + \delta_b}[-k_S S(t) - \sigma_\upsilon|\upsilon(t)|S(t) - \sigma_f|\hat{f}(x,t)|S(t) - \upsilon(t) - \hat{f}(x,t)], \tag{8.61}$$

where $k_S, \sigma_\upsilon, \sigma_f, \delta_b > 0$ are design parameters. The parameters $\theta_{fi}(t)$, $\theta_{gi}(t)$ in (8.59) are updated as follows:

$$\dot{\theta}_{fi}(t) = \gamma_{fi} S_\Delta \phi_{fi}(x), \quad i = 1, 2, \dots, l_f, \tag{8.62}$$

$$\dot{\theta}_{gi}(t) = \gamma_{gi}[S_\Delta u\phi_{gi}(x) + \rho(t)\sigma_b \operatorname{sgn}(g(x))|S_\Delta|(|u'| + |u|)\phi_{gi}(x)], \quad i = 1, 2, \ldots, l_g, \tag{8.63}$$

where

$$S_\Delta(t) = S - \Phi \operatorname{sat}\left(\frac{S}{\Phi}\right),$$

$$u' = \left\{\frac{1}{[(\hat{g})^2 + \delta_b]}\right\}[-k_S S - \sigma_v|v|S - \sigma_f|\hat{f}|S - v - \hat{f}],$$

$\gamma_{fi}, \gamma_{gi} > 0$ are the adaptive gains; $\sigma_b > 0$ is a small design parameter; Φ is a positive design constant; $\operatorname{sgn}(\cdot)$ is the sign function ($\operatorname{sgn}(x) = 1$ if $x \geq 0$ and $\operatorname{sgn}(x) = -1$ otherwise); $\operatorname{sat}(\cdot)$ is the saturation function ($\operatorname{sat}(x) = x$ if $|x| \leq 1$ and $\operatorname{sat}(x) = \operatorname{sgn}(x)$ otherwise); and $\rho(t)$ is a continuous switching function given by

$$\rho(t) = \begin{cases} 0 & \text{if } |\hat{g}(x,t)| \geq \bar{b} - \psi_g, \\ \frac{\bar{b} - \psi_g - |\hat{g}(x,t)|}{\Delta} & \text{if } \bar{b} - \psi_g - \Delta < |\hat{g}(x,t)| < \bar{b} - \psi_g, \\ 1 & \text{if } |\hat{g}(x,t)| \leq \bar{b} - \psi_g - \Delta, \end{cases} \tag{8.64}$$

where $\Delta > 0$ is a design parameter. The continuous switching function $\rho(t)$, instead of a discontinuous one, is used in order to guarantee the existence and uniqueness of solutions of the closed-loop system. By design, the control law (8.61) will never involve division by zero since $(\hat{g}(x,t))^2 + \delta_b > \delta_b > 0 \, \forall x, t$. Therefore, the proposed controller overcomes the difficulty encountered in implementing many adaptive control laws, where the identified model becomes uncontrollable at some points in time. It is also interesting to note that $u \to 0$ with the same speed as $\hat{g}(x,t) \to 0$. Thus, when the estimate $\hat{g}(x,t) \to 0$, the control input remains bounded and also reduces to zero. In other words, in such cases it is pointless to control what appears to the controller as an uncontrollable plant. The above control law is complex and was developed by trial and error using a Lyapunov synthesis approach. The stability and performance properties of the closed-loop system are described by the following theorem.

Theorem 8.7.3. *Consider the system* (8.55), *with the adaptive control law* (8.61)–(8.64). *Assume that the lower bound of* $|g(x)| \geq \bar{b}$ *satisfies the condition*

$$\bar{b} > \sqrt{\delta_b} + \Delta + 3\psi_g \tag{8.65}$$

for some small positive design constants δ_b *and* Δ. *For any given arbitrary small positive number* Φ, *there exist positive constants*

$$\delta_1 = \frac{\psi_g}{\bar{b} - \psi_g - \Delta} + \frac{\delta_b}{(\bar{b} - \psi_g - \Delta)^2 + \delta_b} < 1,$$
$$\delta_2 = \frac{2\delta_b}{\bar{b} - \psi_g}, \qquad \delta_3 = \frac{\psi_g}{\bar{b} - \psi_g} \tag{8.66}$$

such that if

$$k_S \geq \frac{\psi_f}{[(1-\delta_1)\Phi]}, \qquad \sigma_v \geq \frac{\delta_1}{[(1-\delta_1)\Phi]}, \qquad \sigma_f \geq \frac{\delta_1}{[(1-\delta_1)\Phi]}, \qquad \sigma_b \geq \max\{\delta_2, \delta_3\}, \tag{8.67}$$

and $\{\theta_f(0), \theta_g(0)\} \in \Omega_\Theta$, $x(0) \in \Omega_x \subset \Omega$, *where* $\Omega_\Theta \subset \Re^{l_f+l_g}$ *is a compact set in the space of* θ_f, θ_g, *then all signals in the closed-loop system are bounded. Furthermore, the tracking error and its derivatives satisfy*

$$\lim_{t\to\infty} |e^{(i)}(t)| \le 2^i \lambda^{i-n+1}\Phi, \quad i = 0, 1, \ldots, n-1. \tag{8.68}$$

The proof of Theorem 8.7.3 is based on the use of the Lyapunov-like function

$$V(t) = \frac{S_\Delta^2}{2} + \sum_{i=1}^{l_f} \frac{\tilde{\theta}_{fi}^2}{2\gamma_{fi}} + \sum_{i=1}^{l_g} \frac{\tilde{\theta}_{gi}^2}{2\gamma_{gi}},$$

where $\tilde{\theta}_{fi} = \theta_{fi} - \theta_{fi}^*$, $\tilde{\theta}_{gi} = \theta_{gi} - \theta_{gi}^*$ are the parameter errors, and is long and technical. The details of the proof are given in [177].

The importance of the neuroadaptive control law (8.61)–(8.64) is that it guarantees stability and steady-state tracking error, which depends solely on design parameters as indicated by (8.67), provided that the design parameters are chosen to satisfy the inequalities (8.65), (8.66). The design parameters may be chosen to satisfy (8.65)–(8.67) as follows:

1. Using the upper bound of the approximation error ψ_g, check whether the lower bound of $g(x)$ satisfies $\bar{b} > 3\psi_g$. If so, choose the design parameters δ_b, Δ such that $\sqrt{\delta_b} + \Delta + 3\psi_b < \bar{b}$. If not, then the number of nodes l_g of the neural network for $g(x)$ has to be increased in order to obtain a better approximation. It is worth noting that the ratio $\frac{\psi_g}{\bar{b}} < \frac{1}{3}$ gives us an upper bound for the approximation error $d_g(x)$ that can be tolerated by the closed-loop system.

2. Calculate δ_i, $i = 1, 2, 3$, using (8.66) and the knowledge of $\bar{b}$, ψ_g, δ_b, Δ.

3. Set the desired upper bound for the tracking error at steady state equal to $\lambda^{-n+1}\Phi$ and choose λ, Φ to satisfy (8.67).

4. Choose σ_b in the adaptive law such that $\sigma_b \ge \max(\delta_2, \delta_3)$. From $\frac{\psi_g}{\bar{b}} < \frac{1}{3}$, we have $\delta_2 < \frac{3\delta_b}{\bar{b}}$, $\delta_3 < \frac{1}{2}$.

5. Choose k_S, σ_v, σ_f such that $k_S \ge \frac{\psi_f}{1-\delta_1}\Phi$, $\sigma_v \ge \frac{\delta_1}{1-\delta_1}\Phi$, $\sigma_f \ge \frac{\delta_1}{1-\delta_1}\Phi$.

We demonstrate the properties of the proposed adaptive control law using the second-order nonlinear system

$$\ddot{x} = -4\left[\sin\frac{4\pi x}{\pi x}\right]\left[\sin\frac{\pi\dot{x}}{\pi\dot{x}}\right]^2 + (2 + \sin(2\pi x) + 0.1\sin(100t))u.$$

The output $y(t) = x(t)$ is required to track the trajectory $y_d = \sin(\pi t)$. The magnitude of the tracking error at steady state is required to be less than 0.05. A one-hidden-layer radial Gaussian network with a basis function

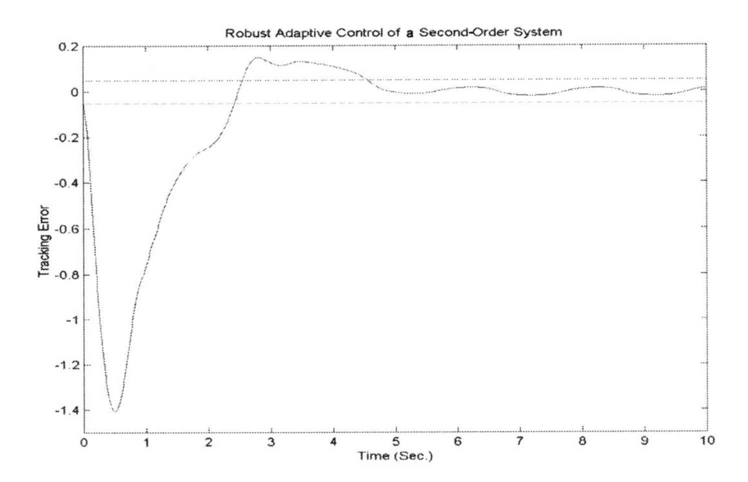

Figure 8.5. *Tracking error during the first* 10 *seconds. (The dashed lines indicate the required error bound.)*

$$\phi_i(x) = \exp[-\pi\sigma^2(x-\xi_i)^T(x-\xi_i)] \tag{8.69}$$

is used to approximate

$$f(x,\dot{x}) = -4\left[\sin\frac{4\pi x}{\pi}x\right]\left[\sin\frac{\pi\dot{x}}{\pi}\dot{x}\right]^2, \qquad g(x) = 2+\sin(2\pi x)+0.1\sin(100t)$$

on a compact set $\Omega = \Omega_x \times \Omega_{\dot{x}}$, where $\Omega_x = \{x|x \in (-3,3)\}$, $\Omega_{\dot{x}} = \{\dot{x}|\dot{x} \in (-5,5)\}$. The design parameters in (8.69) are chosen as $\xi_i = 0.125$, $\sigma^2 = 4\pi$. The approximation error parameters are assumed to satisfy $\psi_f \leq 0.05$, $\psi_g \leq 0.12$. Furthermore, $\bar{b} = 0.9 > 3\psi_b = 0.36$. The values of δ_b, Δ are chosen to be 0.01, 0.04, respectively, such that $\sqrt{\delta_b} + \Delta + 3\psi_g = 0.5 < \bar{b} = 0.9$ is satisfied. Using (8.66), we calculate $\delta_1 = 0.18$, $\delta_2 = 0.026$, $\delta_3 = 0.154$. By selecting $\lambda = 1$ and $\Phi = 0.05$, the tracking error is bounded from above by 0.05 at steady state. The other design parameters are chosen as $k_S = 2$, $\sigma_v = 4.4$, $\sigma_f = 4.4$, and $\sigma_b = 0.16$. The initial values for the estimated parameters are chosen as $\theta_{fi}(0) = 0, \theta_{gi}(0) = 0$. Figures 8.5 and 8.6 show the simulation results. Figure 8.5 shows the convergence of the tracking error to within the prespecified bound. Figure 8.6 demonstrates that the switching of $\rho(t)$ stops after an initial learning stage and $\rho(t)$ converges to a constant with no further switching.

8.8 Case Study: Adaptive Nonlinear Control of a Path-Tracking Vehicle

We consider the control of an autonomous ground vehicle designed to carry out missions involving area coverage. The vehicle is assumed to be equipped with an accurate navigation system so that at any instant it is able to measure its location and orientation in the navigation coordinate frame. Many times, the path is composed of straight-line back and forth sweeps,

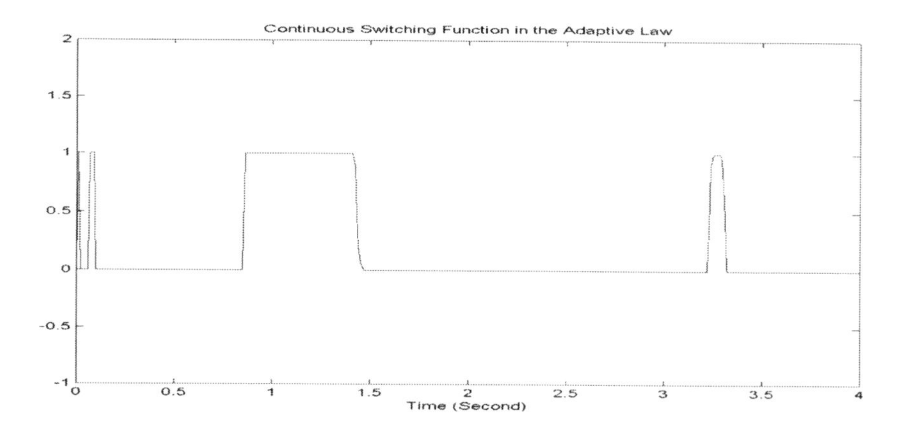

Figure 8.6. *The continuous switching function $\rho(t)$ in the adaptive law during the first 4 seconds.*

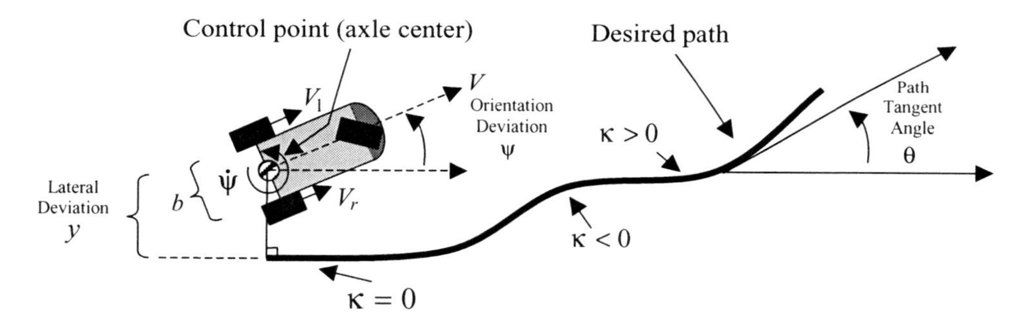

Figure 8.7. *The path-tracking vehicle.*

but this is not necessarily the case. The presence of obstacles or other physical constraints as well as work assignments may require the vehicle to track paths with curvatures, etc.

Suppose now that the vehicle is differential drive with a front castor wheel. The vehicle is capable of turning in place (zero-turn radius) by keeping the velocity at zero with a nonzero heading rate. The kinematic equations for such a vehicle are listed below, with some of the variables presented in Figure 8.7. The linear velocity of the vehicle, V, is the average of V_r and V_l, the linear speeds of the right wheel and left wheel, respectively.

The kinematic relationship between the path deviation and the control input is given by [184]

$$\begin{aligned} \dot{y} &= V \sin \psi, \\ \dot{\psi} &= u - \frac{\kappa}{1 - \kappa y} V \cos \psi, \end{aligned} \tag{8.70}$$

where $\kappa(s) = \frac{d\theta}{ds}$ is the curvature of the desired path, y is the path lateral deviation, ψ is the path orientation deviation, and u is the control input, as illustrated in Figure 8.7. There is a singularity at $y = \frac{1}{\kappa}$, which corresponds to the case where the robot location is the center of the osculating circle; in practice, this condition must be monitored (i.e., this is very poor path-tracking performance, or poor initial conditions, so the system should take special measures when $y \to \frac{1}{\kappa}$) and avoided by not planning paths where the radius of curvature is of the same order as the achievable path-tracking accuracy (i.e., don't plan paths of high curvature).

The path-tracking control objective is to choose the control input u to regulate the lateral error y to zero. Using u and the equations

$$V_r = V + \frac{b\dot{\psi}}{2}, \qquad V_l = V - \frac{b\dot{\psi}}{2}, \qquad V = \frac{V_r + V_l}{2},$$

we can calculate the desired speeds of the wheels V_r and V_l, which are passed to a lower controller which adjusts the appropriate actuators so that the wheels achieve the desired speeds that would lead to y going to zero. In this study we are concerned with the choice of u, and we assume that the wheel actuators will achieve the desired speeds based on the given commands. Assuming small deviations from the desired path, we can use the approximation $\sin\psi \approx \psi$ and express (8.70) as

$$\begin{aligned} \dot{y} &= V\psi + d_0^*, \\ \dot{\psi} &= u + d_1^* V \cos\psi + d_2^*, \end{aligned} \tag{8.71}$$

where d_i^*, $i = 0, 1, 2$, are unknown disturbance terms to account for the modeling inaccuracies as well as curvature calculation inaccuracies. In our derivation of the control we will assume that these terms are constant. The control law, however, will be applied to the more complicated model (8.70) for evaluation using simulations.

We use adaptive backstepping to meet the control objective as follows: If ψ was the control variable, then $\psi = \frac{(-\hat{d}_0 - \alpha y)}{V}$, where $\hat{d}_0$ is the online estimate of the disturbance term d_0^* and $\alpha > 0$ is a design constant, could be used. Following the backstepping procedure, we define the new state variable

$$z = \psi - \frac{(-\hat{d}_0 - \alpha y)}{V} \tag{8.72}$$

and transform (8.71) into

$$\begin{aligned} \dot{y} &= -\alpha y + Vz - (\hat{d}_0 - d_0^*), \\ \dot{z} &= u + d_1^* V \cos\psi + d_2^* + \frac{\dot{\hat{d}}_0}{V} + \alpha\psi + \frac{\alpha}{V} d_0^*, \end{aligned} \tag{8.73}$$

which motivates the feedback linearizing control law

$$u = -cz - \hat{d}_1 V \cos\psi - \hat{d}_2 - \frac{\dot{\hat{d}}_0}{V} - \alpha\psi - \frac{\alpha}{V}\hat{d}_0, \tag{8.74}$$

where $\hat{d}_l$ is the online estimate of d_i^*, $i = 0, 1, 2$, to be generated by an adaptive law and $c > 0$ is a design constant. The closed-loop system becomes

$$\begin{aligned} \dot{y} &= -\alpha y + Vz - \tilde{d}_0, \\ \dot{z} &= -cz - \tilde{d}_1 V \cos\psi - \tilde{d}_2 - \frac{\alpha}{V}\tilde{d}_0, \end{aligned} \tag{8.75}$$

where $\tilde{d}_i = \hat{d}_i - d_i^*$, $i = 0, 1, 2$, is the parameter error. The adaptive law for the parameter estimates $\hat{d}_i$ is generated using the following Lyapunov function candidate:

$$V_a = c_0\frac{y^2}{2} + \frac{z^2}{2} + c_0\frac{\tilde{d}_0^2}{2\gamma_0} + \frac{\tilde{d}_1^2}{2\gamma_1} + \frac{\tilde{d}_2^2}{2\gamma_2},$$

where $\gamma_i > 0$, $i = 0, 1, 2$, are design constants and $c_0 > 0$ is an arbitrary constant to be selected. The time derivative of V_a along the solution of (8.75) is given by

$$\dot{V}_a = -c_0\alpha y^2 - cz^2 + c_0 Vzy - c_0\tilde{d}_0\left(y + \frac{\alpha z}{V}\right) - \tilde{d}_1 zV\cos\psi - \tilde{d}_2 z + c_0\frac{\tilde{d}_0\dot{\hat{d}}_0}{\gamma_0} + \frac{\tilde{d}_1\dot{\hat{d}}_1}{\gamma_1} + \frac{\tilde{d}_2\dot{\hat{d}}_2}{\gamma_2}.$$

Choosing the adaptive laws

$$\dot{\hat{d}}_0 = \gamma_0\left(y + \frac{\alpha z}{V}\right), \qquad \dot{\hat{d}}_1 = \gamma_1 zV\cos\psi, \qquad \dot{\hat{d}}_2 = \gamma_2 z, \tag{8.76}$$

we obtain

$$\dot{V}_a = -c_0\alpha y^2 - cz^2 + c_0 Vzy = -\frac{c_0\alpha y^2}{2} - \frac{cz^2}{2} - \frac{c_0\alpha}{2}\left(y - \frac{zV}{\alpha}\right)^2 - z^2\left(\frac{c}{2} - \frac{c_0V^2}{2\alpha}\right).$$

Choosing the arbitrary constant $c_0 > 0$ as

$$0 < c_0 \le \frac{\alpha c}{V^2},$$

we obtain

$$\dot{V}_a \le -\frac{[c_0\alpha y^2 + cz^2]}{2},$$

which implies that all signals are bounded and that y, z are square integrable. Since from (8.75), $\dot{y}, \dot{z}$ are bounded it follows by applying Barbalat's lemma that z, y converge to zero.

For simulation testing of the nonlinear adaptive control law (8.74), (8.76) we use the model

$$\begin{aligned} \dot{y} &= V\sin\psi + n_s + d_y, \\ \dot{\psi} &= u - \frac{\kappa}{1-\kappa y}V\cos\psi + n_s + d_\psi, \end{aligned} \tag{8.77}$$

where n_s is white Gaussian noise with zero mean and variance 0.01 and $d_y = .1\sin(0.001t)$, $d_\psi = 0.08\sin(0.002t)$ are unknown deterministic disturbances. The path to be followed is a straight line for about 30 meters (10 sec), then a part of a semicircle of radius 100 meters of about 27 meters length (9 sec) in the counterclockwise direction, then a part of a semicircle in the clockwise direction of radius 100 meters for about another 27 meters (9 sec), then in a straight line for about 30 meters (10 sec). The changing curvature of the path is shown in Figure 8.8.

Let us first assume that the curvature of the path is known and use the nonadaptive control law

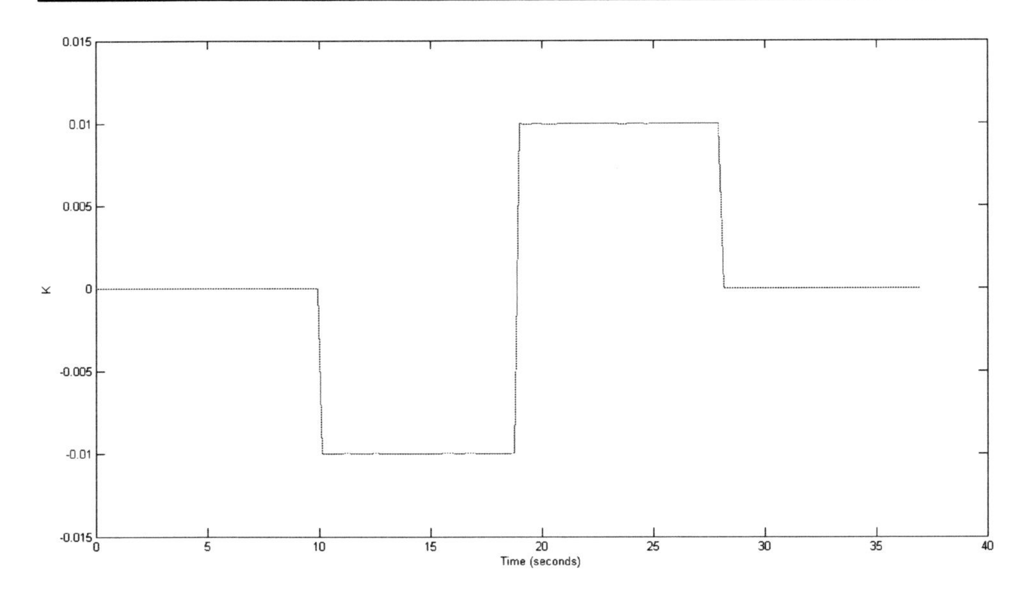

Figure 8.8. *Changing curvature of the path.*

$$u = -cz + \frac{\kappa}{1-\kappa y} V \cos \psi - \alpha \psi, \quad z = \psi + \frac{\alpha y}{V}, \tag{8.78}$$

which can be shown to meet the control objective in the absence of noise and disturbances. Figure 8.9 shows the time response of y, ψ as the vehicle follows the path. There is an initial transient due to initial deviations from the direction of the path, but y converges close to zero with some steady-state oscillations due to the effect of noise and disturbances. The heading deviation ψ converges to -2π, which is equivalent to zero in the polar coordinates, and it is also affected by noise and the disturbances at steady state. Instead of (8.78) let us apply the nonlinear adaptive control law (8.74), (8.76), where the adaptive laws (8.76) are modified for robustness using projection so that the estimated parameters stay within certain bounds, i.e., $\hat{d}_0 \in [-4, 4]$, $\hat{d}_1 \in [-20, 20]$, $\hat{d}_2 \in [-3, 3]$. In this case the curvature is assumed to be unknown. Figure 8.10 shows the results for y, ψ. It is clear that the adaptation, in addition to handling the unknown curvature, also attenuates the effects of noise and disturbances leading to smaller steady-state errors and therefore more accurate tracking of the path.

Problems

1. Consider the nonlinear system

$$\begin{aligned}
\dot{x}_1 &= -x_1 + x_3^2, \\
\dot{x}_2 &= x_3, \\
\dot{x}_3 &= x_1 x_3 + u, \\
y &= x_2.
\end{aligned}$$

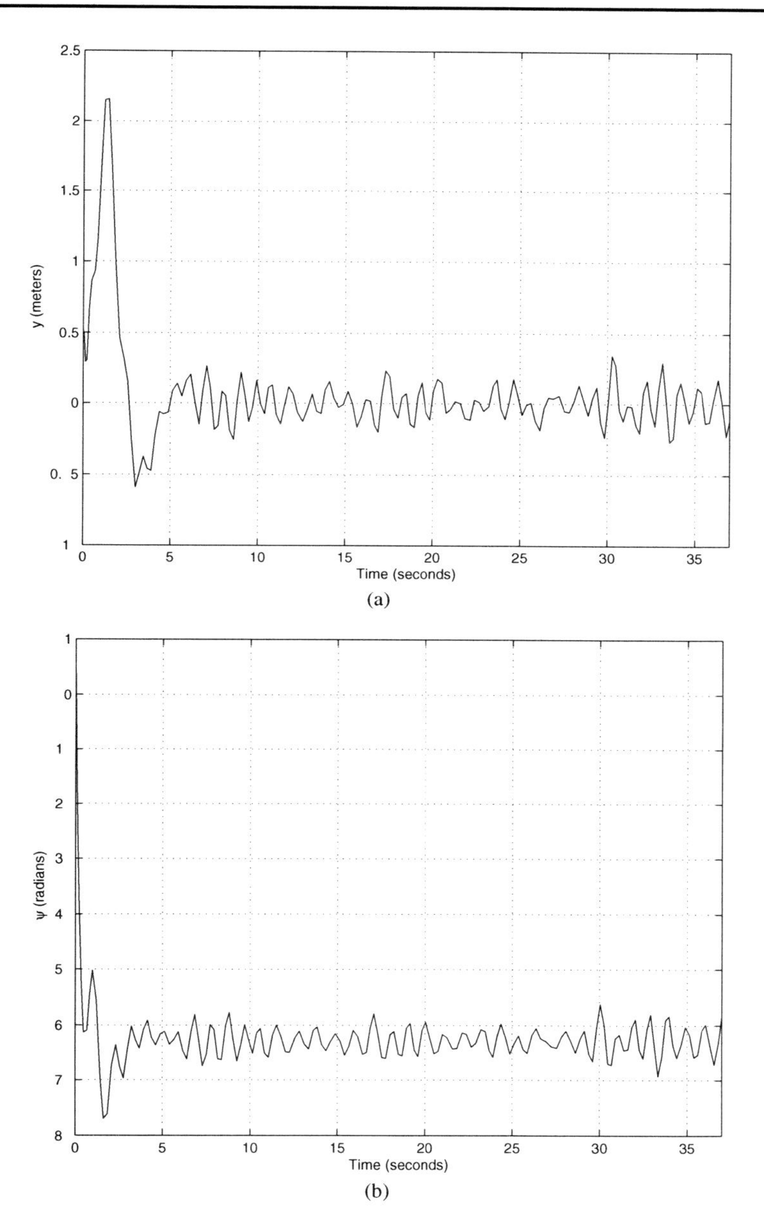

Figure 8.9. *Time response of* (a) *lateral and* (b) *heading deviations in nonadaptive control case.*

Design a control law of your choice to establish stability and convergence of x_1, x_3 to zero.

2. Consider the nonlinear system

$$\dot{x}_1 = u + \theta^* g(x_1),$$

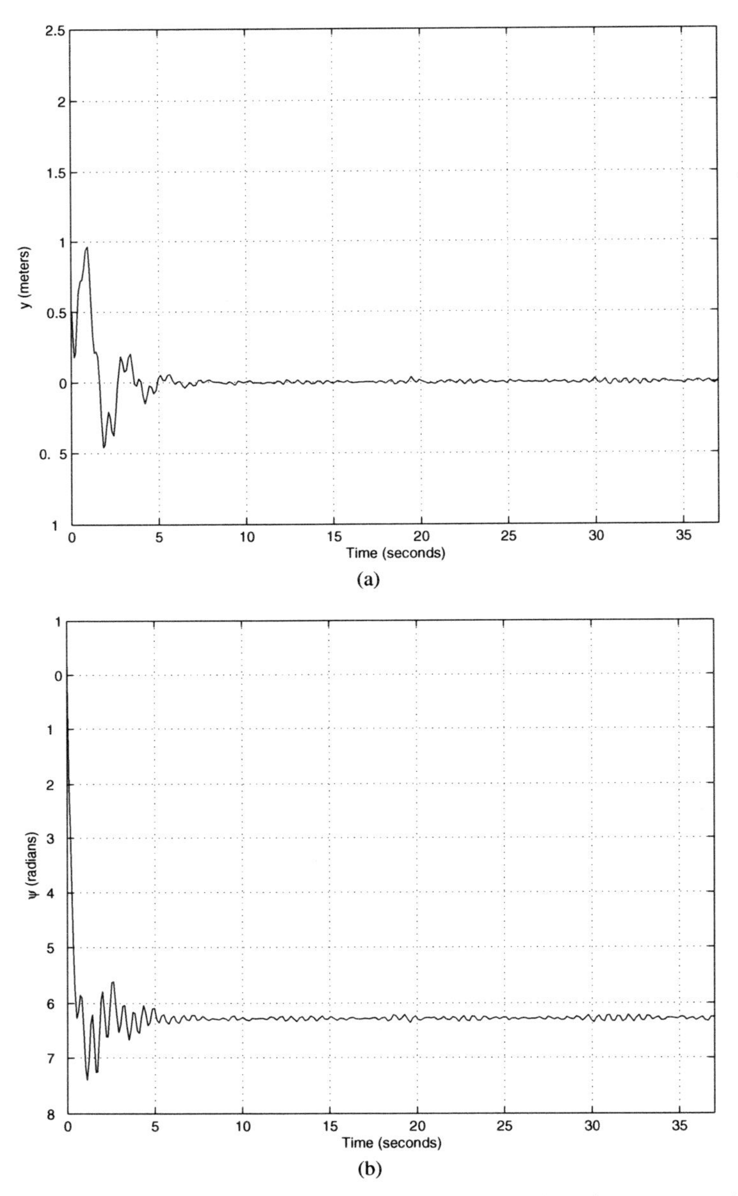

Figure 8.10. *Time response of* (a) *lateral and* (b) *heading deviations in the adaptive control case.*

where g is a continuous nonlinear function and θ^* is an unknown scalar.

(a) Show that the control law

$$u = -cx_1 - kx_1 g^2(x_1),$$

where $c, k > 0$ are design constants, guarantees that x_1 is bounded. Obtain an upper bound for x_1 and comment on how the design constants c, k affect the bound.

(b) Design a control law of your choice to guarantee closed-loop stability and convergence of x_1 to zero.

3. Consider the system

$$\dot{x}_1 = x_2 + \theta^* g(x_1),$$
$$\dot{x}_2 = u,$$

where g is a nonlinear differentiable function with the property $g(0) = 0$ and u is the control input.

(a) Design a stabilizing control law, assuming that θ^* is known.

(b) Design a nonlinear control without adaptation and show that it guarantees signal boundedness for any unknown parameter θ^*.

(c) Design an adaptive control law and show that it guarantees signal boundedness and convergence of the states to zero.

4. Consider the system

$$\dot{x}_1 = x_2 + \theta_1^{*T} g_1(x_1),$$
$$\dot{x}_2 = x_3,$$
$$\dot{x}_3 = u + \theta_2^{*T} g_2(x_1, x_2, x_3),$$

where g_1, g_2 are known nonlinear differentiable functions and u is the control input.

(a) Design a stabilizing control law assuming that θ_1^*, θ_2^* are known.

(b) Design an adaptive control law and show that it guarantees signal boundedness and convergence of the states to zero when θ_1^*, θ_2^* are unknown.

5. Consider the scalar plant

$$\dot{x} = x^5 + u + xd,$$

where d is an unknown bounded function of time acting as a disturbance to the plant. Design a control law to guarantee that all signals are bounded no matter how large the disturbance d is.

6. Consider the scalar plant

$$\dot{x} = f(x) + g(x)u,$$

where f, g are unknown smooth nonlinear functions with $g > 0 \; \forall x$. Design a one-hidden-layer neural network to generate an estimate of f, g online. Simulate your results for $f(x) = -x(2 + \sin x)$, $g(x) = (5 + \sin x^2)(1 + 0.5 \cos 2x)$, and an input of your choice.

7. In Problem 6, design a control law to stabilize the plant and force x to zero. Demonstrate your results using simulations.

8. Consider the scalar plant

$$\dot{x} = f(x) + g(x)u,$$

where f, g are unknown smooth nonlinear functions with $g > 0\ \forall x$. The control objective is to choose the control input u so that x follows the state of the reference model

$$\dot{x}_m = -a_m x_m + b_m r,$$

where r is a bounded reference input and $a_m > 0$, b_m are constants. Design a neuroadaptive control scheme to meet the control objective. Demonstrate your results using simulations with $f(x) = -x(1.2 + \sin x^2)$, $g(x) = (2 + \cos x^2)(1.5 + \cos 2x)$.

9. Consider the nth-order plant in the normal form with no zero dynamics,

$$\begin{aligned} \dot{x}_1 &= x_2, \\ \dot{x}_2 &= x_3, \\ &\vdots \\ \dot{x}_n &= f(x) + g(x)u, \quad y = x_1, \end{aligned}$$

where x is the full state and f, g are unknown smooth nonlinear functions. The control objective is to design the control law u so that y tracks the output of the reference model

$$y_m = \frac{1}{(s + a_m)^n} r,$$

where $a_m > 0$ and r is the reference input.

(a) Design a control law to meet the control objective when f, g are known.

(b) Repeat part (a) when f, g are unknown.

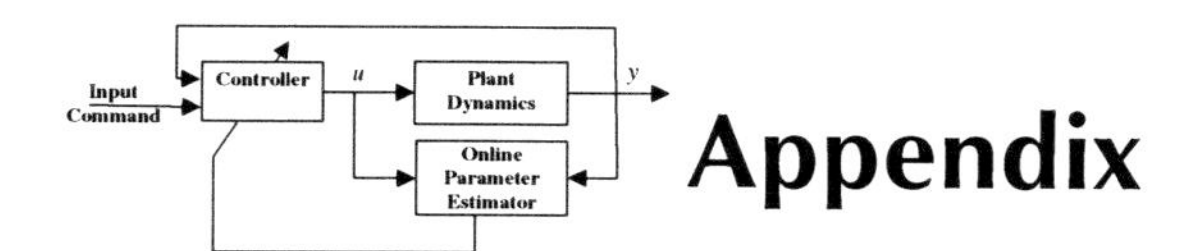

Appendix

A.1 Systems Theory

A *dynamical system* or a *plant* is defined as a structure or process that receives an input $u_p(t)$ at each time t and produces an output $y_p(t)$ as shown in Figure A.1.

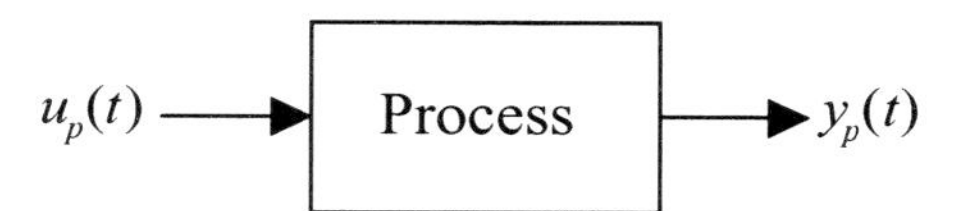

Figure A.1. *Dynamical system.*

Many systems of interest in the area of controls have the structure shown in Figure A.1 and are described by ordinary differential equations of the form

$$\begin{aligned} \dot{x}(t) &= f(x(t), u(t), t), \quad x(t_0) = x_0, \\ y(t) &= g(x(t), u(t), t), \end{aligned} \tag{A.1}$$

where t is the time variable; $x(t) \in \mathcal{R}^n$, $u(t) \in \mathcal{R}^r$, and $y(t) \in \mathcal{R}^l$ denote, respectively, the *state* of the system, the input variables or *control input* of the system, and the *output* variables that can be measured; f and g are real vector-valued functions; n is the dimension of the state x, called the *order* of the system; and $x(t_0)$ denotes the value of $x(t)$ at the initial time $t = t_0 \geq 0$.

When f, g are linear functions of x, u, (A.1) takes the form

$$\begin{aligned} \dot{x} &= A(t)x + B(t)u, \quad x(t_0) = x_0, \\ y &= C^T(t)x + D(t)u, \end{aligned} \tag{A.2}$$

where $A(t) \in \mathcal{R}^{n\times n}$, $B(t) \in \mathcal{R}^{n\times r}$, $C(t) \in \mathcal{R}^{nxl}$, and $D(t) \in \mathcal{R}^{l\times r}$ are matrices with time-varying (TV) elements. If A, B, C, D do not depend on time t, we have

$$\begin{aligned} \dot{x} &= Ax + Bu, \quad x(0) = x_0, \\ y &= C^T x + Du, \end{aligned} \tag{A.3}$$

where A, B, C, and D are matrices of the same dimension as in (A.2) but with constant elements. We refer to (A.3) as the *finite-dimensional LTI system.*

The solution $x(t)$, $y(t)$ of (A.2) is given by

$$\begin{aligned} x(t) &= \Phi(t, t_0)x(t_0) + \int_{t_0}^{t} \Phi(t, \tau)B(\tau)u(\tau)d\tau, \\ y(t) &= C^T(t)x(t) + D(t)u(t), \end{aligned} \tag{A.4}$$

where $\Phi(t, t_0)$ is the state transition matrix defined as the $n \times n$ matrix that satisfies the linear homogeneous matrix equation,

$$\frac{\partial \Phi(t, t_0)}{\partial t} = A(t)\Phi(t, t_0), \quad \Phi(t_0, t_0) = I.$$

It can be shown that

$$\Phi(t, t_0) = \exp\left[\int_{t_0}^{t} A(\tau)d\tau\right].$$

For the LTI system (A.3), $\Phi(t, t_0)$ depends only on the difference $t - t_0$, i.e.,

$$\Phi(t, t_0) = \Phi(t - t_0) = e^{A(t-t_0)},$$

and the solution $x(t)$, $y(t)$ of (A.3) is given by

$$\begin{aligned} x(t) &= e^{A(t-t_0)}x_0 + \int_{t_0}^{t} e^{A(t-\tau)}Bu(\tau)d\tau, \\ y(t) &= C^T x(t) + Du(t), \end{aligned} \tag{A.5}$$

where e^{At} can be identified to be

$$e^{At} \triangleq \mathcal{L}^{-1}[(sI - A)^{-1}],$$

where $\mathcal{L}^{-1}$ denotes the inverse Laplace transform and s is the Laplace variable.

If we take the Laplace transform on both sides of (A.3) and set the initial conditions $x(0) = 0$, we obtain

$$G(s) \triangleq \frac{Y(s)}{U(s)} = C^T(sI - A)^{-1}B + D, \tag{A.6}$$

where $Y(s)$, $U(s)$ are the Laplace transforms of y, u, respectively; $G(s)$ is the *transfer function* of (A.3) and may be expressed as

$$G(s) = \frac{b_m s^m + b_{m-1}s^{m-1} + \cdots + b_0}{s^n + a_{n-1}s^{n-1} + \cdots + a_0} = \frac{Z(s)}{R(s)}, \tag{A.7}$$

where $n \geq m$ and $Z(s)$, $R(s)$ are polynomials of degree n, m, respectively.

The inverse Laplace $g(t)$ of $G(s)$, i.e.,

$$g(t) \triangleq \mathcal{L}^{-1}[G(s)],$$

is known as the *impulse response* of the system (A.3) and

$$y(t) = g(t) * u(t),$$

where $*$ denotes convolution. When $u(t) = \delta_\Delta(t)$, where $\delta_\Delta(t)$ is the delta function defined as

$$\delta_\Delta(t) = \lim_{\varepsilon \to 0} \frac{u_s(t) - u_s(t-\varepsilon)}{\varepsilon},$$

where $u_s(t)$ is the unit step function, we have

$$y(t) = g(t) * \delta_\Delta(t) = g(t).$$

We say that $G(s)$ is *proper* if $G(\infty)$ is finite, i.e., $n \geq m$; *strictly proper* if $G(\infty) = 0$, i.e., $n > m$; and *biproper* if $n = m$. The relative degree n^* of $G(s)$ is defined as $n^* = n - m$, i.e., $n^* =$ degree of denominator of $G(s) -$ degree of numerator of $G(s)$. The *characteristic equation* of the system (A.6) or (A.7) is defined as the equation $s^n + a_{n-1}s^{n-1} + \cdots + a_0 = 0$ or $\det(sI - A) = 0$.

From (A.6) it follows that

$$G(s) = \frac{C^T(\operatorname{adj}(sI - A))B}{\det(sI - A)} + D, \tag{A.8}$$

where $\operatorname{adj}(Q)$ denotes the adjoint[20] of the square matrix $Q \in \mathcal{R}^{n\times n}$. It is obvious from (A.8) that the poles of $G(s)$ are included in the eigenvalues of A. We say that A is *stable* if all its eigenvalues lie in $\Re[s] < 0$. In this case $G(s)$ is referred to as a stable transfer function.

We should note that for zero initial conditions, the Laplace variable s can also be considered as the differential operator $p(x) \stackrel{\Delta}{=} \dot{x}$. For example, the system

$$Y(s) = \frac{s}{s+a}U(s)$$

may be written as $(p+a)y(t) = pu(t)$ or $y(t) = \frac{p}{p+a}u(t)$ with p as the differential operator. Due to the similarity of the two forms, we will use s to denote both the differential operator and Laplace variable and write

$$y = \frac{Z(s)}{R(s)}u, \tag{A.9}$$

where y and u denote $Y(s)$ and $U(s)$, respectively, when s is taken to be the Laplace operator, and y and u denote $y(t)$ and $u(t)$, respectively, when s is taken to be the differential operator. We will also refer to $G(s) = \frac{Z(s)}{R(s)}$ in (A.9) as the filter with input $u(t)$ and output $y(t)$.

A.2 Coprime Polynomials

The I/O properties of most of the systems studied in this book are represented by proper transfer functions expressed as the ratio of two polynomials in s as indicated in (A.7). In this

[20]The (i, j) element q_{ij} of $\operatorname{adj}(Q)$ is given by $q_{ij} = (-1)^{i+j}\det(Q_{ji})$, $i, j = 1, 2, \ldots, n$, where $Q_{ji} \in \mathcal{R}^{(n-1)\times(n-1)}$ is a submatrix of Q obtained by eliminating the jth row and the ith column of Q.

section, we review some of the general properties of polynomials that are used for analysis and control design in subsequent chapters.

Definition A.2.1. *Consider the polynomial* $X(s) = \alpha_n s^n + \alpha_{n-1}s^{n-1} + \cdots + \alpha_0$. *We say that* $X(s)$ *is* monic *if* $\alpha_n = 1$, *and* $X(s)$ *is* Hurwitz *if all the roots of* $X(s) = 0$ *are located in* $\Re[s] < 0$. *We say that the* degree *of* $X(s)$ *is* n *if the coefficient* $\alpha_n \neq 0$.

Definition A.2.2. *A system with a transfer function given by* (A.7) *is referred to as* minimum-phase *if* $Z(s)$ *is Hurwitz; it is referred to as* stable *if* $R(s)$ *is Hurwitz.*

Definition A.2.3. *Two polynomials* $a(s)$ *and* $b(s)$ *are said to be* coprime (*or* relatively prime) *if they have no common factors other than a constant.*

An important characterization of coprimeness of two polynomials is given by the following lemma.

Lemma A.2.4 (Bezout identity [185, 186]). *Two polynomials* $a(s)$ *and* $b(s)$ *are coprime if and only if there exist polynomials* $c(s)$ *and* $d(s)$ *such that*

$$c(s)a(s) + d(s)b(s) = 1.$$

The Bezout identity may have an infinite number of solutions $c(s)$ and $d(s)$ for a given pair of coprime polynomials $a(s)$ and $b(s)$ (see Problem 1).

Coprimeness is an important property that is often used in the design of control schemes for LTI systems. An important theorem used for control design and analysis is the following.

Theorem A.2.5 (see [56, 94]). *If* $a(s)$ *and* $b(s)$ *are coprime and of degree* n_a *and* n_b, *respectively, where* $n_a > n_b$, *then for any given arbitrary polynomial* $a^*(s)$ *of degree* $n_{a^*} \geq n_a$, *the polynomial equation*

$$a(s)l(s) + b(s)p(s) = a^*(s) \tag{A.10}$$

has a unique solution $l(s)$ *and* $p(s)$ *whose degrees* n_l *and* n_p, *respectively, satisfy the constraints* $n_p < n_a$, $n_l \leq \max(n_{a^*} - n_a, n_b - 1)$.

If no constraints are imposed on the degrees of $l(s)$ and $p(s)$, (A.10) has an infinite number of solutions. Equations of the form (A.10) are referred to as *Diophantine equations* and are widely used in the algebraic design of controllers for LTI plants.

Theorem A.2.6 (Sylvester's theorem [56, 94]). *Two polynomials* $a(s) = a_n s^n + a_{n-1}s^{n-1} + \cdots + a_0$, $b(s) = b_n s^n + b_{n-1}s^{n-1} + \cdots + b_0$ *are coprime if and only if their Sylvester matrix*

S_e is nonsingular, where S_e is defined to be the following $2n \times 2n$ matrix:

$$S_e \triangleq \begin{bmatrix} a_n & 0 & \cdots & 0 & b_n & 0 & \cdots & 0 \\ \vdots & \ddots & \ddots & \vdots & \vdots & \ddots & \ddots & \vdots \\ \vdots & \ddots & \ddots & 0 & \vdots & \ddots & \ddots & 0 \\ a_1 & \cdots & \cdots & a_n & b_1 & \cdots & \cdots & b_n \\ a_0 & a_1 & \cdots & a_{n-1} & b_0 & b_1 & \cdots & b_{n-1} \\ 0 & \ddots & \ddots & \vdots & 0 & \ddots & \ddots & \vdots \\ \vdots & \ddots & \ddots & a_1 & \vdots & \ddots & \ddots & b_1 \\ 0 & \cdots & 0 & a_0 & 0 & \cdots & 0 & b_0 \end{bmatrix}. \tag{A.11}$$

The determinant of S_e is known as the *Sylvester resultant* and may be used to examine the coprimeness of a given pair of polynomials. If the polynomials $a(s)$ and $b(s)$ in Theorem A.2.6 have different degrees, say $n_b < n_a$, then $b(s)$ is expressed as a polynomial of degree n_a by augmenting it with the additional powers in s whose coefficients are taken to be equal to zero.

A.3 Norms and $\mathcal{L}_p$ Spaces

For many of the arguments for scalar equations to be extended and remain valid for vector equations, we need an analogue for vectors of the absolute value of a scalar. This is provided by the norm of a vector.

Definition A.3.1. *The* norm $|x|$ *of a vector x is a real-valued function with the following properties:*

(i) $|x| \geq 0$ *with* $|x| = 0$ *if and only if* $x = 0$.

(ii) $|\alpha x| = |\alpha||x|$ *for any scalar* α.

(iii) $|x + y| \leq |x| + |y|$ (triangle inequality).

The norm $|x|$ of a vector x can be thought of as the size or length of the vector x. Similarly, $|x - y|$ can be thought of as the distance between the vectors x and y.

An $m \times n$ matrix A represents a linear mapping from the n-dimensional space $\mathcal{R}^n$ into the m-dimensional space $\mathcal{R}^m$. We define the induced norm of A as follows.

Definition A.3.2. *Let $|\cdot|$ be a given vector norm. Then for each matrix $A \in \mathcal{R}^{m\times n}$, the quantity $\|A\|$ defined by*

$$\|A\| \triangleq \sup_{x\neq 0, x\in\mathcal{R}^n} \frac{|Ax|}{|x|} = \sup_{|x|\leq 1} = \sup_{|x|=1} |Ax|$$

is called the induced (matrix) norm *corresponding to the vector norm $|\cdot|$.*

Table A.1. *Commonly used vector and induced norms.*

Vector norm on $\mathcal{R}^n$	Induced norm on $\mathcal{R}^{m\times n}$
$\lvert x\rvert_\infty = \max_{i\in\{1,2,\dots,n\}} \lvert x_i\rvert$ (infinity norm)	$\|A\|_\infty = \max_{i\in\{1,2,\dots,m\}} \left\{\sum_{j=1}^n \lvert a_{ij}\rvert\right\}$ (row sum)
$\lvert x\rvert_1 = \sum_{i=1}^n \lvert x_i\rvert$	$\|A\|_1 = \max_{j\in\{1,2,\dots,n\}} \left\{\sum_{i=1}^m \lvert a_{ij}\rvert\right\}$ (column sum)
$\lvert x\rvert_2 = \left(\sum_{i=1}^n \lvert x_i\rvert^2\right)^{1/2}$ (Euclidean norm)	$\|A\|_2 = [\lambda_{\max}(A^T A)]^{1/2}$, where $\lambda_{\max}(M)$ is the maximum eigenvalue of M.

The induced matrix norm satisfies the properties (i) to (iii) of Definition A.3.1. Some of the properties of the induced norm that we often use in this book are summarized as follows:

(i) $|Ax| \leq \|A\||x| \; \forall x \in \mathcal{R}^n$,

(ii) $\|A + B\| \leq \|A\| + \|B\|$,

(iii) $\|AB\| \leq \|A\|\|B\|$,

where A, B are arbitrary matrices of compatible dimensions. Table A.1 shows some of the most commonly used vector and induced norms on $\mathcal{R}^n$. In general, for $p \in [1, \infty)$, $x \in \mathcal{R}^n$, and $A \in \mathcal{R}^{m\times n}$ the vector norm

$$|x|_p \triangleq \left(\sum_{i=0}^{n} |x_n|^p\right)^{1/p}$$

and the corresponding induced norm

$$\|A\|_p \triangleq \sup_{x\neq 0, x\in\mathcal{R}^n} \frac{|Ax|_p}{|x|_p} = \sup_{|x|_p=1} |Ax|_p$$

are called the *vector p-norm* of x and the *induced p-norm* of A, respectively. The *vector ∞-norm* $|x|_\infty$ of x and the *induced ∞-norm* $\|A\|_\infty$ of A are defined as given in Table A.1.

For functions of time, we define the *$\mathcal{L}_p$ norm*

$$\|x\|_p \triangleq \left(\int_0^\infty |x(\tau)|^p d\tau\right)^{1/p}$$

for $p \in [1, \infty)$ and say that $x \in \mathcal{L}_p$ when $\|x\|_p$ exists (i.e., when $\|x\|_p$ is finite). The *$\mathcal{L}_\infty$ norm* is defined as

$$\|x\|_\infty \triangleq \sup_{t\geq 0} |x(t)|.$$

We say that $x \in \mathcal{L}_\infty$ when $\|x\|_\infty$ exists. In the above $\mathcal{L}_p$, $\mathcal{L}_\infty$ norm definitions, $x(t)$ can be a scalar or a vector function. If x is a scalar function, then $|\cdot|$ denotes the absolute value. If x is a vector function in $\mathcal{R}^n$, then $|\cdot|$ denotes any norm in $\mathcal{R}^n$. Similarly, for sequences or discrete-time signals $x : \mathcal{Z}^+ \to \mathcal{R}$, where $\mathcal{Z}^+$ denotes the set of nonnegative integers, we define the ℓ_p *norm* as

$$\|x\|_p \triangleq \left(\sum_{i=0}^{\infty} |x(i)|^p\right)^{1/p}, \quad 1 \le p < \infty,$$

and the ℓ_∞ *norm* as

$$\|x\|_\infty \triangleq \sup_{I \in \mathcal{Z}^+} |x(i)|.$$

We say $x \in \ell_p$ (respectively, $x \in \ell_\infty$) if $\|x\|_p$ (respectively, $\|x\|_\infty$) exists.

We are usually concerned with classes of functions of time that do not belong to $\mathcal{L}_p$. To handle such functions we define the $\mathcal{L}_{pe}$ *norm*

$$\|x_t\|_p \triangleq \left(\int_0^t |x(\tau)|^p d\tau\right)^{1/p}$$

for $p \in [1, \infty)$. We say that $x \in \mathcal{L}_{pe}$ when $\|x_t\|_p$ exists for any finite t. Similarly, the $\mathcal{L}_{\infty e}$ norm is defined as

$$\|x_t\|_\infty \triangleq \sup_{0 \le \tau \le t} |x(\tau)|.$$

The function t^2 does not belong to $\mathcal{L}_p$, but $t^2 \in \mathcal{L}_{pe}$. Similarly, any continuous function of time belongs to $\mathcal{L}_{pe}$, but it may not belong to $\mathcal{L}_p$.

We define the ℓ_{pe} *norm* ($p \in [1, \infty)$) and $\ell_{\infty e}$ *norm* of the sequence $x : \mathcal{Z}^+ \to \mathcal{R}$ as

$$\|x_k\|_p \triangleq \left(\sum_{i=0}^{k} |x(i)|^p\right)^{1/p}, \quad 1 \le p < \infty,$$

$$\|x_k\|_\infty \triangleq \sup_{I \in \{0,1,\dots,k\}} |x(i)|.$$

We say $x \in \ell_{pe}$ if $\|x_k\|_p$ exists for any finite $k \in \mathcal{Z}^+$.

For each $p \in [1, \infty]$, the set of functions that belong to $\mathcal{L}_p$ (respectively, $\mathcal{L}_{pe}$) form a linear vector space called the $\mathcal{L}_p$ *space* (respectively, the $\mathcal{L}_{pe}$ *space*) [187]. If we define the truncated function f_t as

$$f_t(\tau) = \begin{cases} f(\tau), & 0 \le \tau \le t, \\ 0, & \tau > t, \end{cases}$$

$\forall t \in [0, \infty)$, then it is clear that for any $p \in [1, \infty]$, $f \in \mathcal{L}_{pe}$ implies that $f_t \in \mathcal{L}_p$ for any finite t. The $\mathcal{L}_{pe}$ space is called the *extended* $\mathcal{L}_p$ *space* and is defined as the set of all functions f such that $f_t \in \mathcal{L}_p$. It can be easily verified that the $\mathcal{L}_p$ and $\mathcal{L}_{pe}$ norms satisfy the properties of the norm given by Definition A.3.1. The following lemmas present some of the properties of $\mathcal{L}_p$ and $\mathcal{L}_{pe}$ norms that we use later in the book.

Lemma A.3.3 (Hölder's inequality). *If $p, q \in [1, \infty]$ and $\frac{1}{p} + \frac{1}{q} = 1$, then $f \in \mathcal{L}_p$, $g \in \mathcal{L}_q$ imply that $fg \in \mathcal{L}_1$ and*

$$\|fg\|_1 \le \|f\|_p \|g\|_q.$$

When $p = q = 2$, the Hölder's inequality becomes the *Schwarz inequality*, i.e.,

$$\|fg\|_1 \leq \|f\|_2 \|g\|_2. \tag{A.12}$$

Lemma A.3.4 (Minkowski inequality). *For $p \in [1, \infty]$, $f, g \in \mathcal{L}_p$ imply that $f + g \in \mathcal{L}_p$ and*

$$\|f + g\|_p \leq \|f\|_p + \|g\|_p. \tag{A.13}$$

The proofs of Lemmas A.3.3 and A.3.4 can be found in any standard book on real analysis such as [188, 189]. We should note that the above lemmas also hold for the truncated functions f_t, g_t of f, g, respectively, provided $f, g \in \mathcal{L}_{pe}$. For example, if f and g are continuous functions, then $f, g \in \mathcal{L}_{pe}$, i.e., $f_t, g_t \in \mathcal{L}_p$ for any finite $t \in [0, \infty)$, and from the Schwarz inequality we have $\|(fg)_t\|_1 \leq \|f_t\|_2 \|g_t\|_2$, i.e.,

$$\int_0^t |f(\tau)g(\tau)| d\tau \leq \left(\int_0^t |f(\tau)|^2 d\tau \right)^{1/2} \left(\int_0^t |g(\tau)|^2 d\tau \right)^{1/2}, \tag{A.14}$$

which holds for any finite $t \geq 0$. We use the above Schwarz inequality extensively throughout the book.

In the remaining chapters of the book, we adopt the following notation regarding norms unless stated otherwise. We will drop the subscript 2 from $|\cdot|_2$, $\|\cdot\|_2$ when dealing with the Euclidean norm, the induced Euclidean norm, and the $\mathcal{L}_2$ norm. If $x : [0, \infty) \to \mathcal{R}^n$, then $|x(t)|$ represents the vector norm in $\mathcal{R}^n$ at each time t, $\|x_t\|_p$ represents the $\mathcal{L}_{pe}$ norm of the function $x(t)$, and $\|x\|_p$ represents the $\mathcal{L}_p$ norm of $x(t)$. If $A \in \mathcal{R}^{m \times n}$, then $\|A\|_i$ represents the induced matrix norm corresponding to the vector norm $|\cdot|_i$. If $A : [0, \infty) \to \mathcal{R}^{m \times n}$ has elements that are functions of time t, then $\|A(t)\|_i$ represents the induced matrix norm corresponding to the vector norm $|\cdot|_i$ at time t.

A.4 Properties of Functions and Matrices

Let us start with some definitions.

Definition A.4.1 (continuity). *A function $f : [0, \infty) \to \mathcal{R}$ is* continuous *on $[0, \infty)$ if for any given $\varepsilon_0 > 0$ there exists a $\delta(\varepsilon_0, t_0)$ such that $\forall t_0, t \in [0, \infty)$ for which $|t - t_0| < \delta(\varepsilon_0, t_0)$, we have $|f(t) - f(t_0)| < \varepsilon_0$.*

Definition A.4.2 (uniform continuity). *A function $f : [0, \infty) \to \mathcal{R}$ is* uniformly continuous *on $[0, \infty)$ if for any given $\varepsilon_0 > 0$ there exists a $\delta(\varepsilon_0)$ such that $\forall t_0, t \in [0, \infty)$ for which $|t - t_0| < \delta(\varepsilon_0)$, we have $|f(t) - f(t_0)| < \varepsilon_0$.*

Definition A.4.3 (piecewise continuity). *A function $f : [0, \infty) \to \mathcal{R}$ is* piecewise continuous *on $[0, \infty)$ if f is continuous on any finite interval $[t_0, t_1] \subset [0, \infty)$ except for a finite number of points.*

Definition A.4.4 (Lipschitz). *A function* $f : [a, b] \to \mathcal{R}$ *is* Lipschitz *on* $[a, b]$ *if* $|f(x_1) - f(x_2)| \le k|x_1 - x_2|\ \forall x_1, x_2 \in [a, b]$, *where* $k \ge 0$ *is a constant referred to as the* Lipschitz constant.

The following fact is important in understanding some of the stability arguments often made in the analysis of adaptive systems.

Fact $\lim_{t\to\infty} \dot{f}(t) = 0$ *does not imply that* $f(t)$ *has a limit as* $t \to \infty$. For example, consider the function $f(t) = (1+t)^{1/2}\cos(\ln(1+t))$, which goes unbounded as $t \to \infty$. Yet $\dot{f}(t) = \frac{\cos(\ln(1+t))}{2(1+t)^{1/2}} - \frac{\sin(\ln(1+t))}{(1+t)^{1/2}} \to 0$ as $t \to \infty$.

Lemma A.4.5 (see [56, 94]). *The following statements hold for scalar-valued functions:*

(i) *A function* $f(t)$ *that is bounded from below and is nonincreasing has a limit as* $t \to \infty$.

(ii) *Consider the nonnegative scalar functions* $f(t)$, $g(t)$ *defined* $\forall t \ge 0$. *If* $f(t) \le g(t)$ $\forall t \ge 0$ *and* $g \in \mathcal{L}_p$, *then* $f \in \mathcal{L}_p\ \forall p \in [1, \infty]$.

A special case of Lemma A.4.5 that we often use in this book is when $f \ge 0$ and $\dot{f} \le 0$, which according to Lemma A.4.5(i) implies that $f(t)$ converges to a limit as $t \to \infty$.

Lemma A.4.6. *Let* $f, V : [0, \infty) \to \mathcal{R}$. *Then*

$$\dot{V} \le -\alpha V + f \quad \forall t \ge t_0 \ge 0$$

implies that

$$V(t) \le e^{-\alpha(t-t_0)}V(t_0) + \int_{t_0}^{t} e^{-\alpha(t-\tau)} f(\tau)d\tau \quad \forall t \ge t_0 \ge 0$$

for any finite constant α.

Proof. Let $w(t) \triangleq \dot{V} + \alpha V - f$. We have $w(t) \le 0$, and

$$\dot{V} = -\alpha V + f + w$$

implies that

$$V(t) = e^{-\alpha(t-t_0)}V(t_0) + \int_{t_0}^{t} e^{-\alpha(t-\tau)} f(\tau)d\tau + \int_{t_0}^{t} e^{-\alpha(t-\tau)} w(\tau)d\tau.$$

Since $w(t) \le 0\ \forall t \ge t_0 \ge 0$, we have

$$V(t) \le e^{-\alpha(t-t_0)}V(t_0) + \int_{t_0}^{t} e^{-\alpha(t-\tau)} f(\tau)d\tau. \qquad \square$$

Lemma A.4.7. *If $f, \dot{f} \in \mathcal{L}_\infty$ and $f \in \mathcal{L}_p$ for some $p \in [1, \infty)$, then $f(t) \to 0$ as $t \to \infty$.*

The result of Lemma A.4.7 is a special case of a more general result given by Barbalat's lemma stated below.

Lemma A.4.8 (Barbalat's lemma [94, 190]). *If $\lim_{t\to\infty} \int_0^t f(\tau)d\tau$ exists and is finite, and $f(t)$ is a uniformly continuous function, then $\lim_{t\to\infty} f(t) = 0$.*

The proof of Lemma A.4.7 follows directly from that of Lemma A.4.8 by noting that the function $f^p(t)$ is uniformly continuous for any $p \in [1, \infty)$ because $f, \dot{f} \in \mathcal{L}_\infty$. The condition that $f(t)$ is uniformly continuous is crucial for the results of Lemma A.4.8 to hold. An example presented in [56] demonstrates a case where $f(t)$ satisfies all the conditions of Lemma A.4.8 except that of uniform continuity but does not converge to zero with time.

A square matrix $A \in \mathcal{R}^{n\times n}$ is called *symmetric* if $A = A^T$. A symmetric matrix A is called *positive semidefinite* if $x^T Ax \geq 0$ for every $x \in \mathcal{R}^n$ and *positive definite* if $x^T Ax > 0$ $\forall x \in \mathcal{R}^n$ with $|x| \neq 0$. It is called *negative semidefinite* (*negative definite*) if $-A$ is positive semidefinite (positive definite).

The definition of a positive definite matrix can be generalized to nonsymmetric matrices. In this book we will always assume that the matrix is symmetric when we consider positive or negative definite or semidefinite properties. We write $A \geq 0$ if A is positive semidefinite, and $A > 0$ if A is positive definite. We write $A \geq B$ and $A > B$ if $A - B \geq 0$ and $A - B > 0$, respectively.

A symmetric matrix $A \in \mathcal{R}^{n\times n}$ is positive definite if and only if any one of the following conditions holds:

(i) $\lambda_i(A) > 0$, $i = 1, 2, \ldots, n$, where $\lambda_i(A)$ denotes the ith eigenvalue of A, which is real because $A = A^T$.

(ii) There exists a nonsingular matrix A_1 such that $A = A_1 A_1^T$.

(iii) Every principal minor of A is positive.

(iv) $x^T Ax \geq \alpha |x|^2$ for some $\alpha > 0$ and $\forall x \in \mathcal{R}^n$.

The decomposition $A = A_1 A_1^T$ in (ii) is unique when A_1 is also symmetric. In this case, A_1 is positive definite, it has the same eigenvectors as A, and its eigenvalues are equal to the square roots of the corresponding eigenvalues of A. We specify this unique decomposition of A by denoting A_1 as $A^{1/2}$, i.e., $A = A^{1/2}A^{T/2}$, where $A^{1/2}$ is a positive definite matrix and $A^{T/2}$ denotes the transpose of $A^{1/2}$.

A symmetric matrix $A \in \mathcal{R}^{n\times n}$ has n orthogonal eigenvectors and can be decomposed as

$$A = U^T \Lambda U, \tag{A.15}$$

where U is a unitary (orthogonal) matrix (i.e., $U^T U = I$) with the eigenvectors of A, and Λ is a diagonal matrix composed of the eigenvalues of A. Using (A.15), it follows that if $A \geq 0$, then for any vector $x \in \mathcal{R}^n$,

$$\lambda_{\min}(A)|x|^2 \leq x^T Ax \leq \lambda_{\max}(A)|x|^2,$$

where $\lambda_{\max}(A)$ and $\lambda_{\min}(A)$ are the maximum and minimum eigenvalues of A, respectively. Furthermore, if $A \geq 0$, then

$$\|A\|_2 = \lambda_{\max}(A),$$

and if $A > 0$, we also have

$$\|A^{-1}\|_2 = \frac{1}{\lambda_{\min}(A)}.$$

We should note that if $A > 0$ and $B \geq 0$, then $A + B > 0$, but it is not true in general that $AB \geq 0$.

A.5 Input/Output Stability

The systems encountered in this book can be described by an I/O mapping that assigns to each input a corresponding output, or by a state variable representation. In this section we present some basic results concerning I/O stability. These results are based on techniques from functional analysis [187], and most of them can be applied to both continuous- and discrete-time systems.

We consider an LTI system described by the convolution of two functions $u, h : [0, \infty) \to \mathcal{R}$ defined as

$$y(t) = u * h \triangleq \int_0^t h(t-\tau)u(\tau)d\tau = \int_0^t u(t-\tau)h(\tau)d\tau, \tag{A.16}$$

where u, y are the input and output of the system, respectively. Let $H(s)$ be the Laplace transform of the I/O operator $h(\cdot)$. $H(s)$ is called the transfer function, and $h(t)$ is called the impulse response of the system (A.16). The system (A.16) may also be represented in the form

$$Y(s) = H(s)U(s), \tag{A.17}$$

where $Y(s)$, $U(s)$ are the Laplace transforms of y, u, respectively. We say that the system represented by (A.16) or (A.17) is *$\mathcal{L}_p$ stable* if $u \in \mathcal{L}_p \Rightarrow y \in \mathcal{L}_p$ and $\|y\|_p \leq c\|u\|_p$ for some constant $c \geq 0$ and any $u \in \mathcal{L}_p$. When $p = \infty$, $\mathcal{L}_p$ stability, i.e., $\mathcal{L}_\infty$ stability, is also referred to as *bounded-input bounded-output* (*BIBO*) *stability*. The following results hold for the system (A.16).

Theorem A.5.1. *If $u \in \mathcal{L}_p$ and $h \in \mathcal{L}_1$, then*

$$\|y\|_p \leq \|h\|_1 \|u\|_p, \tag{A.18}$$

where $p \in [1, \infty]$.

When $p = 2$ we have a sharper bound for $\|y\|_p$ than that of (A.18), given by the following lemma.

Lemma A.5.2. *If $u \in \mathcal{L}_2$ and $h \in \mathcal{L}_1$, then*

$$\|y\|_2 \leq \sup_\omega |H(j\omega)| \|u\|_2. \tag{A.19}$$

Proof. For the proofs of Theorem A.5.1 and Lemma A.5.2, see [187]. □

Remark A.5.3 We should also note that (A.18) and (A.19) hold for the truncated functions of u, y, i.e.,

$$\|y_t\|_p \le \|h\|_1 \|u_t\|_p$$

for any $t \in [0, \infty)$, provided $u \in \mathcal{L}_{pe}$. Similarly,

$$\|y_t\|_2 \le \sup_\omega |H(j\omega)| \|u_t\|_2$$

for any $t \in [0, \infty)$, provided $u \in \mathcal{L}_{2e}$. These results follow by noticing that $u \in \mathcal{L}_{pe} \Rightarrow u_t \in \mathcal{L}_p$ for any finite $t \ge 0$.

The induced $\mathcal{L}_2$ norm in (A.19) is referred to as the $\mathcal{H}_\infty$ *norm* of the transfer function $H(s)$ and is denoted by

$$\|H(s)\|_\infty \triangleq \sup_{\omega \in \mathcal{R}} |H(j\omega)|.$$

Theorem A.5.4. *Let $H(s)$ be a strictly proper rational function of s. Then $H(s)$ is analytic in $\Re[s] \ge 0$ if and only if $h \in \mathcal{L}_1$.*

Corollary A.5.5. *Consider* (A.16). *If $h \in \mathcal{L}_1$, then we have the following:*

(i) *h decays exponentially, i.e., $|h(t)| \le \alpha_1 e^{-\alpha_0 t}$ for some $\alpha_1, \alpha_0 > 0$.*

(ii) *$u \in \mathcal{L}_1 \Rightarrow y \in \mathcal{L}_1 \cap \mathcal{L}_\infty$, $\dot{y} \in \mathcal{L}_1$, y is continuous, and $\lim_{t\to\infty} |y(t)| = 0$.*

(iii) *$u \in \mathcal{L}_2 \Rightarrow y \in \mathcal{L}_2 \cap \mathcal{L}_\infty$, $\dot{y} \in \mathcal{L}_2$, y is continuous, and $\lim_{t\to\infty} |y(t)| = 0$.*

(iv) *For $p \in [1, \infty]$, $u \in \mathcal{L}_p \Rightarrow y, \dot{y} \in \mathcal{L}_p$, and y is continuous.*

Proof. For proofs of Theorem A.5.4 and Corollary A.5.5, see [187]. □

Corollary A.5.6 (see [56, 94]). *Consider* (A.17) *and let $H(s)$ be proper and analytic in $\Re[s] \ge 0$. Then $u \in \mathcal{L}_2 \cap \mathcal{L}_\infty$ and $\lim_{t\to\infty} |u(t)| = 0$ imply that $y \in \mathcal{L}_2 \cap \mathcal{L}_\infty$ and $\lim_{t\to\infty} |y(t)| = 0$.*

Definition A.5.7 (μ-small in the mean square sense (m.s.s.)). *Let $x : [0, \infty) \to \mathcal{R}^n$, where $x \in \mathcal{L}_{2e}$, and consider the set*

$$\mathcal{S}(\mu) = \left\{ x : [0, \infty) \to \mathcal{R}^n \mid \int_t^{t+T} x^T(\tau) x(\tau) d\tau \le c_0 \mu T + c_1 \; \forall t, T \ge 0 \right\}$$

*for a given constant $\mu \ge 0$, where $c_0, c_1 \ge 0$ are some finite constants, and c_0 is independent of μ. We say that x is μ-*small in the m.s.s. *if $x \in \mathcal{S}(\mu)$.*

Using the above definition, we can obtain a result similar to that of Corollary A.5.5(iii) in the case where $u \notin \mathcal{L}_2$ but $u \in \mathcal{S}(\mu)$ for some constant $\mu \geq 0$.

Corollary A.5.8 (see [56, 94]). *Consider the system* (A.16). *If* $h \in \mathcal{L}_1$, *then* $u \in S(\mu)$ *implies that* $y \in S(\mu)$ *and* $y \in \mathcal{L}_\infty$ *for any finite* $\mu \geq 0$. *Furthermore,*

$$|y(t)|^2 \leq \frac{\alpha_1^2}{\alpha_0} \frac{e^{\alpha_0}}{(1-e^{-\alpha_0})}(c_0\mu + c_1) \quad \forall t \geq t_0 \geq 0,$$

where α_0, α_1 *are the parameters in the bound for* h *in Corollary* A.5.5(i).

The *exponentially weighted* $\mathcal{L}_2$ *norm*

$$\|x_t\|_{2\delta} \triangleq \left(\int_0^t e^{-\delta(t-\tau)} x^T(\tau)x(\tau)d\tau\right)^{1/2},$$

where $\delta \geq 0$ is a constant, is found to be useful in analyzing the stability properties of adaptive systems. We say that $x \in \mathcal{L}_{2\delta}$ if $\|x_t\|_{2\delta}$ exists. When $\delta = 0$ we omit it from the subscript and use the notation $x \in \mathcal{L}_{2e}$. We refer to $\|(\cdot)\|_{2\delta}$ as the $\mathcal{L}_{2\delta}$ norm. For any finite time t, the $\mathcal{L}_{2\delta}$ norm satisfies the properties of the norm given by Definition A.3.1.

Let us consider the LTI system given by (A.17), where $H(s)$ is a rational function of s, and examine $\mathcal{L}_{2\delta}$ stability; i.e., given $u \in \mathcal{L}_{2\delta}$, what can we say about the $\mathcal{L}_p, \mathcal{L}_{2\delta}$ properties of the output $y(t)$ and its upper bound.

Lemma A.5.9 (see [56, 94]). *Let* $H(s)$ *in* (A.17) *be proper and analytic in* $\Re[s] \geq -\frac{\delta}{2}$ *for some* $\delta \geq 0$. *If* $u \in \mathcal{L}_{2e}$, *then*

(i) $\|y_t\|_{2\delta} \leq \|H(s)\|_{\infty\delta}\|u_t\|_{2\delta}$, *where*

$$\|H(s)\|_{\infty\delta} \triangleq \sup_\omega \left|H\left(j\omega - \frac{\delta}{2}\right)\right|.$$

(ii) *Furthermore, if* $H(s)$ *is strictly proper, then*

$$|y(t)| \leq \|H(s)\|_{2\delta}\|u_t\|_{2\delta},$$

where

$$\|H(s)\|_{2\delta} \triangleq \frac{1}{\sqrt{2\pi}}\left(\int_{-\infty}^{\infty}\left|H\left(j\omega - \frac{\delta}{2}\right)\right|^2 d\omega\right)^{1/2}.$$

The norms $\|H(s)\|_{2\delta}$, $\|H(s)\|_{\infty\delta}$ *are related by the inequality*

$$\|H(s)\|_{2\delta} \leq \frac{1}{\sqrt{2p-\delta}}\|(s+p)H(s)\|_{\infty\delta}$$

for any $p > \frac{\delta}{2} \geq 0$.

Lemma A.5.10 (see [56, 94]). *Consider the LTV system given by*

$$\begin{aligned} \dot{x} &= A(t)x + B(t)u, \quad x(0) = x_0, \\ y &= C^T(t)x + D(t)u, \end{aligned} \tag{A.20}$$

where $x \in \mathcal{R}^n$, $y \in \mathcal{R}^r$, $u \in \mathcal{R}^m$, *and the elements of the matrices* A, B, C, *and* D *are bounded continuous functions of time. If the state transition matrix* $\Phi(t,\tau)$ *of* (A.20) *satisfies*

$$\|\Phi(t,\tau)\| \le \lambda_0 e^{-\alpha_0(t-\tau)} \tag{A.21}$$

for some $\lambda_0, \alpha_0 > 0$, *and* $u \in \mathcal{L}_{2e}$, *then for any* $\delta \in [0, \delta_1)$, *where* $0 < \delta_1 < 2\alpha_0$ *is arbitrary, we have*

(i) $|x(t)| \le \frac{c\lambda_0}{\sqrt{2\alpha_0 - \delta}} \|u_t\|_{2\delta} + \varepsilon_t$,

(ii) $\|x_t\|_{2\delta} \le \frac{c\lambda_0}{\sqrt{(\delta_1 - \delta)(2\alpha_0 - \delta_1)}} \|u_t\|_{2\delta} + \varepsilon_t$,

(iii) $\|y_t\|_{2\delta} \le c_0 \|u_t\|_{2\delta} + \varepsilon_t$,

where

$$c_0 = \frac{c\lambda_0}{\sqrt{(\delta_1 - \delta)(2\alpha_0 - \delta_1)}} \sup_t \|C(t)\| + \sup_t \|D(t)\|, \quad c = \sup_t \|B(t)\|,$$

and ε_t *is a term exponentially decaying to zero due to* $x_0 \ne 0$.

A useful extension of Lemma A.5.10, applicable to the case where $A(t)$ is not necessarily stable and $\delta = \delta_0 > 0$ is a given fixed constant, is given by the following lemma that makes use of the following definition.

Definition A.5.11. *The pair* $(C(t), A(t))$ *in* (A.20) *is* uniformly completely observable (UCO) *if there exist constants* $\beta_1, \beta_2, \nu > 0$ *such that* $\forall t_0 \ge 0$

$$\beta_2 I \ge N(t_0, t_0 + \nu) \ge \beta_1 I,$$

where $N(t_0, t_0 + \nu) \triangleq \int_{t_0}^{t_0+\nu} \Phi^T(\tau, t_0) C(\tau) C^T(\tau) \Phi(\tau, t_0) d\tau$ *is the so-called* observability Gramian [87, 191] *and* $\Phi(t,\tau)$ *is the state transition matrix associated with* $A(t)$.

Lemma A.5.12 (see [56, 94]). *Consider the LTV system* (A.20). *If* $(C(t), A(t))$ *is UCO, then for any given finite constant* $\delta_0 > 0$ *we have*

(i) $|x(t)| \le \frac{\lambda_1}{\sqrt{2\alpha_1 - \delta_0}} (c_1 \|u_t\|_{2\delta_0} + c_2 \|y_t\|_{2\delta_0}) + \varepsilon_t$,

(ii) $\|x(t)\|_{2\delta_0} \le \frac{\lambda_1}{\sqrt{(\delta_1 - \delta_0)(2\alpha_1 - \delta_1)}} (c_1 \|u_t\|_{2\delta_0} + c_2 \|y_t\|_{2\delta_0}) + \varepsilon_t$,

(iii) $\|y_t\|_{2\delta_0} \le \|x_t\|_{2\delta_0} \sup_t \|C(t)\| + \|u_t\|_{2\delta_0} \sup_t \|D(t)\|$,

where $c_1, c_2 \ge 0$ *are some finite constants,* δ_1, α_1 *satisfy* $\delta_0 < \delta_1 < 2\alpha_1$, *and* ε_t *is a term exponentially decaying to zero due to* $x_0 \ne 0$.

Consider the LTI, SISO system

$$\dot{x} = Ax + Bu, \quad x(0) = x_0,$$
$$y = C^T x + Du, \tag{A.22}$$

whose transfer function is given by

$$y = [C^T(sI - A)^{-1}B + D]u = H(s)u. \tag{A.23}$$

Lemma A.5.13 (see [56, 94]). *Consider the LTI system* (A.22), *where A is a stable matrix and $u \in \mathcal{L}_{2e}$. Let α_0, λ_0 be positive constants satisfying $\|e^{A(t-\tau)}\| \le \lambda_0 e^{-\alpha_0(t-\tau)}$. Then for any constant $\delta \in [0, \delta_1)$, where $0 < \delta_1 < 2\alpha_0$, and any finite t satisfying $t \ge t_1 \ge 0$, we have the following:*

(i) $|x(t)| \le \lambda_0 e^{-\alpha_0(t-t_1)}|x(t_1)| + c_0\|B\|\|u_{t,t_1}\|_{2\delta}$.

(ii) $\|x_{t,t_1}\|_{2\delta} \le c_0 e^{-\delta(t-t_1)/2}|x(t_1)| + c_1\|u_{t,t_1}\|_{2\delta}$.

(iii) $\|y_{t,t_1}\|_{2\delta} \le c_0\|C\|e^{-\delta(t-t_1)/2}|x(t_1)| + \|H(s)\|_{\infty\delta}\|u_{t,t_1}\|_{2\delta}$.

(iv) *Furthermore, if $D = 0$, i.e., $H(s)$ is strictly proper, then $|y(t)| \le \lambda_0\|C\|e^{-\alpha_0(t-t_1)}|x(t_1)| + \|H(s)\|_{2\delta}\|u_{t,t_1}\|_{2\delta}$, where $c_0 = \frac{\lambda_0}{\sqrt{2\alpha_0-\delta}}$, $c_1 = \frac{\lambda_0\|B\|}{\sqrt{(\delta_1-\delta)(2\alpha_0-\delta_1)}}$.*

Considering the case $t_1 = 0$, Lemma A.5.13 also shows the effect of the initial condition $x(0) = x_0$ of the system (A.22) on the bounds for $|y(t)|$ and $\|y_t\|_{2\delta}$. We can obtain a result similar to that in Lemma A.5.12 over the interval $[t_1, t]$ by extending Lemma A.5.13 to the case where A is not necessarily a stable matrix and $\delta = \delta_0 > 0$ is a given fixed constant, provided that (C, A) is an observable pair.

A.6 Bellman–Gronwall Lemma

A key lemma often used in the analysis of adaptive control schemes is the following.

Lemma A.6.1 (Bellman–Gronwall Lemma 1 [94, 192]). *Let $\lambda(t)$, $g(t)$, $k(t)$ be nonnegative piecewise continuous functions of time t. If the function $y(t)$ satisfies the inequality*

$$y(t) \le \lambda(t) + g(t)\int_{t_0}^{t} k(s)y(s)ds \quad \forall t \ge t_0 \ge 0, \tag{A.24}$$

then

$$y(t) \le \lambda(t) + g(t)\int_{t_0}^{t} \lambda(s)k(s)e^{\int_s^t k(\tau)g(\tau)d\tau}ds \quad \forall t \ge t_0 \ge 0. \tag{A.25}$$

In particular, if $\lambda(t) \equiv \lambda$ is a constant and $g(t) \equiv 1$, then

$$y(t) \le \lambda e^{\int_{t_0}^{t} k(s)ds} \quad \forall t \ge t_0 \ge 0.$$

Other useful forms of the Bellman–Gronwall (B–G) lemma are given by Lemmas A.6.2 and A.6.3.

Lemma A.6.2 (B–G Lemma 2 [56, 94]). *Let $\lambda(t)$, $k(t)$ be nonnegative piecewise continuous functions of time t and let $\lambda(t)$ be differentiable. If the function $y(t)$ satisfies the inequality*

$$y(t) \le \lambda(t) + \int_{t_0}^{t} k(s)y(s)ds \quad \forall t \ge t_0 \ge 0,$$

then

$$y(t) \le \lambda(t_0)e^{\int_{t_0}^{t} k(s)ds} + \int_{t_0}^{t} \dot{\lambda}(s)e^{\int_{s}^{t} k(\tau)d\tau}ds \quad \forall t \ge t_0 \ge 0.$$

Lemma A.6.3 (B–G Lemma 3 [56, 94]). *Let c_0, c_1, c_2, α be nonnegative constants and $k(t)$ a nonnegative piecewise continuous function of time. If $y(t)$ satisfies the inequality*

$$y(t) \le c_0 e^{-\alpha(t-t_0)} + c_1 + c_2 \int_{t_0}^{t} e^{-\alpha(t-\tau)}k(\tau)y(\tau)d\tau \quad \forall t \ge t_0,$$

then

$$y(t) \le (c_0 + c_1)e^{-\alpha(t-t_0)}e^{c_2\int_{t_0}^{t} k(s)ds} + c_1\alpha \int_{t_0}^{t} e^{-\alpha(t-\tau)}e^{c_2\int_{\tau}^{t} k(s)ds}d\tau \quad \forall t \ge t_0.$$

A.7 Lyapunov Stability

A.7.1 Definition of Stability

We consider systems described by ordinary differential equations of the form

$$\dot{x} = f(t, x), \quad x(t_0) = x_0, \tag{A.26}$$

where $x \in \mathcal{R}^n$, $f : \mathcal{J} \times \mathcal{B}(r) \to \mathcal{R}$, $\mathcal{J} = [t_0, \infty)$, and $\mathcal{B}(r) = \{x \in \mathcal{R}^n | \, |x| < r\}$. We assume that f is of such a nature that for every $x_0 \in \mathcal{B}(r)$ and every $t_0 \in [0, \infty)$, (A.26) possesses one and only one solution $x(t; t_0, x_0)$.

Definition A.7.1. *A state x_e is said to be an* equilibrium state *of the system described by* (A.26) *if $f(t, x_e) \equiv 0 \; \forall t \ge t_0$.*

Definition A.7.2. *An equilibrium state x_e is called an* isolated equilibrium state *if there exists a constant $r > 0$ such that $\mathcal{B}(x_e, r) \triangleq \{x | x - x_e| < r\} \subset \mathcal{R}^n$ contains no equilibrium state of* (A.26) *other than x_e.*

Definition A.7.3. *The equilibrium state x_e is said to be* stable (in the sense of Lyapunov) *if for arbitrary t_0 and $\varepsilon > 0$ there exists a $\delta(\varepsilon, t_0)$ such that $|x_0 - x_e| < \delta$ implies $|x(t; t_0, x_0) - x_e| < \varepsilon \; \forall t \ge t_0$.*

Definition A.7.4. *The equilibrium state x_e is said to be* uniformly stable (u.s.) *if it is stable and if $\delta(\varepsilon, t_0)$ in Definition* A.7.3 *does not depend on t_0.*

Definition A.7.5. *The equilibrium state x_e is said to be* asymptotically stable (a.s.) *if* (i) *it is stable, and* (ii) *there exists a $\delta(t_0)$ such that $|x_0 - x_e| < \delta(t_0)$ implies* $\lim_{t\to\infty} |x(t; t_0, x_0) - x_e| = 0$.

Definition A.7.6. *The set of all $x_0 \in \mathcal{R}^n$ such that $x(t; t_0, x_0) \to x_e$ as $t \to \infty$ for some $t_0 \geq 0$ is called the* region of attraction *of the equilibrium state x_e. If condition* (ii) *of Definition* A.7.5 *is satisfied, then the equilibrium state x_e is said to be* attractive.

Definition A.7.7. *The equilibrium state x_e is said to be* uniformly asymptotically stable (u.a.s.) *if* (i) *it is uniformly stable,* (ii) *for every $\varepsilon > 0$ and any $t_0 \in [0, \infty)$, there exist a $\delta_0 > 0$ independent of t_0 and ε and a $T(\varepsilon) > 0$ independent of t_0 such that $|x(t; t_0, x_0) - x_e| < \varepsilon \ \forall t \geq t_0 + T(\varepsilon)$ whenever $|x_0 - x_e| < \delta_0$.*

Definition A.7.8. *The equilibrium state x_e is* exponentially stable (e.s.) *if there exists an $\alpha > 0$, and for every $\varepsilon > 0$ there exists a $\delta(\varepsilon) > 0$, such that*

$$|x(t; t_0, x_0) - x_e| \leq \varepsilon e^{-\alpha(t-t_0)} \quad \forall t \geq t_0$$

whenever $|x_0 - x_e| < \delta(\varepsilon)$.

Definition A.7.9. *The equilibrium state x_e is said to be* unstable *if it is not stable.*

When (A.26) possesses a unique solution for each $x_0 \in \mathcal{R}^n$ and $t_0 \in [0, \infty)$, we need the following definitions for the global characterization of solutions.

Definition A.7.10. *A solution $x(t; t_0, x_0)$ of* (A.26) *is* bounded *if there exists a $\beta > 0$ such that $|x(t; t_0, x_0)| < \beta \ \forall t \geq t_0$, where β may depend on each solution.*

Definition A.7.11. *The solutions of* (A.26) *are* uniformly bounded (u.b.) *if for any $\alpha > 0$ and $t_0 \in [0, \infty)$ there exists a $\beta = \beta(\alpha)$ independent of t_0 such that if $|x_0| < \alpha$, then $|x(t; t_0, x_0)| < \beta \ \forall t \geq t_0$.*

Definition A.7.12. *The solutions of* (A.26) *are* uniformly ultimately bounded (u.u.b.) *(with bound B) if there exists a $B > 0$ and if corresponding to any $\alpha > 0$ and $t_0 \in [0, \infty)$ there exists a $T = T(\alpha) > 0$ (independent of t_0) such that $|x_0| < \alpha$ implies $|x(t; t_0, x_0)| < B$ $\forall t \geq t_0 + T$.*

Definition A.7.13. *The equilibrium point x_e of* (A.26) *is* a.s. in the large *if it is stable and every solution of* (A.26) *tends to x_e as $t \to \infty$ (i.e., the region of attraction of x_e is all of $\mathcal{R}^n$).*

Definition A.7.14. *The equilibrium point x_e of* (A.26) *is* u.a.s. in the large *if* (i) *it is uniformly stable,* (ii) *the solutions of* (A.26) *are u.b., and* (iii) *for any $\alpha > 0$, any $\varepsilon > 0$, and $t_0 \in [0, \infty)$ there exists $T(\varepsilon, \alpha) > 0$ independent of t_0 such that if $|x_0 - x_e| < \alpha$, then $|x(t; t_0, x_0) - x_e| < \varepsilon \ \forall t \geq t_0 + T(\varepsilon, \alpha)$.*

Definition A.7.15. *The equilibrium point* x_e *of* (A.26) *is* e.s. in the large *if there exists* $\alpha > 0$ *and for any* $\beta > 0$ *there exists* $k(\beta) > 0$ *such that*

$$|x(t; t_0, x_0)| \leq k(\beta) e^{-\alpha(t-t_0)} \quad \forall t \geq t_0$$

whenever $|x_0| < \beta$.

Definition A.7.16. *If* $x(t; t_0, x_0)$ *is a solution of* $\dot{x} = f(t, x)$, *then the trajectory* $x(t; t_0, x_0)$ *is said to be* stable (u.s., a.s., u.a.s., e.s., unstable) *if the equilibrium point* $z_e = 0$ *of the differential equation*

$$\dot{z} = f(t, z + x(t; t_0, x_0)) - f(t, x(t; t_0, x_0))$$

is stable (u.s., a.s., u.a.s., e.s., unstable, respectively).

A.7.2 Lyapunov's Direct Method

The stability properties of the equilibrium state or solution of (A.26) can be studied using the so-called direct method of Lyapunov (also known as Lyapunov's second method) [193, 194]. The objective of this method is to answer questions of stability by using the form of $f(t, x)$ in (A.26) rather than the explicit knowledge of the solutions. We start with the following definitions [195].

Definition A.7.17. *A continuous function* $\varphi : [0, r] \to [0, \infty)$ *(or a continuous function* $\varphi : [0, \infty] \to [0, \infty)$*) is said to belong to* class $\mathcal{K}$, *i.e.,* $\varphi \in \mathcal{K}$, *if*

(i) $\varphi(0) = 0$.

(ii) φ *is strictly increasing on* $[0, r]$ (or on $[0, \infty)$).

Definition A.7.18. *A continuous function* $\varphi : [0, \infty) \to [0, \infty)$ *is said to belong to* class $\mathcal{KR}$, *i.e.,* $\varphi \in \mathcal{KR}$, *if*

(i) $\varphi(0) = 0$.

(ii) φ *is strictly increasing on* $[0, \infty)$.

(iii) $\lim_{r \to \infty} \varphi(r) = \infty$.

Definition A.7.19. *Two functions* $\varphi_1, \varphi_2 \in \mathcal{K}$ *defined on* $[0, r]$ (or on $[0, \infty)$) *are said to be* of the same order of magnitude *if there exist positive constants* k_1, k_2 *such that* $k_1\varphi_1(r_1) \leq \varphi_2(r_1) \leq k_2\varphi_1(r_1)$ $\forall r_1 \in [0, r]$ *(or* $\forall r_1 \in [0, \infty)$*).*

Definition A.7.20. *A function* $V(t, x) : [0, \infty) \times \mathcal{B}(r) \to \mathcal{R}$ *with* $V(t, 0) = 0$ $\forall t \in [0, \infty)$ *is* positive definite *if there exists a continuous function* $\varphi \in \mathcal{K}$ *such that* $V(t, x) \geq \varphi(|x|)$ $\forall t \in [0, \infty)$, $x \in \mathcal{B}(r)$, *and some* $r > 0$. $V(t, x)$ *is called* negative definite *if* $-V(t, x)$ *is positive definite.*

Definition A.7.21. *A function $V(t,x) : [0,\infty) \times \mathcal{B}(r) \to \mathcal{R}$ with $V(t,0) = 0\ \forall t \in [0,\infty)$ is said to be* positive (negative) semidefinite *if $V(t,x) \geq 0$ ($V(t,x) \leq 0$) $\forall t \in [0,\infty)$ and $x \in \mathcal{B}(r)$ for some $r > 0$.*

Definition A.7.22. *A function $V(t,x) : [0,\infty) \times \mathcal{B}(r) \to \mathcal{R}$ with $V(t,0) = 0\ \forall t \in [0,\infty)$ is said to be* decrescent *if there exists $\varphi \in \mathcal{K}$ such that $|V(t,x)| \leq \varphi(|x|)\ \forall t \geq 0$ and $\forall x \in B(r)$ for some $r > 0$.*

Definition A.7.23. *A function $V(t,x) : [0,\infty) \times \mathcal{R}^n \to \mathcal{R}$ with $V(t,0) = 0\ \forall t \in [0,\infty)$ is said to be* radially unbounded *if there exists $\varphi \in \mathcal{KR}$ such that $V(t,x) \geq \varphi(|x|)\ \forall x \in \mathcal{R}^n$ and $t \in [0,\infty)$.*

It is clear from Definition A.7.23 that if $V(t,x)$ is radially unbounded, it is also positive definite for all $x \in \mathcal{R}^n$, but the converse is not true. The reader should be aware that in some textbooks "positive definite" is used for radially unbounded functions, and "locally positive definite" is used for our definition of positive definite functions.

Let us assume (without loss of generality) that $x_e = 0$ is an equilibrium point of (A.26) and define $\dot{V}$ to be the time derivative of the function $V(t,x)$ along the solution of (A.26), i.e.,

$$\dot{V} = \frac{\partial V}{\partial t} + (\nabla V)^T f(t,x), \tag{A.27}$$

where $\nabla V = [\frac{\partial V}{\partial x_1}, \frac{\partial V}{\partial x_2}, \dots, \frac{\partial V}{\partial x_n}]^T$ is the gradient of V with respect to x. The second method of Lyapunov is summarized by the following theorem.

Theorem A.7.24. *Suppose there exists a positive definite function $V(t,x) : [0,\infty) \times \mathcal{B}(r) \to \mathcal{R}$ for some $r > 0$ with continuous first-order partial derivatives with respect to x, t, and $V(t,0) = 0\ \forall t \in [0,\infty)$. Then the following statements are true:*

(i) *If $\dot{V} \leq 0$, then $x_e = 0$ is stable.*

(ii) *If V is decrescent and $\dot{V} \leq 0$, then $x_e = 0$ is u.s.*

(iii) *If V is decrescent and $\dot{V} < 0$, then x_e is u.a.s.*

(iv) *If V is decrescent and there exist $\varphi_1, \varphi_2, \varphi_3 \in \mathcal{K}$ of the same order of magnitude such that*

$$\varphi_1(|x|) \leq V(t,x) \leq \varphi_2(|x|), \quad \dot{V}(t,x) \leq -\varphi_3(|x|)$$

$\forall x \in \mathcal{B}(r)$ and $t \in [0,\infty)$, then $x_e = 0$ is e.s.

Proof. For the proof, the reader is referred to [193, 196–199]. □

In the above theorem, the state x is restricted to be inside the ball $\mathcal{B}(r)$ for some $r > 0$. Therefore, the results (i) to (iv) of Theorem A.7.24 are referred to as local results. Statement (iii) is equivalent to the existence of functions $\varphi_1, \varphi_2, \varphi_3 \in \mathcal{K}$, where $\varphi_1, \varphi_2, \varphi_3$ do *not* have to be of the same order of magnitude, such that $\varphi_1(|x|) \leq V(t,x) \leq \varphi_2(|x|)$, $\dot{V}(t,x) \leq -\varphi_3(|x|)$.

Theorem A.7.25. *Assume that* (A.26) *possesses unique solutions for all* $x_0 \in \mathcal{R}^n$. *Suppose there exists a positive definite, decrescent, and radially unbounded function* $V(t, x) : [0, \infty) \times \mathcal{R}^n \to [0, \infty)$ *with continuous first-order partial derivatives with respect to* t, x, *and* $V(t, 0) = 0\ \forall t \in [0, \infty)$. *Then the following statements are true:*

(i) *If* $\dot{V} < 0$, *then* $x_e = 0$ *is u.a.s. in the large.*

(ii) *If there exist* $\varphi_1, \varphi_2, \varphi_3 \in \mathcal{KR}$ *of the same order of magnitude such that*

$$\varphi_1(|x|) \leq V(t, x) \leq \varphi_2(|x|), \qquad \dot{V}(t, x) \leq -\varphi_3(|x|),$$

then $x_e = 0$ *is e.s. in the large.*

Proof. For the proof, the reader is referred to [193, 196–199]. □

Statement (i) of Theorem A.7.25 is also equivalent to saying that there exist $\varphi_1, \varphi_2 \in \mathcal{K}$ and $\varphi_3 \in \mathcal{KR}$ such that

$$\varphi_1(|x|) \leq V(t, x) \leq \varphi_2(|x|), \quad \dot{V}(t, x) \leq -\varphi_3(|x|) \quad \forall x \in \mathcal{R}^n.$$

Theorem A.7.26. *Assume that* (A.26) *possesses unique solutions for all* $x_0 \in \mathcal{R}^n$. *If there exists a function* $V(t, x)$ *defined on* $|x| \geq \mathcal{R}$ *(where* $\mathcal{R}$ *may be large) and* $t \in [0, \infty)$ *with continuous first-order partial derivatives with respect to* x, t *and if there exist* $\varphi_1, \varphi_2 \in \mathcal{KR}$ *such that*

(i) $\varphi_1(|x|) \leq V(t, x) \leq \varphi_2(|x|)$,

(ii) $\dot{V}(t, x) \leq 0$

$\forall |x| \geq \mathcal{R}$ *and* $t \in [0, \infty)$, *then the solutions of* (A.26) *are u.b. If, in addition, there exists* $\varphi_3 \in \mathcal{K}$ *defined on* $[0, \infty)$ *and*

(iii) $\dot{V}(t, x) \leq -\varphi_3(|x|)\ \forall |x| \geq R$ *and* $t \in [0, \infty)$,

then the solutions of (A.26) *are u.u.b.*

Proof. For the proof, the reader is referred to [193, 196–199]. □

The system (A.26) is referred to as *nonautonomous*. When the function f in (A.26) does not depend explicitly on time t, the system is referred to as *autonomous*. In this case, we write

$$\dot{x} = f(x). \tag{A.28}$$

Theorems A.7.24–A.7.26 hold for (A.28) as well because it is a special case of (A.26). In the case of (A.28), however, $V(t, x) = V(x)$; i.e., V does not depend explicitly on time t, and all references to "decrescent" and "uniform" could be deleted. This is because $V(x)$ is always decrescent, and the stability (respectively, a.s.) of the equilibrium $x_e = 0$ of (A.28) implies u.s. (respectively, u.a.s.).

For the system (A.28), we can obtain a stronger result than Theorem A.7.25 for a.s. as indicated below.

Definition A.7.27. *A set Ω in $\mathcal{R}^n$ is* invariant *with respect to* (A.28) *if every solution of* (A.28) *starting in Ω remains in Ω $\forall t$.*

Theorem A.7.28. *Assume that* (A.28) *possesses unique solutions for all $x_0 \in \mathcal{R}^n$. Suppose there exists a positive definite and radially unbounded function $V(x) : \mathcal{R}^n \to [0, \infty)$ with continuous first-order derivative with respect to x and $V(0) = 0$. If*

(i) $\dot{V} \leq 0 \, \forall x \in \mathcal{R}^n$,

(ii) *the origin $x = 0$ is the only invariant subset of the set*

$$\Omega = \{x \in \mathcal{R}^n | \dot{V} = 0\},$$

then the equilibrium $x_e = 0$ of (A.28) *is a.s. in the large.*

Proof. For the proof, the reader is referred to [193, 196–199]. □

Theorems A.7.24–A.7.28 are referred to as Lyapunov-type theorems. The function $V(t, x)$ or $V(x)$ that satisfies any Lyapunov-type theorem is referred to as the *Lyapunov function*. Lyapunov functions can also be used to predict the instability properties of the equilibrium state x_e. Several instability theorems based on the second method of Lyapunov are given in [192].

A.7.3 Lyapunov-Like Functions

The choice of an appropriate Lyapunov function and the use of Theorems A.7.24–A.7.26 and A.7.28 to establish stability for a large class of adaptive control schemes may not be obvious or possible in many cases. However, a function that resembles a Lyapunov function, but does not possess all the properties needed to apply these theorems, can be used to analyze the stability properties of certain classes of adaptive systems. We refer to such functions as the *Lyapunov-like functions*. The following example illustrates the use of Lyapunov-like functions.

Example A.7.29 Consider the third-order differential equation

$$\begin{aligned} \dot{x}_1 &= -x_1 - x_2 x_3, & x_1(0) &= x_{10}, \\ \dot{x}_2 &= x_1 x_3, & x_2(0) &= x_{20}, \\ \dot{x}_3 &= x_1^2, & x_3(0) &= x_{30}, \end{aligned} \tag{A.29}$$

which has the nonisolated equilibrium points in $\mathcal{R}^3$ defined by $x_1 = 0, x_2 =$ constant, $x_3 = 0$ or $x_1 = 0$, $x_2 = 0$, $x_3 =$ constant. We would like to analyze the stability properties of the solutions of (A.29) by using an appropriate Lyapunov function and applying Theorems A.7.24–A.7.26 and A.7.28. If we follow Theorems A.7.24–A.7.26 and A.7.28, then we should start with a function $V(x_1, x_2, x_3)$ that is positive definite in $\mathcal{R}^3$. Instead of doing so let us consider the simple quadratic function

$$V(x_1, x_2) = \frac{x_1^2}{2} + \frac{x_2^2}{2},$$

which is positive semidefinite in $\mathcal{R}^3$ and, therefore, does not satisfy the positive definite condition in $\mathcal{R}^3$ of Theorems A.7.24–A.7.26 and A.7.28. The time derivative of V along the solution of the differential equation (A.29) satisfies

$$\dot{V} = -x_1^2 \leq 0, \tag{A.30}$$

which implies that V is a nonincreasing function of time. Therefore,

$$V(x_1(t), x_2(t)) \leq V(x_1(0), x_2(0)) \stackrel{\Delta}{=} V_0$$

and hence $V, x_1, x_2 \in \mathcal{L}_\infty$. Furthermore, V has a limit as $t \to \infty$, i.e., $\lim_{t\to\infty} V(x_1(t), x_2(t)) = V_\infty$ and (A.30) implies that

$$\int_0^t x_1^2(\tau)d\tau = V_0 - V(t) \quad \forall t \geq 0$$

and

$$\int_0^\infty x_1^2(\tau)d\tau = V_0 - V_\infty < \infty,$$

i.e., $x_1 \in \mathcal{L}_2$. Since $x_1 \in \mathcal{L}_2$, integrating both sides of (A.29), we see that $x_3 \in \mathcal{L}_\infty$ and from $x_1, x_2, x_3 \in \mathcal{L}_\infty$ that $\dot{x}_1 \in \mathcal{L}_\infty$. Using $\dot{x}_1 \in \mathcal{L}_\infty$, $x_1 \in \mathcal{L}_2$ and applying Lemma A.4.7, we have $x_1(t) \to 0$ as $t \to \infty$. By using the properties of the positive semidefinite function $V(x_1, x_2)$, we have established that the solution of (A.29) is u.b. and $x_1(t) \to 0$ as $t \to \infty$ for any finite initial condition $x_1(0), x_2(0), x_3(0)$. Because the approach we follow resembles the Lyapunov function approach, we are motivated to refer to $V(x_1, x_2)$ as the Lyapunov-like function. ∎

In the above analysis we assumed that (A.29) has a unique solution. For discussion and analysis on existence and uniqueness of solutions of (A.29) the reader is referred to [200]. We use Lyapunov-like functions and similar arguments as in the above example to analyze the stability of a wide class of adaptive schemes considered throughout this book.

A.7.4 Lyapunov's Indirect Method

Under certain conditions, conclusions can be drawn about the stability of the equilibrium of a nonlinear system by studying the behavior of a certain linear system obtained by linearizing (A.26) around its equilibrium state. This method is known as the *first method of Lyapunov* or as *Lyapunov's indirect method* and is given as follows [192, 196]: Let $x_e = 0$ be an equilibrium state of (A.26) and assume that $f(t, x)$ is continuously differentiable with respect to x for each $t \geq 0$. Then, in the neighborhood of $x_e = 0$, f has a Taylor series expansion written as

$$\dot{x} = f(t, x) = A(t)x + f_1(t, x), \tag{A.31}$$

where $A(t) = \nabla f|_{x=0}$ is referred to as the Jacobian matrix of f evaluated at $x = 0$ and $f_1(t, x)$ represents the remaining terms in the series expansion.

Theorem A.7.30. *Assume that $A(t)$ is u.b. and that*

$$\lim_{|x|\to 0} \sup_{t\geq 0} \frac{|f_1(t, x)|}{|x|} = 0.$$

Let $z_e = 0$ be the equilibrium of

$$\dot{z}(t) = A(t)z(t).$$

The following statements are true for the equilibrium $x_e = 0$ of (A.31):

(i) *If $z_e = 0$ is u.a.s., then $x_e = 0$ is u.a.s.*

(ii) *If $z_e = 0$ is unstable, then $x_e = 0$ is unstable.*

(iii) *If $z_e = 0$ is u.s. or stable, no conclusions can be drawn about the stability of $x_e = 0$.*

Proof. For a proof of Theorem A.7.30, see [192]. □

A.8 Stability of Linear Systems

Equation (A.31) indicates that certain classes of nonlinear systems may be approximated by linear ones in the neighborhood of an *equilibrium point* or, as it is often called in practice, an *operating point*. For this reason we are interested in studying the stability of linear systems of the form

$$\dot{x}(t) = A(t)x(t), \tag{A.32}$$

where the elements of $A(t)$ are piecewise continuous $\forall t \geq t_0 \geq 0$, as a special class of the nonlinear system (A.26) or as an approximation of the linearized system (A.31). The solution of (A.32) is given [193] as

$$x(t; t_0, x_0) = \Phi(t, t_0)x_0$$

$\forall t \geq t_0$, where $\Phi(t, t_0)$ is the *state transition matrix* and satisfies the matrix differential equation

$$\frac{\partial}{\partial t}\Phi(t, t_0) = A(t)\Phi(t, t_0) \quad \forall t \geq t_0,$$
$$\Phi(t_0, t_0) = I.$$

Some additional useful properties of $\Phi(t, t_0)$ are

(i) $\Phi(t, t_0) = \Phi(t, \tau)\Phi(\tau, t_0)\ \forall t \geq \tau \geq t_0$;

(ii) $\Phi(t, t_0)^{-1} = \Phi(t_0, t)$;

(iii) $\frac{\partial}{\partial t_0}\Phi(t, t_0) = -\Phi(t, t_0)A(t_0)$.

Necessary and sufficient conditions for the stability of the equilibrium state $x_e = 0$ of (A.32) are given by the following theorems.

Theorem A.8.1. *Let $\|\Phi(t, \tau)\|$ denote the induced matrix norm of $\Phi(t, \tau)$ at each time $t \geq \tau$. The equilibrium state $x_e = 0$ of* (A.32) *is*

(i) *stable if and only if the solutions of* (A.32) *are bounded or, equivalently,*

$$c(t_0) \triangleq \sup_{t \geq t_0} \|\Phi(t, t_0)\| < \infty;$$

(ii) *u.s. if and only if*

$$c_0 \triangleq \sup_{t_0 \geq 0} c(t_0) = \sup_{t_0 \geq 0}\left(\sup_{t \geq t_0} \|\Phi(t, t_0)\|\right) < \infty;$$

(iii) *a.s. if and only if*

$$\lim_{t \to \infty} \|\Phi(t, t_0)\| = 0 \quad \forall t_0 \in [0, \infty);$$

(iv) *u.a.s. if and only if there exist positive constants α and β such that*

$$\|\Phi(t, t_0)\| \leq \alpha e^{-\beta(t-t_0)} \quad \forall t \geq t_0 \geq 0;$$

(v) *e.s. if and only if it is u.a.s. Moreover, x_e is a.s., u.a.s., e.s. in the large if and only if it is a.s., u.a.s., e.s., respectively.*

Proof. The results follow by applying Definitions A.7.3–A.7.15 to the solution of (A.32). □

Theorem A.8.2 (see [191]). *Assume that the elements of $A(t)$ are u.b. $\forall t \in [0, \infty)$. The equilibrium state $x_e = 0$ of the linear system* (A.32) *is u.a.s. if and only if, given any positive definite matrix $Q(t)$, which is continuous in t and satisfies*

$$0 < c_1 I \leq Q(t) \leq c_2 I < \infty$$

$\forall t \geq t_0$ and some constants $c_1, c_2 > 0$, the scalar function defined by

$$V(t, x) = x^T \left[\int_t^\infty \Phi^T(\tau, t)Q(\tau)\Phi(\tau, t)d\tau\right] x \tag{A.33}$$

exists (i.e., the integral defined by (A.33) *is finite for finite values of x and t) and is a Lyapunov function for* (A.32) *with*

$$\dot{V}(t,x) = -x^T Q(t)x.$$

It follows using the properties of $\Phi(t,t_0)$ that $P(t) \triangleq \int_t^\infty \Phi^T(\tau,t)Q(\tau)\Phi(\tau,t)d\tau$ satisfies the equation

$$\dot{P}(t) = -Q(t) - A^T(t)P(t) - P(t)A(t); \tag{A.34}$$

i.e., the Lyapunov function (A.33) can be rewritten as $V(t,x) = x^T P(t)x$, where $P(t) = P^T(t)$ satisfies (A.34). Hence the following theorem follows as a direct consequence of Theorem A.8.2.

Theorem A.8.3. *A necessary and sufficient condition for the u.a.s. of the equilibrium $x_e = 0$ of* (A.32) *is that there exists a symmetric matrix $P(t)$ such that*

$$\gamma_1 I \le P(t) \le \gamma_2 I,$$
$$\dot{P}(t) + A^T(t)P(t) + P(t)A(t) + \nu C(t)C^T(t) \le 0$$

are satisfied $\forall t \ge 0$ and some constant $\nu > 0$, where $\gamma_1 > 0$, $\gamma_2 > 0$ are constants and $C(t)$ is such that $(C(t), A(t))$ is a UCO pair.

Combining Theorem A.8.3 with the following lemmas has been found useful in establishing stability in many adaptive control systems.

Lemma A.8.4 (see [56, 94]). *Consider a TV output injection gain $K(t) \in \mathcal{R}^{n\times l}$. Assume that there exist constants $\nu > 0$, $k_\nu \ge 0$ such that $\forall t_0 \ge 0$, $K(t)$ satisfies the inequality*

$$\int_{t_0}^{t_0+\nu} |K(\tau)|^2 d\tau \le k_\nu.$$

Then (C, A), where $A(t) \in \mathcal{R}^{n\times n}$, $C(t) \in \mathcal{R}^{n\times l}$, is a UCO pair if and only if $(C, A + KC^\top)$ is a UCO pair.

Lemma A.8.5 (see [56, 94]). *Consider the system*

$$\dot{x}_1 = Ax_1 - B\phi^T x_2,$$
$$\dot{x}_2 = 0,$$
$$y_0 = C^T x_1,$$

where $x_1 \in \mathcal{R}^{n_1}$, $x_2, \phi \in \mathcal{R}^{n_2}$, A is a stable matrix, (C, A) is observable, and $\phi \in \mathcal{L}_\infty$. If

$$\bar{\phi} \triangleq C^T(sI - A)^{-1}B\phi$$

satisfies

$$\alpha_1 I \le \frac{1}{T}\int_t^{t+T} \bar{\phi}(\tau)\bar{\phi}^T(\tau)d\tau \le \alpha_2 I \quad \forall t \ge 0$$

for some constants $\alpha_1, \alpha_2, T > 0$, *then* $(\bar{C}, \bar{A})$, *where*

$$\bar{A}(t) = \begin{bmatrix} A & -B\phi^T(t) \\ 0 & 0 \end{bmatrix} \in \mathcal{R}^{(n_1+n_2)\times(n_1+n_2)}, \qquad \bar{C}(t) = \begin{bmatrix} C \\ 0 \end{bmatrix} \in \mathcal{R}^{n_1+n_2},$$

is a UCO pair.

When $A(t) = A$ is a constant matrix, the conditions for stability of the equilibrium $x_e = 0$ of

$$\dot{x} = Ax \tag{A.35}$$

are given by the following theorem.

Theorem A.8.6. *The equilibrium state* $x_e = 0$ *of* (A.35) *is stable if and only if*

(i) *all the eigenvalues of* A *have nonpositive real parts,*

(ii) *for each eigenvalue* λ_i *with* $Re\{\lambda_i\} = 0$, λ_i *is a simple zero of the minimal polynomial of* A *(i.e., of the monic polynomial* $\psi(\lambda)$ *of least degree such that* $\psi(A) = 0$).

Theorem A.8.7. *A necessary and sufficient condition for* $x_e = 0$ *to be a.s. in the large is that* any one *of the following conditions is satisfied:*

(i) *All the eigenvalues of* A *have negative real parts.*

(ii) *For every positive definite matrix* Q, *the Lyapunov matrix equation* $A^T P + PA = -Q$ *has a unique solution* P *that is also positive definite.*

(iii) *For any given matrix* C *with* (C, A) *observable, the equation* $A^T P + PA = -C^T C$ *has a unique solution* P *that is positive definite.*

Proof. The results follow by applying Theorem A.8.1 to (A.35). □

It is easy to verify that for the LTI system given by (A.35), if $x_e = 0$ is stable, it is also u.s. If $x_e = 0$ is a.s., it is also u.a.s. and e.s. in the large. In the rest of the book we will abuse the notation and call the matrix A in (A.35) *stable* when the equilibrium $x_e = 0$ is a.s., i.e., when all the eigenvalues of A have negative real parts. We call A *marginally stable* when $x_e = 0$ is stable, i.e., A satisfies Theorem A.8.6(i)–(ii).

Theorem A.8.8 (see [56, 94]). *Let the elements of* $A(t)$ *in* (A.32) *be differentiable and bounded functions of time and assume that* $\Re\{\lambda_i(A(t))\} \leq -\sigma_s$ $\forall t \geq 0$ *and for* $i = 1, 2, \ldots, n$, *where* $\sigma_s > 0$ *is some constant. If any one of the conditions*

(a) $\int_t^{t+T} \|\dot{A}(\tau)\| d\tau \leq \mu T + \alpha_0$, *i.e.,* $(\|\dot{A}\|)^{\frac{1}{2}} \in \mathcal{S}(\mu)$,

(b) $\int_t^{t+T} \|\dot{A}(\tau)\|^2 d\tau \leq \mu^2 T + \alpha_0$, $\|\dot{A}\| \in \mathcal{S}(\mu^2)$,

(c) $\|\dot{A}(t)\| \leq \mu$

is satisfied for some $\alpha_0, \mu \in [0, \infty)$ *and* $\forall t, T \geq 0$, *then there exists a* $\mu^* > 0$ *such that if* $\mu \in [0, \mu^*)$, *the equilibrium state* x_e *of* (A.32) *is u.a.s. in the large.*

Corollary A.8.9. *Let the elements of* $A(t)$ *in* (A.32) *be differentiable and bounded functions of time and assume that* $\Re\{\lambda_i(A(t))\} \leq -\sigma_s \ \forall t \geq 0$ *and for* $i = 1, 2, \ldots, n$, *where* $\sigma_s > 0$ *is some constant. If* $\|\dot{A}\| \in \mathcal{L}_2$, *then the equilibrium state* $x_e = 0$ *of* (A.32) *is u.a.s. in the large.*

Theorem A.8.8 states that if the eigenvalues of $A(t)$ for each fixed time t have negative real parts and if $A(t)$ varies sufficiently slowly most of the time, then the equilibrium state $x_e = 0$ of (A.32) is u.a.s. Note that we cannot conclude stability by simply considering the location of the eigenvalues of $A(t)$ at each time t (see Problem 13).

A.9 Positivity and Stability

The concept of *positive real (PR)* and *strictly positive real (SPR)* transfer functions plays an important role in the stability analysis of a large class of nonlinear systems, which also includes adaptive systems. The definition of PR and SPR transfer functions is derived from network theory. That is, a PR (SPR) rational transfer function can be realized as the driving point impedance of a passive (dissipative) network. Conversely, a passive (dissipative) network has a driving point impedance that is rational and PR (SPR). A passive network is one that does not generate energy, i.e., a network consisting only of resistors, capacitors, and inductors. A dissipative network dissipates energy, which implies that it is made up of resistors, capacitors, and inductors that are connected in parallel with resistors. In [201, 202], the following equivalent definitions have been given for PR transfer functions by an appeal to network theory.

Definition A.9.1. *A rational function* $G(s)$ *of the complex variable* $s = \sigma + j\omega$ *is called* PR *if*

(i) $G(s)$ *is real for real* s,

(ii) $\Re[G(s)] \geq 0 \ \forall \Re[s] > 0$.

Lemma A.9.2. *A rational proper transfer function* $G(s)$ *is PR if and only if*

(i) $G(s)$ *is real for real* s;

(ii) $G(s)$ *is analytic in* $\Re[s] > 0$, *and the poles on the* $j\omega$*-axis are simple and such that the associated residues are real and positive;*

(iii) *for all* $\omega \in \mathcal{R}$ *for which* $s = j\omega$ *is not a pole of* $G(s)$, *one has* $\Re[G(j\omega)] \geq 0$.

For SPR transfer functions we have the following definition.

Definition A.9.3 ([201]). *Assume that $G(s)$ is not identically zero for all s. Then $G(s)$ is SPR if $G(s-\varepsilon)$ is PR for some $\varepsilon > 0$.*

The following theorem gives necessary and sufficient conditions in the frequency domain for a transfer function to be SPR.

Theorem A.9.4 (see [203]). *Assume that a rational function $G(s)$ of the complex variable $s = \sigma + j\omega$ is real for real s and is not identically zero for all s. Let n^* be the relative degree of $G(s) = Z(s)/R(s)$ with $|n^*| \leq 1$. Then $G(s)$ is SPR if and only if*

(i) *$G(s)$ is analytic in $\Re[s] \geq 0$;*

(ii) *$\Re[G(j\omega)] > 0 \; \forall \omega \in (-\infty, \infty)$;*

(iii) (a) *when $n^* = 1$, $\lim_{|\omega|\to\infty} \omega^2 \Re[G(j\omega)] > 0$;*
(b) *when $n^* = -1$, $\lim_{|\omega|\to\infty} \frac{G(j\omega)}{j\omega} > 0$.*

It should be noted that when $n^* = 0$, (i) and (ii) in Theorem A.9.4 are necessary and sufficient for $G(s)$ to be SPR. This, however, is not true for $n^* = 1$ or -1. For example,

$$G(s) = (s + \alpha + \beta)/[(s + \alpha)(s + \beta)],$$

$\alpha, \beta > 0$, satisfies (i) and (ii) of Theorem A.9.4, but is not SPR because it does not satisfy (iii)(a). It is, however, PR.

Some useful properties of SPR functions are given by the following corollary.

Corollary A.9.5.

(i) *$G(s)$ is PR (SPR) if and only if $\frac{1}{G(s)}$ is PR (SPR).*

(ii) *If $G(s)$ is SPR, then $|n^*| \leq 1$, and the zeros and the poles of $G(s)$ lie in $\Re[s] < 0$.*

(iii) *If $|n^*| > 1$, then $G(s)$ is not PR.*

A necessary condition for $G(s)$ to be PR is that the Nyquist plot of $G(j\omega)$ lie in the right half complex plane, which implies that the phase shift in the output of a system with transfer function $G(s)$ in response to a sinusoidal input is less than 90°.

The relationship between PR, SPR transfer functions and Lyapunov stability of corresponding dynamic systems leads to the development of several stability criteria for feedback systems with LTI and nonlinear parts. These criteria include the Popov's criterion and its variations [190]. The vital link between PR, SPR transfer functions or matrices and the existence of a Lyapunov function for studying stability can be established by using the following lemmas.

Lemma A.9.6 (Kalman–Yakubovich–Popov (KYP) lemma [186, 190]). *Given a square matrix A with all eigenvalues in the closed left half complex plane, a vector B such that (A, B) is controllable, a vector C, and a scalar $d \geq 0$, the transfer function defined by*

$$G(s) = d + C^T(sI - A)^{-1}B$$

is PR if and only if there exist a symmetric positive definite matrix P and a vector q such that

$$A^T P + PA = -qq^T,$$
$$PB - C = \pm(\sqrt{2d})q.$$

Lemma A.9.7 (Lefschetz–Kalman–Yakubovich (LKY) lemma [203, 204]). *Given a stable matrix A, a vector B such that (A, B) is controllable, a vector C, and a scalar $d \geq 0$, the transfer function defined by*

$$G(s) = d + C^T(sI - A)^{-1}B$$

is SPR if and only if for any positive definite matrix L there exist a symmetric positive definite matrix P, a scalar $\nu > 0$, and a vector q such that

$$A^T P + PA = -qq^T - \nu L,$$
$$PB - C = \pm q\sqrt{2d}.$$

The above lemmas are applicable to LTI systems that are controllable. This controllability assumption is relaxed by the following lemma [86, 205].

Lemma A.9.8 (Meyer–Kalman–Yakubovich (MKY) lemma). *Given a stable matrix A, vectors B, C, and a scalar $d \geq 0$, if*

$$G(s) = d + C^T(sI - A)^{-1}B$$

is SPR, then for any given $L = L^T > 0$ there exist a scalar $\nu > 0$, a vector q, and a $P = P^T > 0$ such that

$$A^T P + PA = -qq^T - \nu L,$$
$$PB - C = \pm q\sqrt{2d}.$$

In many applications of SPR concepts to adaptive systems, the transfer function $G(s)$ involves stable zero-pole cancellations, which implies that the system associated with the triple (A, B, C) is uncontrollable or unobservable. In these situations the MKY lemma is the appropriate lemma to use.

A.10 Optimization Techniques

An important part of every adaptive control scheme is the online estimator or adaptive law used to provide an estimate of the plant or controller parameters at each time t. Most of these adaptive laws are derived by minimizing certain cost functions with respect to the estimated parameters. The type of the cost function and method of minimization determines the properties of the resulting adaptive law as well as the overall performance of the adaptive scheme.

In this section we introduce some simple optimization techniques that include the method of *steepest descent*, referred to as the *gradient method* and the *gradient projection method* for constrained minimization problems.

A.10.1 Notation and Mathematical Background

A real-valued function $f : \mathcal{R}^n \to \mathcal{R}$ is said to be continuously differentiable if the partial derivatives $\frac{\partial f(x)}{\partial x_1}, \ldots, \frac{\partial f(x)}{\partial x_n}$ exist for each $x \in \mathcal{R}^n$ and are continuous functions of x. In this case, we write $f \in \mathcal{C}^1$. More generally, we write $f \in \mathcal{C}^m$ if all partial derivatives of order m exist and are continuous functions of x. If $f \in \mathcal{C}^1$, the *gradient* of f at a point $x \in \mathcal{R}^n$ is defined to be the column vector

$$\nabla f(x) \triangleq \begin{bmatrix} \frac{\partial f(x)}{\partial x_1} \\ \vdots \\ \frac{\partial f(x)}{\partial x_n} \end{bmatrix}.$$

If $f \in \mathcal{C}^2$, the *Hessian* of f at x is defined to be the symmetric $n \times n$ matrix having $\frac{\partial^2 f(x)}{\partial x_i \partial x_j}$ as the ijth element, i.e.,

$$\nabla^2 f(x) \triangleq \left[\frac{\partial^2 f(x)}{\partial x_i \partial x_j} \right]_{n \times n}.$$

A subset $\mathcal{S}$ of $\mathcal{R}^n$ is said to be *convex* if for every $x, y \in \mathcal{S}$ and $\alpha \in [0, 1]$ we have $\alpha x + (1 - \alpha) y \in \mathcal{S}$. A function $f : \mathcal{S} \to \mathcal{R}$ is said to be *convex over the convex set* $\mathcal{S}$ if for every $x, y \in \mathcal{S}$ and $\alpha \in [0, 1]$ we have

$$f(\alpha x + (1 - \alpha) y) \leq \alpha f(x) + (1 - \alpha) f(y).$$

Let $f \in \mathcal{C}^1$ over an open convex set $\mathcal{S}$; then f is convex over $\mathcal{S}$ if and only if

$$f(y) \geq f(x) + (\nabla f(x))^T (y - x) \quad \forall x, y \in \mathcal{S}.$$

If $f \in \mathcal{C}^2$ over $\mathcal{S}$ and $\nabla^2 f(x) \geq 0 \; \forall x \in \mathcal{S}$, then f is convex over $\mathcal{S}$.

Let us now consider the unconstrained minimization problem

$$\begin{array}{ll} \text{minimize} & J(\theta) \\ \text{subject to} & \theta \in \mathcal{R}^n, \end{array} \tag{A.36}$$

where $J : \mathcal{R}^n \to \mathcal{R}$ is a given function. We say that the vector θ^* is a global minimum for (A.36) if

$$J(\theta^*) \leq J(\theta) \quad \forall \theta \in \mathcal{R}^n.$$

A necessary and sufficient condition satisfied by the global minimum θ^* is given by the following lemma.

Lemma A.10.1 (see [206, 207]). *Assume that $J \in \mathcal{C}^1$ is convex over $\mathcal{R}^n$. Then θ^* is a global minimum for* (A.36) *if and only if*

$$\nabla J(\theta^*) = 0.$$

A vector $\bar{\theta}$ is called a *regular point* of the surface $\mathcal{S}_\theta = \{\theta \in \mathcal{R}^n | g(\theta) = 0\}$ if $\nabla g(\bar{\theta}) \neq 0$. At a regular point $\bar{\theta}$, the set

$$M(\bar{\theta}) = \{\theta \in \mathcal{R}^n | \theta^T \nabla g(\bar{\theta}) = 0\}$$

is called the *tangent plane* of g at $\bar{\theta}$.

A.10.2 The Method of Steepest Descent (Gradient Method)

This is one of the oldest and most widely known methods for solving the unconstrained minimization problem (A.36). It is also one of the simplest for which a satisfactory analysis exists. More sophisticated methods are often motivated by an attempt to modify the basic steepest descent technique for better convergence properties [206, 208]. The method of steepest descent proceeds from an initial approximation θ_0 for the minimum θ^* to successive points $\theta_1, \theta_2, \dots$ in $\mathcal{R}^n$ in an iterative manner until some stopping condition is satisfied. Given the current point θ_k, the point θ_{k+1} is obtained by a linear search in the direction d_k, where

$$d_k = -\nabla J(\theta_k).$$

It can be shown [206] that d_k is the direction from θ_k in which the initial rate of decrease of $J(\theta)$ is the greatest. Therefore, the sequence $\{\theta_k\}$ is defined by

$$\theta_{k+1} = \theta_k + \lambda_k d_k = \theta_k - \lambda_k \nabla J(\theta_k) \quad (k = 0, 1, 2, \dots), \tag{A.37}$$

where θ_0 is given and λ_k, known as the *step size* or *step length*, is determined by the linear search method, so that θ_{k+1} minimizes $J(\theta)$ in the direction d_k from θ_k. A simpler expression for θ_{k+1} can be obtained by setting $\lambda_k = \lambda \ \forall k$, i.e.,

$$\theta_{k+1} = \theta_k - \lambda \nabla J(\theta_k). \tag{A.38}$$

In this case, the linear search for λ_k is not required, though the choice of the step length λ is a compromise between accuracy and efficiency. Considering infinitesimally small step lengths, (A.38) can be converted to the continuous-time differential equation

$$\dot{\theta} = -\nabla J(\theta(t)), \quad \theta(t_0) = \theta_0, \tag{A.39}$$

whose solution $\theta(t)$ is the descent path in the time domain starting from $t = t_0$.

The direction of steepest descent $d = -\nabla J$ can be scaled by a constant positive definite matrix $\Gamma = \Gamma^T$ as follows: We let $\Gamma = \Gamma_1 \Gamma_1^T$, where Γ_1 is an $n \times n$ nonsingular matrix, and consider the vector $\bar{\theta} \in \mathcal{R}^n$ given by

$$\Gamma_1 \bar{\theta} = \theta.$$

Then the minimization problem (A.36) is equivalent to

$$\begin{array}{ll} \text{minimize} & \bar{J}(\bar{\theta}) \triangleq J(\Gamma_1 \bar{\theta}) \\ \text{subject to} & \bar{\theta} \in \mathcal{R}^n. \end{array} \tag{A.40}$$

If $\bar{\theta}^*$ is a minimum of $\bar{J}$, the vector $\theta^* = \Gamma_1 \bar{\theta}^*$ is a minimum of J. The steepest descent for (A.40) is given by

$$\bar{\theta}_{k+1} = \bar{\theta}_k - \lambda \nabla \bar{J}(\bar{\theta}_k). \tag{A.41}$$

Because $\nabla \bar{J}(\bar{\theta}) = \frac{\partial J(\Gamma_1 \bar{\theta})}{\partial \bar{\theta}} = \Gamma_1^T \nabla J(\theta)$ and $\Gamma_1 \bar{\theta} = \theta$, it follows from (A.41) that

$$\theta_{k+1} = \theta_k - \lambda \Gamma_1 \Gamma_1^T \nabla J(\theta_k).$$

Setting $\Gamma = \Gamma_1 \Gamma_1^T$, we obtain the scaled version for the steepest descent algorithm

$$\theta_{k+1} = \theta_k - \lambda \Gamma \nabla J(\theta_k). \tag{A.42}$$

The continuous-time version of (A.42) is given by

$$\dot{\theta} = -\Gamma \nabla J(\theta). \tag{A.43}$$

The convergence properties of (A.37), (A.38), (A.42) for different step lengths are given in any standard book on optimization.

A.10.3 Gradient Projection Method

In section A.10.2, the search for the minimum of the function $J(\theta)$ given in (A.36) was carried out for all $\theta \in \mathcal{R}^n$. In some cases, θ is constrained to belong to a certain convex set

$$\mathcal{S} \triangleq \{\theta \in \mathcal{R}^n | g(\theta) \leq 0\} \tag{A.44}$$

in $\mathcal{R}^n$, where $g(\cdot)$ is a scalar-valued function if there is only one constraint, and a vector-valued function if there is more than one constraint. In this case, the search for the minimum is restricted to the convex set defined by (A.44) instead of $\mathcal{R}^n$.

Let us first consider the simple case where we have an equality constraint, i.e., the problem

$$\begin{aligned} &\text{minimize } J(\theta) \\ &\text{subject to } g(\theta) = 0, \end{aligned} \tag{A.45}$$

where $g(\theta)$ is a scalar-valued function. One of the most common techniques for handling constraints is to use a descent method in which the direction of descent is chosen to reduce the function $J(\theta)$ while remaining within the constrained region. Such a method is usually referred to as the *gradient projection method*.

We start with a point θ_0 satisfying the constraint, i.e., $g(\theta_0) = 0$. To obtain an improved vector θ_1, we project the negative gradient of J at θ_0, i.e., $-\nabla J(\theta_0)$, onto the tangent plane $M(\theta_0) = \{\theta \in \mathcal{R}^n | \nabla g^T(\theta_0)\theta = 0\}$, obtaining the direction vector $\Pr(\theta_0)$. Then θ_1 is taken as $\theta_0 + \lambda_0 \Pr(\theta_0)$, where λ_0 is chosen to minimize $J(\theta_1)$. The general form of this iteration is given by

$$\theta_{k+1} = \theta_k + \lambda_k \Pr(\theta_k), \tag{A.46}$$

where λ_k is chosen to minimize $J(\theta_k)$ and $\Pr(\theta_k)$ is the new direction vector after projecting $-\nabla J(\theta_k)$ onto $M(\theta_k)$. The explicit expression for $\Pr(\theta_k)$ can be obtained as follows: The vector $-\nabla J(\theta_k)$ can be expressed as a linear combination of the vector $\Pr(\theta_k)$ and the normal vector $N(\theta_k) = \nabla g(\theta_k)$ to the tangent plane $M(\theta_k)$ at θ_k, i.e.,

$$-\nabla J(\theta_k) = \alpha \nabla g(\theta_k) + \Pr(\theta_k) \tag{A.47}$$

for some constant α. Because $\Pr(\theta_k)$ lies on the tangent plane $M(\theta_k)$, we also have $\nabla g^\top(\theta_k) \Pr(\theta_k) = 0$, which together with (A.47) implies that

$$-\nabla g^T \nabla J = \alpha \nabla g^T \nabla g,$$

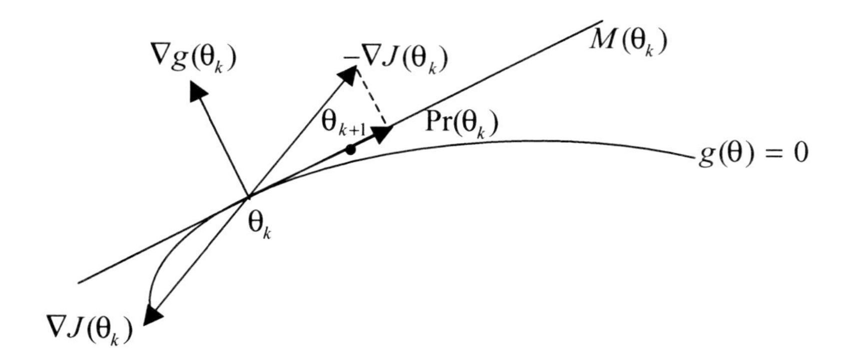

Figure A.2. *Projection.*

i.e.,

$$\alpha = -(\nabla g^T \nabla g)^{-1} \nabla g^T \nabla J.$$

Hence, from (A.47) we obtain

$$\Pr(\theta_k) = -[I - \nabla g(\nabla g^T \nabla g)^{-1} \nabla g^T] \nabla J. \tag{A.48}$$

We refer to $\Pr(\theta_k)$ as the *projected direction* onto the tangent plant $M(\theta_k)$. The gradient projection method is illustrated in Figure A.2.

It is clear from Figure A.2 that when $g(\theta)$ is not a linear function of θ, the new vector θ_{k+1} given by (A.46) may not satisfy the constraint, so it must be modified. There are several successive approximation techniques that can be employed to move θ_{k+1} from $M(\theta_k)$ to the constraint surface $g(\theta) = 0$. One special case, which is often encountered in adaptive control applications, is when θ is constrained to stay inside a ball with a given center and radius, i.e., $g(\theta) = (\theta - \theta_0)^T(\theta - \theta_0) - M^2$, where θ_0 is a fixed constant vector and $M > 0$ is a scalar. In this case, the discrete projection algorithm which guarantees that $\theta_k \in \mathcal{S}\ \forall k$ is

$$\begin{aligned} \bar{\theta}_{k+1} &= \theta_k + \lambda_k \nabla J, \\ \theta_{k+1} &= \begin{cases} \bar{\theta}_{k+1} & \text{if } |\bar{\theta}_{k+1} - \theta_0| \le M, \\ \theta_0 + \frac{\bar{\theta}_{k+1} - \theta_0}{|\bar{\theta}_{k+1} - \theta_0|} M & \text{if } |\bar{\theta}_{k+1} - \theta_0| > M. \end{cases} \end{aligned} \tag{A.49}$$

Letting the step length λ_k become infinitesimally small, we obtain the continuous-time version of (A.49), i.e.,

$$\dot{\theta} = \Pr(\theta) = -[I - \nabla g(\nabla g^T \nabla g)^{-1} \nabla g^T] \nabla J. \tag{A.50}$$

Because of the sufficiently small step length, the trajectory $\theta(t)$, if it exists, will satisfy $g(\theta(t)) = 0\ \forall t \ge 0$, provided that $\theta(0) = \theta_0$ satisfies $g(\theta_0) = 0$.

The scaled version of the gradient projection method can be obtained by using the change of coordinates $\Gamma_1 \bar{\theta} = \theta$, where Γ_1 is a nonsingular matrix that satisfies $\Gamma = \Gamma_1 \Gamma_1^T$ and Γ is the scaling positive definite constant matrix. Following a similar approach as in section A.11, the scaled version of (A.50) is given by

$$\dot{\theta} = \overline{\Pr}(\theta),$$

where

$$\overline{\Pr}(\theta) = -[I - \Gamma\nabla g(\nabla g^T \Gamma \nabla g)^{-1} \nabla g^T]\Gamma\nabla J.$$

The minimization problem (A.45) can now be extended to

$$\begin{aligned} &\text{minimize } J(\theta) \\ &\text{subject to } g(\theta) \leq 0, \end{aligned} \tag{A.51}$$

where $\mathcal{S} = \{\theta \in \mathcal{R}^n | g(\theta) \leq 0\}$ is a convex subset of $\mathcal{R}^n$.

The solution to (A.51) follows directly from that of the unconstrained problem and (A.45). We start from an initial point $\theta_0 \in \mathcal{S}$. If the current point is in the interior of $\mathcal{S}$, defined as $\mathcal{S}_0 \stackrel{\Delta}{=} \{\theta \in \mathcal{R}^n | g(\theta) < 0\}$, then the unconstrained algorithm is used. If the current point is on the boundary of $\mathcal{S}$, defined as $\delta(S) \stackrel{\Delta}{=} \{\theta \in \mathcal{R}^n | g(\theta) = 0\}$, and the direction of search given by the unconstrained algorithm is pointing away from $\mathcal{S}$, then we use the gradient projection algorithm. If the direction of search is pointing inside $\mathcal{S}$, then we keep the unconstrained algorithm. In view of the above, the solution to the constrained optimization problem (A.51) is given by

$$\dot{\theta} = \begin{cases} -\Gamma\nabla J(\theta) & \text{if } \theta \in \mathcal{S}_0 \\ & \text{or if } \theta \in \delta(S) \text{ and } -(\Gamma\nabla J)^T \nabla g \leq 0, \\ -\Gamma\nabla J + \Gamma\frac{\nabla g \nabla g^T}{\nabla g^T \Gamma \nabla g}\Gamma\nabla J & \text{otherwise,} \end{cases}$$

where $\theta(0) \in \mathcal{S}$ and $\Gamma = \Gamma^T > 0$ is the scaling matrix.

A.11 Swapping Lemmas

The following lemmas are useful in establishing stability in most of the adaptive control schemes presented in this book.

Lemma A.11.1 (Swapping Lemma 1). *Consider two functions $\theta, \omega : [0, \infty) \to \mathcal{R}^n$, where θ is differentiable. Let $W(s)$ be a proper stable rational transfer function with a minimal realization (A, B, C, d), i.e.,*

$$W(s) = C^T(sI - A)^{-1}B + d.$$

Then

$$W(s)[\theta^T \omega] = \theta^T W(s)\omega + W_c(s)[(W_b(s)\omega^T)\dot{\theta}],$$

where

$$W_c(s) = -C^T(sI - A)^{-1}, \qquad W_b(s) = (sI - A)^{-1}B.$$

Proof. For the proof see [56] or the web resource [94]. $\square$

Lemma A.11.2 (Swapping Lemma 2). *Consider two differentiable functions $\theta, \omega : [0, \infty) \to \mathcal{R}^n$. Then*

$$\theta^T \omega = F_1(s, \alpha_0)[\dot{\theta}^T \omega + \theta^T \dot{\omega}] + F(s, \alpha_0)[\theta^T \omega],$$

where $F(s, \alpha_0) \triangleq \frac{\alpha_0^k}{(s+\alpha_0)^k}$, $F_1(s, \alpha_0) \triangleq \frac{1-F(s,\alpha_0)}{s}$, $k \geq 1$, and $\alpha_0 > 0$ is an arbitrary constant. Furthermore, for $\alpha_0 > \delta$, where $\delta \geq 0$ is any given constant, $F_1(s, \alpha_0)$ satisfies

$$\|F_1(s, \alpha_0)\|_{\infty\delta} \leq \frac{c}{\alpha_0}$$

for a finite constant $c \in [0, \infty)$ which is independent of α_0.

Proof. For the proof see [56] or the web resource [94]. □

The following two swapping lemmas can be used to interchange the sequence of TV polynomial operators that appear in the proof of indirect adaptive control schemes.

Lemma A.11.3 (Swapping Lemma 3). *Consider two arbitrary polynomials*

$$A(s, t) = a_n(t)s^n + a_{n-1}(t)s^{n-1} + \cdots + a_1(t)s + a_0(t) = a^T(t)\alpha_n(s),$$
$$B(s, t) = b_m(t)s^m + b_{m-1}(t)s^{m-1} + \cdots + b_1(t)s + b_0(t) = b^T(t)a_m(s)$$

with differentiable TV coefficients, where $a \triangleq [a_n, a_{n-1}, \ldots, a_0]^T$, $b \triangleq [b_m, b_{m-1}, \ldots, b_0]^T$, and $\alpha_i(s) \triangleq [s^i, s^{i-1}, \ldots, s, 1]^T$. Let

$$C(s, t) \triangleq A(s, t) \cdot B(s, t) = B(s, t) \cdot A(s, t)$$

be the algebraic product of $A(s, t)$ and $B(s, t)$. Then for any function $f(t)$ such that $C(s, t)f$, $A(s, t)(B(s, t)f)$, $B(s, t)(A(s, t)f)$ are well defined we have

(i) $A(s, t)(B(s, t)f) = C(s, t)f + a^T(t)D_{n-1}(s)[\alpha_{n-1}(s)(\alpha_m^T(s)f)\dot{b}]$,

(ii) $A(s, t)(B(s, t)f) = B(s, t)(A(s, t)f) + G(s, t)((H(s)f)[{}^{\dot{a}}_{\dot{b}}])$,

where

$$G(s, t) \triangleq [a^T D_{n-1}(s), b^T D_{m-1}(s)];$$

$D_i(s)$, $H(s)$ are matrices of dimension $(i + 2) \times (i + 1)$ and $(n + m) \times (n + 1 + m + 1)$, respectively, defined as

$$D_i(s) \triangleq \begin{bmatrix} 1 & s & s^2 & \cdots & s^i \\ 0 & 1 & s & \cdots & s^{i-1} \\ 0 & 0 & \ddots & \ddots & \vdots \\ \vdots & \ddots & 1 & s & \\ 0 & \cdots & 0 & 1 & \\ 0 & \cdots & 0 & 0 & \end{bmatrix}, \qquad H(s) \triangleq \begin{bmatrix} 0 & \alpha_{n-1}(s)\alpha_m^T(s) \\ -\alpha_{m-1}(s)\alpha_n^T(s) & 0 \end{bmatrix}.$$

Proof. For the proof see [56] or the web resource [94]. □

Lemma A.11.4 (Swapping Lemma 4). *Consider the polynomials*

$$A(s,t) = a_n(t)s^n + a_{n-1}(t)s^{n-1} + \cdots + a_1(t)s + a_0(t) = a^T(t)\alpha_n(s),$$
$$\bar{A}(s,t) = s^n a_n(t) + s^{n-1}a_{n-1}(t) + \cdots + sa_1(t) + a_0(t) = \alpha_n^T(s)a(t),$$
$$B(s,t) = b_m(t)s^m + b_{m-1}(t)s^{m-1} + \cdots + b_1(t)s + b_0(t) = b^T(t)\alpha_m(s),$$
$$\bar{B}(s,t) = s^m b_m(t) + s^{m-1}b_{m-1}(t) + \cdots + sb_1(t) + b_0(t) = \alpha_m^T(s)b(t),$$

with differentiable TV coefficients, where $s \triangleq \frac{d}{dt}$, $\alpha_i(s) \triangleq [s^i, s^{i-1}, \ldots, s, 1]^T$, $a \triangleq [a_n, a_{n-1}, \ldots, a_0]^T$, $b \triangleq [b_m, b_{m-1}, \ldots, b_0]^T$, *and* $a, b \in \mathcal{L}_\infty$. *Let*

$$A(s,t)\cdot B(s,t) \triangleq \sum_{i=0}^{n}\sum_{j=0}^{m} a_i b_j s^{i+j}, \qquad \overline{A(s,t)\cdot B(s,t)} \triangleq \sum_{i=0}^{n}\sum_{j=0}^{m} s^{i+j} a_i b_j$$

be the algebraic product of $A(s,t)$, $B(s,t)$ *and* $\bar{A}(s,t)$, $\bar{B}(s,t)$, *respectively, and let* f *be any function for which* $\overline{A(s,t)\cdot B(s,t)}f$, $\bar{A}(s,t)(B(s,t)f)$, *and* $\bar{B}(s,t)(A(s,t)f)$ *are defined. Then*

(i) $\bar{A}(s,t)(B(s,t)f) = \bar{B}(s,t)(A(s,t)f) + \alpha_{\bar{n}}^T(s)F(a,b)\alpha_{\bar{n}}(s)f$, *where* $\bar{n} = \max\{n,m\} - 1$ *and* $F(a,b)$ *satisfies* $\|F(a,b)\| \le c_1|\dot{a}| + c_2|\dot{b}|$ *for some constants* c_1, c_2;

(ii) $\bar{A}(s,t)(B(s,t)\frac{1}{\Lambda_0(s)}f) = \frac{1}{\Lambda_0(s)}\overline{A(s,t)\cdot B(s,t)}f + \alpha_n^T(s)G(s,f,a,b)$ *for any Hurwitz polynomial* $\Lambda_0(s)$ *of order greater or equal to* m, *where* $G(s,f,a,b)$ *is defined as*

$$G(s,f,a,b) \triangleq [g_n, \ldots, g_1, g_0], \quad g_j = -\sum_{j=0}^{m} W_{jc}(s)((W_{jb}(s)f)(\dot{a}_i b_j + a_i \dot{b}_j)),$$

and W_{jc}, W_{jb} *are strictly proper transfer functions that have the same poles as* $\frac{1}{\Lambda_0(s)}$.

Proof. For the proof see [56] or the web resource [94]. □

A.12 Discrete-Time Systems

A.12.1 Lyapunov Stability Theory

In this section, we discuss the stability of a class of discrete-time systems, described by the difference equation

$$x(k+1) = f(k, x(k)), \quad x(k_0) = x_0, \tag{A.52}$$

where $x \in \mathcal{R}^n$, $k \in \mathcal{Z}^+$, $\mathcal{Z}^+$ is the set of all nonnegative integers, and $f : \mathcal{Z}^+ \times \mathcal{B}(r) \to \mathcal{R}^n$ is a single-valued function which is continuous in x. $\mathcal{B}(r) \triangleq \{x | x \in \mathcal{R}^n, |x| \le r\}$. Thus, for every $k_0 \in \mathcal{Z}^+$, $x_0 \in \mathcal{B}(r)$, there exists one and only one solution $x(k; k_0, x_0)$ that satisfies (A.52).

Definition A.12.1. *A state x_e is said to be an* equilibrium state *of* (A.52) *if*

$$x_e \equiv f(k, x_e)$$

$\forall k \in \mathcal{Z}^+$.

Definition A.12.2. *An equilibrium state x_e is an* isolated equilibrium state *if there exists a constant $r > 0$ (which can be sufficiently small) such that $\mathcal{B}(x_e, r) \triangleq \{x \,|\, |x - x_e| < r\} \subset \mathcal{R}^n$ contains no equilibrium state of* (A.52) *other than x_e.*

Definition A.12.3. *An equilibrium state x_e is said to be* stable *(in the sense of Lyapunov) if for any given k_0 and $\varepsilon > 0$ there exists $\delta(\varepsilon, k_0)$ which depends on ε and k_0 such that $|x_0 - x_e| < \delta$ implies $|x(k; k_0, x_0) - x_e| < \varepsilon \ \forall k \geq k_0$.*

Definition A.12.4. *An equilibrium state x_e is said to be* uniformly stable (u.s.) *if it is stable according to Definition* A.12.3 *and $\delta(\varepsilon, k_0)$ does not depend on k_0.*

Definition A.12.5. *An equilibrium state x_e is said to be* asymptotically stable (a.s.) *if* (i) *it is stable and* (ii) *there exists a $\delta(k_0)$ such that $|x_0 - x_e| < \delta(k_0)$ implies* $\lim_{k\to\infty} |x(k; k_0, x_0) - x_e| = 0$.

Definition A.12.6. *The set of all $x_0 \in \mathcal{R}^n$ such that $x(k; k_0, x_0) \to x_e$ as $k \to \infty$ for some $k_0 \geq 0$ is called the* region of attraction *of the equilibrium state x_e. If condition* (ii) *of Definition* A.12.5 *is satisfied for an equilibrium state x_e, then x_e is said to be* attractive.

Definition A.12.7. *The equilibrium state x_e is said to be* uniformly asymptotically stable (u.a.s.) *if* (i) *it is u.s. by Definition* A.12.4, (ii) *for every $\varepsilon > 0$ and $k_0 \in \mathcal{Z}^+$, there exists a $\delta_0 > 0$ which is independent of k_0 and a $K(\varepsilon) > 0$ which is independent of k_0 but may be dependent on ε such that $|x(k; k_0, x_0) - x_e| < \varepsilon \ \forall k \geq k_0 + K(\varepsilon)$ whenever $|x_0 - x_e| < \delta_0$.*

Definition A.12.8. *An equilibrium state x_e is said to be* exponentially stable (e.s.) *if there exists an $\alpha (0 < \alpha < 1)$ and for every $\varepsilon > 0$ there exists a $\delta(\varepsilon)$ such that*

$$|x(k; k_0, x_0) - x_e| \leq \varepsilon \alpha^{k-k_0}$$ [21]

$\forall k \geq k_0$ *whenever* $|x_0 - x_e| < \delta(\varepsilon)$.

Definition A.12.9. *An equilibrium state x_e is said to be* unstable *if it is not stable.*

Definition A.12.10. *The solutions of* (A.52) *are* bounded *if there exists a $\beta > 0$ such that $|x(k; k_0, x_0)| \leq \beta \ \forall k \geq k_0$, where β may depend on each solution.*

Definition A.12.11. *The solutions of* (A.52) *are* uniformly bounded (u.b.) *if for any $\alpha > 0$ and $k_0 \in \mathcal{Z}^+$ there exists a $\beta = \beta(\alpha)$ which is independent of k_0 such that if $|x_0| < \alpha$, then $|x(k; k_0, x_0)| < \beta \ \forall k \geq k_0$.*

[21] Note that $\varepsilon \alpha^{k-k_0} = \varepsilon e^{-\beta(k-k_0)}$, where $\beta = -\ln \alpha$.

Definition A.12.12. *The solutions of* (A.52) *are* uniformly ultimately bounded (u.u.b.) *(with bound B) if there exists a $B > 0$ and if corresponding to any $\alpha > 0$ and $x_0 \in \mathcal{R}^n$ there exists a $K = K(\alpha) > 0$ which is independent of k_0 such that $|x_0| < \alpha$ implies $|x(k; k_0, x_0)| < B$ $\forall k \geq k_0 + K$.*

Definition A.12.13. *An equilibrium state x_e of* (A.52) *is* a.s. in the large *if it is stable and every solution of* (A.52) *tends to x_e as $k \to \infty$ (i.e., the region of attraction of x_e is all of $\mathcal{R}^n$).*

Definition A.12.14. *An equilibrium state x_e of* (A.52) *is* u.a.s. in the large *if* (i) *it is u.s.,* (ii) *the solutions of* (A.52) *are u.b., and* (iii) *for any $\alpha > 0$, $\varepsilon > 0$, and $k_0 \in \mathcal{Z}^+$ there exists $K(\varepsilon, \alpha) > 0$ ($K \in \mathcal{Z}^+$) which is dependent on ε and α but independent of k_0 such that if $|x_0 - x_e| < \alpha$, then $|x(k; k_0, x_0) - x_e| < \varepsilon$ $\forall k \geq k_0 + K$.*

Definition A.12.15. *An equilibrium state x_e of* (A.52) *is* e.s. in the large *if there exists $\alpha \in (0, 1)$ and for every $\beta > 0$ there exists $\gamma(\beta) > 0$ which is independent of k such that*

$$|x(k; k_0, x_0) - x_e| \leq \gamma(\beta)\alpha^{k-k_0}$$

$\forall k \geq k_0$ whenever $|x_0 - x_e| < \beta$.

Definition A.12.16. *If $x(k; k_0, x_0)$ is a solution of* (A.52)*, then the trajectory $x(k; k_0, x_0)$ is said to be* stable (*or* unstable, u.s., *etc.*) *if the equilibrium state $z_e = 0$ of the difference equation*

$$z(k+1) = f(k, z(k) + x(k; k_0, x_0)) - f(k, x(k; k_0, x_0)) \triangleq f_1(k, z(k))$$

is stable (or unstable, u.s., etc., respectively).

In order to study the stability properties of an equilibrium state using the so-called Lyapunov direct method (or the second method, as it is referred to in some references), we need the following definitions about the positive definiteness of a function.

Definition A.12.17. *A function $V(k, x) : \mathcal{Z}^+ \times \mathcal{B}(r) \to \mathcal{R}$ with $V(k, 0) = 0$ $\forall k \in \mathcal{Z}^+$ and $V(k, x)$ continuous at $x = 0$ for any fixed $k \in \mathcal{Z}^+$ is said to be* positive definite *if there exists a continuous function $\phi : [0, \infty) \to [0, \infty)$, with the properties* (i) *$\phi(0) = 0$,* (ii) *ϕ is strictly increasing such that $V(k, x) \geq \phi(|x|)$ $\forall k \in \mathcal{Z}^+$, $x \in \mathcal{B}(r)$, and some $r > 0$. $V(k, x)$ is called* negative definite *if $-V(k, x)$ is positive definite.*

Definition A.12.18. *A function $V(k, x) : \mathcal{Z}^+ \times \mathcal{B}(r) \to \mathcal{R}$ with $V(k, 0) = 0$ $\forall k \in \mathcal{Z}^+$ is said to be* positive (negative) semidefinite *if $V(k, x) \geq 0$ (or $V(k, x) \leq 0$, respectively) $\forall k \in \mathcal{Z}^+$ and $x \in \mathcal{B}(r)$ for some $r > 0$.*

Definition A.12.19. *A function $V(k, x) : \mathcal{Z}^+ \times \mathcal{R}^n \to \mathcal{R}$ with $V(k, 0) = 0$, $\forall k \in \mathcal{Z}^+$ is said to be* decrescent *if there exists a continuous strictly increasing function $\phi : [0, \infty) \to [0, \infty)$ with $\phi(0) = 0$ such that $|V(k, x)| \leq \phi(|x|)$ $\forall k \in \mathcal{Z}^+$, $x \in \mathcal{B}(r)$, and some $r > 0$.*

Definition A.12.20. *A function* $V(k,x) : \mathcal{Z}^+ \times \mathcal{B}(r) \to \mathcal{R}$ *with* $V(k,0) = 0\ \forall k \in \mathcal{Z}^+$ *is said to be* radially unbounded *if there exists a continuous function* $\phi : [0,\infty) \to [0,\infty)$ *with the properties* (i) $\phi(0) = 0$, (ii) ϕ *is strictly increasing on* $[0,\infty)$, *and* (iii) $\lim_{s\to\infty}\phi(s) = \infty$, *such that* $V(k,x) \geq \phi(|x|)\ \forall x \in \mathcal{R}^n$ *and* $k \in \mathcal{Z}^+$.

When we discuss the stability of an equilibrium state x_e of (A.52), we will assume, without loss of generality, that $x_e = 0$. In fact, if $x_e \neq 0$ is an equilibrium state of (A.52), it is trivial to show that the stability properties of x_e are equivalent to the stability properties of the equilibrium state $y_e = 0$ of the system

$$y(k+1) = f(k, y(k) + x_e) - x_e \triangleq f_1(k, y(k)).$$

Let us consider a positive definite function $V(k,x)$ and compute the change of $V(k,x)$ along the solution of (A.52) as

$$\Delta V(k) \triangleq V(k+1, x(k+1)) - V(k, x(k)) = V(k+1, f(k, x(k))) - V(k, x(k)). \quad \text{(A.53)}$$

Then we have the following Lyapunov stability theorem based on the properties of V and ΔV.

Theorem A.12.21. *Suppose that there exists a positive definite function* $V(k,x) : \mathcal{Z}^+ \times \mathcal{B}(r) \to [0,\infty)$ *for some* $r > 0$ *and* $V(k,0) = 0\ \forall k \in \mathcal{Z}^+$. *Consider the following conditions for* $V(k,x)$:

(a) ΔV *is negative semidefinite.*

(b) $V(k,x)$ *is decrescent.*

(c) (1) ΔV *is negative definite or*

 (2) ΔV *is not identically zero along any trajectory other than* $x \equiv 0$.

(d) $V(k,x)$ *is radially unbounded.*

(e) *There exist constants* $c_1, c_2, c_3 > 0$ *such that* $V(k,x)$ *satisfies*

 (1) $c_1|x|^2 \leq V(k,x) \leq c_2|x|^2$ *and*

 (2) $\Delta V \leq -c_3|x|^2$, *where* $c_3 < c_2$.

Then

(i) (a) $\Rightarrow x_e = 0$ *is stable;*

(ii) (a), (b) $\Rightarrow x_e = 0$ *is u.s.;*

(iii) (a), (c) $\Rightarrow x_e = 0$ *is a.s.;*

(iv) (a), (b), (c) $\Rightarrow x_e = 0$ *is u.a.s.;*

(v) (a), (c), (d) $\Rightarrow x_e = 0$ *is a.s. in the large;*

(vi) (a), (b), (c), (d) $\Rightarrow x_e = 0$ *is u.a.s. in the large;*

(vii) (e) $\Rightarrow x_e = 0$ *is e.s.;*

(viii) (d), (e) $\Rightarrow x_e = 0$ *is e.s. in the large.*

Proof. For the proof see the web resource [94]. □

For a linear system given by

$$x(k+1) = A(k)x(k), \tag{A.54}$$

where $A \in \mathcal{R}^{n \times n}$ is a matrix, the stability of the equilibrium state $x_e = 0$ depends on the properties of the matrix $A(k)$.

Theorem A.12.22. *Consider the linear difference equation given by* (A.54). *Suppose that there exists a positive definite symmetric constant matrix P such that*

$$A^T(k)PA(k) - P = -N(k)N^T(k) \tag{A.55}$$

for some matrix sequence $\{N(k)\}$ and all k. Then the equilibrium state $x_e = 0$ of (A.54) *is stable in the sense of Lyapunov. If, further, the pair $(A(k), N(k))$ is* uniformly completely observable (UCO), *i.e., there exist constants $\alpha, \beta > 0$ and $l > 0$ such that for all k*

$$0 < \alpha I \le \sum_{i=0}^{l-1} \Phi^T(k+i,k)N(k+i)N^T(k+i)\Phi(k+i,k) \le \beta I < \infty, \tag{A.56}$$

where $\Phi(k+i,k) = A(k+i-1)A(k+i-2)\cdots A(k+1)A(k)$ is the transition matrix of (A.54), *then $x_e = 0$ is e.s. If the condition of UCO on $(A(k), N(k))$ is replaced by the UCO condition on $(A(k) - K(k)N^T, N(k))$ for any bounded $K(k)$, then $x_e = 0$ is e.s.*

Proof. Consider the Lyapunov function $V(k) = x^T(k)Px(k)$. From (A.55), we can compute the change of V along the trajectory of (A.54) as

$$V(k+1) - V(k) = -x^T(k)N(k)N^T(k)x(k) \le 0,$$

which implies u.s. Let us consider

$$\begin{aligned} V(k+l) - V(k) &= \sum_{i=0}^{l-1}[V(k+i+1) - V(k+i)] \\ &= -\sum_{i=o}^{l-1} x^T(k+i)N(k+i)N^T(k+i)x(k+i). \end{aligned}$$

Since $\Phi(i, j)$ is the transition matrix of (A.54), we have

$$x(k+i) = \Phi(k+i,k)x(k).$$

Therefore, using (A.56), we get

$$\begin{aligned} V(k+l) - V(k) &= -x^T(k)\left[\sum_{i=0}^{l-1} \Phi^T(k+i,k)N(k+i)N^T(k+i)\Phi(k+i,k)\right]x(k) \\ &\le -\alpha x^T(k)x(k) \le -\frac{\alpha}{\lambda_{\max}(P)}V(k). \end{aligned}$$

The last inequality follows from the fact $V(k) = x^T(k)Px(k) \le x^T(k)x(k)\lambda_{\max}(P)$. Thus

$$V(k+l) \le \left(1 - \frac{\alpha}{\lambda_{\max}(P)}\right)V(k). \tag{A.57}$$

Since $V(k) \ge 0$, $\forall k$ we have

$$0 \le 1 - \frac{\alpha}{\lambda_{\max}(P)} < 1.$$

Therefore, (A.57) implies that

$$V(nl) \le \left(1 - \frac{\alpha}{\lambda_{\max}(P)}\right)^n V(0)$$

and $V(nl) \to 0$ as $n \to \infty$. We know that $V(k)$ is a nonincreasing function, $V(k) \le V(nl)$ $\forall nl \le k \le (n+1)l$, and hence

$$V(k) \le \left(1 - \frac{\alpha}{\lambda_{\max}(P)}\right)^n V(0) \le \left(1 - \frac{\alpha}{\lambda_{\max}(P)}\right)^{\left(\frac{k}{l}-1\right)} V(0).$$

This implies that $x(k)$ and $V(k)$ converge to zero exponentially fast as $k \to \infty$.

The last conclusion follows from the fact that the observability property is invariant to output feedback. Consider the fictitious system

$$\begin{aligned} x(k+1) &= A(k)x(k) + K(k)u(k), \\ y(k) &= N^T(k)x(k). \end{aligned}$$

The observability property from the measurement of $u(k)$ and $y(k)$ is given completely by the UCO property of the pair $(A(k), N(k))$ as defined in Theorem A.12.22. If we use output feedback, i.e.,

$$u(k) = v(k) - y(k),$$

we get

$$\begin{aligned} x(k+1) &= (A(k) - K(k)N^T(k))x(k) + K(k)v(k), \\ y(k) &= N^T(k)x(k), \end{aligned}$$

and the observability from the measurement of $v(k)$ and $y(k)$ is given completely by the UCO property of the pair $(A(k) - K(k)N^T(k), N(k))$. Because the output feedback does not change the observability properties, we conclude that $(A(k), N(k))$ is UCO if and only if $(A(k) - K(k)N^T(k), N(k))$ is UCO. Therefore, the condition placed on the observability Gramian of $(A(k), N(k))$ can be replaced by the same condition on the observability

Gramian of $(A(k) - K(k)N^T(k), N(k))$; i.e., $\Phi(k, j)$ in (A.56) can be replaced by the transition matrix $\bar{\Phi}(k, j)$ of $x(k+1) = (A(k) - K(k)N^T(k))x(k)$. □

For an LTI system, i.e., when A is a constant matrix, we have the following theorem.

Theorem A.12.23.

(i) *The equilibrium state* $x_e = 0$ *of* (A.54) *is e.s. in the large if and only if* $|\lambda_i(A)| < 1$, $i = 1, 2, \ldots, n$, *where* $\lambda_i(A)$ *is the* i*th eigenvalue of the matrix* A.

(ii) *The equilibrium state* $x_e = 0$ *of system* (A.54) *is stable if and only if* A *has no eigenvalue outside the unit circle and the eigenvalues of* A *which are on the unit circle correspond to Jordan blocks of order* 1.

(iii) *Given any* $Q \triangleq GG^T \geq 0$ *such that* (A, G) *is completely observable, the equilibrium state* $x_e = 0$ *of the system* (A.54) *is a.s. if and only if there exists a unique symmetric positive definite matrix* P *such that*

$$A^T PA - P = Q. \tag{A.58}$$

(iv) *Given any* $Q = Q^T > 0$, *the equilibrium state* $x_e = 0$ *of the system* (A.54) *is a.s. if and only if there exists a unique symmetric positive definite matrix* P *such that* (A.58) *is satisfied.*

Proof. The proof of Theorem A.12.23 can be found in any standard book on linear system theory. □

For most control problems, the system is described by the difference equation

$$x(k+1) = Ax(k) + Bu(k), \tag{A.59}$$

where $u \in \mathcal{R}^m$ is a function of $k \in \mathcal{Z}^+$.

Theorem A.12.24. *Consider the system* (A.59). *Let the following conditions be satisfied:*

(a) $|\lambda_i(A)| \leq 1$, $i = 1, 2, \ldots, n$.

(b) *All controllable modes of (A,B) are inside the unit circle.*

(c) *The eigenvalues of* A *on the unit circle have a Jordan block of size* 1.

Then

(i) *there exist constants* $k_1, k_2 > 0$ *which are independent of* N *such that*

$$\sum_{k=0}^{N} |x(k)|^2 \leq k_1 \sum_{k=0}^{N} |u(k)|^2 + k_2 \quad \forall N \geq 0,$$

and the system is said to be finite gain stable in this case;

(ii) *there exist constants* $m_1, m_2 > 0$ *which are independent of k, N such that*

$$|x(k)| \leq m_1 + m_2 \max_{0 \leq \tau < N} |u(\tau)| \quad \forall 0 \leq k \leq N.$$

Proof. For the proof see the web resource [94]. □

For a discrete-time system, the so-called Lyapunov indirect method, or the first Lyapunov method, can also be used to study the stability of an equilibrium state of a nonlinear system by linearizing the system around the equilibrium state.

Let $x_e = 0$ be an isolated equilibrium state of (A.52) and assume that $f(k, x)$ is continuously differentiable with respect to x for each k. Then in the neighborhood of $x_e = 0$, f has a Taylor series expansion which can be written as

$$x(k+1) = A(k)x(k) + f_1(k, x(k)),$$

where $A(k)$ is the Jacobian matrix of f evaluated at $x = 0$ and $f_1(k, x)$ represents the remaining terms in the series expansion. The following theorem gives the condition under which the stability of the nonlinear system can be concluded from that of its linearized approximated part.

Theorem A.12.25. *If* $x_e = 0$ *is an a.s. equilibrium state of* $x(k+1) = A(k)x(k)$, *then* $x_e = 0$ *is an a.s. equilibrium state of* (A.52). *If* $x_e = 0$ *is an unstable equilibrium state of* $x(k+1) = A(k)x(k)$, *it is also an unstable equilibrium state of* (A.52).

Proof. The proof of Theorem A.12.25 can be found in any standard book on nonlinear control, e.g., [192]. □

Note that if the equilibrium $x_e = 0$ of $x(k+1) = A(k)x(k)$ is only stable, no conclusion can be made about the stability of $x_e = 0$ of (A.52).

A.12.2 Positive Real Functions

The concept of positive real (PR) and strictly positive real (SPR) functions originated from network theory and plays an important role in the stability analysis of a class of nonlinear systems which also includes adaptive control systems. The definitions and some useful properties of PR and SPR functions for continuous-time systems have been discussed in detail in section A.9. Similar definitions and results can be drawn for discrete-time systems as well.

In order to define the PR and SPR for a discrete-time system, we need the following definition for a Hermitian matrix.

Definition A.12.26. *A matrix function* $H(z)$ *of the complex variable* z *is a* Hermitian matrix (*or* Hermitian) *if*

$$H(z) = H^T(z^*),$$

where the asterisk denotes the complex conjugate.

Definition A.12.27. *An $n \times n$ discrete transfer matrix $H(z)$ whose elements are rational functions of z is said to be* PR *if*

(i) *all elements of $H(z)$ are analytic outside the unit circle;*

(ii) *the eventual poles (the poles left after pole-zero cancellation) of any element of $H(z)$ on the unit circle $|z| = 1$ are simple and the associated residue matrix is a positive semidefinite Hermitian;*

(iii) *the matrix*

$$H(e^{j\omega}) + H^T(e^{-j\omega})$$

is a positive semidefinite Hermitian for all real values of ω for which $z = e^{j\omega}$ and $z = e^{-j\omega}$ are not poles of any element of $H(z)$.

Definition A.12.28. *An $n \times n$ discrete transfer matrix $H(z)$ whose elements are real rational functions of z is* SPR *if*

(i) *all elements of $H(z)$ are analytic in $|z| \geq 1$;*

(ii) *the matrix*

$$H(e^{j\omega}) + H(e^{-j\omega})$$

is a positive definite Hermitian for all real ω.

Consider the discrete-time system

$$\begin{aligned} x(k+1) &= Ax(k) + Bu(k), \\ y(k) &= C^T x(k) + Du(k), \end{aligned}$$

where $A \in R^{nxn}$, $B \in R^{nxm}$, $C \in R^{nxm}$, D^{mxm}. The transfer function matrix of the system is given by

$$H(z) = C^T(zI - A)^{-1}B + D. \tag{A.60}$$

We have the following theorem.

Theorem A.12.29. *The following statements are equivalent to each other:*

(i) *$H(z)$ given by* (A.60) *is a PR discrete transfer function.*

(ii) *There exist a symmetric positive definite matrix P, a symmetric positive semidefinite matrix Q, and matrices S and R of proper dimensions such that*

$$\begin{aligned} A^T PA - P &= -Q, \\ B^T PA + S^T &= C^T, \\ D + D^T - B^T PB &= R, \end{aligned}$$

$$\begin{bmatrix} Q & S \\ S^T & R \end{bmatrix} \geq 0.$$

(iii) (discrete-time positive real lemma *or* Kalman–Szogo–Popov lemma) *There exist a symmetric positive definite matrix P and matrices K, L such that*

$$A^T PA - P = -LL^T,$$
$$B^T PA + K^T L^T = C^T,$$
$$D + D^T - B^T PB = K^T K.$$

Theorem A.12.30. *The discrete-time transfer matrix $H(z)$ given by* (A.60) *is SPR if there exist a symmetric positive definite matrix P, a symmetric positive definite matrix Q, and matrices K, L such that*

$$A^T PA - P = -LL^T - Q,$$
$$B^T PA + K^T L^T = C^T,$$
$$D + D^T - B^T PB = K^T K.$$

Proof. The proofs of Theorems A.12.29 and A.12.30 can be found in [209]. □

It should be noted that the definitions of PR and SPR discrete transfer matrix $H(z)$ can be related to the definitions of the PR and SPR continuous transfer matrix $H(s)$, respectively, by the transformation $s = \frac{z-1}{z+1}$. Therefore, we can study the properties of a PR or SPR discrete transfer matrix by studying the corresponding continuous transfer function using the lemmas and theorems given in section A.9.

A.12.3 Stability of Perturbed Systems

In this section, we discuss the stability of a class of linear systems which has a perturbation term as follows:

$$x(k+1) = A(k)x(k) + f(k, x(k)), \quad x(k_0) = x_0 \in \mathcal{R}^n.$$

In proving the stability properties of such perturbed linear system, we often need the following B–G lemma.

Lemma A.12.31 (discrete-time B–G lemma). *Let $\{u_k\}$, $\{f_k\}$, $\{h_k\}$ be real-valued sequences defined on $\mathcal{Z}^+$. If u_k satisfies the condition*

$$u_k \leq f_k + \sum_{i=0}^{k-1} h_i u_i, \quad k = 0, 1, 2, \ldots,$$

and $h_k \geq 0 \ \forall k \in \mathcal{Z}^+$, then

$$u_k \leq f_k + \sum_{i=0}^{k-1} \left(\prod_{i<j<k} (1 + h_j) h_i f_i \right),$$

where $\prod_{i<j<k}(1 + h_j)$ is set equal to 1 when $i = k - 1$. The following special cases hold:

(i) *If for some h_M, $h_k \leq h_M \ \forall k \in \mathcal{Z}^+$, then*

$$u_k \leq f_k + h_M \sum_{i=0}^{k-1} (1 + h_M)^{k-i-1} f_i.$$

(ii) *If for some constant f_M, $f_k \leq f_M$ $\forall k \in \mathcal{Z}^+$, then*

$$u_k \leq f_M \prod_{i=0}^{k-1}(1+h_i).$$

Proof. The proof can be found in [187]. □

Lemma A.12.32. *Consider the TV difference equation*

$$x(k+1) = A(k)x(k) + f(k, x(k)), \tag{A.61}$$

where $x \in \mathcal{R}^n$ and $|f(k,x)| \leq \gamma_0(k)|x| + \gamma_1(k)$ for some bounded γ_0, γ_1, where γ_0 satisfies

$$\frac{1}{N}\sum_{k=k_0}^{k_0+N-1} \gamma_0(k) \leq \mu + \frac{a_0}{N}$$

for some positive constants μ and a_0. If the unperturbed part

$$x(k+1) = A(k)x(k) \tag{A.62}$$

is e.s., i.e., for some constants $0 \leq \lambda < 1$ and $\beta > 0$ the state transition matrix $\Phi(k_2, k_1)$ of (A.62) *satisfies*

$$\|\Phi(k_2, k_1)\| \leq \beta\lambda^{(k_2-k_1)} \quad \forall k_2 > k_1 > k_0,$$

then there exists a $\mu^* > 0$ such that for $\mu \in [0, \mu^*)$ we have (i) *x is uniformly bounded and* (ii) $\gamma_1(k) = 0\ \forall k \Rightarrow$ (A.61) *is e.s.*

Proof. For the proof see the web resource [94]. □

A.12.4 I/O Stability

Similar to the continuous-time case, the discrete-time systems considered in this book can be described by I/O mappings that assign to each input a corresponding output. In this section we present some basic definitions and results concerning I/O stability of discrete-time systems. These definitions and results, which can be thought of as the discrete-time counterparts of the ones in section A.5, are particularly useful in the stability and robustness analysis of discrete-time adaptive systems.

We consider an LTI system described by the convolution of two functions $u, h : \mathcal{Z}^+ \to \mathcal{R}$ defined as

$$y(k) = u * h \triangleq \sum_{i=0}^{k} h(k-i)u(i) = \sum_{i=0}^{k} u(k-i)h(i), \tag{A.63}$$

where u, y are the input and output of the system, respectively. Let $H(z)$ be the z-transform of the I/O operator $h(\cdot)$. $H(z)$ is called the *transfer function* and $h(k)$ the *impulse response* of the system (A.63). The system (A.63) may also be represented in the form

$$Y(z) = H(z)U(z), \tag{A.64}$$

where $Y(z)$, $U(z)$ are the z-transforms of the sequences $y(k)$, $u(k)$, respectively, or as

$$y = H(z)u \tag{A.65}$$

by treating the complex variable z as the discrete-time shift operator, i.e., $zy(k) = y(k+1)$.

We say that the system represented by (A.63), (A.64), or (A.65) is *ℓ_p stable* if $u \in \ell_p \Rightarrow y \in \ell_p$ and $\|y\|_p \leq c\|u\|_p$ for some constant $c \geq 0$ and any $u \in \ell_p$ (note that the *ℓ_p norm* $\|\cdot\|_p$ is defined in section A.3). When $p = \infty$, ℓ_∞ stability is also referred to as *bounded-input bounded-output (BIBO) stability*.

For $p = 2$, in addition to the *ℓ_2 norm* we consider another norm that is used to simplify the stability and robustness analysis of adaptive systems, the *exponentially weighted ℓ_2 norm* defined [111] as

$$\|x_k\|_{2\delta} \triangleq \left(\sum_{i=0}^{k} \delta^{k-i} x^T(i)x(i) \right)^{1/2},$$

where $0 < \delta \leq 1$ is a constant and $k \in \mathcal{Z}^+$. We say that $x \in \ell_{2\delta}$ if $\|x_k\|_{2\delta}$ exists. We refer to $\|(\cdot)\|_{2\delta}$ as the *$\ell_{2\delta}$ norm*. Note that, for $\delta = 1$, as $k \to \infty$ the $\ell_{2\delta}$ and ℓ_2 become equal, i.e.,

$$\lim_{k\to\infty} \|x_k\|_{2\delta} = \|x\|_2 \quad \text{for } \delta = 1 \quad \text{and} \quad x \in \ell_2.$$

Note also that for any finite step k, the $\ell_{2\delta}$ norm satisfies the following properties similar to the $\mathcal{L}_{2\delta}$ norm:

(i) $\|x_k\|_{2\delta} \geq 0$.

(ii) $\|\alpha x_k\|_{2\delta} = |\alpha| \|x_k\|_{2\delta}$ for any constant scalar α.

(iii) $\|(x+y)_k\|_{2\delta} \leq \|x_k\|_{2\delta} + \|y_k\|_{2\delta}$.

(iv) $\|\alpha x_k\|_{2\delta} \leq \|x_k\|_{2\delta} \sup_k |\alpha(k)|$ for any $\alpha \in \ell_\infty$.

Let us examine the $\ell_{2\delta}$ stability of the LTI system (A.65); i.e., given $u \in \ell_{2\delta}$ for $0 < \delta \leq 1$, what can we say about the ℓ_p, $\ell_{2\delta}$ properties of the output $y(k)$ and its upper bound?

Lemma A.12.33. *Let $H(z)$ in* (A.63) *be proper. If $H(z)$ is analytic in $|z| \geq \sqrt{\delta}$ for some $0 < \delta \leq 1$ and $u \in \ell_{2\delta}$, then $\forall k \geq 0$,*

(i) $\|y_k\|_{2\delta} \leq \|H(z)\|_{\infty\delta} \|u_k\|_{2\delta}$, *where*

$$\|H(z)\|_{\infty\delta} \triangleq \sup_{\omega\in[0,2\pi]} |H(\sqrt{\delta}e^{j\omega})|;$$

(ii) $|y(k)| \leq \|H(z)\|_{2\delta}\|u_k\|_{2\delta}$, *where*

$$\|H(z)\|_{2\delta} \triangleq \sqrt{\int_0^{2\pi} |H(\sqrt{\delta}e^{j\omega})|^2 \frac{d\omega}{2\pi}};$$

(iii) *furthermore, when* $H(z)$ *is strictly proper, we have*

$$|y(k)| \leq \|zH(z)\|_{2\delta}\|u_{k-1}\|_{2\delta}.$$

Proof. Lemma A.12.33 is a special case of [111, Lemma 2.1]. □

Corollary A.12.34. *Let* $H(z)$ *in* (A.63) *be a proper stable transfer function. If* $u \in \ell_2$, *then* $\forall k \geq 0$,

(i) $\|y\|_2 \leq \|H(z)\|_\infty \|u\|_2$, *where*

$$\|H(z)\|_\infty \triangleq \sup_{\omega \in [0, 2\pi]} |H(e^{j\omega})|;$$

(ii) $|y(k)| \leq \|H(z)\|_2 \|u_k\|_2$, *where*

$$\|H(z)\|_2 \triangleq \sqrt{\int_0^{2\pi} |H(e^{j\omega})|^2 \frac{d\omega}{2\pi}};$$

(iii) *furthermore, if* $H(z)$ *is strictly proper, we have*

$$|y(k)| \leq \|zH(z)\|_2 \|u_{k-1}\|_2.$$

Proof. The results follow from Lemma A.12.33 by letting $\delta = 1$ and taking limits where necessary. □

A.12.5 Swapping Lemmas

The following lemma is useful in establishing stability in classes of discrete-time adaptive control systems considered in this book.

Lemma A.12.35 (Discrete-Time Swapping Lemma 1). *Let* $\theta, \omega : \mathcal{Z}^+ \to \mathcal{R}^n$. *Let* $W(z)$ *be a proper stable rational scalar transfer function with a minimal realization* (A, B, C, d), *i.e.,*

$$W(z) = C^T(zI - A)^{-1}B + d.$$

Then

$$W(z)[\theta^T\omega]|_k = \theta^T W(z)[\omega]|_k + W_c(z)[(W_b(z)[\omega^T])\Delta\theta]|_k,$$

where $\Delta\theta(k) \triangleq \theta(k+1) - \theta(k)$ *and*

$$W_c(z) = -C^T(zI - A)^{-1}, \qquad W_b(z) = z(zI - A)^{-1}B.$$

Proof. We have

$$\begin{aligned}
W(z)[\theta^T\omega]|_k &= W(z)[\omega^T\theta]|_k = d\theta^T(k)\omega(k) + C^T\sum_{i=0}^{k-1} A^{k-1-i}B\omega^T(i)\theta(i)\\
&= d\theta^T(k)\omega(k) + C^T\left(\sum_{i=0}^{k-1} A^{k-1-i}B\omega^T(i)\left(\theta(k) - \sum_{j=i}^{k-1}\Delta\theta(j)\right)\right)\\
&= \theta^T(k)W(z)[\omega]|_k - C^T\left(\sum_{i=0}^{k-1}\sum_{j=i}^{k-1} A^{k-1-i}B\omega^T(i)\Delta\theta(j)\right)\\
&= \theta^T(k)W(z)[\omega]|_k - C^T\left(\sum_{i=0}^{k-1} A^{k-1-i}\sum_{j=0}^{i} A^{i-j}B\omega^T(j)\Delta\theta(i)\right)\\
&= \theta^T(k)W(z)[\omega]|_k - C^T\left(\sum_{i=0}^{k-1} A^{k-1-i}(W_b(z)[\omega^T]|_i)\Delta\theta(i)\right)\\
&= \theta^T(k)W(z)[\omega]|_k + W_c(z)[(W_b(z)[\omega^T])\Delta\theta]|_k. \qquad \square
\end{aligned}$$

Lemma A.12.36 (Discrete-Time Swapping Lemma 2). *Let* $\theta, \omega : \mathcal{Z}^+ \to \mathcal{R}^n$. *Then*

$$\theta^T\omega = F_1(z,\alpha_0)[\Delta\theta^T\omega + \theta^T\Delta\omega + \Delta\theta^T\Delta\omega] + F(z,\alpha_0)[\theta^T\omega],$$

where $F(z,\alpha_0) \triangleq \frac{\alpha_0^k}{(z-1+\alpha_0)^k}$, $F_1(z,\alpha_0) \triangleq \frac{1-F(z,\alpha_0)}{z-1}$, $k \geq 1$, *and* $1 < \alpha_0$ *is an arbitrary constant. Furthermore, for* $\alpha_0 \geq c_1(\sqrt{\delta}+1)$, *where* $0 < \delta < 1$ *and* $c_1 > 1$ *are any given constants,* $F_1(z,\alpha_0)$ *satisfies*

$$\|F_1(z,\alpha_0)\|_{\infty\delta} \leq \frac{c}{\alpha_0}$$

for a finite constant $c \in [0,\infty)$ *which is independent of* α_0.

Proof. For the proof see the web resource [94]. $\square$

Problems

1. Consider the coprime polynomials $a(s) = s + 2$ and $b(s) = s + 6$. Show that the Bezout identity in Lemma A.2.4 has an infinite number of solutions.

2. Consider the feedback system

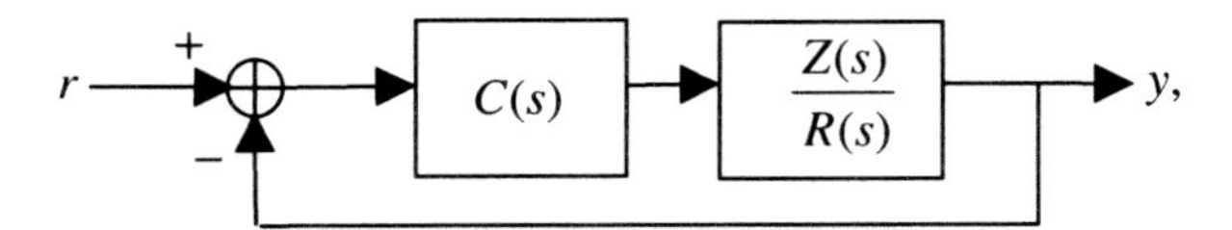

where $C(s) = \frac{p(s)}{l(s)}$, $Z(s) = s$, $R(s) = (s-1)^2$. Use Theorem A.2.5 to choose the degree and elements of the polynomials $p(s)$, $l(s)$ of the controller transfer function $C(s)$ so that the closed-loop characteristic equation is $(s+2)^3 = 0$.

3. Find the Sylvester matrix of the polynomials $a(s) = s^2 + 2s + 2$ and $b(s) = s - 1$, and show that $a(s)$ and $b(s)$ are coprime.

4. (a) Compute the ∞-, 2-, and 1-norms of $x = [1, 2, -1, 0]^T$.

   (b) Compute the induced ∞-, 2-, and 1-norms of $A = \begin{bmatrix} 0 & 5 \\ 1 & 0 \\ 0 & -10 \end{bmatrix}$, $B = \begin{bmatrix} -1 & 5 \\ 0 & 2 \end{bmatrix}$.

   (c) In (b), compute $\|AB\|_1$, $\|AB\|_2$, $\|AB\|_\infty$ and verify the inequality $\|AB\| \leq \|A\|\|B\|$. Comment.

5. Show that the function $f(t) = \frac{1}{t+2}$, $t \geq 0$, has the property $f \in \mathcal{L}_2 \cap \mathcal{L}_\infty$, $f \in \mathcal{L}_{1e}$, but $f \notin \mathcal{L}_1$.

6. Show that the functions $f(t) = t + 2$, $g(t) = \frac{1}{t+2}$ satisfy the Schwarz inequality for each $t \in [0, \infty)$.

7. According to Lemma A.4.5, a function $f(t)$ that is bounded from below and is nonincreasing has a limit as $t \to \infty$. Does this imply that $f(t)$ is bounded from above? (*Hint*: Consider $f(t) = \frac{1}{t}$ for $t \in (0, \infty)$.)

8. Consider the system

$$\dot{x} = Ax + \mu Bx,$$

where A is a stable matrix, i.e., all the eigenvalues of A have negative real parts, B is a constant matrix, and μ is a small constant. Use the B–G lemma to show that for some μ^* and any $\mu \in [0, \mu^*]$, $x(t) \to 0$ as $t \to \infty$ exponentially fast.

9. Find the equilibrium points of the following differential equations. In each case, specify whether the equilibrium is isolated or not.

   (a) $\dot{x}_1 = x_1 x_2$, $\dot{x}_2 = x_1^3$.

   (b) $\dot{x} = (x-2)^3 x^2$.

10. Specify the stability properties of the equilibrium x_e of the differential equation

$$\dot{x} = -x^3.$$

11. Consider the system

$$\dot{x}_1 = x_2 + c x_1 (x_1^2 + x_2^2),$$
$$\dot{x}_2 = -x_1 + c x_2 (x_1^2 + x_2^2),$$

where c is a constant.

   (a) Find the equilibrium points of the system.

(b) Use the Lyapunov candidate

$$V(x) = \frac{x_1^2 + x_2^2}{2}$$

to examine the stability properties of the equilibrium points in (a) for $c = 0$, $c > 0$, and $c < 0$.

12. Consider the system

$$\dot{x}_1 = x_2,$$
$$\dot{x}_2 = -x_2 - e^{-t}x_1.$$

(a) Find the equilibrium points.

(b) Examine stability of the equilibrium points using $V(x) = \frac{x_1^2+x_2^2}{2}$.

(c) Repeat (b) with $V(x) = x_1^2 + e^t x_2^2$.

13. Consider the system

$$\dot{x} = A(t)x, \tag{A.66}$$

where

$$A(t) = \begin{bmatrix} -1 + 1.5\cos^2 t & 1 - 1.5\sin t\cos t \\ -1 - 1.5\sin t\cos t & -1 + 1.5\sin^2 t \end{bmatrix}.$$

(a) Find the eigenvalues of $A(t)$ at each fixed time t.

(b) Verify that

$$x(t) = \begin{bmatrix} e^{0.5t}\cos t & e^{-t}\sin t \\ -e^{0.5t}\sin t & e^{-t}\cos t \end{bmatrix} x(0)$$

satisfies the differential equation (A.66).

(c) Comment on your results in (a), (b).

14. Examine whether the following transfer functions are PR or SPR:

(a) $G_1(s) = \frac{s-2}{(s+5)^2}$.

(b) $G_2(s) = \frac{1}{(s+5)^3}$.

(c) $G_3(s) = \frac{s+4}{(s+1)(s+3)}$.

(d) $G_4(s) = \frac{s+\alpha}{(s+1)(s+2)}$.

15. Use the KYP and LKY lemmas to check whether the system

$$\dot{x} = \begin{bmatrix} 0 & 1 \\ -2 & -3 \end{bmatrix} x + \begin{bmatrix} 0 \\ 1 \end{bmatrix} u,$$
$$y = [3,\ 1]x$$

is PR or SPR.

16. Consider the discrete-time system

$$x_1(k+1) = |x_1(k)x_2(k)|^{1/2},$$
$$x_2(k+1) = \gamma x_2(k) + \delta x_1(k),$$

where γ, δ are some nonzero constants.

(a) Find the equilibrium state of the system.

(b) Use a Lyapunov function to analyze stability and specify γ, δ for different types of stability.

17. Consider the LTI system

$$x(k+1) = \begin{bmatrix} .1 & \lambda \\ .5 & .01 \end{bmatrix} x(k).$$

(a) Find the value of λ for stability.

(b) Repeat (a) by using the Lyapunov equation $A^T PA - P = Q$ when Q is positive semidefinite and when it is positive definite. (*Hint*: Use Theorem A.12.23.) In each case specify the numerical values of P, Q.

18. Examine whether the following transfer functions are PR or SPR:

(a) $G(z) = \frac{(z+.01)}{(z+.02)(z-.03)}$.

(b) $G(z) = \frac{(z+.01)(z-.04)}{(z+.02)(z-.03)}$.

19. Use the discrete-time B–G lemma to find a bound for the norm of the matrix $B(k)$ so that the equilibrium of the system is e.s.,

$$x(k+1) = (A(k) + B(k))x(k),$$

assuming that the zero equilibrium of the system

$$x(k+1) = A(k)x(k)$$

is e.s., i.e., its state transition matrix $\Phi(k, k_0)$ satisfies $\|\Phi(k, k_0)\| \le \beta\lambda^{k-k_0}$, where $0 \le \lambda < 1$ and $\beta > 0$.

Bibliography

[1] J. A. Aseltine, A. R. Mancini, and C. W. Sartune, *A survey of adaptive control systems*, IEEE Trans. Automat. Control, 3 (1958), pp. 102–108.

[2] W. I. Caldwell, *Control System with Automatic Response Adjustment*, U. S. patent 2,517,081, 1950 (filed April 25, 1947).

[3] D. McRuer, I. Ashkenas, and D. Graham, *Aircraft Dynamics and Automatic Control*, Princeton University Press, Princeton, NJ, 1973.

[4] G. Kreisselmeier, *An indirect adaptive controller with a self-excitation capability*, IEEE Trans. Automat. Control, 34 (1989), pp. 524–528.

[5] K. S. Tsakalis and P. A. Ioannou, *Linear Time Varying Systems: Control and Adaptation*, Prentice–Hall, Englewood Cliffs, NJ, 1993.

[6] G. Stein, *Adaptive flight control: A pragmatic view*, in Applications of Adaptive Control, K. S. Narendra and R. V. Monopoli, eds., Academic Press, New York, 1980.

[7] N. Andreiev, *A process controller that adapts to signal and process conditions*, Control Eng., 38 (1977), pp. 38–40.

[8] B. Martensson, *The order of any stabilizing regulator is sufficient a prior information for adaptive stabilization*, Systems Control Lett., 6 (1985), pp. 87–91.

[9] M. Fu and B. R. Barmish, *Adaptive stabilization of linear systems via switching control*, IEEE Trans. Automat. Control, 31 (1986), pp. 1097–1103.

[10] D. E. Miller and E. J. Davison, *An adaptive controller which provides an arbitrarily good transient and steady-state response*, IEEE Trans. Automat. Control, 36 (1991), pp. 68–81.

[11] B. Anderson, T. Brinsmead, D. Liberzon, and A. S. Morse, *Multiple model adaptive control with safe switching*, Internat. J. Adapt. Control Signal Process., 15 (2001), pp. 445–470.

[12] S. Fuji, J. P. Hespanha, and A. S. Morse, *Supervisory control of families of noise suppressing controllers*, in Proceedings of the 37th IEEE Conference on Decision and Control, Vol. 2, IEEE Press, New York, 1998, pp. 1641–1646.

[13] J. P. HESPANHA AND A. S. MORSE, *Stability of switched systems with average dwell-time*, in Proceedings of the 38th IEEE Conference on Decision and Control, Vol. 3, IEEE Press, New York, 1999, pp. 2655–2660.

[14] J. P. HESPANHA, D. LIBERZON, AND A. S. MORSE, *Towards the supervisory control of uncertain nonholonomic systems*, in Proceedings of the American Control Conference, San Diego, CA, 1999, pp. 3520–3524.

[15] J. P. HESPANHA, D. LIBERZON, AND A. S. MORSE, *Bounds on the number of switchings with scale-independent hysteresis: Applications to supervisory control*, in Proceedings of the 39th IEEE Conference on Decision and Control, Vol. 4, IEEE Press, New York, 2000, pp. 3622–3627.

[16] J. P. HESPANHA AND A. S. MORSE, *Switching between stabilizing controllers*, Automatica, 38 (2002), pp. 1905–1917.

[17] J. P. HESPANHA, D. LIBERZON, AND A. S. MORSE, *Hysteresis-based switching algorithms for supervisory control of uncertain systems*, Automatica, 39 (2003), pp. 263–272.

[18] J. P. HESPANHA, D. LIBERZON, AND A. S. MORSE, *Overcoming the limitations of adaptive control by means of logic-based switching*, Systems Control Lett., 49 (2003), pp. 49–65.

[19] A. S. MORSE, *Supervisory control of families of linear set-point controllers—Part 1: Exact matching*, IEEE Trans. Automat. Control, 41 (1996), pp. 1413–1431.

[20] A. S. MORSE, *Supervisory control of families of linear set-point controllers—Part 2: Robustness*, IEEE Trans. Automat. Control, 42 (1997), pp. 1500–1515.

[21] F. M. PAIT AND A. S. MORSE, *A cyclic switching strategy for parameter-adaptive control*, IEEE Trans. Automat. Control, 39 (1994), pp. 1172–1183.

[22] A. PAUL AND M. G. SAFONOV, *Model reference adaptive control using multiple controllers and switching*, in Proceedings of the 42nd IEEE Conference on Decision and Control, Vol. 4, IEEE Press, New York, 2003, pp. 3256–3261.

[23] R. WANG AND M. G. SAFONOV, *Stability of unfalsified adaptive control using multiple controllers*, in Proceedings of the American Control Conference, 2005, pp. 3162–3167.

[24] K. S. NARENDRA AND J. BALAKRISHNAN, *Improving transient response of adaptive control systems using multiple models and switching*, IEEE Trans. Automat. Control, 39 (1994), pp. 1861–1866.

[25] K. S. NARENDRA, J. BALAKRISHNAN, AND M. K. CILIZ, *Adaptation and learning using multiple models, switching, and tuning*, IEEE Control Systems Magazine, 15 (1995), pp. 37–51.

[26] K. S. NARENDRA AND J. BALAKRISHNAN, *Adaptive control using multiple models*, IEEE Trans. Automat. Control, 42 (1997), pp. 171–187.

[27] B. D. O. ANDERSON, T. S. BRINSMEAD, F. DE BRUYNE, J. HESPANHA, D. LIBERZON, AND A. S. MORSE, *Multiple model adaptive control, Part 1: Finite controller coverings*, Internat. J. Robust Nonlinear Control, 10 (2000), pp. 909–929.

[28] J. P. HESPANHA, D. LIBERZON, A. S. MORSE, B. D. O. ANDERSON, T. S. BRINSMEAD, AND F. DE BRUYNE, *Multiple model adaptive control, Part 2: Switching*, Internat. J. Robust Nonlinear Control, 11 (2001), pp. 479–496.

[29] M. ATHANS, S. FEKRI, AND A. PASCOAL, *Issues on robust adaptive feedback control*, in Proceedings of the 16th IFAC World Congress, Prague, 2005.

[30] H. P. WHITAKER, J. YAMRON, AND A. KEZER, *Design of Model Reference Adaptive Control Systems for Aircraft*, Report R-164, Instrumentation Laboratory, MIT, Cambridge, MA, 1958.

[31] P. V. OSBURN, A. P. WHITAKER, AND A. KEZER, *New Developments in the Design of Model Reference Adaptive Control Systems*, Paper 61-39, Institute of the Aerospace Sciences, Easton, PA, 1961.

[32] R. E. KALMAN, *Design of a self optimizing control system*, Trans. ASME, 80 (1958), pp. 468–478.

[33] K. J. ASTROM, *Theory and applications of adaptive control: A survey*, Automatica, 19 (1983), pp. 471–486.

[34] L. W. TAYLOR AND E. J. ADKINS, *Adaptive control and the X-15*, in Proceedings of the Princeton University Conference on Aircraft Flying Qualities, Princeton University Press, Princeton, NJ, 1965.

[35] R. E. BELLMAN, *Dynamic Programming*, Princeton University Press, Princeton, NJ, 1957.

[36] R. E. BELLMAN, *Adaptive Control Processes: A Guided Tour*, Princeton University Press, Princeton, NJ, 1961.

[37] A. A. FELDBAUM, *Optimal Control Systems*, Academic Press, New York, 1965.

[38] K. J. ASTROM AND P. EYKHOF, *System identification: A survey*, Automatica, 7 (1971), pp. 123–162.

[39] Y. Z. TSYPKIN, *Adaptation and Learning in Automatic Systems*, Academic Press, New York, 1971.

[40] P. C. PARKS, *Lyapunov redesign of model reference adaptive control systems*, IEEE Trans. Automat. Control, 11 (1966), pp. 362–367.

[41] B. EGARDT, *Stability of Adaptive Controllers*, Lecture Notes in Control and Inform. Sci. 20, Springer-Verlag, New York, 1979.

[42] A. S. MORSE, *Global stability of parameter adaptive control systems*, IEEE Trans. Automat. Control, 25 (1980), pp. 433–439.

[43] K. S. Narendra, Y. H. Lin, and L. S. Valavani, *Stable adaptive controller design, Part* II: *Proof of stability*, IEEE Trans. Automat. Control, 25 (1980), pp. 440–448.

[44] G. C. Goodwin, P. J. Ramadge, and P. E. Caines, *Discrete-time multi-variable adaptive control*, IEEE Trans. Automat. Control, 25 (1980), pp. 449–456.

[45] I. D. Landau, *Adaptive Control: The Model Reference Approach*, Marcel Dekker, New York, 1979.

[46] G. C. Goodwin and K. S. Sin, *Adaptive Filtering Prediction and Control*, Prentice–Hall, Englewood Cliffs, NJ, 1984.

[47] C. J. Harris and S. A. Billings, eds., *Self-Tuning and Adaptive Control: Theory and Applications*, Peter Peregrinus, London, 1981.

[48] K. S. Narendra and R. V. Monopoli, eds., *Applications of Adaptive Control*, Academic Press, New York, 1980.

[49] H. Unbehauen, ed., *Methods and Applications in Adaptive Control*, Springer-Verlag, Berlin, 1980.

[50] P. A. Ioannou and P. V. Kokotovic, *Adaptive Systems with Reduced Models*, Lecture Notes in Control and Inform. Sci. 47, Springer-Verlag, New York, 1983.

[51] P. A. Ioannou and P. V. Kokotovic, *Instability analysis and improvement of robustness of adaptive control*, Automatica, 20 (1984), pp. 583–594.

[52] C. E. Rohrs, L. Valavani, M. Athans, and G. Stein, *Robustness of continuous-time adaptive control algorithms in the presence of unmodeled dynamics*, IEEE Trans. Automat. Control, 30 (1985), pp. 881–889.

[53] P. A. Ioannou and J. Sun, *Theory and design of robust direct and indirect adaptive control schemes*, Internat. J. Control, 47 (1988), pp. 775–813.

[54] P. A. Ioannou and A. Datta, *Robust adaptive control: Design, analysis and robustness bounds*, in Grainger Lectures: Foundations of Adaptive Control, P. V. Kokotovic, ed., Springer-Verlag, New York, 1991.

[55] P. A. Ioannou and A. Datta, *Robust adaptive control: A unified approach*, Proc. IEEE, 79 (1991), pp. 1736–1768.

[56] P. A. Ioannou and J. Sun, *Robust Adaptive Control*, Prentice–Hall, Englewood Cliffs, NJ, 1996; also available online from http://www-rcf.usc.edu/~ioannou/Robust_Adaptive_Control.htm.

[57] L. Praly, *Robust model reference adaptive controllers, Part* I: *Stability analysis*, in Proceedings of the 23rd IEEE Conference on Decision and Control, IEEE Press, New York, 1984.

[58] L. PRALY, S. T. HUNG, AND D. S. RHODE, *Towards a direct adaptive scheme for a discrete-time control of a minimum phase continuous-time system*, in Proceedings of the 24th IEEE Conference on Decision and Control, Ft. Lauderdale, FL, IEEE Press, New York, 1985, pp. 1188–1191.

[59] P. A. IOANNOU AND K. S. TSAKALIS, *A robust direct adaptive controller*, IEEE Trans. Automat. Control, 31 (1986), pp. 1033–1043.

[60] K. S. TSAKALIS AND P. A. IOANNOU, *Adaptive control of linear time-varying plants*, Automatica, 23 (1987), pp. 459–468.

[61] K. S. TSAKALIS AND P. A. IOANNOU, *Adaptive control of linear time-varying plants: A new model reference controller structure*, IEEE Trans. Automat. Control, 34 (1989), pp. 1038–1046.

[62] K. S. TSAKALIS AND P. A. IOANNOU, *A new indirect adaptive control scheme for time-varying plants*, IEEE Trans. Automat. Control, 35 (1990), pp. 697–705.

[63] K. S. TSAKALIS AND P. A. IOANNOU, *Adaptive control of linear time-varying plants: The case of "jump" parameter variations*, Internat. J. Control, 56 (1992), pp. 1299–1345.

[64] I. KANELLAKOPOULOS, P. V. KOKOTOVIC, AND A. S. MORSE, *Systematic design of adaptive controllers for feedback linearizable systems*, IEEE Trans. Automat. Control, 36 (1991), pp. 1241–1253.

[65] I. KANELLAKOPOULOS, *Adaptive Control of Nonlinear Systems*, Ph.D. thesis, Report UIUC-ENG-91-2244, DC-134, Coordinated Science Laboratory, University of Illinois at Urbana-Champaign, Urbana, IL, 1991.

[66] M. KRSTIC, I. KANELLAKOPOULOS, AND P. KOKOTOVIC, *Nonlinear and Adaptive Control Design*, Wiley, New York, 1995.

[67] M. KRSTIC, I. KANELLAKOPOULOS, AND P. V. KOKOTOVIC, *A new generation of adaptive controllers for linear systems*, in Proceedings of the 31st IEEE Conference on Decision and Control, Tucson, AZ, IEEE Press, New York, 1992.

[68] M. KRSTIC, P. V. KOKOTOVIC, AND I. KANELLAKOPOULOS, *Transient performance improvement with a new class of adaptive controllers*, Systems Control Lett., 21 (1993), pp. 451–461.

[69] M. KRSTIC AND P. V. KOKOTOVIC, *Adaptive nonlinear design with controller-identifier separation and swapping*, IEEE Trans. Automat. Control, 40 (1995), pp. 426–441.

[70] A. DATTA AND P. A. IOANNOU, *Performance improvement versus robust stability in model reference adaptive control*, in Proceedings of the 30th IEEE Conference on Decision and Control, IEEE Press, New York, 1991, pp. 1082–1087.

[71] J. SUN, *A modified model reference adaptive control scheme for improved transient performance*, IEEE Trans. Automat. Control, 38 (1993), pp. 1255–1259.

[72] Y. Zhang, B. Fidan, and P. A. Ioannou, *Backstepping control of linear time-varying systems with known and unknown parameters*, IEEE Trans. Automat. Control, 48 (2003), pp. 1908–1925.

[73] B. Fidan, Y. Zhang, and P. A. Ioannou, *Adaptive control of a class of slowly time-varying systems with modeling uncertainties*, IEEE Trans. Automat. Control, 50 (2005), pp. 915–920.

[74] A. L. Fradkov, *Continuous-time model reference adaptive systems: An east-west review*, in Proceedings of the IFAC Symposium on Adaptive Control and Signal Processing, Grenoble, France, 1992.

[75] R. Ortega and T. Yu, *Robustness of adaptive controllers: A survey*, Automatica, 25 (1989), pp. 651–678.

[76] B. D. O. Anderson, R. R. Bitmead, C. R. Johnson, P. V. Kokotovic, R. L. Kosut, I. Mareels, L. Praly, and B. Riedle, *Stability of Adaptive Systems*, MIT Press, Cambridge, MA, 1986.

[77] K. J. Astrom and B. Wittenmark, *Adaptive Control*, Addison–Wesley, Reading, MA, 1989.

[78] R. R. Bitmead, M. Gevers, and V. Wertz, *Adaptive Optimal Control*, Prentice–Hall, Englewood Cliffs, NJ, 1990.

[79] V. V. Chalam, *Adaptive Control Systems: Techniques and Applications*, Marcel Dekker, New York, 1987.

[80] A. L. Fradkov, *Adaptive Control in Large Scale Systems*, Nauka, Moscow, 1990 (in Russian).

[81] P. J. Gawthrop, *Continuous-Time Self-Tuning Control: Volume* I: *Design*, Research Studies Press, New York, 1987.

[82] M. M. Gupta, ed., *Adaptive Methods for Control System Design*, IEEE Press, Piscataway, NJ, 1986.

[83] C. R. Johnson, Jr., *Lectures on Adaptive Parameter Estimation*, Prentice–Hall, Englewood Cliffs, NJ, 1988.

[84] P. V. Kokotovic, ed., *Foundations of Adaptive Control*, Springer-Verlag, New York, 1991.

[85] K. S. Narendra, ed., *Adaptive and Learning Systems: Theory and Applications*, Plenum Press, New York, 1986.

[86] K. S. Narendra and A. M. Annaswamy, *Stable Adaptive Systems*, Prentice–Hall, Englewood Cliffs, NJ, 1989.

[87] S. Sastry and M. Bodson, *Adaptive Control: Stability, Convergence and Robustness*, Prentice–Hall, Englewood Cliffs, NJ, 1989.

[88] I. Landau, R. Lozano, and M. M'Saad, *Adaptive Control*, Springer-Verlag, New York, 1997.

[89] G. Tao and P. V. Kokotovic, *Adaptive Control of Systems with Actuator and Sensor Nonlinearities*, Wiley, New York, 1996.

[90] G. Tao and F. Lewis, eds., *Adaptive Control of Nonsmooth Dynamic Systems*, Springer-Verlag, New York, 2001.

[91] G. Tao, *Adaptive Control Design and Analysis*, Wiley, New York, 2003.

[92] G. Tao, S. Chen, X. Tang, and S. M. Joshi, *Adaptive Control of Systems with Actuator Failures*, Springer-Verlag, New York, 2004.

[93] A. Datta, *Adaptive Internal Model Control*, Springer-Verlag, New York, 1998.

[94] http://www.siam.org/books/dc11.

[95] J. S. Yuan and W. M. Wonham, *Probing signals for model reference identification*, IEEE Trans. Automat. Control, 22 (1977), pp. 530–538.

[96] P. A. Ioannou, E. B. Kosmatopoulos, and A. Despain, *Position error signal estimation at high sampling rates using data and servo sector measurements*, IEEE Trans. Control Systems Tech., 11 (2003), pp. 325–334.

[97] P. Eykhoff, *System Identification: Parameter and State Estimation*, Wiley, New York, 1974.

[98] L. Ljung and T. Söderström, *Theory and Practice of Recursive Identification*, MIT Press, Cambridge, MA, 1983.

[99] I. M. Y. Mareels and R. R. Bitmead, *Nonlinear dynamics in adaptive control: Chaotic and periodic stabilization*, Automatica, 22 (1986), pp. 641–655.

[100] A. S. Morse, *An adaptive control for globally stabilizing linear systems with unknown high-frequency gains*, in Proceedings of the 6th International Conference on Analysis and Optimization of Systems, Nice, France, 1984.

[101] A. S. Morse, *A model reference controller for the adaptive stabilization of any strictly proper, minimum phase linear system with relative degree not exceeding two*, in Proceedings of the 1985 MTNS Conference, Stockholm, 1985.

[102] D. R. Mudgett and A. S. Morse, *Adaptive stabilization of linear systems with unknown high-frequency gains*, IEEE Trans. Automat. Control, 30 (1985), pp. 549–554.

[103] R. D. Nussbaum, *Some remarks on a conjecture in parameter adaptive-control*, Systems Control Lett., 3 (1983), pp. 243–246.

[104] J. C. Willems and C. I. Byrnes, *Global adaptive stabilization in the absence of information on the sign of the high frequency gain*, in Proceedings of the INRIA Conference on Analysis and Optimization of Systems, Vol. 62, 1984, pp. 49–57.

[105] B. D. O. ANDERSON, *Adaptive systems, lack of persistency of excitation and bursting phenomena*, Automatica, 21 (1985), pp. 247–258.

[106] P. A. IOANNOU AND G. TAO, *Dominant richness and improvement of performance of robust adaptive control*, Automatica, 25 (1989), pp. 287–291.

[107] K. S. NARENDRA AND A. M. ANNASWAMY, *Robust adaptive control in the presence of bounded disturbances*, IEEE Trans. Automat. Control, 31 (1986), pp. 306–315.

[108] G. KREISSELMEIER, *Adaptive observers with exponential rate of convergence*, IEEE Trans. Automat. Control, 22 (1977), pp. 2–8.

[109] G. LUDERS AND K. S. NARENDRA, *An adaptive observer and identifier for a linear system*, IEEE Trans. Automat. Control, 18 (1973), pp. 496–499.

[110] G. LUDERS AND K. S. NARENDRA, *A new canonical form for an adaptive observer*, IEEE Trans. Automat. Control, 19 (1974), pp. 117–119.

[111] A. DATTA, *Robustness of discrete-time adaptive controllers: An input-output approach*, IEEE Trans. Automat. Control, 38 (1993), pp. 1852–1857.

[112] L. PRALY, *Global stability of a direct adaptive control scheme with respect to a graph topology*, in Adaptive and Learning Systems: Theory and Applications, K. S. Narendra, ed., Plenum Press, New York, 1985.

[113] P. A. IOANNOU AND K. S. TSAKALIS, *Robust discrete-time adaptive control*, in Adaptive and Learning Systems: Theory and Applications, K. S. Narendra, ed., Plenum Press, New York, 1985.

[114] K. S. TSAKALIS, *Robustness of model reference adaptive controllers: An input-output approach*, IEEE Trans. Automat. Control, 37 (1992), pp. 556–565.

[115] K. S. TSAKALIS AND S. LIMANOND, *Asymptotic performance guarantees in adaptive control*, Internat. J. Adapt. Control Signal Process., 8 (1994), pp. 173–199.

[116] L. HSU AND R. R. COSTA, *Bursting phenomena in continuous-time adaptive systems with a σ-modification*, IEEE Trans. Automat. Control, 32 (1987), pp. 84–86.

[117] B. E. YDSTIE, *Stability of discrete model reference adaptive control—revisited*, Systems Control Lett., 13 (1989), pp. 1429–1438.

[118] P. A. IOANNOU AND Z. XU, *Throttle and brake control systems for automatic vehicle following*, IVHS J., 1 (1994), pp. 345–377.

[119] H. ELLIOTT, R. CRISTI, AND M. DAS, *Global stability of adaptive pole placement algorithms*, IEEE Trans. Automat. Control, 30 (1985), pp. 348–356.

[120] T. KAILATH, *Linear Systems*, Prentice–Hall, Englewood Cliffs, NJ, 1980.

[121] G. KREISSELMEIER, *An approach to stable indirect adaptive control*, Automatica, 21 (1985), pp. 425–431.

[122] G. KREISSELMEIER, *A robust indirect adaptive control approach*, Internat. J. Control, 43 (1986), pp. 161–175.

[123] R. H. MIDDLETON, G. C. GOODWIN, D. J. HILL, AND D. Q. MAYNE, *Design issues in adaptive control*, IEEE Trans. Automat. Control, 33 (1988), pp. 50–58.

[124] P. DE LARMINAT, *On the stabilizability condition in indirect adaptive control*, Automatica, 20 (1984), pp. 793–795.

[125] E. W. BAI AND S. SASTRY, *Global stability proofs for continuous-time indirect adaptive control schemes*, IEEE Trans. Automat. Control, 32 (1987), pp. 537–543.

[126] G. KREISSELMEIER AND G. RIETZE-AUGST, *Richness and excitation on an interval—with application to continuous-time adaptive control*, IEEE Trans. Automat. Control, 35 (1990), pp. 165–171.

[127] A. S. MORSE AND F. M. PAIT, *MIMO design models and internal regulators for cyclicly-switched parameter-adaptive control systems*, IEEE Trans. Automat. Control, 39 (1994), pp. 1809–1818.

[128] R. WANG, M. STEFANOVIC, AND M. SAFONOV, *Unfalsified direct adaptive control using multiple controllers*, in Proceedings of the AIAA Guidance, Navigation, and Control Exhibit, Vol. 2, Providence, RI, 2004.

[129] F. GIRI, F. AHMED-ZAID, AND P. A. IOANNOU, *Stable indirect adaptive control of continuous-time systems with no apriori knowledge about the parameters*, IEEE Trans. Automat. Control, 38 (1993), pp. 766–770.

[130] B. D. O. ANDERSON AND R. M. JOHNSTONE, *Global adaptive pole positioning*, IEEE Trans. Automat. Control, 30 (1985), pp. 11–22.

[131] R. CRISTI, *Internal persistency of excitation in indirect adaptive control*, IEEE Trans. Automat. Control, 32 (1987), pp. 1101–1103.

[132] F. GIRI, J. M. DION, M. M'SAAD, AND L. DUGARD, *A globally convergent pole placement indirect adaptive controller*, IEEE Trans. Automat. Control, 34 (1989), pp. 353–356.

[133] F. GIRI, J. M. DION, L. DUGARD, AND M. M'SAAD, *Parameter estimation aspects in adaptive control*, Automatica, 27 (1991), pp. 399–402.

[134] R. LOZANO-LEAL, *Robust adaptive regulation without persistent excitation*, IEEE Trans. Automat. Control, 34 (1989), pp. 1260–1267.

[135] R. LOZANO-LEAL AND G. C. GOODWIN, *A globally convergent adaptive pole placement algorithm without a persistency of excitation requirement*, IEEE Trans. Automat. Control, 30 (1985), pp. 795–797.

[136] J. W. POLDERMAN, *Adaptive Control and Identification: Conflict or Conflux?*, Ph.D. thesis, Rijksuniversiteit, Groningen, The Netherlands, 1987.

[137] J. W. Polderman, *A state space approach to the problem of adaptive pole placement*, Math. Control Signals Systems, 2 (1989), pp. 71–94.

[138] B. L. Stevens and F. L. Lewis, *Aircraft Control and Simulation*, Wiley, New York, 1992.

[139] L. T. Nguyen, M. E. Ogburn, W. P. Gilbert, K. S. Kibler, P. W. Brown, and P. L. Deal, *Simulator Study of Stall/Post-Stall Characteristics of a Fighter Airplane with Relaxed Longitudinal Static Stability*, Technical Paper 1538, NASA, Washington, DC, 1979.

[140] A. Ajami, N. Hovakimyan, and E. Lavretsky, *Design examples: Adaptive tracking control in the presence of input constraints for a fighter aircraft and a generic missile*, AIAA Guidance, Navigation, and Control Conference, AIAA 2005-6444, San Francisco, 2005.

[141] K. Wise, E. Lavretsky, and N. Hovakimyan, *Adaptive control of flight: Theory, applications, and open problems*, in Proceedings of the American Control Conference, 2006, pp. 5966–5971.

[142] S. Ferrari and R. F. Stengel, *Online adaptive critical flight control*, AIAA J. Guidance Control Dynam., 27 (2004), pp. 777–786.

[143] K. S. Narendra and Y. H. Lin, *Stable discrete adaptive control*, IEEE Trans. Automat. Control, 25 (1980), pp. 456–461.

[144] R. Ortega, L. Praly, and I. D. Landau, *Robustness of discrete-time direct adaptive controllers*, IEEE Trans. Automat. Control, 30 (1985), pp. 1179–1187.

[145] S. Limanond and K. S. Tsakalis, *Model reference adaptive and nonadaptive control of linear time-varying plants*, IEEE Trans. Automat. Control, 45 (2000), pp. 1290–1300.

[146] P. V. Kokotović and M. Arcak, *Constructive nonlinear control: A historical perspective*, Automatica, 37 (2001), pp. 637–662.

[147] S. Sastry, *Nonlinear Systems: Analysis, Stability, and Control*, Springer-Verlag, New York, 1999.

[148] H. K. Khalil, *Nonlinear Systems*, 2nd ed., Prentice–Hall, Upper Saddle River, NJ, 1996.

[149] A. Isidori, *Nonlinear Control Systems*, Springer-Verlag, Berlin, 1989.

[150] H. Nijmeijer and A. van der Schaft, *Nonlinear Dynamical Control Systems*, Springer-Verlag, Berlin, 1990.

[151] B. Kosko, *Neural Networks and Fuzzy Systems: A Dynamical Systems Approach to Machine Intelligence*, Prentice–Hall, Englewood Cliffs, NJ, 1991.

[152] G. BEKEY AND K. GOLDBERG, EDS., *Neural Networks in Robotics*, Kluwer Academic Publishers, Dordrecht, The Netherlands, 1992.

[153] P. ANTSAKLIS AND K. PASSINO, EDS., *An Introduction to Intelligent and Autonomous Control*, Kluwer Academic Publishers, Dordrecht, The Netherlands, 1992.

[154] D. A. WHITE AND D. A. SOFGE, *Handbook of Intelligent Control*, Van Nostrand Reinhold, New York, 1992.

[155] K. S. NARENDRA AND K. PARTHASARATHY, *Identification and control of dynamical systems using neural networks*, IEEE Trans. Neural Networks, 1 (1990), pp. 4–27.

[156] M. M. POLYCARPOU, *Stable adaptive neural control scheme for nonlinear systems*, IEEE Trans. Automat. Control, 41 (1996), pp. 447–451.

[157] M. M. POLYCARPOU AND M. J. MEARS, *Stable adaptive tracking of uncertain systems using nonlinearly parametrized on-line approximators*, Internat. J. Control, 70 (1998), pp. 363–384.

[158] M. M. POLYCARPOU AND P. A. IOANNOU, *Modeling, identification and stable adaptive control of continuous-time nonlinear dynamical systems using neural networks*, in Proceedings of the American Control Conference, 1992, pp. 36–40.

[159] S. SESHAGIRI AND H. KHALIL, *Output feedback control of nonlinear systems using RBF neural networks*, IEEE Trans. Automat. Control, 11 (2000), pp. 69–79.

[160] G. A. ROVITHAKIS, *Stable adaptive neuro-control design via Lyapunov function derivative estimation*, Automatica, 37 (2001), pp. 1213–1221.

[161] F.-C. CHEN AND H. K. KHALIL, *Adaptive control of nonlinear systems using neural networks*, Internat. J. Control, 55 (1992), pp. 1299–1317.

[162] F.-C. CHEN AND C.-C. LIU, *Adaptively controlling nonlinear continuous-time systems using multilayer neural networks*, IEEE Trans. Automat. Control, 39 (1994), pp. 1306–1310.

[163] J. T. SPOONER AND K. M. PASSINO, *Stable adaptive control using fuzzy systems and neural networks*, IEEE Trans. Fuzzy Syst., 4 (1996), pp. 339–359.

[164] R. SANNER AND J.-J. E. SLOTINE, *Gaussian networks for direct adaptive control*, IEEE Trans. Neural Networks, 3 (1992), pp. 837–863.

[165] H. XU AND P. A. IOANNOU, *Robust adaptive control of linearizable nonlinear single input systems*, in Proceedings of the 15th IFAC World Congress, Barcelona, 2002.

[166] A. YESILDIREK AND F. L. LEWIS, *Feedback linearization using neural networks*, Automatica, 31 (1995), pp. 1659–1664.

[167] S. JAGANNATHAN AND F. L. LEWIS, *Multilayer discrete-time neural-net controller with guaranteed performance*, IEEE Trans. Neural Networks, 7 (1996), pp. 107–130.

[168] C. Kwan, and F. L. Lewis, *Robust backstepping control of nonlinear systems using neural networks*, IEEE Trans. Systems, Man and Cybernet. A, 30 (2000), pp. 753–766.

[169] Y. H. Kim and F. L. Lewis, *Neural network output feedback control of robot manipulators*, IEEE Trans. Robotics Automation, 15 (1999), pp. 301–309.

[170] F. L. Lewis, A. Yesildirek, and K. Liu, *Multilayer neural-net robot controller with guaranteed tracking performance*, IEEE Trans. Neural Networks, 7 (1996), pp. 388–399.

[171] F. L. Lewis, K. Liu, and A. Yesildirek, *Neural-net robot controller with guaranteed tracking performance*, IEEE Trans. Neural Networks, 6 (1995), pp. 703–715.

[172] C.-C. Liu and F.-C. Chen, *Adaptive control of nonlinear continuous-time systems using neural networks: General relative degree and MIMO cases*, Internat. J. Control, 58 (1993), pp. 317–335.

[173] R. Ordonez and K. M. Passino, *Stable multi-input multi-output adaptive fuzzy/neural control*, IEEE Trans. Fuzzy Syst., 7 (1999), pp. 345–353.

[174] E. B. Kosmatopoulos, M. Polycarpou, M. Christodoulou, and P. A. Ioannou, *High-order neural network structures for identification of dynamical systems*, IEEE Trans. Neural Networks, 6 (1995), pp. 422–431.

[175] K. Funahashi, *On the approximate realization of continuous mappings by neural networks*, Neural Networks, 2 (1989), pp. 183–192.

[176] J. Park and I. W. Sandberg, *Approximation and radial-basis-function networks*, Neural Comput., 5 (1993), pp. 305–316.

[177] H. Xu and P. A. Ioannou, *Robust adaptive control of linearizable nonlinear single input systems with guaranteed error bounds*, Automatica, 40 (2003), pp. 1905–1911.

[178] H. Xu and P. A. Ioannou, *Robust adaptive control for a class of MIMO nonlinear systems with guaranteed error bounds*, IEEE Trans. Automat. Control, 48 (2003), pp. 728–743.

[179] E. B. Kosmatopoulos and P. A Ioannou, *A switching adaptive controller for feedback linearizable systems*, IEEE Trans. Automat. Control, 44 (1999), pp. 742–750.

[180] K. Hornik, M. Stinchcombe, and H. White, *Multilayer feedforward networks are universal approximators*, Neural Networks, 2 (1989), pp. 359–366.

[181] G. Cybenko, *Approximation by superpositions of a sigmoidal function*, Math. Control Signals Systems, 2 (1989), pp. 303–314.

[182] E. J. Hartman, J. D. Keeler, and J. M. Kowalski, *Layered neural networks with Gaussian hidden units as universal approximators*, Neural Comput., 2 (1990), pp. 210–215.

[183] J. Park and I. W. Sandberg, *Universal approximation using radial basis function networks*, Neural Comput., 3 (1991), pp. 246–257.

[184] A. De Luca, G. Oriolo, and C. Samson, *Feedback Control of a Nonholonomic Car-Like Robot*, in Robot Motion Planning and Control, J.-P. Laumond, ed., Laboratoire d'Analyse et d'Architecture des Systèmes, Toulouse, France, 1997, pp. 187–188; also available online from http://www.laas.fr/~jpl/book-toc.html.

[185] W. A. Wolovich, *Linear Multivariable Systems*, Springer-Verlag, New York, 1974.

[186] B. D. O. Anderson and S. Vongpanitlerd, *Network Analysis and Synthesis: A Modern Systems Theory Approach*, Prentice–Hall, Englewood Cliffs, NJ, 1973.

[187] C. A. Desoer and M. Vidyasagar, *Feedback Systems: Input-Output Properties*, Academic Press, New York, 1975.

[188] H. L. Royden, *Real Analysis*, Prentice–Hall, Englewood Cliffs, NJ, 1988.

[189] W. Rudin, *Real and Complex Analysis*, 2nd ed., McGraw–Hill, New York, 1986.

[190] V. M. Popov, *Hyperstability of Control Systems*, Springer-Verlag, New York, 1973.

[191] B. D. O. Anderson, *Exponential stability of linear equations arising in adaptive identification*, IEEE Trans. Automat. Control, 22 (1977), pp. 83–88.

[192] M. Vidyasagar, *Nonlinear Systems Analysis*, 2nd ed., SIAM, Philadelphia, 2002.

[193] J. P. LaSalle and S. Lefschetz, *Stability by Lyapunov's Direct Method with Application*, Academic Press, New York, 1961.

[194] J. P. LaSalle, *Some extensions of Lyapunov's second method*, IRE Trans. Circuit Theory, 7 (1960), pp. 520–527.

[195] A. Michel and R. K. Miller, *Ordinary Differential Equations*, Academic Press, New York, 1982.

[196] W. A. Coppel, *Stability and Asymptotic Behavior of Differential Equations*, D. C. Heath, Boston, 1965.

[197] W. Hahn, *Theory and Application of Lyapunov's Direct Method*, Prentice–Hall, Englewood Cliffs, NJ, 1963.

[198] J. K. Hale, *Ordinary Differential Equations*, Wiley–Interscience, New York, 1969.

[199] R. E. Kalman and J. E. Bertram, *Control systems analysis and design via the "second method" of Lyapunov*, J. Basic Engrg., 82 (1960), pp. 371–392.

[200] M. Polycarpou and P. A. Ioannou, *On the existence and uniqueness of solutions in adaptive control systems*, IEEE Trans. Automat. Control, 38 (1993), pp. 474–479.

[201] K. S. Narendra and J. H. Taylor, *Frequency Domain Criteria for Absolute Stability*, Academic Press, New York, 1973.

[202] D. D. Siljak, *New algebraic criteria for positive realness*, J. Franklin Inst., 291 (1971), pp. 109–120.

[203] P. A. Ioannou and G. Tao, *Frequency domain conditions for strictly positive real functions*, IEEE Trans. Automat. Control, 32 (1987), pp. 53–54.

[204] S. Lefschetz, *Stability of Nonlinear Control Systems*, Academic Press, New York, 1965.

[205] K. R. Meyer, *On the existence of Lyapunov functions for the problem on Lur'e*, SIAM J. Control, 3 (1966), pp. 373–383.

[206] S. S. Rao, *Engineering Optimization: Theory and Practice*, Wiley, New York, 1996.

[207] D. G. Luenberger, *Optimization by Vector Space Methods*, Wiley, New York, 1969.

[208] D. P. Bertsekas, *Dynamic Programming: Deterministic and Stochastic Models*, Prentice–Hall, Englewood Cliffs, NJ, 1987.

[209] G. Tao and P. A. Ioannou, *Necessary and sufficient conditions for strictly positive real matrices*, IEE Proc. G: Circuits, Devices, Systems, 137 (1990), pp. 360–366.

[210] J. S. Shamma and M. Athans, *Gain scheduling: Potential hazards and possible remedies*, IEEE Control Systems Magazine, 12 (1992), pp. 101–107.

[211] W. J. Rugh and J. S. Shamma, *Research on gain scheduling*, Automatica, 36 (2000), pp. 1401–1425.

[212] Y. Huo, M. Mirmirani, P. Ioannou, and R. Colgren, *Adaptive linear quadratic design with application to F-16 fighter aircraft*, in Proceedings of the AIAA Guidance, Navigation, and Control Conference, AIAA-2004-5421, 2004.

Index